AF869491

Jozef Šamaj • Jay J. Thelen (Editors)

Plant Proteomics

Jozef Šamaj
Jay J. Thelen (Editors)

Plant Proteomics

With 38 Figures, 7 in color and 13 Tables

PILON

B.U. ORLÉANS
SCIENCES
INV. S09-2410
Cote Mag. E12375
Cote Salle 572.6 PLA

Springer

Dr. Jozef Šamaj
Institute of Cellular and Molecular Botany
Rheinische Friedrich-Wilhelms-University Bonn
Department of Plant Cell Biology
Kirschallee 1
D-53115 Bonn
Germany

Institute of Plant Genetics and Biotechnology
Slovak Academy of Sciences
Akademicka 2
95007 Nitra
Slovak Republic

Dr. Jay J. Thelen
University of Missouri-Columbia
Department of Biochemistry
Columbia, MO 65211
USA

Library of Congress Control Number: 2007929273

ISBN-13 978-3-540-72616-6 Springer-Verlag Berlin Heidelberg New York

This work is subject to copyright. All rights are reserved, whether the whole or part of the material is concerned, specifically the rights of translation, reprinting, reuse of illustrations, recitation, broadcasting, reproduction on microfilm or in any other way, and storage in data banks. Duplication of this publication or parts thereof is permitted only under the provisions of the German Copyright Law of September 9, 1965, in its current version, and permission for use must always be obtained from Springer. Violations are liable to prosecution under the German Copyright Law.

Springer-Verlag is a part of Springer Science + Business Media

springer.com

© Springer-Verlag Berlin Heidelberg 2007

The use of general descriptive names, registered names, trademarks, etc. in this publication does not imply, even in the absence of a specific statement, that such names are exempt from the relevant protective laws and regulations and therefore free for general use.

Editor: Dr. Christina Eckey, Heidelberg
Desk Editor: Anette Lindqvist, Heidelberg
Production: SPi
Typesetting: SPi
Cover Design: WMXDesign Gmbh, Heidelberg

Printed on acid-free paper 149/3152-HM 5 4 3 2 1 0

Editors

Jozef Šamaj

Jay Thelen

Jozef Šamaj received his Ph.D. degree in Plant Physiology from the Comenius University in Bratislava, Slovakia. He completed three post-doctoral programmes supported by Eurosilva, the Alexander von Humboldt Foundation, and the EU Marie Curie Programme in the highly regarded laboratories of Alain Boudet in Toulouse, Dieter Volkmann in Bonn, and Heribert Hirt in Vienna. He worked on the cell biology of somatic embryogenesis, lignification in tree species, arabinogalactan proteins, the cytoskeleton, and signalling proteins. Jozef Šamaj has co-edited three books and co-authored more than 75 research papers, reviews, and book chapters. He is a senior lecturer and group leader at the Institute of Cellular and Molecular Botany in Bonn, Germany, and senior researcher at the Institute of Plant Genetics and Biotechnology, Slovak Academy of Sciences, Nitra, Slovakia. His current research is focussed on the role of signalling components and the cytoskeleton in relation to the vesicular trafficking during plant development and stress responses using integrated cell-biological and functional proteomics approaches.

Jay Thelen received his B.Sc. degree in Biology and Biochemistry from the University of Nebraska-Lincoln in 1993. He earned his Ph.D. from the University of Missouri-Columbia (UMC) studying the structure and regulation of plant mitochondrial pyruvate dehydrogenase complexes under the guidance of Douglas Randall. In 1999 he started a 3-year postdoctoral position in John Ohlrogge's lab at Michigan State University investigating the plastid acetyl-CoA carboxylase protein complex. He returned to UMC in 2002 as the Associate Director of a campus Proteomics Center. In 2004, he was promoted to Assistant Professor in the Biochemistry Department, a position he currently holds. He has authored or co-authored 35 research and review articles since 1994. His research interests are centered around the regulation of plant metabolism, particularly carbon assimilation in oilseeds, and multienzyme metabolic complexes. He is currently studying seed filling in numerous crop oilseeds, using various quantitative proteomics approaches. He is also investigating global phosphoprotein networks involved in seed development and is developing improved strategies for quantitative proteomics.

Contents

Preface

Plant proteomics is a relatively new research field focused on the large-scale functional analysis of proteins extracted from intact plants, particular plant organs, tissues, individual cells, subcellular organelles and/or separated suborganellar structures. Rapidly increasing numbers of excellent publications on plant proteomics, both in primary and applied research, demonstrate the imense potential and importance of this research field for current and future plant science. One of the main aims of plant proteomics is to study the assembly and functional interactions of plant proteins. Proteins often function as molecular machines organized into multiprotein complexes localized within specialized subcellular compartments. Enormous methodological and technical developments have moved recent proteomics towards the large-scale study of post-translational modifications of proteins involved in cellular signalling (regulated by reversible phosphorylation), protein turnover (ubiquitinylation) and membrane association (palmitoylation and myristylation).

This book highlights this rapid progress in plant proteomics with emphasis on model species, subcellular organelles as well as specific aspects such as signalling, plant reproduction, stress biology and/or pathogen/symbiotic interactions between plants and microorganisms. Additionally, brief historical overviews on plant proteomics and two-dimensional gel electrophoresis as well as an introduction to bioinfomatics are provided here. Thus, this monograph represents a synthesis of the most current knowledge in this field, including the most important biological aspects as well as new methodological approaches such as high-resolution two-dimensional electrophoresis, protein microchips, MudPIT (multidimensional protein identification technology), fluorescent DIGE (difference gel electrophoresis) alone and/or in combination with stable isotope reagents such as ICAT (isotope-coded affinity tag) and iTRAQ (isobaric tag for relative and absolute quantitation), which allow relative protein quantification. The reader is provided with an up-to date view on plant proteomes in carefully selected model plant species such as Arabidopsis, cereals, legumes and oil seed plants. One chapter focuses on the cell division model represented by suspension cultured tobacco BY-2 cells. Several chapters are devoted to the proteomics of plant organelles and compartments. Among the latter, special attention is paid to the cell wall, plasma membrane, plastids, mitochondria and nucleolus. Two chapters focus on proteomic approaches used to study plant reproduction, namely pollen proteomics and the proteomics of seed development in oilseed crops. Finally, four chapters describe proteomes during pathogenic and symbiotic interactions between plants and microrganisms, and during plant stress responses. Regarding future perspectives, it is very important that diverse integrated approaches including advanced proteomic techniques combined with functional genomics, bioinformatics, metabolomics and/or with advanced molecular cell biology are nicely presented in several chapters. Thus, this book not only covers the rapid progress in the field of plant proteomics but also delivers this recent knowledge to a broad spectrum of readers including advanced students, teachers and researchers.

At this point I would like to thank my co-editor Jay Thelen and all the authors for their great job and excellent contributions to this book. Last but not least, my special thanks goes to my family, my wife Olinka and sons Matejko and Tomáško, for their encouragement and patience with me during this book project.

Bonn, April 2007 Jozef Šamaj

List of Abbreviations

Chapter 1

PAGE	polyacrylamide gel electrophoresis
SDS-PAGE	sodium dodecyl sulfate polyacrylamide gel electrophoresis
LC	liquid chromatography
2-D	two-dimensional
IPG	immobilized pH gradient
IEF	isoelectric focusing
CBB	Coomassie Brilliant Blue
ESI	Electrospray Ionization
MALDI	Matrix Assisted Laser Desorption Ionization
TOF	Time of Flight
PMF	peptide mass fingerprinting
EST	expressed sequence tag
ICAT	Isotope-Coded Affinity Tag
MS	Mass spectrometry

Chapter 2

2-DE	Two-dimensional gel electrophoresis
MS	mass spectrometry
IPG	immobilized pH gradients
NEPHGE	non-equilibrium pH gradient electrophoresis
mABC1	mitochondrial ATP-binding cassette protein 1
SB 3-10	N-decyl-N, N-dimethyl-3-ammonio-1-propane sulfonate
TBP	tributyl phosphine
DIGE	difference gel electrophoresis
MALDI	Matrix-assisted laser desorption/ionization
TOF	Time of flight
CID	collision-induced
IEF	isoelectric focusing
SDS-PAGE	sodium dodecyl sulfate polyacrylamide

	gel electrophoresis
EST	expressed sequence tag
CBB	Coomassie Brilliant Blue

Chapter 3

CID	collision-induced dissociation
PTM	post-translational modification
GPM	global proteome machine

Chapter 4

MAPK	mitogen-activated protein kinase
MAPKKKs	MAPK kinase kinases
MAPKKs	MAPK kinases
PP2C	Protein Ser/Thr phosphatase 2C
DsPs	Ser/Thr/Tyr phosphatases
PTP	protein Tyr phosphatase
MS	mass spectrometry
LC-MS	liquid chromatography-mass spectrometry
IMAC	immobilised metal ion affinity chromatography
SCX	strong cationic exchange
SRPK	SR protein-specific kinase
SILAC	Stable Isotope Labelling by Amino acids in Cell culture
TiO_2	Titanium dioxide
EST	expressed sequence tag

Chapter 5

2-DE	two-dimensional gel electrophoresis
MS	mass-spectrometry
UPA	universal protein array
PMAs	protein microarrays
AMAs	antibody microarrays
RPMAs	reverse protein microarrays
HMG	high mobility group
ACS-6	1-aminocyclopropane-1-carboxylic acid synthase-6
MKS1	MAPK substrate 1
VSP1	vegetative storage protein1
Pro-Q DPS	ProQ-Diamond phosphoprotein strain
CK2α	casein kinase2α
MPK	mitogen-activated kinase
PKA	Protein kinase A

Chapter 6

ECM	extracellular matrix
ER	endoplasmic reticulum

GAPDH	glyceraldehyde-3-phosphate dehydrogenase
2D-DiGE	2-dimensional difference gel electrophoresis
2-DE	2-dimensional gel electrophoresis
MS	mass-spectrometry
GFP	green fluorescent protein
FB1	fumonisin B1

Chapter 7

2-DE	2-dimensional gel electrophoresis
BN-PAGE	blue native-polyacrylamide gel electrophoresis
CK2α	casein kinase 2α
DIGE	difference gel electrophoresis
DLC	diamond-like carbon coated stainless steel
ESI	electrospray ionization
FT	Fourier transform ion cyclotron resonance
GFP	green fluorescent protein
ICAT	isotope-coded affinity tag
IMAC	immobilized metal affinity chromatography
IT	ion trap
iTRAQ	isobaric tag for relative and absolute quantitation
LC	liquid chromatography
MALDI	matrix-assisted laser desorption ionization
MS	mass spectrometry
SDS-PAGE	sodium dodecyl sulfate-polyacrylamide gel electrophoresis
PTM	post-translational modification
Q	quadrupole
TAP	tandem affinity purification
TOF	time-of-flight
2-DE	two-dimensional electrophoresis

Chapter 8

MS	mass-spectrometry
MALDI	matrix-assisted laser desorption ionization
Q	quadrupole
TOF	time-of-flight
ESI	electrospray ionization
eSLDB	eukaryotic sub-cellular localisation database
SUBA	subcellular location database for Arabidopsis proteins
GFP	green fluorescent protein
AMPDB	Arabidopsis Mitochondrial Protein Database
IMAC	metal ion affinity chromatography
TRX	thioredoxin
TAP	tandem affinity purification

TEV	tobacco etch virus
ICAT	isotope-coded affinity tag
iTRAQ	isobaric tag for relative and absolute quantitation
SILAC	stable isotope labeling with amino acids in cell culture
MRM	multiple reaction monitoring
HPLC	high performance liquid chromatography
LC	liquid chromatography

Chapter 9

2-DE	two-dimensional electrophoresis
Mb	Megabase
EST	expressed sequence tag
TC	tentative consensus
ppm	parts per million
IEF	isoelectric focusing
AM	arbuscular mycorrhizal
dai	days after inoculation
PR	pathogenesis-related
ABA	abscisic acid
Ado-Met	S-adenosyl-Met
LEA	late embryogenesis abundant
A. euteiches	Aphanomyces euteiches
NSF	N-ethylmaleimide-sensitive fusion
ABA	abscisic acid
G. mosseae	Glomus mosseae
G. intraradices	Glomus intraradices
iTRAQ	isobaric tags for relative and absolute quantitation
ICAT	isotope-coded affinity tags
IPG	immobilized pH gradient
LC-MS/MS	liquid chromatography coupled to tandem mass spectrometry
MALDI TOF-MS	matrix assisted laser desorption/ionization time-of-flight mass spectrometry
MS	mass spectrometry
MS/MS	tandem mass spectrometry
M. truncatula	Medicago truncatula
MudPIT	multidimensional protein identification technology
pI	isoelectric point
PMF	peptide mass fingerprinting
PR	pathogenesis-related
RuBisCO	ribulose 1,5-bisphosphate carboxylase/oxygenase
S. meliloti	Sinorhizobium meliloti.

Chapter 10

WAF	weeks after flowering

2-DE	two-dimensional gel electrophoresis
PTM	post-translational modifications
PMF	peptide mass fingerprint
TC	tentative consensus
IPG	immobilized pH gradient
MALDI-TOF	Matrix Assisted Laser Desorption Ionization-Time of Flight
EST	expressed sequence tag
SSP	seed storage proteins
LOX	lipoxygenases
SuSy	sucrose synthase
SBP	sucrose-binding protein
PEP	phosphoenolpyruvate
PDC	pyruvate dehydrogenase complex
FBA	fructose bisphosphate aldolase
RuBisCO	ribulose-1,5-bisphosphate carboxylase/oxygenase 3-PGA 3-phosphoglycerate
GAPDH	glyceraldehyde-3-phosphate dehydrogenase
PGK	phosphoglycerate kinase
PGM	phosphoglucomutase
PGI	phosphoglucose isomerase
TPI	triose-phosphate isomerases
iPGAM	2,3-bisphosphoglycerate-independent phopsphoglycerate mutase
PK	pyruvate kinase
LEA	late embryogenesis
IEF	isoelectric focusing
TAG	triacylglycerol
ER	endoplasmic reticulum
BiP	luminal binding protein
MS/MS	tandem mass spectrometry
CBB	Coomassie Brilliant Blue

Chapter 11

EST	expressed sequence tag
BY-2	Bright Yellow-2
2-DE	two dimensional gel-electrophoresis
PTM	post-translational modification
MS	mass spectroscopy
PMF	peptide mass fingerprint
DiGE	difference in gel electrophoresis
iTRAQ	isobaric tags for relative and absolute quantification
LC	liquid chromatography
IPG	immobilised pH gradient
SDS-PAGE	sodium dodecyl sulfate polyacrylamide

	gel electrophoresis
IEF	isoelectric focusing
LC-ESI-Q-TOF MS	liquid chromatography–electrospray ionisation–quadrupole–time-of-flight mass spectrometry
MALDI	matrix-assisted laser desorption/ionization
MS/MS	tandem MS
LPS	lipopolysaccharides
RuBP	ruthenium II tris bathophenanthroline disulfonate
ROS	reactive oxygen species
BN	blue native
RuBisCO	ribulose 1,5-bisphosphate carboxylase/oxygenase

Chapter 12

CWP	cell wall protein
HRGPs	Hyp-rich glycoproteins
H/PRPs	Hyp/Pro-rich proteins
GRPs	Gly-rich proteins
GH	glycoside hydrolases
XTH	xyloglucan endotransglucosylase/hydrolases
PMEs	pectin methylesterases
LRX	leucine-rich repeat-extensins
AGPs	arabinogalactan proteins
MS	mass spectrometry
PTM	post-translational modifications
CWME	cell wall modifying enzymes
CWMEI	inhibitors of cell wall modifying enzymes
PG	polygalaturonase
XEGIPs	xyloglucan endoglucanase inhibiting proteins
GPI	glycosylphosphatidylinositol
MALDI-TOF	Matrix-assisted laser desorption/ionization-time of flight
TDIF	tracheary differentiation inhibitory factor
PGIPs	inhibitors of polygalacturonases
PMEI	Inhibitor of PME
Nt-CIF	inhibitor of tobacco invertase
ER	endoplasmic reticulum

Chapter 13

MS	mass spectroscopy
RLKs	receptor-like protein kinases
IEF	iso-electrofocusing
CTAB	cationic trimethyl ammonium bromide
BN-PAGE	blue-native electrophoresis
MuDPIT	multidimensional protein identification technique

MALDI	matrix assisted laser desorption/ionisation
ESI	electrospray ionisation
PI-PLC	phosphatidylinositol phospholipase C
IMAC	immobilised metal ion affinity chromatography
LDS	lithium dodecyl sulphate
LOPIT	localisation of organelle proteins by isotope tagging
GPI	glycosylphosphatidylinositol
iTRAQ	isobaric tags for relative and absolute quantification
ER	endoplasmic reticulum
2-DE	two-dimensional gel electrophoresis
16-BAC	benzyldimethyl-n-hexadecylammonium chloride
CNBr	cyanogen bromide
GO	gene ontology
DIGE	difference in gel electrophoresis
ESTs	expressed sequence tags

Chapter 14

MS	mass spectroscopy
PTM	post-translational modification
TIC	Translocon at the Inner envelope membrane of Chloroplasts
TOC	Translocon at the Outer envelope membrane of Chloroplasts
2-DE	two-dimensional gel electrophoresis
TAT	twin-arginine translocation
TPR	tetratricopeptide
PPR	pentatricopeptide
PAP	plastid lipid-associated protein
ER	endoplasmic reticulum
ceQORH	chloroplast envelope quinone oxidoreductase homologue
IEP32	inner envelope protein of 32 kDa
CAH1	carbonic anhydrase 1
Clp	caseinolytic protease
IEF	isoelectric focusing

Chapter 15

BSA	bovine serum albumin
PVP	polyvinylpyrrolidone
FW	fresh weight
PAGE	Polyacrylamide gel electrophoresis
IEF	isoelectric focusing
SDS-PAGE	sodium dodecyl sulfate PAGE

MS	mass spectrometry
BN	blue native
TOM	translocase of the outer membrane
HSPs	heat shock proteins
CMS	cytoplasmic male sterility
PTM	post-translational modification
PDC	pyruvate dehydrogenase complex
TCA cycle	tricarboxylic acid cycle
AOS	active oxygen species
ROS	reactive oxygen species
HNE	4-hydroxy-2-nonenal
TRX	thioredoxin
Y2H	yeast two-hybrid technique
FRET	fluorescence resonance energy transfer
BRET	bioluminescence resonance energy transfer
GRAVY	grand average of hydrophobicity

Chapter 16

NoLS	Nucleolar localisation sequences
PLRV	potato leaf-roll virus
GFP	green fluorescent protein
2-DE	2D polyacrylamide gel electrophoresis
LC	liquid chromatography
MS	mass spectrometry
Mr	relative molecular mass
DiGE	difference gel electrophoresis
MuDPIT	multidimensional protein identification technique
MALDI	matrix assisted laser desorption/ionisation
ESI	electrospray ionisation
MS/MS	tandem MS
TOF	time-of-flight
FT-ICR-MS	Fourier-transform ion-cyclotron resonance mass spectrometer
snRNP	small nuclear RNP
PTMs	Post-translational modifications
GPI	glycosylphosphatidylinositol
SILAC	Stable Isotope Labelling by Amino acids in Cell culture
ICAT	isotope coded affinity tagging
LOPIT	Localisation of organelle proteins by isotope tagging
ER	endoplasmic reticulum
ITRAQ	Isobaric tag for relative and absolute quantification
TAP	tandem affinity purification
TEV	tobacco etch-virus
CBP	calmodulin-binding protein

EF	elongation factor
EST	expressed sequence tag

Chapter 17

PTMs	Post-translational modifications
MS	mass spectrometry
MS/MS	tandem mass spectrometry
2-DE	two-dimensional gel electrophoresis
GO	gene ontology
EST	expressed sequence tag
Q	quadrupole
ESI	electrospray ionisation
MALDI-TOF	matrix assisted laser desorption/ionisation–time-of-flight

Chapter 18

PAMPs	pathogen-associated molecular patterns
PRs	pathogenesis-related proteins
SAR	systemic acquired resistance
ROS	reactive oxygen species
PR	pathogenesis-related
SA	salicylic acid
JA	jasmonic acid
ET	ethylene
MS	mass spectrometry
AM	arbuscular mycorrhizal
2D-LC	two dimensional liquid chromatography
MS/MS	tandem mass spectrometry
Nep1	necrosis- and ethylene-inducing peptide
HR	hypersensitive response
MAPKs	mitogen-activated protein kinases
EST	expressed sequence tag
CSI	cross-species identification
AVR	avirulence
R	resistance
RLP	receptor-like protein
CITRX	Cf-9-interacting thioredoxin
NBS-LRR	nucleotide binding site leucine-rich repeat
GIP	glucanase inhibitor proteins
PI	protease inhibitors
TGases	transglutaminases
CBEL	cellulose binding elicitor lectin
PAMP	pathogen-associated molecular pattern
NLPs	Nep1-like proteins
CBD	cellulose-binding domain

PRLs	PR-like proteins
PAL	phenylalanine ammonia-lyase
TMV	tobacco mosaic virus
FHB	Fusarium head blight
SOD	superoxide dismutase
HRGP	hydroxyproline-rich glycoprotein
SAR	systemic acquired resistance

Chapter 19

2-DE	two-dimensional electrophoresis
MALDI	matrix assisted laser desorption ionisation
PMF	peptide mass fingerprinting
CDPK	calmodulin-like domain protein kinase
NDPK	Nucleoside diphosphate kinase

Chapter 20

AM	arbuscular mycorrhizal
1D-SDS PAGE	one dimensional sodium dodecyl sulfate polycacrylamide gel
2-DE	two-dimensional gel electrophoresis
LC	liquid chromatography
ESI-MS/MS	electrospray ionisation tandem mass spectrometry
ESI-Q-TOF	electrospray ionisation quadrupole time of flight
ESTs	expressed sequence tags
GPI	glycosylphosphatidylinositol
HPLC	high performance liquid chromatography
IPG	immobilized pH gradient
MALDI-TOF	matrix assisted laser desorption ionisation time of flight
PCR	polymerase chain reaction
PM	plasma membrane
RNAi	RNA interference
TILLING	targeting induced local lesions in genomes
PMF	peptide mass fingerprint
MS/MS	tandem mass spectrometry
HPLC	high performance liquid chromatography
DIGE	2-D difference gel electrophoresis
ICAT	isotope coded affinity tag
MUDPIT	multidimensional protein identification technology
RNAi	RNA interference

Chapter 21

MS	Mass spectrometry
MALDI-TOF	matrix assisted laser desorption ionisation time of flight

ABA	Abscisic acid
RuBisCO	ribulose 1,5-bisphosphate carboxylase/oxygenase
HSP	heat shock protein
2-DE	two-dimensional electrophoresis
PR	pathogenesis-related
ASR	ABA/stress/ripening responsive protein
COMT	caffeate-O-methyltransferase
SAM	S-adenosyl-L-methionine
QTLs	quantitative trait loci
PQL	protein quantity locus
LEA	late embryogenesis abundant
CV	coefficient of variation
ROS	reactive oxygen species
DIGE	Difference in-gel electrophoresis
PMF	peptide mass fingerprint
EIFs	eukaryotic initiation factors
SOD	superoxide dismutase
PG	plastoglobule
PC	phytochelatin
GSH	glutathione
BN	Blue native
GST	Glutathione S-transferase

Chapter 1
Introduction to Proteomics: a Brief Historical Perspective on Contemporary Approaches

Jay J. Thelen

Abstract The field of proteomics has experienced numerous milestones over the course of the past 35–40 years. As an introductory chapter to this larger review text on plant proteomics, this article provides a cursory historical perspective on protein separation and identification techniques widely used in plant biochemistry laboratories today. In the past 10 years alone, advancements in techniques such as two-dimensional gel electrophoresis, mass spectrometry, and mass spectral data mining have made previously intractable proteomics problems almost routine by today's standards. In analyzing these various proteomics approaches I also discuss and project their utility for the next generation of proteomics research.

1.1 Introduction

Proteomics, or the high-throughput identification and analysis of proteins, is an emerging field of research facilitated by numerous advancements over the past 35–40 years in protein separation, mass spectrometry, genome sequencing/annotation, and protein search algorithms. Recognizing this trend in the physical and life sciences, the term "proteome" was first used by Wilkins et al. (1995) to describe the protein complement to the genome. Since the first use of this term its meaning and scope have narrowed. The host of post-translational modifications, alternative splice products, and proteins intractable to conventional separation techniques has each presented a challenge towards the achievement of the classic definition of the word (Chapman 2000; Westermeier and Naven 2002; Wilkins and Gooley 1998). The broad dynamic range of protein expression has also contributed to difficulties in efforts towards identifying every protein expressed in the life cycle of any given organism (Corthals et al. 2000). For example, identification of every protein expressed in plant leaves would never reveal proteins that are specifically expressed

Jay J. Thelen: University of Missouri-Columbia, Department of Biochemistry, Columbia, MO 65211, E-mail: thelenj@missouri.edu

J. Šamaj and J. Thelen (eds.), *Plant Proteomics*
© Springer 2007

in roots. Despite these limitations, hundreds if not thousands of proteins can be resolved, profiled and identified using the latest methods – a remarkable achievement given the recent genesis of this discipline.

The purpose of this chapter is to briefly introduce and provide a historical perspective on established proteomics concepts and methods that are being used in many plant biology laboratories today to comparatively profile protein expression and identify proteins. I will also attempt to provide a perspective on the future outlook of each of these approaches. This introduction will hopefully be useful for non-experts in the field of proteomics as an aid to comprehension of most of the terminology and jargon used in this highly technical field of life sciences research. The varied approaches to proteomics research can be generally classified as having one of two major objectives: (1) protein or peptide separation, and (2) identification and characterization of resolved proteins or peptides, typically by mass spectrometry. I will address these two aspects of proteomics research in the first two sections in this introductory chapter and then discuss general strategies for quantitative protein profiling.

1.2 Protein Separation and Detection for Proteome Investigations

Currently, there are three preferred methods for separation of complex protein or peptide samples: (1) denaturing polyacrylamide gel electrophoresis (PAGE) also referred to as sodium dodecyl sulfate polyacrylamide (SDS-PAGE); (2) two-dimensional (2-D) gel electrophoresis; and (3) liquid chromatography (LC) a general term that includes all forms of ion exchange, affinity, and reversed-phase chromatography (Hunter et al. 2002). There are of course other forms of protein separation, including preparative isoelectric focusing (protein separation according to native charge) and native or blue-native PAGE, to name but a few alternative techniques. Due to space constraints however, only SDS-PAGE and 2-D gel electrophoresis will be discussed here.

1.2.1 Sodium Dodecyl Sulfate–Polyacrylamide Gel Electrophoresis

No protein separation technique is more widely used than SDS-PAGE, first reported by Laemmli in 1970. It would not be an exaggeration to state that nearly all contemporary laboratories performing life sciences research employ this technique. The widespread use of SDS-PAGE to separate proteins according to size can be attributed to its ease, reproducibility, and modest consumable and instrument expenses. Although an easy technique to perform, the resolving power of SDS-PAGE is somewhat limited. Mass spectrometry (MS) analysis of any single discreet SDS-PAGE protein band from a complex protein sample consistently reveals multiple proteins, frequently greater than ten (Phinney and Thelen 2005). However, for highly enriched samples of low complexity (<10 unique proteins) SDS-PAGE may be suitable. In general, accurate quantitative analysis of SDS-PAGE protein bands from a complex sample is not feasible as the volume of any band is the collective composition of each unique protein in that band. However, as a pre-fractionation technique for alternative quantification strategies including chemical labeling (using stable isotope conjugates; Ramus et al. 2006) and perhaps label-free quantification using recently developed software tools (SIEVE, DeCyder MS), SDS-PAGE may find a new niche as a rapid, reproducible separation technique prior to MS quantification.

1.2.2 Two-Dimensional Gel Electrophoresis

Around the time at which SDS-PAGE was introduced, O'Farrell applied isoelectric focusing (IEF) to protein samples prior to SDS-PAGE to pioneer the concept of two-dimensional (2-D) gel electrophoresis (O'Farrell 1975). Although extremely powerful in its resolving capabilities, this method suffered from reproducibility issues owing to the casting, focusing, and extrusion of the fragile tube gels used for IEF. Over the years this procedure has been improved through the introduction in 1978 (Görg et al. 1978) and recent commercialization (Görg et al. 2000) of the immobilized pH gradient (IPG) strip, to replace IEF tube gels, which has resulted in a major resurgence in this technique.

Reproducibility, sample loading and resolution for 2-D gel electrophoresis have significantly improved with the introduction of the IPG strip in conjunction with commercial Peltier-cooled programmable focusing units for IEF (Görg et al. 2000). These advancements have made 2-D electrophoresis an attractive method for the separation of complex protein samples. Besides the impressive separation capabilities, another reason 2-DE is frequently preferred to LC-based approaches for protein separation is that a reproducible 2-DE proteome reference map is a static, visual entity. A fully annotated 2-DE reference map for a specific organ, tissue, cell, or organelle of interest is a valuable tool that can save time and money when 'landmarking' differentially expressed proteins in response to a treatment, mutation, or transgene introduction. Although 2-D electrophoresis suffers from well-publicized limitations, such as under-representation of membrane proteins (Wilkins et al. 1998; Santoni et al. 2000), this time-honored method is presently one of the preferred approaches for quantitative characterization of complex protein samples. The popularity of 2-DE will no doubt continue with recent technical developments such as sensitive and quantitative pre- and post-electrophoretic stains for total proteins, as well as phospho- and glycoproteins, as discussed herein.

1.2.3 Extracting Proteins From Plant Samples

Performing 2-D electrophoresis with plant samples can be a challenging endeavor, in part due to the high carbohydrate:protein ratio in most plant tissues. Direct grinding of samples in IEF extraction media, while generally sufficient for non-plant cells, results in extensive streaking in the IEF dimension with plant tissues. Therefore, obtaining high quality 2D gels from plant tissues requires the removal of carbohydrates prior to IEF. Mooney et al. (2004) found that phenol partitioning of proteins followed by ammonium acetate/methanol precipitation was one consistent strategy with which to isolate proteins free of complex polysaccharides in mature soybean seeds. This approach was first employed for 2-DE by Hurkman and Tanaka (1986) for plant membrane proteins. This protein isolation procedure has also been successfully employed with developing seeds from soybean, castor, rapeseed, and Arabidopsis as well as purified oil bodies from rapeseed (Hajduch et al. 2005, 2006; Agrawal and Thelen 2006; Katavic et al. 2006). A comparison of phenol/methanol precipitation versus TCA/acetone extraction in grape berries revealed greater protein yield and spot resolution using the phenol extraction procedure (Vincent et al. 2006). Another investigation compared phenol/methanol extraction with two variations of TCA/acetone extraction in a range of tomato tissues and the fruits of banana, avocado, and orange (Saravanan and Rose 2004). This latter investigation reported that phenol extraction gave higher protein yields and greater spot resolution and intensity, particularly from fruits rich in polysaccharides. However, it was noted that the spot patterns were different amongst these extraction procedures, in part due to enhanced glycoprotein extraction with the phenol procedure. Despite the nearly universal success of the phenol extraction procedure for recalcitrant protein samples, other less time-consuming procedures have also proven successful with plant protein samples (discussed further in Chap. 2 by Hurkman and Tanaka).

1.2.4 In-Gel Detection of Proteins

Protein quantitation and detection is an area of proteomics research that has changed dramatically over the past 10 years. Although several methods for protein detection have been reported (for a comprehensive treatise, see Allen and Budowle 1999), silver and Coomassie Brilliant Blue (CBB) staining methods have historically been the preferred methods for in-gel protein detection. The two chemical forms of CBB, G-250 and R-250, differ in their sensitivity, quantitative linear range, and destaining properties. Since CBB G-250, also referred to as colloidal CBB, outperforms the R-250 variant on all counts it is generally recommended for proteomics applications. Silver staining is at least 10-fold more sensitive than colloidal CBB, with a reported detection range of 0.1–1 ng (Ocbs et al. 1981; Shevchenko et al. 1996). However, silver staining is plagued by problems such as inferior reproducibility, poor linear dynamic range, and non-quantitative negative-staining of some modified proteins (Westermeier and Naven 2002; Wilkins and Gooley 1998; Görg et al. 2000), all of which complicate downstream quantitation and spot matching. Broad dynamic range fluorescent protein stains including SyproRuby™, Deep Purple™, and ruthenium II, which have detection sensitivities around 10–20 ng (Rabilloud et al. 2000, 2001; Steinberg et al. 2000; Chevalier et al. 2004) are promising, but expensive, alternatives to Coomassie and silver staining as general protein stains. Other, more specific fluorescent stains include Pro-Q Diamond and Pro-Q Emerald (both manufactured by Invitrogen), which are specific for phosphoproteins and glycoproteins, respectively (Steinberg et al. 2000, 2001). As with any of these specialized commercial stains, expense is a concern when implementing their use. Modified protocols for both Sypro Ruby and Pro-Q Diamond have demonstrated that multi-fold dilutions of these commercial stains are possible without compromising linear dynamic range or sensitivity (Krieg et al. 2003; Agrawal and Thelen 2005). With any of these fluorescent stains, manual spot picking from gels can be a challenging task. Generally, "over-staining" of these gels with Coomassie or silver is required to accurately determine spot location (Agrawal and Thelen 2006). For further discussion on staining, see Chap. 2 by Hurkman and Tanaka.

1.3 Protein Identification using Mass Spectrometry

Analysis of any analyte by MS requires ionization of that molecule and its entrance into the gas phase. For large macromolecules such as proteins and peptides this was long considered a Herculean task, analogous to making elephants fly (Fenn 2003). The development and commercialization of two different "soft" ionization approaches, electrospray ionization (ESI) and matrix assisted laser desorption ionization (MALDI) enabled large macromolecules such as proteins to be analyzed either in flowing, liquid solution or in a dry, crystalline state, respectively (Fenn et al. 1989; Tanaka et al. 1988; Karas and Hillenkamp 1988). The importance of these developments was appreciated by the scientific community and in 2002 led to John Fenn and Koichi Tanaka sharing the Nobel prize in Chemistry. In ESI, high voltage is applied to a flowing solution containing the analyte as it passes through a narrow bore needle. As the solution of charged molecules evaporates, the emitted droplets shrink into smaller droplets and shortly thereafter enter the gas phase. In MALDI, a laser is fired at a stainless steel plate under vacuum. Dried on that plate is the analyte, co-crystallized with "matrix" – a small organic molecule that absorbs the pulsed UV laser light. Although the mechanism is not completely understood, the laser light pulsed on the co-crystallized matrix-analyte layer results in the vaporization of matrix and the associated analyte. Early work with MALDI necessitated high vacuum conditions, which were easily configured with time of flight (TOF) mass analyzers; however, the emergence of atmospheric pressure MALDI has allowed this ionization approach to be configured with many other types of mass analyzers. For an in-depth discussion of mass analyzers a recent review is available (see Domon and Aebersold 2005).

1.3.1 Peptide Mass Fingerprint Protein Identification

Identification of proteins using MS has multiple levels of complexity and rigor, each with specific instrumentation requirements. Minimally, accurate mass determination of at least four proteolytic peptides, and comparison to a database of predicted masses from in silico-digested proteins containing the authentic gene or protein, can determine the identity of a protein (Courchesne and Patterson 1999; Mann et al. 2001). This is referred to as peptide mass fingerprinting (PMF) and, while being the least rigorous approach to protein identification, it has been utilized widely for proteome investigations from model as well as crop plants (Porubleva et al. 2001; Mathesius et al. 2001; Kruft et al. 2001; Maltman et al. 2002; Bardel et al. 2002; Mooney et al. 2004; Hajduch et al. 2005). Protein identification by PMF analyses places a premium on mass accuracy and thus necessitates the use of mass spectrometers with sufficient resolution for monoisotopic mass determination. Since peptide fragmentation is not necessary for PMF identification, a MALDI-TOF mass spectrometer is generally the instrument of choice for this approach. PMF identification is a rapid and simple approach to protein identification and is best suited for high-throughput proteome projects in which the complete, annotated genome is available. In the absence of genome information, large expressed sequence tag (EST) datasets can be useful as evidenced by recent proteome investigations of model and crop plants including Medicago, maize, and soybean (Mathesius et al. 2001; Porubleva et al. 2001; Mooney et al. 2004). However, the use of non-redundant, contiguous EST databases, also referred to as Unigene databases (Pontius et al. 2002), yields higher confidence assignments than an unannotated, redundant EST database (Mooney et al. 2004). Although an inexpensive and rapid approach to assign protein identity, PMF is a poor approach to analyze samples that contain multiple proteins (e.g. SDS-PAGE bands). Therefore PMF is most effectively employed when identifying resolved spots from 2D gels, i.e., low complexity samples in which one protein is prominent. Protein samples that do not yield conclusive assignments by PMF must be further analyzed by tandem MS for a more accurate assignment.

1.3.2 Tandem Mass Spectrometry Protein Identification

Proceeding beyond PMF analysis requires fragmentation of individual peptides (precursor ions), which is most reliably performed within the collision cell of a tandem mass spectrometer using an inert gas such as nitrogen. Precursor ion fragmentation is followed by a second round of mass analysis, to characterize the fragment ions. Based upon the fragmentation spectrum of the precursor ion (proteolytic peptide), a partial or full sequence of the peptide can be obtained. Sequence construction of an unknown peptide based solely on the tandem MS spectrum is referred to as de novo peptide sequencing. One of the first uses of ESI-MS/MS for peptide sequencing was reported by Hunt et al. (1989). However, manual interpretation of tandem mass spectra can be a tedious and frustrating endeavor, particularly when the MS/MS spectrum provides only partial sequence information (as is frequently the case). In these instances the partial sequence information, plus the precursor mass and knowledge of the proteolytic cleavage rules together are generally sufficient information to query a gene or protein database and find a matching peptide. Specifically, automated peptide matching using MS/MS ion spectra is typically initiated by matching the precursor ion mass according to specified mass tolerance followed by comparison with in silico generated fragment ion series to find the closest match. This is the concept of what is now referred to as an MS/MS ions search (Perkins et al. 1999; Clauser et al. 1999). If the introduction of soft ionization approaches has revolutionized peptide mass spectrometry, the development of mass spectral search algorithms has arguably had a similar impact on the high-throughput analysis and interpretation of peptide mass spectral data. One only needs to visually analyze a tandem mass spectrum once to realize that manual de novo sequencing, though under-appreciated and critical for post-translational site mapping, does not lend itself to large-scale proteomics research.

Proteomics is strongly dependent upon the bioinformatic tools currently available to the research community. The three original proteomics search algorithms were Sequest (Ducret et al. 1998), Protein Prospector (Clauser et al. 1999), and Mascot (Perkins et al. 1999). However, mass spectral search tools are rapidly increasing in number; at least four new PMF or MS/MS ions search algorithms have been introduced in the past few years. Although most of these search engines are available for use at no-charge via the Internet, licensed copies for most of these programs are available and ensure faster, secure searches. Customized or proprietary databases can also be utilized with licensed copies of these programs. This is particularly important since the vast array of Unigene databases (composed of annotated, archived ESTs) are infrequently uploaded onto the workstations that serve these programs at no-charge. Using these databases, search times are reduced and hit accuracy improved when compared to searches of entire EST collections (Mooney et al. 2004).

Although highly useful, MS/MS ions search tools are arguably not as accurate at protein assignments as de novo peptide sequencing, particularly for plants that lack genome or EST resources for database comparisons. As an alternative to MS/MS ions searches, at least five software programs capable of de novo sequencing from tandem mass spectra data are available (Pevtsov et al. 2006). The efficacy of these programs varies considerably, also depending on the choice of tandem mass spectrometer (Pevtsov et al. 2006), and therefore further refinement of these algorithms is clearly needed. However, computer-assisted de novo sequencing appears a promising possibility for proteomics research on plants that lack genome or EST resources, for assisting post-translational site mapping, and for an alternative or parallel protein assignment workflow for any proteomics project. For an in-depth discussion of the status of proteomics bioinformatics tools I recommend the chapter in this book by Curtis Wilkerson (Chap. 3).

1.4 Differential Protein Profiling Approaches

The majority of proteomics research projects are profiling studies designed to identify differentially expressed proteins in response to a stimulus or physiological treatment. Like microarray profiling, protein profiling is almost entirely relative in scope; only rarely are the absolute levels of proteins determined. The methods employed and the statistical analyses performed ultimately determine the confidence and success of the survey. Fortunately, multiple approaches are now available for differential protein profiling. Comparison of protein expression among 2-D gels is generally performed by imaging gels post-stained with quantitative protein dyes, quantifying the volumes of defined protein spots, subtracting background and normalizing spot volumes, all using 2-D image analysis software, and finally comparing these values to the same protein spots observed in gels from other 'treated' samples (Blackstock and Weir 1999; Raman et al. 2002). This is a simple approach requiring only an imaging system and 2-D quantitation software. Using this method, sensitivity and accuracy in protein spot detection and quantitation is strongly dependent upon the choice of stain (and staining method), imaging devices and analysis platform (Raman et al. 2002; Nishihara and Champion 2002). For global investigations, sophisticated 2-D analysis software with sensitive spot detection and matching algorithms greatly simplify the analysis (Lopez 2000; Hajduch et al. 2005).

1.4.1 Difference Gel Electrophoresis

Although comparative 2-D electrophoresis is perhaps the most widely used protein profiling approach, technical reproducibility can be problematic due to variability with focusing of IPG strips and laboratory-cast polyacrylamide gels. To control these variables, a recently developed method for differential protein profiling utilizes N-hydroxysuccinimide-ester conjugated cyanine fluorophores to react with free primary amine groups on proteins prior to 2-D electrophoresis separation (Mustafa et al. 1997; Tonge et al. 2001). Internal standards, control and experimental samples are separately labeled with three different cyanine dyes (Cy2, Cy3 and Cy5), pooled, and resolved on the same 2-D gel. Imaging is typically performed with a laser imager with dual photomultiplier tubes for increased throughput and imaging accuracy. Overlaying of colored images allows relative expression level comparisons to be visualized directly. Quantification can be performed by most image analysis software, particularly since spot matching is easy under these conditions. This approach, which is beginning to be used by plant biologists, promises to be particularly valuable for profiling samples with minor differences in protein expression (e.g., transgenic germplasm or stimulus-treated plants; Casati et al. 2005) as well as highly diverse protein samples that would otherwise present a challenge for protein spot matching (Mooney et al. 2006).

1.4.2 Chemical Labeling Approaches for Relative Quantification

An improved approach for quantitative expression profiling of polypeptides from whole tissues takes advantage of the high resolution capabilities of current mass spectrometers. The isotope-coded affinity tag (ICAT) procedure pioneered by Aebersold and colleagues allows comparative analyses to be performed on protein samples using stable isotopic labeling to discriminate two, pooled, samples for relative quantitation (Gygi et al. 1999). This method is facilitated by the development of iodoacetamide derivatives containing zero or eight deuterium ions on a linker region separating the cysteine reactive end from a biotin affinity tag. Conjugation to whole protein samples is followed by trypsin cleavage, ion exchange chromatography, affinity purification of labeled peptides and subsequent biotin cleavage/elution prior to LC-MS/MS to "mass discriminate" between control and experimental samples. Relative expression levels are determined by comparing the abundance of each Cys-labeled peptide from the two samples. Identifications are made from the MS/MS spectra of individual peptides. The availability of kits (commercialized by Applied Biosystems, Foster City, CA) for the reagents and columns as well as software to automate identification and quantitation promises to simplify this approach considerably.

The concept of chemical labeling peptides for quantitation can be further expanded from Cys labeling (ICAT) to consideration of other residues. This is particularly important as Cys residues are absent in some proteins and peptides, rendering them undetectable with this approach. A variation on the chemical labeling principle targets free primary amines (Lys) with N-hydroxysuccinimide conjugation chemistry, referred to as iTRAQ. However, unlike the case of ICAT, the chemical labels contain labile, isobaric mass tags that easily fragment during a round of tandem mass spectrometry. Thus, with iTRAQ chemical labeling, the peptides are quantified in the low mass range of the MS/MS spectrum instead of the MS spectrum. Quantifying the peptides in the MS/MS spectrum which also contains the peptide fragmentation information promises greater quantitative confidence. Also, ICAT offers only pairwise comparisons whereas iTRAQ reagents are capable of four sample multiplexing. Therefore, iTRAQ chemistry appears to be a significant step forward in the field of commercial chemical labeling strategies and is beginning to be applied to quantitative plant proteome investigations (Jones et al. 2006).

1.5 Outlook

In the 12 years since the first mention of the word 'proteome' the field of proteomics has had numerous breakthrough advances, particularly with respect to protein separation techniques, MS instrumentation and mass spectral search algorithms. I expect this trend to continue as this technology is placed into the hands of more biologists and biochemists. More sensitive, higher resolution mass spectrometers are being introduced on almost an annual basis, and with new, bundled instrumentation packages, new strategies for proteome characterization are becoming a reality for non-technical biologists. For example, top-down proteomics whereby intact protein masses are determined followed by collision fragmentation for protein assignment (Zabrouskov et al. 2003), once considered to be possible only with advanced Fourier transform ion cyclotron resonance mass spectrometers, now seems achievable with the recently released LTQ-Orbitrap mass spectrometer (Scigelova and Makarov 2006). Top-down proteomics offers the possibility of discriminating post-translational modifications in a more systematic manner than traditional bottom-up strategies that start with proteolytic peptides. Also, the introduction of label-free quantitative software programs such as DeCyder MS and SIEVE holds tremendous promise for quantitative profiling without the need for expensive and time-consuming 2-D gel-based or chemical labeling approaches. If these and other hardware and software developments continue to deliver on their promises, the next 12 years of proteomics research will undoubtedly be exciting.

Acknowledgements The author acknowledges Dr. Brian Mooney (University of Missouri, Columbia) for critical reading of this manuscript.

References

Agrawal GK, Thelen JJ (2005) Development of a simplified, economical polyacrylamide gel staining protocol for phosphoproteins. Proteomics 5:4684–4688

Agrawal GK, Thelen JJ (2006) Large-scale identification and quantitative profiling of phosphoproteins expressed during seed filling in oilseed rape. Mol Cell Prot 5:2044–2059

Allen RC, Budowle B (1999) Protein staining and identification techniques. The Biotechniques Series on Molecular Laboratory Methods, Eaton, Manchester

Bardel J, Louwagie M, Jaquinod M, Jourdain A, Luche S, Rabilloud T, Macherel D, Garin J, Bourguignon J (2002) A survey of the plant mitochondrial proteome in relation to development. Proteomics 2:880–898

Blackstock WP, Weir MP (1999) Proteomics: quantitative and physical mapping of cellular proteins. Trends Biotechnol 17:121–127

Bourguignon J (2002) A survey of the plant mitochondrial proteome in relation to development. Proteomics 2:880–898

Casati P, Zhang X, Burlingame AL, Walbot V (2005) Analysis of leaf proteome after UV-B irradiation in maize lines differing in sensitivity. Mol Cell Prot 4:1673–1685

Chapman JR (2000) Mass spectrometry of proteins and peptides. Humana, Totowa, NJ

Chevalier F, Rofidal V, Vanova P, Bergoin A, Rossignol M (2004) Proteomic capacity of recent fluorescent dyes for protein staining. Phytochemistry 65:1499–1506

Clauser KR, Baker PR, Burlingame AL (1999) Role of accurate mass measurement (+/– 10 ppm) in protein identification strategies employing MS or MS/MS and database searching. Anal Chem 71:2871–2882

Corthals GL, Wasinger VC, Hochstrasser DF, Sanchez J-C (2000) The dynamic range of protein expression: a challenge for proteomic research. Electrophoresis 21:1104–1115

Courchesne PL, Patterson SD (1999) Identification of proteins by matrix-assisted laser desorption/ionization mass spectrometry using peptide and fragment ion masses. In: Link A (ed) 2-D proteome analysis protocols. Humana, Totowa, NJ, pp 487–511

Domon B, Aebersold R (2005) Mass spectrometry and protein analysis. Science 312:212–217

Ducret A, Van Oostveen I, Eng JK, Yates JR 3rd, Aebersold R (1998) High throughput protein characterization by automated reverse-phase chromatography/electrospray tandem mass spectrometry. Protein Sci 7:706–719

Fenn JB (2003) Electrospray wings for molecular elephants (Nobel lecture). Angew Chem Int Ed Engl 42:3871–3894

Fenn JB, Mann M, Meng CK, Wong SF, Whitehouse CM (1989) Electrospray ionization for mass spectrometry of large biomolecules. Science 246:64–71

Görg A, Postel W, Westermeier R (1978) Ultrathin-layer isoelectric focusing in polyacrylamide gels on cellophane. Anal Biochem 89:60–70

Görg A, Obermaier C, Boguth G, Harder A, Scheibe B, Wildgruber R, Weiss W (2000) The current state of two-dimensional electrophoresis with immobilized pH gradients. Electrophoresis 21:1037–1053

Gygi SP, Rist B, Gerber SA, Turecek F, Gelb MH, Aebersold R (1999) Quantitative analysis of complex protein mixtures using isotope-coded affinity tags. Nat Biotechnol 17:994–999

Hajduch M, Ganapathy A, Stein JW, Thelen JJ (2005) A systematic proteomic study of seed-filling in soybean: establishment of high-resolution two-dimensional reference maps, expression profiles, and an interactive proteome database. Plant Physiol 137:1397–1419

Hajduch M, Casteel JE, Hurrelmeyer KE, Song Z, Agrawal GK, Thelen JJ (2006) Proteomic analysis of seed filling in *Brassica napus*: developmental characterization of metabolic isozymes using high-resolution two-dimensional gel electrophoresis. Plant Physiol 141:32–46

Hunt DF, Zhu NZ, Shabanowitz J (1989) Oligopeptide sequence analysis by collision-activated dissociation of multiply charged ions. Rapid Commun Mass Spectr 3:122–124

Hunter TC, Andon NL, Koller A, Yates JR III, Haynes PA (2002) The functional proteomics toolbox: methods and applications. J Chromatogr B 782:165–181

Hurkman WJ, Tanaka CK (1986) Solubilization of plant membrane proteins for analysis by two-dimensional gel electrophoresis. Plant Physiol 81:802–806

Jones AM, Bennett MH, Mansfield JW, Grant M (2006) Analysis of the defence phosphoproteome of *Arabidopsis thaliana* using differential mass tagging. Proteomics 6:4155–4165

Karas M, Hillenkamp F (1988) Laser desorption ionization of proteins with molecular masses exceeding 10,000 Daltons. Anal Chem 60:2299–2301

Katavic V, Agrawal GK, Hajduch M, Harris SL, Thelen JJ (2006) Protein and lipid composition analysis of oil bodies from two *Brassica napus* cultivars. Proteomics 6:4586–4598

Krieg RC, Paweletz CP, Liotta LA, Petricoin EF 3rd (2003) Dilution of protein gel stain results in retention of staining capacity. Biotechniques 35:376–378

Kruft V, Eubel H, Jansch L, Werhahn W, Braun H-P (2001) Proteomic approach to identify novel mitochondrial proteins in Arabidopsis. Plant Physiol 127:1694–1710

Laemmli UK (1970) Cleavage of structural proteins during the assembly of the head of bacteriophage T4. Nature 227:680–685

Lopez MF (2000) Better approaches to finding the needle in a haystack: optimizing proteome analysis through automation. Electrophoresis 21:1082–1093

Maltman DJ, Simon WJ, Wheeler CH, Dunn MJ, Wait R, Slabas AR (2002) Proteomic analysis of the endoplasmic reticulum from developing and germinating seed of castor (*Ricinus communis*). Electrophoresis 23:626–639

Mann M, Hendrickson RC, Pandey A (2001) Analysis of proteins and proteomes by mass spectrometry. Annu Rev Biochem 70:437–473

Mathesius U, Keijzers G, Natera SHA, Weinman JJ, Djordjevic MA, Rolfe BG (2001) Establishment of a root proteome reference map for the model legume *Medicago truncatula* using the expressed sequence tag database for peptide mass fingerprinting. Proteomics 1:1424–1440

Mooney BP, Krishnan HB, Thelen JJ (2004) High-throughput peptide mass fingerprinting of soybean seed proteins: automated workflow and utility of Unigene expressed sequence tag databases for protein identification. Phytochemistry 65:1733–1744

Mooney BP, Miernyk JA, Greenlief CM, Thelen JJ (2006) Using quantitative proteomics of Arabidopsis roots and leaves to predict metabolic activity. Physiol Plant 128:237–250

Mustafa U, Morgan ME, Minden JS (1997) Difference gel electrophoresis: a single method for detecting changes in protein extracts. Electrophoresis 18:2071–2077

Nishihara JC, Champion KM (2002) Quantitative evaluation of proteins in one- and two-dimensional polyacrylamide gels using a fluorescent stain. Electrophoresis 23:2203–2215

Ocbs DC, McConkey EH, Sammons DW (1981) Silver stains for proteins in polyacrylamide gels: a comparison of six methods. Electrophoresis 2:304–307

O'Farrell PH (1975) High resolution two-dimensional electrophoresis of proteins. J Biol Chem 250:4007–4021

Perkins DN, Pappin DJ, Creasy DM, Cottrell JS (1999) Probability-based protein identification by searching sequence databases using mass spectrometry data. Electrophoresis 20:3551–3567

Pevtsov S, Fedulova I, Mirzaei H, Buck C, Zhang X (2006) Performance evaluation of existing de novo sequencing algorithms. J Proteome Res 5:3018–3028

Phinney B, Thelen JJ (2005) Proteomic characterization of a Triton X-100 insoluble fraction from chloroplasts defines a novel group of proteins associated with macromolecular structures. J Proteome Res 4:497–450

Pontius JU, Wagner L, Schuler GD (2002) UniGene: a unified view of the transcriptome. In: The NCBI Handbook, National Center for Biotechnology Information, Bethesda, MD, pp 201–212

Porubleva L, VanderVelden K, Kothari S, Oliver DJ, Chitnis PR (2001) The proteome of maize leaves: use of gene sequences and expressed sequence tag data for identification of proteins with peptide mass fingerprints. Electrophoresis 22:1724–1738

Rabilloud T, Strub J-M, Luche S, Girardet JL, van Dorsselaer A, Lunardi J (2000) Ruthenium II tris (bathophenanthroline disulfonate), a powerful fluorescent stain for detection of proteins in gel with minimal interference in subsequent mass spectrometry analysis. Proteome 1:1–14

Rabilloud T, Strub J-M, Luche S, van Dorsselaer A, Lunardi J (2001) A comparison between SyproRuby and ruthenium II tris (bathophenanthroline disulfonate) as fluorescent stains for protein detection in gels. Proteomics 1:699–704

Raman B, Cheung A, Marten MR (2002) Quantitative comparison and evaluation of two commercially available, two-dimensional electrophoresis image analysis software packages, Z3 and Melanie. Electrophoresis 23:2194–2202

Ramus C, Gonzalez de Peredo A, Dahout C, Gallagher M, Garin J (2006) An optimized strategy for ICAT quantification of membrane proteins. Mol Cell Prot 5:68–78

Santoni V, Molloy M, Rabilloud T (2000) Membrane proteins and proteomics: Un amour impossible? Electrophoresis 21:1054–1070

Saravanan RS, Rose JK (2004) A critical evaluation of sample extraction techniques for enhanced proteomic analysis of recalcitrant plant tissues. Proteomics 4:2522–2532

Scigelova M, Makarov A (2006) Orbitrap mass analyzer – overview and applications in proteomics. Proteomics 6:16–21

Shevchenko A, Wilm M, Vorm O, Mann M (1996) Mass spectrometric sequencing of proteins silver-stained polyacrylamide gels. Anal Chem 68:850–858

Steinberg TH, Chernokalskaya E, Berggren K, Lopez MF, Diwu Z, Haugland RP, Patton WF (2000) Ultrasensitive fluorescence protein detection in isoelectric focusing gels using a ruthenium metal chelate stain. Electrophoresis 21:486–496

Steinberg TH, Pretty On Top K, Berggren KN, Kemper C, Jones L, Diwu Z, Haugland RP, Patton WF (2001) Rapid and simple single nanogram detection of glycoproteins in polyacrylamide gels and on electroblots. Proteomics 1:841–855

Tanaka KH, Wake H, Ido Y, Akita S, Yoshida, Y, Yoshida I (1988) Protein and polymer analyses up to m/z 100,000 by laser ionization time-of-flight mass spectrometry. Rapid Commun Mass Spectr 2:151–153

Tonge RP, Shaw J, Middleton B, Rowlinson R, Rayner S, Young J, Pognan F, Hawkins E, Currie I, Davison M (2001) Validation and development of fluorescence two-dimensional difference gel electrophoresis proteomics technology. Proteomics 1:377–396

Vincent D, Wheatley MD, Cramer GR (2006) Optimization of protein extraction and solubilization for mature grape berry clusters. Electrophoresis 27:1853–1865

Westermeier R, Naven T (2002) Proteomics in practice. Wiley-VCH, Weinheim

Wilkins MR, Gooley AA (1998) Protein identification in proteome projects. In: Appel RD, Hochstrasser DF (eds) Proteome research: new frontiers in functional genomics. Springer, New York, pp 35–64

Wilkins MR, Sanchez JC, Gooley AA, Appel RD, Humphery-Smith I, Hochstrasser DF, Williams KL (1995) Progress with proteome projects: why all proteins expressed by a genome should be identified and how to do it. Biotechnol Genet Eng Rev 13:19–50

Wilkins MR, Gasteiger E, Sanchez J-C, Bairoch A, Hochstrasser DF (1998) Two dimensional gel electrophoresis for proteome projects: the effects of protein hydrophobicity and copy number. Electrophoresis 19:1501–1505

Zabrouskov V, Giacomelli L, van Wijk KJ, McLafferty FW (2003) A new approach for plant proteomics: characterization of chloroplast proteins of *Arabidopsis thaliana* by top-down mass spectrometry. Mol Cell Prot 2:1253–1259

Chapter 2
High-Resolution Two-Dimensional Gel Electrophoresis: A Cornerstone of Plant Proteomics

William J. Hurkman and Charlene K. Tanaka

Abstract Two-dimensional gel electrophoresis (2-DE) in combination with mass spectrometry (MS) is a principal tool for plant proteomic studies. Global protein identification together with analysis of the intrinsic patterns and relationships of protein populations provide new insights into the interrelated biochemical processes of specific proteomes. High-resolution 2-D gels in which proteins are clearly resolved facilitate MS identification and pattern analysis. Of the many factors that affect 2-D gel quality, optimization of protein extraction, solubilization, and detection methods ensure high resolution 2-D gels and maximal proteome coverage. Plant proteomic studies most often investigate total protein populations. Protein extraction protocols that utilize TCA or phenol are widely used because protein degradation and non-protein components that interfere with protein separation during electrophoresis are minimized. The high dynamic range of protein abundance in eukaryotic cells makes analysis of low abundance proteins problematic, but a wide array of separation techniques is available for isolating specific protein populations. Protein solubilization requires a buffer that denatures and reduces proteins as well as preventing their aggregation and precipitation. 2-DE gel patterns can differ substantially depending on the combination of chaotropes, reducing agents, and surfactants in the chosen buffer. Protein detection methods must be compatible with MS analysis and have the sensitivity and dynamic range required for pattern analysis. The sensitivity of Coomassie brilliant blue can rival that of silver stains, but the quantitative linear dynamic range for both these stains is only one order of magnitude. Alternatively, fluorescent stains are more sensitive, with a dynamic range covering three to four orders of magnitude. Obtaining high resolution 2-D gels of plant proteins can be challenging, but the technique, when optimized, is invaluable for profiling and comparing plant proteomes.

William J. Hurkman, US Department of Agriculture, Agricultural Research Service, Western Regional Research Center, Albany, CA 94710, USA, E-mail: bhurkman@pw.usda.gov

Charlene K. Tanaka, US Department of Agriculture, Agricultural Research Service, Western Regional Research Center, Albany, CA 94710, USA

J. Šamaj and J. Thelen (eds.), *Plant Proteomics*
© Springer 2007

2.1 Introduction

The past two decades have seen the introduction of large-scale investigation into the characterization of biological systems. Genomic analysis, in particular assembly of complete genomes and compilation of transcription products, has provided an invaluable foundation for this research. However, the translation of this wealth of genomic data into functional analysis of cellular processes will be in large part dependent upon characterization of the proteome it encodes. As with genomic studies, proteomic research has seen the introduction and refinement of a variety of techniques suitable for large-scale analysis. One of the most important of these techniques is high-resolution two-dimensional gel electrophoresis (2-DE) which, in combination with mass spectrometry (MS), provides a central tool for the identification of specific gene products and characterization of proteome-wide profiles (Agrawal and Rakwal 2006; Cánovas et al. 2004; Ephritikhine et al. 2004; Görg et al. 2004; Newton et al. 2004; Rose et al. 2004; Wittmann-Liebold et al. 2006).

From its beginnings nearly half a century ago, true high-resolution 2-DE did not become practical until the mid-1970s (O'Farrell 1975; Klose 1975; Scheele 1975) due to important technical advances that included the addition of carrier ampholytes, urea, and non-ionic detergent to the solubilization buffer and the first dimension gel as well as the addition of SDS and reducing agent to the second dimension gel. This evolution of isoelectric focusing (IEF) in the first dimension and sodium dodecyl sulfate polyacrylamide gel electrophoresis (SDS-PAGE) in the second dimension became the foundation for high-resolution protein mapping studies of complex biological samples. The development of sensitive Edman microsequencing methods in the 1980s made possible the identification of the small amounts of protein present in 2-D gel spots (Aebersold et al. 1986, 1987) and ushered in the age of proteomics.

During the late 1980s and into the 1990s, progress in genome and expressed sequence tag (EST) sequencing projects resulted in databases that served as important resources for predicted protein sequences. Concurrently, refinements in MS instrumentation, including increased sensitivity, selectivity, and mass measurement accuracy coupled with its ability to analyze and identify proteins in gel spots utilizing the rapidly growing databases, led to wide use of MS in the biological sciences (Aebersold and Goodlett 2001, Newton et al. 2004). Refinements in 2-DE methodology, including separation of proteins using immobilized pH gradients (IPG) in the first dimension, and the discovery of increasingly sensitive fluorescent dyes for protein detection, as well as automation of spot picking, in-gel spot digestion, and gel analysis have increased 2-DE sample throughput, continue to make 2-DE the method of choice for protein separations. These advances and many others set the stage for the increasing number of plant 2-DE/MS proteomics studies that are highlighted in a number of recent reviews (Agrawal and Rakwal 2006; Agrawal et al. 2005b, 2005c; Baginsky and Gruissem 2006; Cánovas et al. 2004; Ephritikhine et al. 2004; Holtorf et al. 2002; Peck 2005).

In the 2-DE/MS proteomics approach, proteins from plant samples are extracted, solubilized, and separated in 2-D gels (Fig. 2.1). The gels are scanned and the

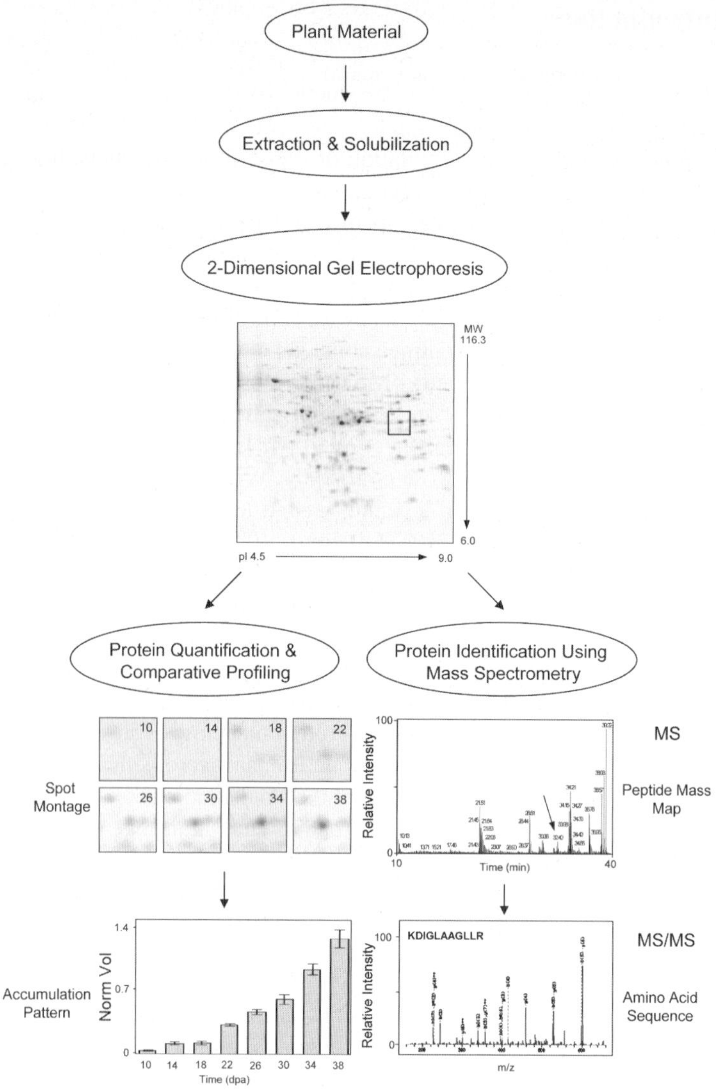

Fig. 2.1 Two-dimensional gel electrophoresis (2-DE) in the proteomics age. Proteins from plant samples are extracted, solubilized, and separated by 2-DE. The 2-D gel displays albumins and globulins extracted from wheat endosperm. The gel is digitized and spots detected and quantified using computer software. The *box* in the 2-D gel outlines the protein in the *spot montage*, which depicts this spot during a developmental time course. The *bar graph* shows that this protein increases significantly during grain development. The spot is excised from the gel, subjected to in-gel digestion with trypsin, and the peptide fragments analyzed by mass spectrometry (MS). The *upper graph* displays the peptide mass map for the excised spot. The *lower graph* displays the amino acid sequence for the indicated peak, which identifies the protein as a peroxidase

digitized images are imported into gel analysis software for spot detection and matching. Following normalization of spot volumes, qualitative and quantitative differences between protein populations are identified. Spots are excised from the gels and subjected to in-gel digestion with site-specific protease(s), most commonly trypsin, to generate peptide fragments. The fragments are analyzed by MS and proteins identified by matching experimental peptide masses and sequence information with theoretical tryptic peptide masses generated in silico from protein sequence or translated nucleotide sequence databases. High-resolution gels are essential to this approach; proteins that are clearly resolved from one another greatly facilitate subsequent computer analysis and MS identification. Obtaining high-resolution 2-D gels is challenging, because there is no universal protocol for all protein samples; 2-DE, often considered a routine technique, is a multi-step procedure that requires as much art as it does skill to master. Many variables affect the resolution of 2-D gels. Major points of optimization of the technique include choice of extraction, solubilization, and equilibration buffers, choice of first dimension pH range and second dimension acrylamide concentration, selection of protocols for reproducible first and second dimension electrophoresis, selection of staining procedure, and source and age of reagents (Dunn and Burghes 1983a, 1983b; Dunbar 1987; Wittmann-Liebold et al. 2006). Although all of these factors are extremely important, this review focuses on protein extraction, solubilization, and detection, because these steps are the key to high-resolution 2-DE and maximal proteome coverage. The review concludes with a brief discussion of comparative proteomics, the quantitative comparison of protein populations for which high-resolution 2-D gels are unrivaled.

2.2 Protein Extraction

2.2.1 Total Proteomes

Plant proteomic studies most often investigate total protein populations. Extracting proteins from plant samples is challenging, because plant cells generally contain low amounts of protein protected by cell walls that require extreme measures to disrupt. A number of methods, individually or in combination, can be used to homogenize plant tissues but one of the most successful is to grind plant samples to a powder with a pestle in a mortar containing liquid nitrogen (e.g., Giavalisco et al. 2003; Hajduch et al. 2005; Hurkman and Tanaka 2004; Sheoran et al. 2005). The ideal extraction procedure for plant proteins would be a rapid one-step procedure utilizing a single buffer. The diversity of protein solubilities and plant tissue composition ensures that no single method can be effective with all samples. Plant cells often contain proteases that remain active in the extraction buffer, reducing and altering protein populations. Plant cells also contain various non-protein components that, when present in protein extracts, cause streaking and smearing of the 2-DE patterns. Among these components are cell wall and storage polysaccharides, lipids, phenolics, salts, nucleic acids, and a broad array of secondary metabolites (Görg et al. 2004). The optimal extraction procedure must minimize protein degradation and eliminate non-protein components that interfere with protein separation during electrophoresis.

One of the most popular extraction protocols utilizes TCA in combination with acetone to extract proteins from plant tissues (Damerval et al. 1986). The low pH of the TCA inactivates proteases and acetone-insoluble proteins are separated from acetone-soluble compounds that interfere with electrophoresis. Another commonly used protocol involves partitioning of proteins into phenol and subsequent precipitation with ammonium acetate in methanol or acetone (Schuster and Davies 1983; Hurkman and Tanaka 1986). As in the TCA/acetone method, the presence of phenol minimizes proteolysis during extraction (Schuster and Davies 1983) and separates proteins from interfering compounds (Hurkman and Tanaka 1986). The phenol method has been used to extract proteins from recalcitrant plant samples, including membrane fractions (DuPont et al. 1988; Hurkman and Tanaka 1986; Hurkman et al. 1988). The phenol method gave a higher protein yield than the TCA/acetone method for tomato fruit pericarp, which is rich in soluble pectins (Saravanan and Rose 2004). Recently, it was shown that sequentially extracting leaves and fruit – organs that contain large amounts of polyphenols – with TCA/acetone followed by phenol was necessary for high-resolution 2-D gels (Wang et al. 2006).

In our laboratory, the TCA/acetone method is a starting point for extraction of proteins from plant samples. It is rapid, inhibits proteases, minimizes compounds that interfere with electrophoresis, precipitates the majority of proteins, and generally results in high-resolution 2-D gels. The phenol method is employed when the TCA/acetone method does not produce high-resolution 2-D gels. One problem with the phenol method is that not all protein classes partition into the phenol phase. We found that some protein classes in wheat endosperm are lost because they are soluble in the ammonium acetate/methanol used to precipitate proteins from the phenol phase (W.J. Hurkman and C.K. Tanaka, unpublished observations). In contrast, Saravanan and Rose (2004) noted that, compared to the TCA/acetone method, the phenol method showed enhanced extraction of glycoproteins from tomato fruit pericarp. These examples emphasize the importance of comparing extraction procedures, monitoring protein amounts, and examining gel patterns for the various fractions generated by sample extraction protocols.

2.2.2 *Subproteomes*

The high dynamic range of protein abundance in eukaryotic cells prevents analysis of low abundance proteins. The concentration of individual proteins in a single cell spans five to six orders of magnitude, e.g. ranging from several million copies/cell for some highly abundant proteins to a few copies/cell for low abundance proteins (Patterson 2004; Patterson and Aebersold 2003). This enormous dynamic range presents a major challenge to the detection, quantification, and identification of low abundance proteins. It has been estimated that only 25% of an expected proteome can be observed in 2-D gels (Patterson 2004). One strategy for improving proteome coverage is the use of multiple, overlapping, narrow-range, first dimension IPG gels (Bak-Jensen et al. 2004; Corthals et al. 2000; Görg et al. 2000; Gygi et al. 2000; Hoving et al. 2000; Rabilloud 2002; Wildgruber et al. 2000). While it is true that this approach increases the numbers of proteins that can be displayed, entire proteome coverage is still not possible. This is not surprising since, in addition to abundance, proteins are highly heterogeneous and structurally complex molecules that exhibit a wide range of molecular mass, charge, solubility properties, cellular locations, and post-translational modifications. In order to analyze low abundance proteins, the most viable strategy is to isolate subproteomes by exploiting these protein properties. For example, leaf and seed proteomes are dominated by one or a few proteins that, when depleted, greatly extend proteome coverage of the remaining proteins. For example, RuBisCO dominates the leaf proteome; Kim et al. (2001) extracted rice leaf proteins with a Mg/Nonidet P-40 buffer and further fractionated this extract using polyethylene glycol to obtain fractions depleted in RuBisCO. This strategy improved proteome coverage 2.7-fold over TCA/acetone extraction. Storage proteins dominate seed proteomes. In total extracts of wheat endosperm, the gliadins and glutenins mask all but the most abundant albumins and globulins. Hurkman and Tanaka (2004) exploited the solubility properties of these proteins in KCl to effectively separate the gliadins and glutenins from the low abundance albumins and globulins. This fractionation method improved proteome coverage twofold and made it possible to analyze developmental patterns of the many previously 'hidden' albumins and globulins (Vensel et al. 2005).

A wide array of techniques is available for isolating protein subpopulations. The advantages of fractionation are that it reduces sample complexity while enriching for specific protein classes. Proteomic studies have been conducted on protein fractions prepared by traditional chromatographic methods (ion exchange chromatography, hydrophobic interaction chromatography, reverse-phase chromatography, size exclusion chromatography, affinity chromatography) as well as electrokinetic methods (e.g. Rotofor); details are provided in earlier reviews (Righetti et al. 2005; Shaw and Riederer 2003). Centrifugation – a classical technique for separating cell proteomes – has been employed extensively in Arabidopsis proteomics for analysis of fractions enriched in amyloplasts, cell walls, chloroplasts, cytoplasmic ribosomes, endoplasmic reticulum, mitochondria, peroxisomes, plasma membrane, and vacuoles. Proteomic analysis of subcellular fractions provides new insights into organelle function, regulation, and intracellular protein trafficking; a number of reviews provide detailed overviews of such studies (Agrawal et al. 2005a; Baginsky and Gruissem 2006; Cánovas et al. 2004; Corthals et al. 2000; Ephritikhine et al. 2004; Peck 2005; Righetti et al. 2005; Rose et al. 2004; Stasyk and Huber 2004).

2.3 Protein Solubilization

When the extraction protocol is optimized, protein solubilization becomes the most critical factor in obtaining high-resolution 2-D gels. In this step, proteins are denatured, disaggregated, and reduced to achieve complete disruption of molecular interactions while maintaining inherent charge properties. Like extraction protocols, optimal solubilization protocols depend on the sample and, as a result, many protocols are available (see Görg et al. 2004; Rose et al. 2004; Shaw and Riederer 2003). Sample solubilization is usually carried out using a buffer that contains chaotropes, reducing agents, and surfactants – reagents that denature and reduce proteins as well as prevent protein aggregation and precipitation. The composition of this buffer must be optimized to solubilize particular protein classes. For example, substitution of 2 M for 9.2 M urea results in more complete extraction and solubilization of wheat seed storage proteins (Payne et al. 1985). Solubilization of hydrophobic membrane proteins is greatly improved by combining thiourea (Rabilloud 1998) with urea, substituting CHAPS and other sulfobetaine surfactants for NP-40, and replacing 2-mercaptoethanol with DTT (Görg et al. 2004; Molloy 2000; Herbert 1999).

In recent years IPG strips have replaced carrier ampholytes in tube gels for first dimension IEF (reviewed in Görg et al. 2004; Wittmann-Liebold et al. 2006). Compared to carrier ampholytes, IPGs generate more stable and reproducible pH gradients that can be adjusted to produce extremely broad or narrow ranges. IPG strips are easier to implement than carrier ampholytes in tube gels in 2-DE, but basic proteins (pI > 10) and membrane proteins continue to be difficult to analyze (Wittmann-Liebold 2006). Arguably, narrow range IPG strips solved the problem of analysis of basic proteins (Shaw and Riederer 2003; Wittmann-Liebold 2006), but non-equilibrium pH gradient electrophoresis (NEPHGE) remains a useful technique (O'Farrell 1975; O'Farrell et al. 1977; Wittmann-Liebold 2006). Transfer of proteins (Méchin et al. 2003), especially membrane proteins (McDonough and Marbán 2005; Molloy 2000), from the IPG strip to the second dimension is problematic due to protein aggregation or precipitation, or tight association with the acrylamide gel matrix of the strip. Méchin et al. (2003) found that a solubilization buffer containing two chaotropes, two reducing agents, two detergents, and two kinds of carrier ampholytes enhanced the transfer of maize endosperm proteins from the IPG strip to the second dimension. McDonough and Marbán (2005) improved transfer of mABC1 (mitochondrial ATP-binding cassette protein 1), a basic integral membrane protein, to the second dimension by increasing the SDS concentration in the equilibration buffer and increasing the volume of this buffer relative to the IPG strip.

Like extraction protocols, no single solubilization method is effective for all samples. Depending on the buffer used to solubilize the protein extracts, 2-DE gels patterns can differ substantially, both qualitatively and quantitatively. Méchin et al. (2003) evaluated the efficiency of 15 different buffers in solubilizing proteins extracted from maize endosperm using TCA/acetone. The number of spots detected in the 2-D gels ranged from 1,033 to 1,826 depending on chaotrope concentration and combination, reducing agent mixture, detergent cocktail, and carrier ampholyte addition. Molloy et al. (1998) sequentially solubilized Tris-insoluble proteins from *Escherichia coli* with a buffer containing urea, CHAPS and DTT and a buffer containing urea, thiourea, CHAPS, SB 3–10 (N-decyl-N, N-dimethyl-3-ammonio-1-propane sulfonate), and TBP (tributyl phosphine). Although 89% of the initial *E. coli* sample mass was solubilized in the first buffer, the second buffer solubilized membrane proteins that were insoluble in the first buffer. These examples illustrate the important role that buffer composition plays in protein solubilization and underlines the fact that, like protein extraction methods, solubilization methods must be optimized.

2.4 Protein Detection

Numerous procedures are available for visualizing proteins in 2-D gels. Important selection criteria include ease of use, reliability, sensitivity, and compatibility with MS. Protocols utilizing organic dyes, silver stains, and fluorescent stains are the most popular. Coomassie Brilliant Blue (CBB) dye has traditionally been used to stain proteins in 2-D gels due to its ease of use and compatibility with subsequent protein identification methods. A principle limitation of CBB staining is sensitivity, which typically ranges from 200 to 500 ng protein/spot for conventional CBB methods utilizing R-250 (Wilson 1979). However, the colloidal CBB G-250 method (Neuhoff et al. 1985, 1988) lowers the detection limit to approximately 30 ng protein/spot, and a recent modification of this method (Candiano et al. 2004) further lowers the detection limit to 1 ng protein/spot. The linear dynamic range of both CBB R-250 and CBB G-250 is limited to one order of magnitude (Patton 2002).

Silver staining has long been the dominant method for protein detection because sensitivity is 1–10 ng protein/spot (Switzer et al. 1979; Rabilloud 1992). Only a few silver staining procedures are compatible with protein digestion and MS (Mortz et al. 2001). Since silver forms complexes with nucleophilic groups, such as the $-NH_2$ group of lysine (Rabilloud 1990), silver staining intensity correlates with protein lysine content (Yüksel and Gracy 1985). In a number of silver staining methods, glutaraldehyde- or formaldehyde-based sensitizers are used to promote the binding of silver to proteins. These aldehydes modify and crosslink lysine residues, reducing the efficiency of peptide extraction from the gel as well as preventing complete trypsin digestion of the protein (Shevchenko et al. 1996). These disadvantages were overcome by replacing the aldehydes with sodium thiosulfate as the sensitizer (Shevchenko et al. 1996) and, additionally, destaining the silver-stained proteins with potassium ferricyanide and sodium thiosulfate (Gharahdaghi et al. 1999). These improvements have been incorporated into silver staining protocols optimized for staining sensitivity, peptide recovery, and compatibility with trypsin digestion and MS analysis (Mortz et al. 2001). Like CBB stains, silver staining has a linear dynamic range of one order of magnitude (Patton 2000).

Methods for the fluorescent detection of proteins have recently become available (Patton 2002), the most popular of which is SYPRO Ruby, a proprietary ruthenium-based metal chelate stain (Mackintosh et al. 2003; Rabilloud et al. 2001). The SYPRO Ruby staining procedure is rapid and compatible with MS identification. Sensitivity is approximately 1–2 ng protein/spot and the linear dynamic range is at least three orders of magnitude (Breggren et al. 2002). Lightning Fast, a fluorescence-based stain containing the fluorophore epicocconone from the fungus *Epicoccum nigrum*, is the most sensitive stain to date (Mackintosh et al. 2003). The detection limit for Lightning Fast is 100 pg protein/spot and the linear dynamic range is more than four orders of magnitude.

An advantage of fluorescence stains is that different protein populations can be labeled with different, spectrally distinct, fluorophores. In difference gel electrophoresis (DIGE), cyanine-based fluorescent dyes (Cy2, Cy3, and Cy5) that have matched masses and charges, but different excitation and emission wavelengths, are employed to covalently label the lysyl residues in different protein populations (Tonge et al. 2001; Ünlü et al. 1997). In a standard protocol (e.g. Amme et al. 2006), proteins from paired samples are labeled with Cy3 or Cy5. The samples are combined prior to separation by 2-DE and the gels scanned consecutively at the two excitation wavelengths. Because proteins in both samples have the same relative mobilities regardless of the dye used to label them, and the protein populations are separated under identical electrophoretic conditions, the digitized images can be overlaid and compared directly. In a refinement of the method, an internal standard consisting of a mixture of equal amounts of the control and treated samples is also labeled with a third dye, Cy2, and combined with the Cy3- or Cy5-labeled samples. The inclusion of this internal standard allows normalization of protein volumes and improves quantitative comparison of protein levels. The detection limit for the cyanine-based fluorescent dyes is less than 1 ng protein/spot and the linear dynamic range is four to five orders of magnitude (Tonge et al. 2001). One requirement for DIGE is that, to maintain protein solubility during electrophoresis, only 1–2% of the lysine residues can be labeled with the dyes (Ünlü et al. 1997). Because the small population of labeled proteins migrates slightly slower than the bulk of unlabeled proteins, the 2-D gels must be post-stained to excise spots for MS identification (Gharbi et al. 2002; Tonge et al. 2001). CBB has been used for post-staining gels containing fluorescent-labeled proteins (Amme et al. 2006; Tonge et al. 2001), but SYPRO Ruby, with its lower detection limit, is the stain of choice (Gharbi et al. 2002; Zhou et al. 2002). DIGE is effective for pairwise comparisons, but is less suited for studies that require quantitative analysis of complex experiments that involve, for example, multiple replicates of multiple treatments over a developmental time course (Patton 2002).

2.5 Comparative Proteomics

A principal goal of proteome analysis is comparative proteomics, the quantitative comparison of protein populations. This approach often requires quantitative analysis of numerous sets of gels across multiple experiments. The first step in the analysis is to scan the 2-D gels and capture them as digital images, which are imported into gel analysis software for spot detection, pattern matching, and spot quantification. Although image analysis is 'automated', results are far from perfect. Images must be inspected and edited manually to verify that spot detection and matching are correctly and uniformly applied across all of the gels. This verification is essential to ensure that normalized spot volumes are correctly associated with properly matched spots. Clearly, high-resolution2-D gels with reproducible patterns facilitate accurate spot detection and matching, reduce manual editing, and ensure accurate quantitative data. With gel analysis complete, the stage is set to mine the data set for the intrinsic patterns and relationships that, together with protein identifications, provide insights into the interrelated biochemical processes associated with specific proteomes.

Comparative proteomics relies on protein identifications, with MS being the method of choice for the global identification of proteins contained in 2-D gel spots (Aebersold and Goodlett 2001; Aebersold and Mann 2003). Matrix-assisted laser desorption/ionization (MALDI) and electrospray ionization (ESI) are the two techniques most commonly used to volatize and ionize proteins or peptides for MS analysis. Prior to analysis by MS, spots are excised from the gels and subjected to in-gel digestion with site-specific protease(s), most commonly trypsin, to generate peptide fragments. Development of robotic spot picking and in-gel spot digestion instrumentation has minimized the labor required in these operations. For MALDI, the digested protein sample is mixed with a crystalline matrix and the sample/matrix is spotted onto a plate to co-crystallize. The plate is placed into the instrument and bombarded by a laser, converting the samples to single charged ions in a gas phase. MALDI is usually coupled to a time-of-flight (TOF) analyzer, which measures the mass of intact peptides. A list of the peptide masses, termed the mass fingerprint, is derived from the mass spectrum for each protein. Through the use of appropriate computer algorithms, proteins are identified by matching the experimentally determined peptide masses with those calculated from entries in sequence databases. For ESI, the tryptic peptides are separated by a micro-HPLC system coupled to the mass spectrometer. The HPLC effluent is sprayed into the MS and the liquid droplets are ionized to the gas phase. ESI is most often coupled to ion traps and triple quadrupole instruments and used to generate fragment ion (collision-induced, CID) spectra of selected precursor ions. Protein identifications using CID spectra are more precise than those achieved by mass mapping because, in addition to the peptide mass, the peak pattern in the CID spectrum also provides information about the peptide sequence (Fig. 2.1). Proteins are identified by matching experimental peptide sequence and peptide mass information with the same information calculated from entries in sequence databases. MS results list protein identifications and confidence matching scores, i.e., an evaluation of the probability that the sample protein is the specific protein in the database. Protein identifications provide the basis for assigning biological function, and peptide sequence information is invaluable for searching for related proteins and nucleotides.

2.6 Conclusions

Two-dimensional gel electrophoresis, with its unparalleled resolving power of complex biological samples and unrivaled ability to globally identify quantitative differences between protein populations, will remain a widely employed tool for proteomics studies. The key to high-resolution 2-DE is tailoring extraction and solubilization methods for optimal proteome coverage. The broad array of cellular functions carried out by structural and enzymatic proteins is reflected in the diverse physical/chemical properties of individual members of the proteome. While this divergence of properties (abundance, solubility, charge, molecular mass, etc.) prevents visualization of the total proteome in a single 2-D gel, appropriate fractionation techniques effectively extend proteome coverage. Some questions, such as the identification of proteins associated with subcellular organelles and the characterization of proteins that undergo post-translational modification, can best be answered by the analysis of subproteomes.

Comparative 2-DE proteomics relies on computer algorithms for spot detection, matching, and quantification. Improved algorithms are needed to fully automate these analyses and reduce user intervention. Data mining tools are making it possible to identify intrinsic patterns and relationships in 2-D gel data sets, but much better tools must be developed. The role of proteomic characterization, particularly 2-DE-based protocols, in large-scale research on complex biological systems cannot be overstated. While genomic/transcriptomic analyses indicate the potential for accumulation of specific enzymatic and structural proteins, these data are insufficient for defining the nature and dynamics of proteome profiles. In addition, as characterization of cellular chemistry enters the arena of large-scale experimentation (e.g., metabolomics), proteomics will play a central role in establishing the essential links between genes and metabolism.

Disclaimer The mention of a trademark or proprietary product does not constitute a guarantee or warranty of the product by the United States Department of Agriculture and does not imply its approval to the exclusion of other products that may be suitable.

References

Aebersold R, Goodlett DR (2001) Mass spectrometry in proteomics. Chem Rev 101:269–295

Aebersold R, Mann M (2003) Mass spectrometry-based proteomics. Nature 422:198–207

Aebersold RH, Teplow DB, Hood LE, Kent SB (1986) Electroblotting onto activated glass. High efficiency preparation of proteins from analytical sodium dodecyl sulfate-polyacrylamide gels for direct sequence analysis. J Biol Chem 261:4229–4238

Aebersold R, Leavitt J, Saavedra RA, Hood LE, Kent SB (1987) Internal amino acid sequence analysis of proteins separated by one- or two-dimensional gel electrophoresis after in situ protease digestion on nitrocellulose. Proc Natl Acad Sci USA 84:6970–6974

Agrawal GK, Rakwal R (2006) Rice proteomics: a cornerstone for cereal food crop proteomics. Mass Spectrom Rev 25:1–53

Agrawal GK, Yonekura M, Iwahashi Y, Iwahashi H, Rakwal R (2005a) System, trends and perspectives of proteomics in dicot plants. Part I: Technologies in proteome establishment. J Chromatogr B Anal Technol Biomed Life Sci 815:109–123

Agrawal GK, Yonekura M, Iwahashi Y, Iwahashi H, Rakwal R (2005b) System, trends and perspectives of proteomics in dicot plants. Part II: Proteomes of the complex developmental stages. J Chromatogr B Anal Technol Biomed Life Sci 815:125–136

Agrawal GK, Yonekura M, Iwahashi Y, Iwahashi H, Rakwal R (2005c) System, trends and perspectives of proteomics in dicot plants. Part III: Unraveling the proteomes influenced by environment, and at the levels of function and genetic relationships. J Chromatogr B Anal Technol Biomed Life Sci 815:137–145

Amme S, Matros A, Schleseer B, Mock H-P (2006) Proteome analysis of cold stress response in *Arabidopsis thaliana* using DIGE-technology. J Exp Bot 57:1537–1546

Baginsky S, Gruissem W (2006) *Arabidopsis thaliana* proteomics: from proteome to genome. J Exp Bot 57:1485–1491

Bak-Jensen KS, Laugesen S, Roepstorff P, Svensson B (2004) Two-dimensional electrophoresis pattern (pH 6–11) and identification of water-soluble barley seed and malt proteins by mass spectrometry. Proteomics 4:728–742

Breggren KN, Schulenberg B, Lopez MF, Steinberg TH, Bogdanova A, Smejkal G, Wang A, Patton WF (2002) An improved formulation of SYPRO Ruby protein gel stain: comparison with the original formulation and with a ruthenium II tris (bathophenanthroline disulfonate) formulation. Proteomics 2:486–498

Candiano G, Bruschi M, Musante L, Santucci L, Ghiggeri GM, Carnemolla B, Orecchia P, Zardi L, Righetti PG (2004) Blue silver: a very sensitive colloidal Coomassie G-250 staining for proteome analysis. Electrophoresis 25:1327–1333

Cánovas FM, Dumas-Gaudot E, Recorbet G, Jorrin J, Mock H-P, Rossignol M (2004) Plant proteome analysis. Proteomics 4:285–298

Corthals GL, Wasinger VC, Hochstrasser DF, Sanchez J-C (2000) The dynamic range of protein expression: a challenge for proteomic research. Electrophoresis 21:1104–1115

Damerval C, de Vienne D, Zivy M, Thellement H (1986) Technical improvements in two-dimensional electrophoresis increase the level of genetic variation detected in wheat-seedling proteins. Electrophoresis 7:52–54

Dunbar BS (1987) Two-dimensional electrophoresis and immunological techniques. Plenum, New York

Dunn MJ, Burghes AHM (1983a) High resolution two-dimensional polyacrylamide gel electrophoresis I. Methodological procedures. Electrophoresis 4:97–116

Dunn MJ, Burghes AHM (1983b) High resolution two-dimensional polyacrylamide gel electrophoresis II. Analysis and applications. Electrophoresis 4:173–189

DuPont FM, Tanaka CK, Hurkman WJ (1988) Separation and immunological characterization of membrane fractions from barley roots. Plant Physiol 86:717–724

Ephritikhine G, Ferro M, Rolland N (2004) Plant membrane proteomics. Plant Physiol Biochem 42:943–962

Gharahdaghi F, Weinberg CR, Meagher DA, Imai BS, Mische SM (1999) Mass spectrometric identification of proteins from silver-stained polyacrylamide gel: a method for the removal of silver ions to enhance sensitivity. Electrophoresis 20:601–605

Gharbi S, Gaffney P, Yang A, Zvelebil MJ, Cramer R, Waterfield MD, Timms JF (2002) Evaluation of two-dimensional differential gel electrophoresis for proteomic expression analysis of a model breast cancer cell system. Mol Cell Prot 1:91–98

Giavalisco P, Nordhoff E, Lehrach H, Gobom J, Klose J (2003) Extraction of proteins from plant tissues for two-dimensional electrophoresis analysis. Electrophoresis 24:207–216

Görg A, Obermaier C, Boguth G, Harder A, Scheibe B, Wildgruber R, Weiss W (2000) The current state of two-dimensional electrophoresis with immobilized pH gradients. Electrophoresis 21:1037–1053

Görg A, Weiss W, Dunn MJ (2004) Current two-dimensional electrophoresis technology for proteomics. Proteomics 4:3665–3685

Gygi SP, Corthals GL, Zhang Y, Rochon Y, Aebersold R (2000) Evaluation of two-dimensional gel electrophoresis-based proteome analysis technology. Proc Natl Acad Sci USA 97:9390–9395

Hajduch M, Ganapathy A, Stein JW, Thelen JJ (2005) A systematic proteomic study of seed filling in soybean. Establishment of high-resolution two-dimensional reference maps, expression profiles, and an interactive proteome database. Plant Physiol 137:1397–1419

Herbert B (1999) Advances in protein solubilisation for two-dimensional electrophoresis. Electrophoresis 20:660–663

Holtorf H, Guitton M-C, Reski R (2002) Plant functional genomics. Naturwissenschaften 89:235–249

Hoving S, Voshol H, van Oostrum J (2000) Towards high performance two-dimensional gel electrophoresis utilizing ultrazoom gels. Electrophoresis 21:2617–2621

Hurkman WJ, Tanaka CK (1986) Solubilization of plant membrane proteins for analysis by two-dimensional gel electrophoresis. Plant Physiol 81:802–806

Hurkman WJ, Tanaka CK (2004) Improved methods for separation of wheat endosperm proteins and analysis by two-dimensional gel electrophoresis. J Cereal Sci 40:295–299

Hurkman WJ, Tanaka CK, DuPont FM (1988) The effects of salt stress on polypeptides in membrane fractions from barley roots. Plant Physiol 88:1263–1273

Kim ST, Cho KS, Jang YS, Kang KY (2001) Two-dimensional electrophoretic analysis of rice proteins by polyethylene glycol fractionation for protein arrays. Electrophoresis 22:2103–2109

Klose J (1975) Protein mapping by combined isoelectric focusing and electrophoresis of mouse tissues. A novel approach to testing for induced point mutations in mammals. Humangenetik 26:231–243

Mackintosh JA, Choi H-Y, Bae S-H, Veal DA, Bell PJ, Ferrari BC, Van Dyk DD, Verrills NM, Paik Y-K, Karuso P (2003) A fluorescent natural product for ultra sensitive detection of proteins in one-dimensional and two-dimensional gel electrophoresis. Proteomics 3:2273–2288

McDonough J, Marbán E (2005) Optimization of IPG strip equilibration for the basic membrane protein mABC1. Proteomics 5:2892–2895

Méchin V, Consoli L, Le Guillou M, Damerval C (2003) An efficient solubilization buffer for plant proteins focused in immobilized pH gradients. Proteomics 3:1299–1302

Molloy MP (2000) Two-dimensional electrophoresis of membrane proteins using immobilized pH gradients. Anal Biochem 280:1–10

Molloy MP, Herbert BR, Walsh BJ, Tyler MI, Traini M, Sanchez J-C, Hochstrasser DF, Williams KL, Gooley AA (1998) Extraction of membrane proteins by differential solubilization for separation using two-dimensional gel electrophoresis. Electrophoresis 19:837–844

Mortz E, Krogh TN, Vorum H, Görg A (2001) Improved silver staining protocols for high sensitivity protein identification using matrix-assisted laser desorption/ionization–time of flight analysis. Proteomics 1:1359–1363

Neuhoff V, Stamm R, Eibl H (1985) Clear background and highly sensitive protein staining with Coomassie Blue dyes in polyacrylamide gels: a systematic analysis. Electrophoresis 6:427–448

Neuhoff V, Arold N, Taube D, Erhardt W (1988) Improved staining of proteins in polyacrylamide gels including isoelectric focusing gels with clear background at nanogram sensitivity using Coomassie Brilliant Blue G-250 and R-250. Electrophoresis 9:255–262

Newton RP, Brenton AG, Smith CJ, Dudley E (2004) Plant proteome analysis by mass spectrometry: principles, problems, pitfalls and recent developments. Phytochemistry 65:1449–1485

O'Farrell PH (1975) High resolution two-dimensional electrophoresis of proteins. J Biol Chem 250:4007–4021

O'Farrell PZ, Goodman HM, O'Farrell PH (1977) High resolution two-dimensional electrophoresis of basic as well as acidic proteins. Cell 12:1133–1142

Patterson SD (2004) How much of the proteome do we see with discovery-based proteomic methods and how much do we need to see? Curr Proteomics 1:3–12

Patterson SD, Aebersold RH (2003) Proteomics, the first decade and beyond. Nat Genet 33:311–323

Patton WF (2000) A thousand points of light: the application of fluorescence detection technologies to two-dimensional gel electrophoresis and proteomics. Electrophoresis 21:1123–1144

Patton WF (2002) Detection technologies in proteome analysis. J Chromatogr B 771:3–31

Payne PI, Holt LM, Jarvis MG, Jackson EA (1985) Two-dimensional fractionation of the endosperm proteins of bread wheat (*Triticum aestivum*): biochemical and genetic studies. Cereal Chem 62:319–326

Peck SC (2005) Update on proteomics in *Arabidopsis*. Where do we go from here? Plant Physiol 138:591–599

Rabilloud T (1990) Mechanisms of protein silver staining in polyacrylamide gels: a 10-year synthesis. Electrophoresis 11:785–794

Rabilloud T (1992) A comparison between low background silver diammine and silver nitrate protein stains. Electrophoresis 13:429–439

Rabilloud T (1998) Use of thiourea to increase the solubility of membrane proteins in two-dimensional electrophoresis. Electrophoresis 19:758–760

Rabilloud T (2002) Two-dimensional gel electrophoresis in proteomics: old, old fashioned, but it still climbs the mountains. Proteomics 2:3–10

Rabilloud T, Strub J-M, Luche S, van Dorsselaer A, Lunardi J (2001) A comparison between SYPRO Ruby and ruthenium II tris (bathophenanthroline disulfonate) as fluorescent stains for protein detection in gels. Proteomics 1:699–704

Righetti PG, Castagna A, Antonioli P, Boschetti E (2005) Prefractionation techniques in proteome analysis: the mining tools of the third millennium. Electrophoresis 26:297–319

Rose JKC, Bashir S, Giovannoni JJ, Jahn MM, Saravanan, RS (2004) Tackling the plant proteome: practical approaches, hurdles and experimental tools. Plant J 39:715–733

Saravanan, RS, Rose JKC (2004) A critical evaluation of sample extraction techniques for enhanced proteomic analysis of recalcitrant plant tissues. Proteomics 4:2522–2532

Scheele GA (1975) Two-dimensional gel analysis of soluble proteins: characterization of guinea-pig exocrine pancreatic proteins. J Biol Chem 250:5375–5385

Schuster AM, Davies E (1983) Ribonucleic acid and protein metabolism in pea epicotyls I. The aging process. Plant Physiol 76:809–816

Shaw MM, Riederer BM (2003) Sample preparation for two-dimensional gel electrophoresis. Proteomics 3:1408–1417

Sheoran IS, Olson DJH, Ross ARS, Sawhney VK (2005) Proteome analysis of embryo and endosperm from germinating tomato seeds. Proteomics 5:3752–3764

Shevchenko A, Wilm M, Vorm O, Mann M (1996) Mass spectrometric sequencing of proteins from silver-stained polyacrylamide gels. Anal Chem 68:850–858

Stasyk T, Huber LA (2004) Zooming in: fractionation strategies in proteomics. Proteomics 4:3704–3716

Switzer III RC, Merrill CR, Shifrin S (1979) A highly sensitive silver stain for detecting proteins and peptides in polyacrylamide gels. Anal Biochem 98:231–237

Tonge R, Shaw J, Middleton B, Rowlinson R, Rayner S, Young J, Pognan F, Hawkins E, Currie I, Davison M (2001) Validation and development of fluorescence two-dimensional differential gel electrophoresis proteomics technology. Proteomics 1:377–396

Ünlü M, Morgan ME, Minden JS (1997) Difference gel electrophoresis. A single gel method for detecting changes in protein extracts. Electrophoresis 18:2071–2077

Vensel WH, Tanaka CK, Cai N, Wong JH, Buchanan BB, Hurkman WJ (2005) Developmental changes in the metabolic proteins profiles of wheat endosperm. Proteomics 5:1594–1611

Wang W, Vignani R, Scali M, Cresti M (2006) A universal and rapid protocol for protein extraction from recalcitrant plant tissues for proteomic analyses. Electrophoresis 27:2782–2786

Wildgruber R, Hardre A, Obermaier C, Goguth G, Weill W, Fey SJ, Larsen PM, Görg A (2000) Towards higher resolution: two-dimensional electrophoresis of *Saccharomyces cerevisiae* proteins using overlapping narrow immobilized pH gradients. Electrophoresis 21:2610–2616

Wilson, CM (1979) Staining of proteins on gels: comparisons of dyes and procedures. Methods Enzymol 91:236–247

Wittmann-Liebold B, Graack H-R, Pohl T (2006) Two-dimensional gel electrophoresis as tool for proteomics studies in combination with protein identification by mass spectrometry. Proteomics 6:4688–4703

Yüksel KU, Gracy RW (1985) The quantitation of proteins in silver stained polyacrylamide gels. Electrophoresis 6:361–366

Zhou G, Li H, DeCamp D, Chen S, Shu H, Gong Y, Flaig M, Gillespie JW, Hu N, Taylor PR, Emmert-Buck MR, Liotta LA, Petricoin EF III, Zhao Y (2002) 2D differential in-gel electrophoresis for the identification of esophageal scans cell cancer-specific protein markers. Mol Cell Prot 1:117–123

Chapter 3
An Introduction to Proteomics Data Analysis

Curtis G. Wilkerson

Abstract Recent advances in techniques and instrumentation now allow many researchers to acquire large proteomics data sets. Such data sets require more rigorous analysis than was required when single proteins were identified from purified or partially purified samples. This chapter discusses the types of software programs used to analyze these data sets. A detailed examination of the open software program X!-Tandem is used to illustrate how such programs are used to generate reliable protein identifications from mass spectra.

3.1 Introduction

To appreciate the complexity in interpreting the results from a standard proteomics experiment it is useful to understand how the experiments are conducted and how the data is analyzed. Until recently, researchers analyzed simple protein mixtures. This required purifying proteins using various biochemical techniques or using a two dimensional gel electrophoresis (2-DE) technique. In such samples one protein is expected to predominate in abundance. Two methods were successful in analyzing such samples. One method involved digesting the protein with an enzyme such as trypsin and determining the masses of the fragments produced. Since trypsin cleaves at only 2 of the 20 amino acids, it is possible to produce a list of expected fragment masses for every protein in any sample. Software is available that will report the probability of a protein being present in the sample given the mass accuracy of the mass spectrometer, the mass of the fragments observed, and the size of the protein set searched (Cottrell 1994). This method is referred to as peptide mass fingerprinting. Mass fingerprinting is unable to successfully identify

Curtis G. Wilkerson, Michigan State University, Department of Energy Plant Research Laboratory, Michigan State University, East Lansing, MI 48824, USA, E-mail: Wilker13@msu.edu

J. Šamaj and J Thelen (eds.), *Plant Proteomics*
© Springer 2007

proteins in complex mixtures (Thiede et al. 2005) and becomes less successful as the number of proteins searched increases. It is most effective when used to identify proteins from organisms with small proteomes such as bacteria and yeast.

In the second type of analysis, peptides are analyzed in a mass spectrometer that can perform tandem mass analysis. In a technique known as MS/MS analysis, a peptide referred to as the precursor ion is isolated and fragment ions are produced by collision of the isolated ion with neutral gas molecules in a process termed collision-induced dissociation (CID). The masses of the fragment ions produced by CID are then determined. Due to the speed of modern mass spectrometers this technique can analyze moderately complex mixtures of proteins by the sequential analysis of different precursor ions. The data produced by an MS/MS experiment can be used in two different ways to identify proteins. Both rely on the fact that MS/MS spectra resulting from CID have two main ion series resulting from peptide bond cleavage. The b-ion series are fragments containing the amino-terminal amino acid and the y-ion series are fragments containing the carboxy-terminal amino acid. The pattern of sequential amino acid mass loss from the precursor ion in these two ion series can be used to determine the amino acid sequence of the peptide. This is referred to as de novo sequencing and several software programs are available to automate this process (Taylor and Johnson 2001; Ma et al. 2003). The resulting peptide sequences can then be matched to proteins using a modified BLAST sequence comparison program optimized for short peptide sequences that takes into account the ambiguities resulting from amino acids that have similar masses (e.g., Leu/Ile; Shevchenko et al. 2001). The second strategy is to predict an MS/MS spectrum for each of the peptides in a fixed, designated database that would generate a mass nearly identical to the mass of the precursor ion, given the accuracy of the MS instrument. These in silico spectra are then compared to an experimentally observed MS/MS spectrum and a score is assigned to indicate the closeness of the match. In this second method, no sequence comparison is performed and the technique is referred to as an uninterpreted mass spectra analysis (Yates et al. 1995). The second technique is most often used due to the shorter times required for searches and the robustness of the identifications. Both of these techniques for using MS/MS data to identify peptides are complicated by experimental constraints. Not all peptides ionize well, thus many are not observed. Bond breakage responsible for generating the MS/MS spectra is sequence dependent, so gaps in the ion series occur. In addition, ions can have multiple charge states and this can make it difficult to assign an ion to the proper series. Finally, ions that are not produced by the peptide of interest are frequently observed in mass spectra, especially from precursor ions with low abundance (Aebersold and Goodlett 2001). The result of these complications is that it is not always possible to unambiguously assign a single peptide sequence to an MS/MS spectrum. Most programs give multiple sequences for each spectrum with some indication of the confidence of the assignment.

3.2 Search Software

A large number of software programs are available to perform uninterpreted spectral matching. These include both open source and commercial efforts. The open source programs most frequently used are X!-Tandem (Craig and Beavis 2004) and OMSSA (Geer et al. 2004). The more often used commercial programs include SEQUEST (Yates 1998), MASCOT (Perkins et al. 1999), PHENYX (Colinge et al. 2004) and Spectrum Mill. A comparison of these programs is described in Kapp et al. (2005). The choice of program is often determined by the experience of the user and to some extent the type of instrument. Instrument manufacturers generally sell a software package with their instrument. As an example, Thermo-Fisher (Madison, WI) sells SEQUEST with its ion trap and hybrid instruments. An advantage of using the instrument manufacturer's search software is that it is usually integrated with the instrument control software package. This makes it easier to automate the search process. Software only vendors such as Matrix Science (Boston, MA) also have solutions to automate the search process. Many of the commercial programs are available without charge via a web interface. The open source programs have the advantage of low cost and the fact that the algorithm for the search is known. It is possible, if you have sufficient programming talent, to adapt the open source programs to your local needs. The disadvantage is that you cannot expect the same level of support as you would get from a commercial vendor. The choice may depend on the user's experience with search programs, the amount of informatics support locally available, and the amount of time available for experimenting with such programs. I will discuss the X!-Tandem open source search program in detail. Many of the features of data entry and data output are shared between X!-Tandem and other search programs and so it serves as good introduction to using such software.

3.3 X!-Tandem

X!-Tandem is the first widely distributed open source search program. A number of innovative features make X!-Tandem both faster than other programs and able to detect a larger number of post-translational modifications (PTMs). A key feature of X!-Tandem is the recognition that modern searches are likely to involve thousands of MS/MS spectra. It makes sense to optimize the speed of such searches over those having only a few spectra (Craig and Beavis 2003). X!-Tandem processes all the spectra before any searching takes place and creates a lookup table indexed on the precursor ion mass of each spectrum. One pass is made through the database of proteins to be searched and, as each predicted peptide is generated from the database, it is compared against those spectra whose mass is within the designated mass tolerance. This results in a large increase in speed when many spectra are to be searched. Another factor that slows database searches is the need to search for peptides resulting from incomplete enzymatic digestion. If many missed cleavages are allowed then the number of peptides that must be compared with the observed spectra will be large and the search slow. X!-Tandem initially searches the database with the number of allowed missed cleavages set to one. Proteins that are identified in the initial search are used to generate a subset of proteins from the original set of proteins. This set is then used in a new search with the number of missed cleavages set to a higher number. This procedure can be iterated many times. This procedure will work if at least one peptide for the protein of interest is found in the initial search. Additionally, this multiple pass searching strategy increases the ability of X!-Tandem to identify PTMs. PTMs are identified by predictable mass shifts in the tryptic fragments containing the PTM. Modern search programs allow the user to specify which PTM masses to include in the search. Since a protein can be found with and without the modification, each modification included in the search increases by two-fold the number of fragments to be searched for each residue that could contain that modification. This limits the number of PTMs that can be included in a single search due to the geometric expansion of the search space. By restricting the number of PTMs in the first search and increasing the number in subsequent searches with a reduced set of proteins, many PTMs and combinations of PTMs on a single tryptic fragment can be identified.

X!-Tandem includes a number of features that aid in evaluating the quality of the match between observed spectra and peptides. As X!-Tandem executes the search it saves the distribution of scores and uses these data to estimate the distribution of false positive scores. X!-Tandem then estimates the probability of observing a score as high as or higher than the score assigned to any given spectra. This e-value is useful in gauging the significance of the assignment (Fenyo and Beavis 2003). X!-Tandem will also search a sequence reversed database and report the scores in the same report as the original database. The user can scan the output and determine when the reversed database hits begin to appear. This allows the user to determine what e-value to use as a cutoff and gives a robust estimate of the false positive rate for a chosen e-value cutoff.

The user interface for X!-Tandem is web-based and is named the GPM (global proteome machine). If you run the program locally you will still use the web-based interface. The data entry page is shown in Fig. 3.1. Each of the data entry fields is numbered; I will describe the essential items that need to be entered for a search. Field 1 is the file name containing the peak lists generated from the mass spectra. These peak lists are generally produced by software provided by the instrument manufacture. Three file formats are accepted. DTA and PKI files are generated from Thermo Fisher and Waters (Bedford, MA) software respectively. MGF is the Mascot generic file format developed by Matrix Science. Software is available to convert DTA and PKI files to MGF. Once the peak lists have been entered, a file containing protein sequences (often referred to as a database) must be chosen from the list in Field 2. X!-Tandem lets you select as many species as you like from the list although it is best to select as small a set of proteins as possible. X!-Tandem appends a set of proteins that are common contaminants of proteomics samples to each search. This is helpful in correctly identifying peptides such as those resulting from keratin and trypsin in the sample that might otherwise result in false identifications. The user

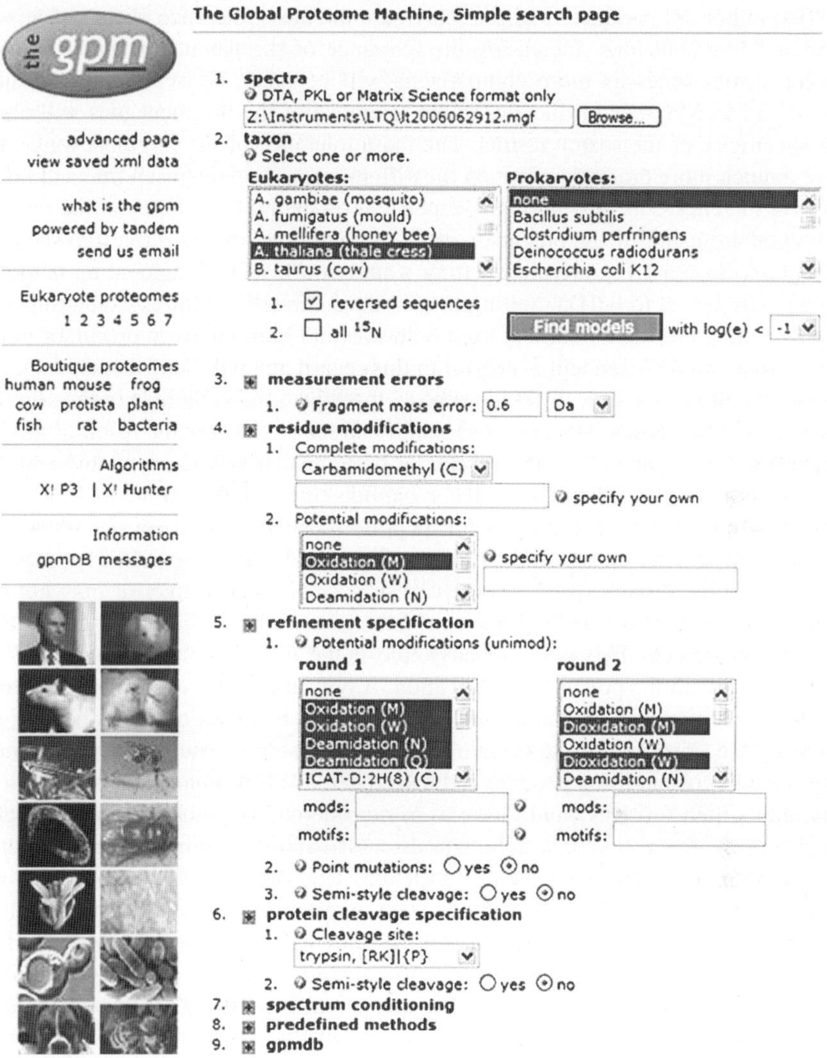

Fig. 3.1 Data entry web page for X!-Tandem. This page allows the user to specify all the parameters necessary to conduct a search of MS/MS data against a protein database

can also choose to append the reversed sequences of the selected proteins. This is very useful in determining the appropriate e-value to use as a cutoff. The user can also chose a upper limit of the e-value for which peptide matches will be reported. Initially, this is set at the default of 0.1 until a better value can be determined using the reversed sequences.

Two important parameters of the search are the mass tolerance of the precursor ions and fragment ions. Increasing the tolerance of the parent ion will result in longer search times as more comparisons will be made between synthetic and observed MS/MS spectra. Increasing the tolerance of the fragment ions will alter the specificity of the search results. The mass tolerance of the fragment ions will have a much more dramatic effect on the difference in score between true and false positive matches. The best setting is dependent on the mass accuracy of the instrument you are using. If the precursor ion is measured in a modern ion trap such as an Thermo Fisher LTQ, a typical setting would be 1 Da, and fragment mass tolerance would be set to 0.6 Da. Setting these values correctly requires examining the results from valid identifications from both low and high intensity precursor ions. The output from X!-Tandem is helpful in this regard and will be discussed later.

Field 4 allows the user to set the peptide modifications to include in the search. These are of two types. The complete modifications are usually the result of modification made by the user to the peptide or protein such as carboxymethylation to prevent disulfide bond formation. If the peptide contains the amino acid then the modification must be present for the peptide to be identified by the software. Such modifications do not add to the time required for the search. The second type of modification is listed as potential or variable. These may or may not be present on the peptide. Adding these modifications can significantly increase the time of the search. This is particularly true if the peptide contains a number of amino acids with the potential modification. A typical example would be phosphorylation, where serine, threonine and tyrosine could be modified. Each peptide in the database containing a large number of these residues would generate a very large number of modified peptides to be searched as all combinations of modified and unmodified residues would have to be considered. The modifications in Field 4 apply to the first round of searching as discussed above. Additional modifications can be added for subsequent rounds in Field 5. The user may also choose to allow single point mutations in the peptides. This is extremely useful when using the proteome of a closely related species for the search. It is also possible to allow for matches to peptides that are the result of a specific cleavage and a non-specific cleavage (semi-style). Field 6 allows the user to set the enzyme specificity. Field 7 allows the user to remove spectra that are not likely to produce reliable peptide matches. Field 8 has a pull-down menu that allows the user to pick data files that set all the X!-Tandem settings automatically for a number of popular instruments.

Field 9 submits the user's search results anonymously to the GPMDB (global proteome machine database), which is a database currently containing over 15 million spectra to peptide mappings. This is an important community resource, which I will discuss later. The output from such a search is shown in Fig. 3.2. Proteins are ranked by the e-value, which is the number of times a match with a score as high or higher would be found randomly given the number of spectra and the size of the searched database. The log (I) column reports the summed ion intensity, the % column reports the percentage of the protein sequence covered by the identified peptides, the # column reports the number of unique peptides identified for each identified protein, the total column reports the total number of spectra that mapped

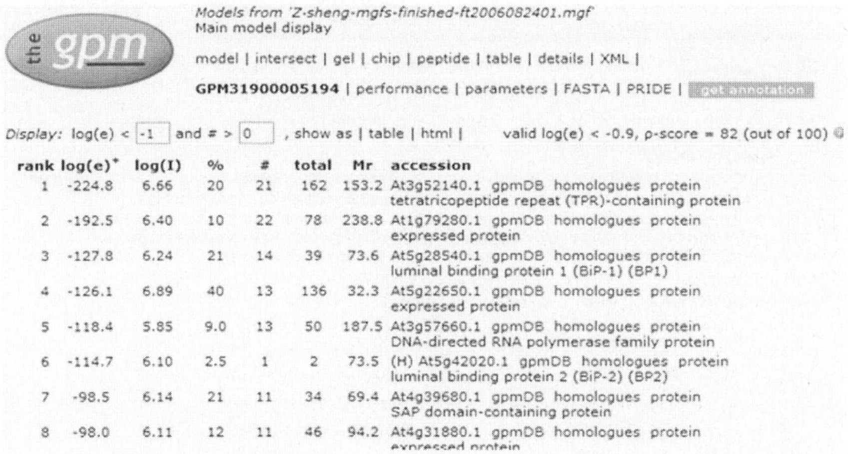

Models from 'Z-sheng-mgfs-finished-ft2006082401.mgf'
Main model display

model | intersect | gel | chip | peptide | table | details | XML |

GPM31900005194 | performance | parameters | FASTA | PRIDE | get annotation

Display: log(e) < -1 and # > 0 , show as | table | html | valid log(e) < -0.9, p-score = 82 (out of 100)

rank	log(e)*	log(I)	%	#	total	Mr	accession
1	-224.8	6.66	20	21	162	153.2	At3g52140.1 gpmDB homologues protein tetratricopeptide repeat (TPR)-containing protein
2	-192.5	6.40	10	22	78	238.8	At1g79280.1 gpmDB homologues protein expressed protein
3	-127.8	6.24	21	14	39	73.6	At5g28540.1 gpmDB homologues protein luminal binding protein 1 (BiP-1) (BP1)
4	-126.1	6.89	40	13	136	32.3	At5g22650.1 gpmDB homologues protein expressed protein
5	-118.4	5.85	9.0	13	50	187.5	At3g57660.1 gpmDB homologues protein DNA-directed RNA polymerase family protein
6	-114.7	6.10	2.5	1	2	73.5	(H) At5g42020.1 gpmDB homologues protein luminal binding protein 2 (BiP-2) (BP2)
7	-98.5	6.14	21	11	34	69.4	At4g39680.1 gpmDB homologues protein SAP domain-containing protein
8	-98.0	6.11	12	11	46	94.2	At4g31880.1 gpmDB homologues protein expressed protein

Fig. 3.2 The initial results page from X!-Tandem showing the top proteins identified. This is the protein summary page. More detailed data is available using hyperlinks on this page

to peptides from the indicated protein, and the Mr column reports the calculated molecular weight of the identified protein. Also shown is annotation for the proteins. If X!-Tandem has not identified the protein in previous searches the annotation may be blank.

Clicking on the "Get annotation" button will retrieve annotation from TIGR (www.tigr.org) for Arabidopsis proteins. Clicking the protein hyperlink displays a more detailed analysis of the evidence for the presence of that protein in the sample, as shown in Fig. 3.3. Most search software programs provide a similar output. Typical of such outputs is the display of the protein sequence with the identified peptides indicated by color highlighting. This is very useful as it can reveal a bias in the coverage that may indicate post-translational processing of the protein such as the removal of signal or transit peptides. X!-Tandem includes a graphical representation of this data above the protein sequence that indicates both the coverage and the confidence in assignment of each peptide. This figure is a very compact way to look at this data and is used to good effect in comparing a large number of proteins as is seen in Fig. 3.4. The display in Fig. 3.4 was generated by clicking on the "validate" link. Because a very large number of protein searches have been archived in the GPMDB it is likely that someone has identified the protein of interest in a previous search. The data in Fig. 3.4 allows the user to determine if a similar collection of peptides for that protein have been identified in other searches. A well-documented fact is that some peptides are frequently observed for a particular protein while others are rarely seen (Craig et al. 2005). If the peptides the user has found for a particular protein match those found by other researchers then it is more likely that the protein is present in the sample. Below the protein sequence is a table showing information for each peptide. This information includes the expected

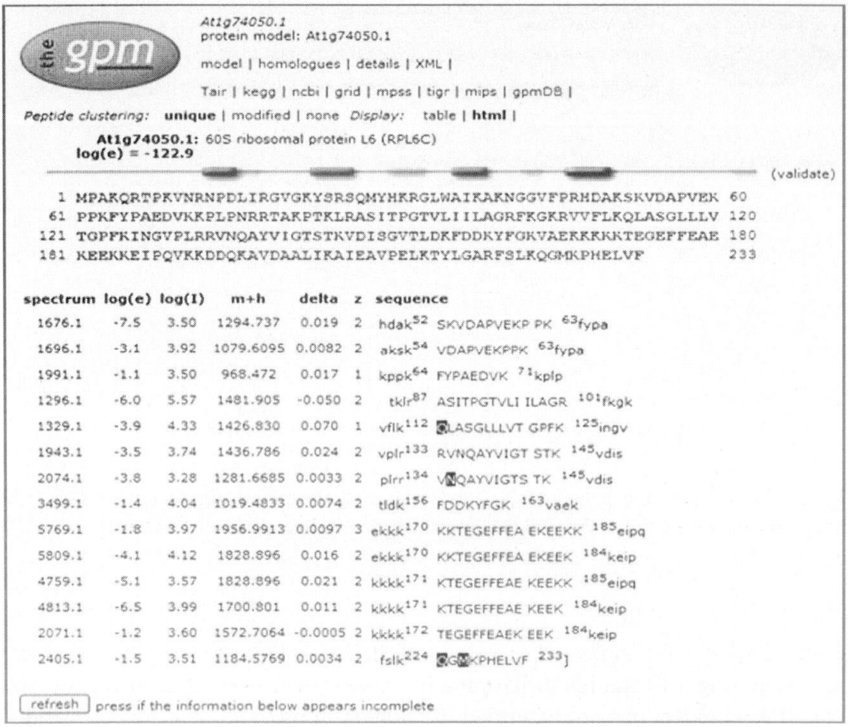

Fig. 3.3 The peptide summary page resulting from clicking the hyperlink-labeled protein in the protein summary page. Each identified peptide is listed with information relating to the quality of the match. More detailed data for each peptide is available using hyperlinks on this page

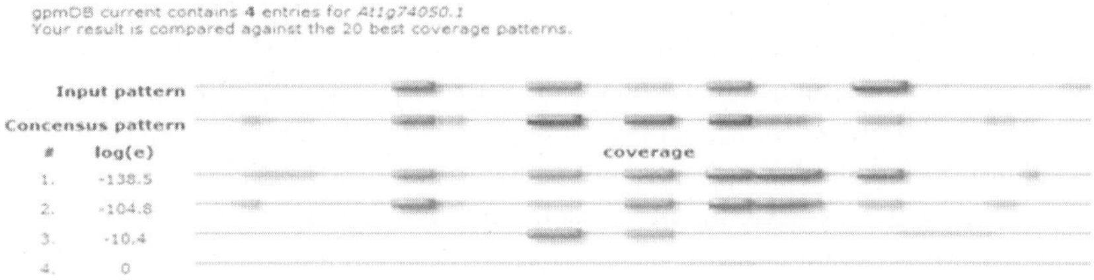

Fig. 3.4 A comparison of the peptides identified in Fig. 3.3 with four proteins present in the global proteome machine database (GPMDB). This output resulted from clicking the *validate* hyperlink shown in Fig. 3.3

score for the peptide, the intensity of the ion, the observed mass corrected for charge, the difference in the observed vs calculated mass (delta), and the sequence of the peptide. The sequence information includes a small amount of the surrounding protein sequence. Clicking on the peptide sequence will show detailed information about that particular spectrum-to-peptide mapping as shown in Fig. 3.5. The

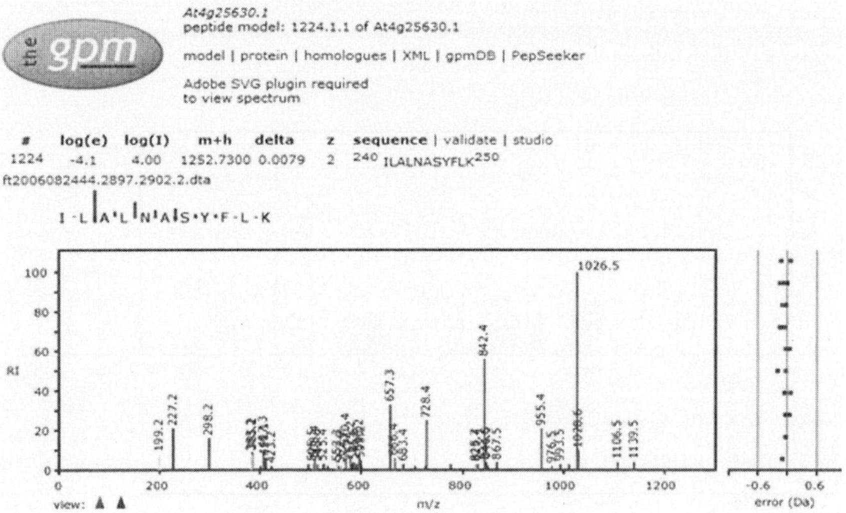

bond	+1y	+1y-17	+1y-18	+1b	+1b-17	+1b-18
I1	+1,2 1139.646	1122.619	1121.635	114.091	97.065	96.081
L2	+1,2 1026.562	1009.535	1008.551	227.175	210.149	209.165
A3	955.525	938.498	937.514	298.213	281.186	280.202
L4	842.441	825.414	824.430	411.297	394.270	393.286
N5	728.398	711.371	710.387	525.339	508.313	507.329
A6	657.361	640.334	639.350	+1,2 596.377	579.350	578.366
S7	570.329	553.302	552.318	683.409	666.382	665.398
Y8	407.265	390.239	389.255	+1,2 846.472	829.445	828.461
F9	260.197	243.170	242.186	993.540	976.514	975.530
L10	147.113	130.086	129.102	+1,2 1106.624	1089.598	1088.614

Fig. 3.5 A detailed peptide report of a single identified peptide is shown. This report was generated by clicking the sequence of a peptide shown on a peptide summary page

display of the fragment ions spectrum allows the user to judge the relative intensities of the observed masses that match predicted fragment ions versus those not matching. The graph to the right of the spectrum indicates the mass errors of matching fragments. The table below the spectrum associates the masses in the spectrum with the ion type. In the example shown there is a complete b and y ion series within the mass range limits of the ion trap, which is indicative of a valid match. The graphic above the spectrum is unique to X!-Tandem and is useful in assessing the relative amount of peptide bond breakage between amino acids. It is well known that the CID spectrum is not random but depends on the amino acid sequence. This graphic allows a very efficient comparison of CID spectra. Clicking the "validate" link generates the data shown in Fig. 3.6. This again uses GPMDB data. If other searches have identified the same peptide you are able to compare their data to yours. You can quickly compare the spectra by looking at the summary graphic. In this example there were 11 similar spectra. This comparison is particularly useful if the spectrum found in the GPMDB for comparison is from a search that yielded more peptides for the protein of interest or if the spectrum was of higher quality.

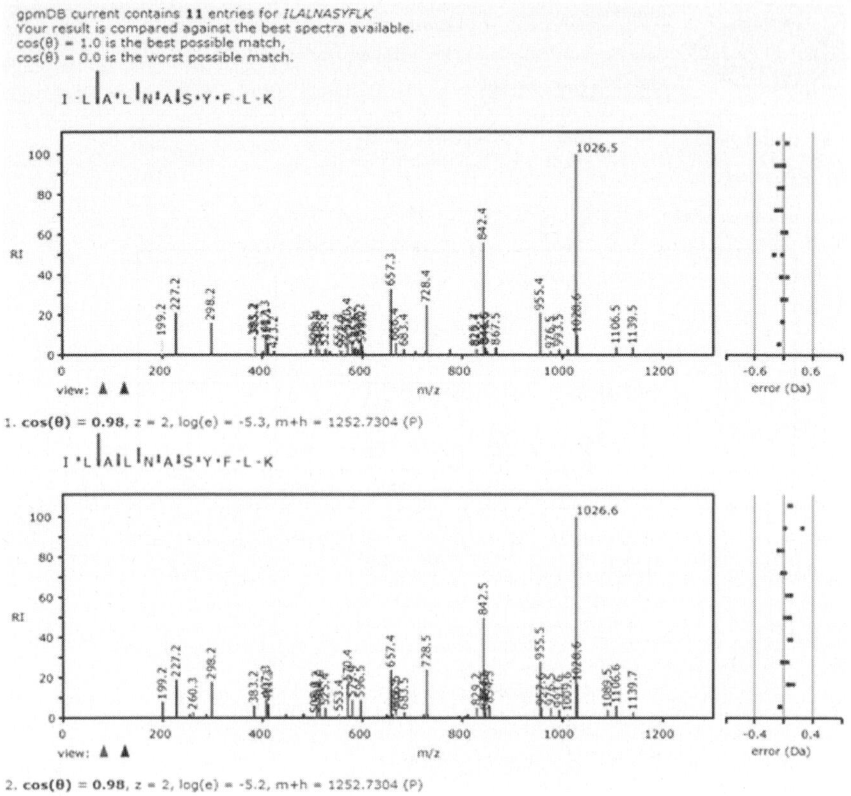

Fig. 3.6 Clicking on the *validate* link on the detailed peptide report allows the user to compare the observed mass spectrum to spectra in the GPMDB that were mapped to the same peptide sequence

3.4 Informatics Considerations

The transition of researchers from using mass spectrometry techniques to obtain the identity of single proteins to larger scale proteomics experiments requires a more robust informatics infrastructure. A few years ago analysis of proteomics data consisted of generating a list of proteins found in the sample. Often these identifications had no information about the confidence of the identification. An improvement was the use of the programs such as MASCOT or X!-Tandem. These programs generated an HTML report for each MS/MS experiment and provided sufficient information for the user to infer the confidence of the identification. A list of the most likely proteins present in the samples was generated and some simple information about each protein was provided by a hyperlink to a public database providing some annotation for the identified protein. This solution suffered from several problems. A major problem was the user often needed to look at a number of experiments in order to draw a conclusion. One solution to this problem is to summarize MASCOT or X!-Tandem data and distribute this summary to biologists as an Excel spreadsheet. Various hyperlinks to annotation can be provided to aid the researcher in analyzing the results. The researcher can add comments or data to the spreadsheet to user-defined categories to allow custom grouping of the proteins based on such characteristics as metabolic pathway or cellular location. One advantage with this approach is that most biologists have familiarity with Excel and are able to sort and filter the data. This approach limits the ways in which the data can be analyzed to the structure of the spreadsheet.

3.5 Relation Database Management of Large Scale Proteomics Data

To overcome the above-mentioned limitation, a number of groups have developed relational database schemata for large scale MS/MS data including SBEAMS (http://tools.proteomecenter.org/SBEAMS.php), PRIDE (http://www.ebi.ac.uk/pride), PRIME (http://www.proteomeconsortium.org/prime.html) and CPAS (https://proteomics.fhcrc.org/CPAS/Project/home/begin.view?). A well constructed relational database allows more flexibility in the types of queries that can be generated than is possible with a spreadsheet. In particular, it is much easier to query for combinations of peptide attributes such as PTMs. With a relational database it is possible to set the e-value for peptides containing modifications that may compromise the MS/MS data quality to a lower value than other peptides. The user can also ask to find proteins that have a specific PTM only in certain samples. Many users wish to assign the proteins found in their samples to classification schemes developed in their own laboratories or elsewhere. This allows them to filter and sort the proteins in ways that optimize the time involved in analyzing these large data sets. A relational database makes it much easier to integrate such information with the proteomics data. A relational database allows for the association of annotation information with specific users. This is very useful when not all users agree on the function of a protein. The combination of modern MS/MS search programs with such relational databases greatly facilitates the analysis of large-scale proteomics data sets.

References

Aebersold R, Goodlett DR (2001) Mass spectrometry in proteomics. Chem Rev 101:269–295

Colinge J, Masselot A, Cusin I, Mahe E, Niknejad A, Argoud-Puy G, Reffas S, Bederr N, Gleizes A, Rey PA, Bougueleret L (2004) High-performance peptide identification by tandem mass spectrometry allows reliable automatic data processing in proteomics. Proteomics 4:1977–1984

Cottrell JS (1994) Protein identification by peptide mass fingerprinting. Pept Res 7:115–124

Craig R, Beavis RC (2003) A method for reducing the time required to match protein sequences with tandem mass spectra. Rapid Commun Mass Spectrom 17:2310–2316

Craig R, Beavis RC (2004) TANDEM: matching proteins with tandem mass spectra. Bioinformatics 20:1466–1467

Craig R, Cortens JP, Beavis RC (2005) The use of proteotypic peptide libraries for protein identification. Rapid Commun Mass Spectrom 19:1844–1850

Fenyo D, Beavis RC (2003) A method for assessing the statistical significance of mass spectrometry-based protein identifications using general scoring schemes. Anal Chem 75:768–774

Geer LY, Markey SP, Kowalak JA, Wagner L, Xu M, Maynard DM, Yang X, Shi W, Bryant SH (2004) Open mass spectrometry search algorithm. J Proteome Res 3:958–964

Kapp EA, Schutz F, Connolly LM, Chakel JA, Meza JE, Miller CA, Fenyo D, Eng JK, Adkins JN, Omenn GS, Simpson RJ (2005) An evaluation, comparison, and accurate benchmarking of several publicly available MS/MS search algorithms: sensitivity and specificity analysis. Proteomics 5:3475–3490

Ma B, Zhang K, Hendrie C, Liang C, Li M, Doherty-Kirby A, Lajoie G (2003) PEAKS: powerful software for peptide de novo sequencing by tandem mass spectrometry. Rapid Commun Mass Spectrom 17:2337–2342

Perkins DN, Pappin DJ, Creasy DM, Cottrell JS (1999) Probability-based protein identification by searching sequence databases using mass spectrometry data. Electrophoresis 20:3551–3567

Shevchenko A, Sunyaev S, Loboda A, Shevchenko A, Bork P, Ens W, Standing KG (2001) Charting the proteomes of organisms with unsequenced genomes by MALDI–quadrupole time-of-flight mass spectrometry and BLAST homology searching. Anal Chem 73:1917–1926

Taylor JA, Johnson RS (2001) Implementation and uses of automated de novo peptide sequencing by tandem mass spectrometry. Anal Chem 73:2594–2604

Thiede B, Hohenwarter W, Krah A, Mattow J, Schmid M, Schmidt F, Jungblut PR (2005) Peptide mass fingerprinting. Methods 35:237–247

Yates JR 3rd (1998) Database searching using mass spectrometry data. Electrophoresis 19:893–900

Yates JR 3rd, Eng JK, McCormack AL, Schieltz D (1995) Method to correlate tandem mass spectra of modified peptides to amino acid sequences in the protein database. Anal Chem 67:1426–1436

Chapter 4
Phosphoproteomics in Plants

Sergio de la Fuente van Bentem, Thomas S. Nühse, and Heribert Hirt

Abstract Reversible modification of proteins by phosphorylation is crucial in regulating signal transduction in plants and other organisms. Research on protein phosphorylation has greatly benefitted from recent developments in mass spectrometry (MS)-based technology. In combination with this technology, different, highly specific phosphopeptide purification methods have been explored to determine hundreds to thousands of phosphorylation sites. Using accurate mass spectrometers, researchers are now able to quantitatively determine phosphopeptide concentrations from different samples on a large scale. Contrasting with studies on yeast and animal systems, phosphoproteomic research on plants has concentrated mainly on the identification of novel phosphorylation sites. Here, we describe recent MS-based approaches that will enable the elucidation of dynamic changes in plant phosphoproteomes induced by environmental signals.

4.1 Introduction

Hundreds of different posttranslational modifications expand the complexity and function of the proteome far beyond that which is encoded in the genome. The regulation of chromatin by histone acetylation and methylation, or of protein turnover and signalling by ubiquitination and SUMOylation, are just a few examples relevant for all eukaryotes. Protein phosphorylation has traditionally been the most intensively studied post-translational mechanism. Since its discovery in the 1950s,

Sergio de la Fuente van Bentem, Department of Plant Molecular Biology, Max F. Perutz Laboratories, Dr. Bohr-Gasse 9, University of Vienna, 1030 Vienna, Austria

Thomas S. Nühse, The Sainsbury Laboratory, Norwich Research Park, Norwich NR4 7UH, UK

Heribert Hirt, Department of Plant Molecular Biology, Max F. Perutz Laboratories, Dr. Bohr-Gasse 9, University of Vienna, 1030 Vienna, Austria, E-mail: heribert.hirt@univie.ac.at

Current address: Heribert Hirt, URGV, 2, Rue Gaston Crémieux, 91057 Evry, France

J. Šamaj and J. Thelen (eds.), *Plant Proteomics*
© Springer 2007

the number of studies focused on protein phosphorylation has expanded tremendously. Protein phosphorylation is important: it can influence multiple characteristics of proteins, including enzymatic activity, turnover, subcellular localisation and protein–protein interactions. Protein phosphorylation is also abundant: as many as 30% of all proteins may be phosphorylated during their life span (Hubbard and Cohen 1993).

Protein kinases catalyse the transfer of one or more phosphate moieties to their substrates, whereas protein phosphatases catalyse the removal of these phosphate groups. The analysis of signalling pathways in plants has focused mostly on protein kinases. Approximately 1,000 protein kinases are encoded by the *Arabidopsis thaliana* genome (The Arabidopsis Genome Initiative 2000) – a much more complex kinase complement than that of humans, with only about 500 kinase genes (Manning et al. 2002). The main factor contributing to this complexity is an enormous expansion of transmembrane receptor-like kinases in plants. Intriguingly, unlike in animal genomes, none of these are protein tyrosine kinases. Many plant protein kinases that are required for developmental processes or that are crucial in resistance to stress such as cold and pathogen invasion have been identified. To fully understand such processes in plants, we need to identify not only which kinases are activated and required for the biological response, but also which proteins are phosphorylated on which residues. Finally, we want to connect kinases and substrates to establish signalling pathways and networks. Each one of these questions holds considerable challenges, and our picture of protein phosphorylation and dephosphorylation events in plants is still fragmented. Half a century after the initial discovery of regulatory protein phosphorylation, novel technology has brought the field into a new era. Mass spectrometry, along with other phosphoproteomic methods, has tackled the major challenges and bottlenecks, most prominently by identifying large numbers of phosphorylation sites in yeast, animal and plant proteins.

4.2 Mitogen-Activated Protein Kinase Modules are Central Components of Plant Signalling Pathways

In plant biology, there is a strong disparity between our knowledge of kinases that are essential in different biological contexts, and our knowledge of which proteins are actually being phosphorylated by those kinases to effect the cellular response. The most striking examples of this are probably the plant mitogen-activated protein kinase (MAPK) modules. A general feature of hormone- and stress-induced pathways in eukaryotes, MAPK modules consist of MAPK kinase kinases (MAPKKKs) that phosphorylate and thereby activate MAPK kinases (MAPKKs), which in their turn phosphorylate and activate MAPKs (Fig. 4.1). Essential for the activation of a protein kinase is phosphorylation on one or several residues in its activation loop (Fig. 4.1). The expression of phosphorylation-mimicking mutants in transient systems such as tobacco mesophyll cells and Arabidopsis protoplasts has enabled us

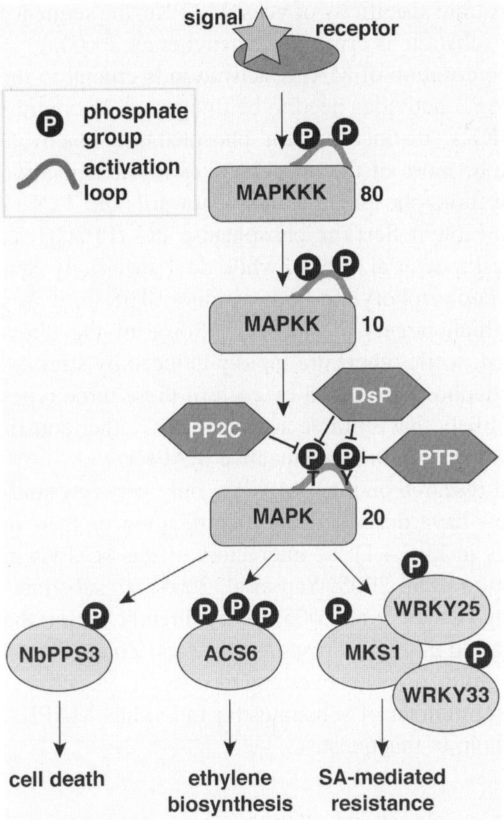

Fig. 4.1 Mitogen-activated protein signalling cascades in plants. Components of mitogen-activated protein kinase (MAPK) modules are activated by phosphorylation on one or two residues, such as a threonine (T) and a tyrosine (Y) for MAPKs, in their activation loops. Three distinct protein phosphatase activities, protein Ser/Thr phosphatase 2C (PP2C), dual specificity phosphatases (DsPs) and protein tyrosine phosphatase (PTP), can dephosphorylate one or both phosphorylated residues. *Numbers* next to the MAPK pathway components indicate the number of members in *Arabidopsis thaliana*. *ACS6* 1-Aminocycloproprane-1-carboxylate synthase 6 *SA* salicylic acid, *MAPKK* MAPK kinase, *MAPKKK* MAPK kinase kinase *NbPPS3 Nicotiana benthamiana* protein phosphorylated by potato MPK1, *MKS1* MAP kinase 4 substrate 1, *WRKY* WRKY motif-containing transcription factor

to discover the constituents and biological function of different MAPK modules (Nakagami et al. 2005; Pedley and Martin 2005). In *A. thaliana*, a potential set of 80 MAPKKKs, 10 MAPKKs and 20 MAPKs exists (Fig. 4.1). These numbers suggested that multiple signals converge through MAPKKKs onto MAPKKs, which could activate different MAPKs. However, it now has become clear that a signal can activate several MAPKKs, which in turn can activate the same or distinct downstream MAPKs. What determines the signalling specificity in plant MAPK cascades is poorly understood – the role of substrate docking domains and scaffolds – key regulatory factors in animals and yeast – remains to be investigated in plants. In vitro studies have suggested that different MAPKs can phosphorylate a partially overlapping set of proteins (Feilner et al. 2005; see also Chap. 5 by Kersten and Feilner in this volume), as is also the case in vivo for some animal MAPK substrates (Qi and Elion 2005). This indicates that MAPK functions are likely to be partially redundant, complicating the analysis of the in vivo function of individual

MAPK members. To enable the substrate specificity of yeast MAPKs, the sequence at which the kinase docks onto the substrate is critical (Remenyi et al. 2005).

Since both the amplitude and the duration of MAPK activation is crucial to the physiological output generated, MAPK activities need to be fine-tuned. In contrast to the activating upstream MAPKKs, distinct protein phosphatases inactivate MAPKs by dephosphorylating one or more of the phospho-sites in the activation loop (Fig. 4.1). There appear to be three phosphatases active towards the TxY (x denotes D or E in plants) motif. Protein Ser/Thr phosphatase 2C (PP2C) can dephosphorylate the threonine (Meskiene et al. 1998), while dual specificity Ser/Thr/Tyr phosphatases (DsPs) can dephosphorylate both residues. The third is a protein Tyr phosphatase (PTP), which targets the tyrosine. Some of the phosphatases are constitutively expressed, while others are rapidly induced by stresses. Although evidence for the in vivo dephosphorylation by each of these three types of phosphatases is still scarce, it is likely that all three act in concert, either constitutively or after induction by the pathway, in order to inactive MAPKs.

In spite of more than a decade of research on plant MAPKs, only very few studies, all using "classical" techniques, have described the identification of their in vivo substrates (see some examples in Fig. 4.1): as interactors of the MAPKs in yeast two-hybrid screens (Andreasson et al. 2005; Yap et al. 2005), as substrates found by in vitro expression cloning (Katou et al. 2005), or by directly testing the key candidate protein in the investigated biological response (Liu and Zhang 2004). Given the fact that many tested stimuli activate overlapping sets of MAPKs, it would be very helpful to find a larger number of substrates for individual MAPKs. Phosphoproteomic techniques can help in this quest.

4.3 Mass Spectrometry-Based Approaches to Identify Phosphorylation Sites

Until the development of mass spectrometric technology suitable for peptides and proteins, the identification of phosphorylation sites has been technically very difficult. A combination of in vivo labelling with ^{32}P, Edman sequencing of radiolabelled protein fragments, and site-directed mutagenesis of candidate phosphorylation sites was a typical procedure. Only with an enormous increase in sensitivity have modern mass spectrometers allowed us to "see" the phosphorylated peptides directly without the aid of radioactivity. This has not only made it easier to find phosphorylation sites on a protein of interest (although still not a trivial task), but has also sparked an entirely different approach. The isolation of phosphopeptides from complex protein digests by affinity chromatography followed by tandem MS (Fig. 4.2) has emerged as a novel tool with which to determine phosphorylation sites in eukaryotes en masse (Chen and White 2004; Laugesen et al. 2004). While most large-scale phosphoproteomic studies have been carried out in yeast or mammalian systems, some recent pioneers have made breakthroughs in the understanding of protein phosphorylation in plants (Heinz et al. 2004; Nühse et al. 2003).

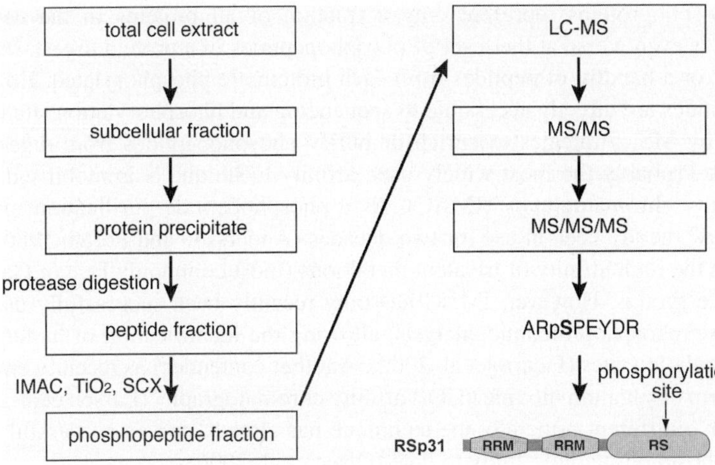

Fig. 4.2 Purification and sequencing of phosphopeptides from complex mixtures. A standard method for large-scale identification of phosphopeptides is depicted. For MS/MS, peptides are fragmented so that the sequence can be deduced from the fragmentation pattern. Since phosphopeptides normally yield poor fragmentation spectra because of the prominent loss of phosphate from the peptide, the peptide that lost the phosphate can be selected for a second round of fragmentation. This usually gives a better spectrum, allowing the sequence to be confidently assigned to the peptide. The phosphoserine is indicated in *bold lettering* and is preceded by a lowercase *p*. *MS* Mass spectrometry, *RRM* RNA recognition motif, *RS* arginine/serine-rich domain

4.3.1 Enrichment Strategies for MS-based Phosphoproteomic Analysis

Several problems are encountered during the examination of phosphorylation sites (Mann et al. 2002). Most importantly, phosphoproteins are generally of extremely low abundance within the proteome. This feature has prompted the development of phosphopeptide or phosphoprotein purification methods from complex mixtures. Affinity purification of phosphoproteins with phospho-specific antibodies prior to MS has been successful in several studies (Grønborg et al. 2002). Anti-phosphotyrosine antibodies are particularly suitable for specific affinity purification. Because of the low occurrence of tyrosine phosphorylation in animal cells (an estimated 1–2% against more than 98% on serine and threonine residues), phosphotyrosine-containing peptides/proteins need to be specifically purified from complex mixtures. Immunoprecipitation with anti-phosphotyrosine antibodies combined with liquid chromatography-mass spectrometry (LC-MS) identified around 1,000 phosphotyrosine sites (Brill et al. 2004; Ficarro et al. 2005; Rush et al. 2005; Zhang et al. 2005). Such a study has never been performed in plants. Given that direct and unambiguous evidence of tyrosine phosphorylation in plants (besides MAPKs and cyclin-dependent kinases) is scarce, a proteomic approach might yield some very interesting results.

If phosphoproteins represent only a fraction of all proteins in the cell, the problem is even worse at the level of phosphopeptides in a protein digest, because only one or a handful of peptides from each protein are phosphorylated. However, only peptides are directly accessible to sequencing and phosphorylation site identification by MS. Strategies to enrich or purify phosphopeptides from digests are essential. Probably the most widely used affinity technique is immobilised metal ion affinity chromatography (IMAC). As a phosphopeptide purification method, IMAC has already been in use for two decades (Andersson and Porath 1986). It is based on the high affinity of trivalent metal ions (most commonly Fe^{3+} or Ga^{3+}) for phosphate groups. However, IMAC has only recently been successfully used for large-scale phosphoproteomic analysis, allowing the identification of hundreds of phosphorylation sites (Ficarro et al. 2002). Another contender has recently emerged in the form of titanium dioxide (TiO_2) affinity chromatography (Larsen et al. 2005). Based on a different principle, the technique has already proven successful in the largest phosphoproteomic study to date (Olsen et al. 2006).

Several studies have described the isolation of phosphopeptides from complex mixtures by strong cationic exchange (SCX) chromatography or strong anionic exchange (SAX) chromatography, mostly in combination with IMAC or TiO_2 (Ballif et al. 2004; Beausoleil et al. 2004; Nühse et al. 2003; Olsen et al. 2006). Prefractionation of protein digests by ion exchange chromatography is, on the one hand, essential to deconvolute the complexity of the proteome and split it up into manageable portions, but on the other hand also makes use of the specific ionic properties conferred on peptides by the phosphate group. Coupling these techniques to mass spectrometric peptide sequencing has proved very successful.

4.3.2 Global Identification of in vivo Phosphorylation Sites using IMAC Coupled to MS Technology

The first successful large-scale phosphoproteomic study used IMAC to enrich phosphopeptides from trypsin-digested yeast protein extracts (Ficarro et al. 2002). MS allowed the identification of 383 phosphorylation sites from this enriched fraction. Since then, ever larger projects have been initiated with variations of the same approach, and recently 6,600 sites on 2,244 proteins were described from a single human cell type (Olsen et al. 2006). Although several thousands of phosphorylation sites can be found in the SwissProt database, the majority of the sites from the latter phosphoproteomic study were novel. This indicates that most of the phosphoproteome in human cells is still unknown.

The power of the combined use of IMAC and MS for plant biology was first shown by Vener et al. (2001) in Arabidopsis thylakoid membranes. A similar large-scale analysis of phosphopeptides from *A. thaliana* plasma membranes (Nühse et al. 2004) identified more than 300 phosphorylation sites, virtually all of them novel. As is the case for trehalose-6-phosphate synthases (Glinski and Weckwerth 2005), sucrose-phosphate synthase isoforms are modified by multiple phosphorylation sites in *A. thaliana* (De la Fuente van Bentem et al. 2006a). The mRNA splicing machinery was revealed as a major target of phosphorylation in this plant by examination of IMAC-purified phosphopeptides from nuclear and cytosolic extracts (De la Fuente van Bentem et al. 2006b). As in animals, a large set of serine-/arginine-rich splicing factors (or SR proteins) are heavily phosphorylated on serine residues in their arginine-/serine-rich (RS) domains. Analysis of conserved phosphorylation motifs in RS domain-containing proteins led to the identification of SR protein-specific kinase (SRPK) as the protein kinase responsible for several in vivo phosphorylation sites. The in vivo sites targetted by each of the four *A. thaliana* SRPKs, and the functional relevance of the individual modifications still remain to be determined.

The first phosphoproteomic studies in plants provided an insight into the general characteristics of protein phosphorylation in plants (Nühse et al. 2004). While only a few sites were conserved within one protein family, practically all them were conserved among putative orthologues, i.e. the likely equivalent of a specific isoform in another plant species. Thus, feature analyses of conserved sequences surrounding the identified phosphorylation sites can provide clues as to the nature of the kinase responsible for targeting the sites and, conversely, sites that show this conservation are good candidates for phosphorylation sites.

4.4 Technologies Applicable to Studying Changes in the Phosphoproteome

The isolation and sequencing of phosphopeptides provides a limited biological insight. The key question is: which sites are regulated by a particular biological stimulus or process? By analogy with the progression from expressed sequence tag (EST) sequencing to transcript profiling, the next stage in studying changes in the phosphoproteome is to obtain quantitative information on those phosphorylation sites. In vivo, the induction of a site by a stimulus means that it is phosphorylated at a higher stoichiometry. After digesting the proteins, a peptide carrying this phosphorylation will be more abundant in a stimulated versus a control sample.

Peptide signals in a mass spectrometer are not inherently quantitative, and the large amount of sample handling involved in phosphopeptide isolation means that signal intensities for the same peptide in sequential analyses can vary. This problem can be solved by introducing stable isotope tags into the different samples and analysing them simultaneously – isotopic variants of the same peptide are chemically identical, co-elute in chromatography and give the same "molar" signal intensities in the mass spectrometer, but yield separate signals because of their mass difference. Several adaptations of such techniques have been developed to study the dynamics of phosphorylation sites in signalling networks. In animal and yeast systems, several such studies have identified many novel stress-induced phosphorylation events (Ballif et al. 2005; Blagoev et al. 2004; Cutillas et al. 2005; Gruhler et al. 2005a; Ibarrola et al. 2004; Kratchmarova et al. 2005; Zhang et al. 2005).

The isotopic tag can be introduced either in vivo by metabolic labelling or after digestion of the proteins. The most widely used in vivo technique is stable isotope labelling by amino acids in cell culture (SILAC; Ong et al. 2002). This method is based on the application of distinct isotopically labelled amino acids to separate cell cultures during their growth. For instance, a normal arginine can be applied to a control cell culture, while a 6 Da heavier $^{13}C_6$-arginine can be applied to a cell culture that is later subjected to a certain treatment. These isotopes are incorporated into proteins and can later be differentiated during mass spectrometric analysis of peptides derived from these proteins. SILAC followed by phosphopeptide purification by IMAC or TiO_2 and MS allows the relative quantification of a peptide derived from the control cell culture and the same peptide from the treated culture. This method has been successfully used for differential phospho-site profiling of human and yeast signalling pathways (Blagoev et al. 2004; Gruhler et al. 2005a; Kratchmarova et al. 2005). SILAC has also been successfully applied to *A. thaliana* for quantitative proteomics (Gruhler et al. 2005b). To quantify all tryptic peptides with SILAC, cells need to be labelled with both arginine and lysine, and lysine labelling was poor in Arabidopsis. This complicates SILAC in plant cells for peptide-based phosphoproteomics because about one-half of all tryptic phosphopeptides have a C-terminal lysine.

An alternative approach to determine relative phosphopeptide differences between samples is the in vitro labelling of all peptides with an isotopic label. A commercially available multiplex set of reagents allows the incorporation of mass labels at the N-termini and lysine side chains of peptides in a digest mixture (Ross et al. 2004). Using four isoforms of this so-called iTRAQ reagent, Zhang et al. (2005) gave the first insights into the dynamic changes of phosphorylation profiles during EGFR tyrosine kinase signalling. The signals containing the quantitative information appear at the low mass end of the fragment mass spectra (m/z = 114–117 Da), which currently precludes the use of iTRAQ in sensitive, fast ion trap mass spectrometers.

The complexity of signalling cascades is reflected by the fact that substrate proteins can be phosphorylated by different protein kinases at multiple sites. Proteins often serve as platforms that integrate signals at different sites, each individually leading to different outputs. Quantitative phosphoproteomics therefore needs to resolve protein phosphorylation down to the individual residue.

4.5 Plant Phosphoproteomics in the Future

The recently developed MS-based methods have established themselves as invaluable tools in the identification of novel phosphorylation sites. There is a clear wish list for the future: we want the largest possible number of phosphorylation sites, reliably identified and accurately quantified during development and in response to multiple stimuli, and ideally with spatial resolution within tissues and cells.

The issue of sheer numbers of identified phosphopeptides touches an inherent problem of protein chemistry: In vivo protein concentrations vary over an enormous range, and we have no amplification technique analagous to PCR for proteins or peptides. The scale of the challenge is illustrated by the recent study of Olsen et al. (2006); this study shows that, as had been speculated before (Hubbard and Cohen 1993), a large portion of the proteome is phosphorylated, and that each phosphoprotein has on average at least three phosphorylation sites, although this ratio is skewed by some very highly phosphorylated proteins. Assuming that the proteome of a higher eukaryote may contain about 20,000–30,000 different proteins, and that about one-third is modified by phosphorylation, each on average on three sites, one arrives at a similar figure of 20,000–30,000 potential phosphorylation sites in the proteome – almost an order of magnitude more than encountered in even the most ambitious studies to date. Similarly, the unbiased approach of Olsen et al. (2006) found 103 phosphotyrosines (1.8% of all sites) compared with a (certainly not complete) inventory of about 1,000 previously published sites in human proteins. Ever faster and more sensitive mass spectrometers push the identification of more peptides, but fractionation of total cellular proteins is an even more important approach. As demonstrated by several studies, the purification of single organelles or prefractionation of complex mixtures by SAX/SCX is essential to increase the coverage of the phosphoproteome. Another limitation is that trypsin may cut proteins into peptides that are too short or too long to be detected by MS and consequently would be overlooked. This limitation can be overcome by the use of a protease panel that increases the sequence coverage of the phosphoproteome (Rush et al. 2005), albeit at the expense of greatly increased instrument time. It seems likely that IMAC or TiO_2 affinity chromatography will remain the tools of the trade for global phosphoproteomics.

Phosphopeptides frequently yield poor quality mass spectra, and the reliability of phosphorylation site identification is continuously improved by mass spectrometers with increasingly high mass accuracy as well as "smart" algorithms that match spectra with database entries based on more spectrum features than are currently used for routine protein identification. As the scale of phosphoproteomic studies expands, better software for efficient structuring of the large amounts of data becomes indispensable.

The recent development of specific labelling techniques, especially SILAC and iTRAQ, has made the quantification of phosphorylation profiles and their stress-induced changes in time possible. We can expect these studies to reveal novel signalling pathways and regulatory processes that are dependent on phosphorylation. Multiplexing, i.e. the possibility to carry out relative quantification of multiple samples, is the key for detailed time courses and the comparison of, for example, different stress signalling pathways or responses in different genotypes. The difficulty of incorporating multiple isotopic "flavours" into reagents for metabolic or post-digest labelling currently remains a challenge.

Our knowledge of plant signalling networks has large gaps, and we are still far from knowing the full phosphoproteome and its dynamics in response to development and environmental changes. This means, however, that the opportunities arising from quantitative phosphoproteomics are all the greater: while, at least the earlier, mammalian studies largely confirmed what was already known about growth factor signalling, plant biology is exploring *terra incognita* as far as targets for protein kinases are concerned. Launching large-scale quantitative studies into phosphorylation in response to key biotic and abiotic stresses or plant hormones will give us an entirely new view of plant biology and allow the generation of new hypotheses. The jury is still out on the ideal approach to quantitative phosphoproteomics in plants, but iTRAQ seems particularly well-suited because it avoids problems with metabolic labelling.

In summary, future studies aiming at global phosphoproteomics will greatly benefit from the recent developments in MS-based technology. These methods are becoming available for plant studies. As studies in animal and yeast systems have demonstrated, quantitative phosphoproteomics can bring our view of plant biology to the next level. The elucidation of plant signalling networks has only just begun.

References

Andersson L, Porath J (1986) Isolation of phosphoproteins by immobilized metal (Fe^{3+}) affinity chromatography. Anal Biochem 154:250–254

Andreasson E, Jenkins T, Brodersen P, Thorgrimsen S, Petersen NH, Zhu S, Qiu JL, Micheelsen P, Rocher A, Petersen M, Newman MA, Bjorn Nielsen H, Hirt H, Somssich I, Mattsson O, Mundy J (2005) The MAP kinase substrate MKS1 is a regulator of plant defense responses. EMBO J 24:2579–2589

Ballif BA, Villen J, Beausoleil SA, Schwartz D, Gygi SP (2004) Phosphoproteomic analysis of the developing mouse brain. Mol Cell Proteomics 3:1093–1101

Ballif BA, Roux PP, Gerber SA, MacKeigan JP, Blenis J, Gygi SP (2005) Quantitative phosphorylation profiling of the ERK/p90 ribosomal S6 kinase-signaling cassette and its targets, the tuberous sclerosis tumor suppressors. Proc Natl Acad Sci USA 102:667–672

Beausoleil SA, Jedrychowski M, Schwartz D, Elias JE, Villen J, Li J, Cohn MA, Cantley LC, Gygi SP (2004) Large-scale characterization of HeLa cell nuclear phosphoproteins. Proc Natl Acad Sci USA 101:12130–12135

Blagoev B, Ong SE, Kratchmarova I, Mann M (2004) Temporal analysis of phosphotyrosine-dependent signaling networks by quantitative proteomics. Nat Biotechnol 22:1139–1145

Brill LM, Salomon AR, Ficarro SB, Mukherji M, Stettler-Gill M, Peters EC (2004) Robust phosphoproteomic profiling of tyrosine phosphorylation sites from human T cells using immobilized metal affinity chromatography and tandem mass spectrometry. Anal Chem 76:2763–2772

Chen WG, White FM (2004) Proteomic analysis of cellular signaling. Expert Rev Proteomics 1:343–354

Cutillas PR, Geering B, Waterfield MD, Vanhaesebroeck B (2005) Quantification of gel-separated proteins and their phosphorylation sites by LC-MS using unlabeled internal standards: analysis of phosphoprotein dynamics in a B cell lymphoma cell line. Mol Cell Proteomics 4:1038–1051

De la Fuente van Bentem S, Roitinger E, Anrather D, Csaszar E, Hirt H (2006a) Phosphoproteomics as a tool to unravel plant regulatory mechanisms. Physiol Planta 126:110–119

De la Fuente van Bentem S, Anrather D, Roitinger E, Djamei A, Hufnagl T, Barta A, Csaszar E, Dohnal I, Lecourieux D, Hirt H (2006b) Phosphoproteomics reveals extensive in vivo phosphorylation of Arabidopsis proteins involved in RNA metabolism. Nucleic Acids Res 34:3267–3278

A, Scheel D, Cahill DJ, Lehrach H, Kreutzberger J, Kersten B (2005) High-throughput identification of potential Arabidopsis MAP kinases substrates. Mol Cell Proteomics 4:1558–1568

Ficarro SB, McCleland ML, Stukenberg PT, Burke DJ, Ross MM, Shabanowitz J, Hunt DF, White FM (2002) Phosphoproteome analysis by mass spectrometry and its application to *Saccharomyces cerevisiae*. Nat Biotechnol 20:301–305

Ficarro SB, Salomon AR, Brill LM, Mason DE, Stettler-Gill M, Brock A, Peters EC (2005) Automated immobilized metal affinity chromatography/nano-liquid chromatography/electrospray ionization mass spectrometry platform for profiling protein phosphorylation sites. Rapid Commun Mass Spectrom 19:57–71

Glinski M, Weckwerth W (2005) Differential multisite phosphorylation of the trehalose-6-phosphate synthase gene family in *Arabidopsis thaliana* – a mass spectrometry-based process for multiparallel peptide library phosphorylation analysis. Mol Cell Proteomics 4:1614–1625

Grønborg M, Kristiansen TZ, Stensballe A, Andersen JS, Ohara O, Mann M, Jensen ON, Pandey A (2002) A mass spectrometry-based proteomic approach for identification of serine/threonine-phosphorylated proteins by enrichment with phospho-specific antibodies: identification of a novel protein, Frigg, as a protein kinase A substrate. Mol Cell Proteomics 1:517–527

Gruhler A, Olsen JV, Mohammed S, Mortensen P, Faergeman NJ, Mann M, Jensen ON (2005a) Quantitative phosphoproteomics applied to the yeast pheromone signaling pathway. Mol Cell Proteomics 4:310–327

Gruhler A, Schulze WX, Matthiesen R, Mann M, Jensen ON (2005b) Stable isotope labeling of *Arabidopsis thaliana* cells and quantitative proteomics by mass spectrometry. Mol Cell Proteomics 4:1697–1709

Heintz D, Wurtz V, High AA, Van Dorsselaer A, Reski R, Sarnighausen E (2004) An efficient protocol for the identification of protein phosphorylation in a seedless plant, sensitive enough to detect members of signalling cascades. Electrophoresis 25:1149–1159

Hubbard MJ, Cohen P (1993) On target with a new mechanism for the regulation of protein phosphorylation. Trends Biochem Sci 18:172–177

Ibarrola N, Molina H, Iwahori A, Pandey A (2004) A novel proteomic approach for specific identification of tyrosine kinase substrates using [^{13}C]tyrosine. J Biol Chem 279:15805–15813

Katou S, Yoshioka H, Kawakita K, Rowland O, Jones JD, Mori H, Doke N (2005) Involvement of PPS3 phosphorylated by elicitor-responsive mitogen-activated protein kinases in the regulation of plant cell death. Plant Physiol 139:1914–1926

Kratchmarova I, Blagoev B, Haack-Sorensen M, Kassem M, Mann M (2005) Mechanism of divergent growth factor effects in mesenchymal stem cell differentiation. Science 308:1472–1477

Larsen MR, Thingholm TE, Jensen ON, Roepstorff P, Jorgensen TJ (2005) Highly selective enrichment of phosphorylated peptides from peptide mixtures using titanium dioxide microcolumns. Mol Cell Proteomics 4:873–886

Laugesen S, Bergoin A, Rossignol M (2004) Deciphering the plant phosphoproteome: tools and strategies for a challenging task. Plant Physiol Biochem 42:929–936

Liu Y, Zhang S (2004) Phosphorylation of 1-aminocyclopropane-1-carboxylic acid synthase by MPK6, a stress-responsive mitogen-activated protein kinase, induces ethylene biosynthesis in Arabidopsis. Plant Cell 16:3386–3399

Mann M, Ong SE, Grønborg M, Steen H, Jensen ON, Pandey A (2002) Analysis of protein phosphorylation using mass spectrometry: deciphering the phosphoproteome. Trends Biotechnol 20:261–268

Manning G, Whyte DB, Martinez R, Hunter T, Sudarsanam S (2002) The protein kinase complement of the human genome. Science 298:1912–1934

Meskiene I, Bogre L, Glaser W, Balog J, Brandstotter M, Zwerger K, Ammerer G, Hirt, H (1998) MP2C, a plant protein phosphatase 2C, functions as a negative regulator of mitogen-activated protein kinase pathways in yeast and plants. Proc Natl Acad Sci USA 95:1938–1943

Nakagami H, Pitzschke A, Hirt H (2005) Emerging MAP kinase pathways in plant stress signalling. Trends Plant Sci 10:339–46

Nühse TS, Stensballe A, Jensen ON, Peck SC (2003) Large-scale analysis of in vivo phosphorylated membrane proteins by immobilized metal ion affinity chromatography and mass spectrometry. Mol Cell Proteomics 2:1234–1243

Nühse TS, Stensballe A, Jensen ON, Peck SC (2004) Phosphoproteomics of the Arabidopsis plasma membrane and a new phosphorylation site database. Plant Cell 16:2394–2405
Olsen JV, Blagoev B, Gnad F, Macek B, Kumar C, Mortensen P, Mann M (2006) Global, in vivo, and site-specific phosphorylation dynamics in signaling networks. Cell 127:635–648
Ong SE, Blagoev B, Kratchmarova I, Kristensen DB, Steen H, Pandey A, Mann M (2002) Stable isotope labeling by amino acids in cell culture, SILAC, as a simple and accurate approach to expression proteomics. Mol Cell Proteomics 1:376–386
Pedley KF, Martin GB (2005) Role of mitogen-activated protein kinases in plant immunity. Curr Opin Plant Biol 8:541–547
Qi M, Elion EA (2005) MAP kinase pathways. J Cell Biol 118:3569–3572
Remenyi A, Good MC, Bhattacharyya RP, Lim WA (2005) The role of docking interactions in mediating signaling input, output, and discrimination in the yeast MAPK network. Mol Cell 20:951–962
Ross PL, Huang YN, Marchese JN, Williamson B, Parker K, Hattan S, Khainovski N, Pillai S, Dey S, Daniels S, Purkayastha S, Juhasz P, Martin S, Bartlet-Jones M, He F, Jacobson A, Pappin DJ (2004) Multiplexed protein quantitation in *Saccharomyces cerevisiae* using amine-reactive isobaric tagging reagents. Mol Cell Proteomics 3:1154–1169
Rush J, Moritz A, Lee KA, Guo A, Goss VL, Spek EJ, Zhang H, Zha XM, Polakiewicz RD, Comb MJ (2005) Immunoaffinity profiling of tyrosine phosphorylation in cancer cells. Nat Biotechnol 23:94–101
The Arabidopsis Genome Initiative (2000) Analysis of the genome sequence of the flowering plant *Arabidopsis thaliana*. Nature 408:796–815
Vener AV, Harms A, Sussman MR, Vierstra RD (2001) Mass spectrometric resolution of reversible protein phosphorylation in photosynthetic membranes of *Arabidopsis thaliana*. J Biol Chem 276:6959–6966
Yap YK, Kodama Y, Waller F, Chung KM, Ueda H, Nakamura K, Oldsen M, Yoda H, Yamaguchi Y, Sano H (2005) Activation of a novel transcription factor through phosphorylation by WIPK, a wound-induced mitogen-activated protein kinase in tobacco plants. Plant Physiol 139:127–137
Zhang Y, Wolf-Yadlin A, Ross PL, Pappin DJ, Rush J, Lauffenburger DA, White FM (2005) Time-resolved mass spectrometry of tyrosine phosphorylation sites in the epidermal growth factor receptor signaling network reveals dynamic modules. Mol Cell Proteomics 4:1240–1250

Chapter 5
High-Throughput Identification of Plant Protein Kinase Substrates

Birgit Kersten and Tanja Feilner

Abstract Proteomic approaches, such as protein microarray technology, play an important role in the study of complex biological systems. Their application in plant science has been strongly supported by the completion of genome sequence projects in the model plants *Arabidopsis thaliana* and rice. This chapter focuses on the identification of substrate phosphorylation by protein kinases using protein microarrays. Protein microarrays are widely used to profile antibodies and sera for their specificity, and/or to screen entire proteomes for new protein interactions. Here, we emphasise the feasibility of using protein microarrays in the discovery of potential protein kinase substrates in plants. Our group has identified potential substrates in two different plant species: first barley and later *Arabidopsis*. A signal quantification and threshold-based selection method was introduced during optimisation of protein microarray technology for substrate identification in order to better evaluate the microarray data, and to provide a preliminary list of candidate substrates for further investigation.

5.1 Introduction

Proteomics facilitates the discovery of new proteins as well as of new protein interaction partners, e.g. other proteins, DNA or small molecules. Proteomic technologies are becoming increasingly efficient and less time consuming. The exceptional achievements of the high-throughput and automated approaches have already contributed to improving the global analysis of proteins (Zhu et al. 2001; Ramachandran et al. 2004).

To give a short overview of the most well-known current proteomic techniques, two-dimensional gel electrophoresis (2-DE) certainly has to be mentioned as the

Birgit Kersten, Max Planck Institute of Molecular Plant Physiology, Bioinformatics Group, Am Mühlenberg 1, 14476 Potsdam-Golm, Germany

Tanja Feilner, Genetics and Biotechnology Laboratory, Department of Biochemistry, University College Cork, Lee Maltings, Cork, Ireland, E-mail: t_feilner@yahoo.de

J. Šamaj and J. Thelen (eds.), *Plant Proteomics*
© Springer 2007

classical method of protein expression studies. This technology has been used for proteomic analysis for more than 30 years (reviewed by Görg et al. 2004) and is often combined with mass-spectrometry (MS) to gain more detailed information concerning the identity, and the amino acid sequence of the separated proteins. Additionally, this combined approach allows the generation of catalogues of expressed proteins in a cell or tissue of interest.

Another, more recent, proteomic approach based on antibody arrays (Angenendt 2005) is used for the profiling of proteins in a manner analogous to that of profiling RNA expression using cDNA arrays. Expression profiling on the protein level is becoming increasingly important as studies on relative mRNA expression cannot provide detailed information about the protein level, not to mention the state or the function of a protein, or whether it exists alone or in a network (Gygi et al. 1999).

For functional studies, expression of proteins of interest in recombinant form is usually required. Two main approaches are currently used to find novel interaction partners: protein microarrays and the two hybrid-system in yeast. The latter approach is used principally to discover novel protein–protein interactions but it may also be used to verify data obtained from the application of protein microarrays or other in vitro methods. Protein microarrays can be also applied to analyse biochemical properties such as enzyme activity and substrate specificity.

In this chapter we will focus on protein microarrays, which have been used mainly to identify potential phosphorylation targets of different plant kinases, first in barley (Kramer et al. 2004) and later in *Arabidopsis thaliana* (Feilner et al. 2005). Protein kinases are enzymes that modify other proteins via the chemical addition of phosphate groups. Phosphorylation of proteins is crucial to the regulation of the signalling pathways that control various biological responses. Phosphorylation is one of the most important posttranslational protein modifications, others being glycosylation, acetylation, methylation, etc. Phosphorylation can influence various features of proteins, for example by altering the conformation state of a protein thus modifying its activity. Furthermore, it also can have an influence on multiple protein interactions. Phosphorylation and dephosphorylation (performed by phosphatases) are reversible biological processes and play a major role in controlling, e.g., metabolism, transcription, translation, cell proliferation and cell growth as well as cell cycling in diverse organisms including plants (Cohen 2000; Knight et al. 2003; Huber and Hardin 2004; Mukherji 2005; Pawson and Scott 2005; Kersten et al. 2006). It is assumed that approximately one-third of all proteins are phosphorylated in vivo at any given time (Cohen 2000; Zolnierowicz and Bollen 2000; Ahn and Resing 2001; Manning et al. 2002; Knight et al. 2003). The outstanding significance of kinases is reflected in the high number of genes predicted to encode protein kinases. Almost 5% of the *Arabidopsis* genome encodes kinases or phosphatases (1,100 and 100 genes each, respectively; Arabidopsis Genome Initiative 2000). Interestingly, the complement of kinases in *Arabidopsis* seems to be more complex than that of humans; the human genome encodes approximately 500 kinases and 100 phosphatases, i.e. accounting for around 2% of the genome (Venter et al. 2001; Manning et al. 2002).

Various proteomics technologies for large-scale functional analysis of signalling pathways, which concentrate on the analyses of phosphoproteins, the identification of novel kinase substrates and their phosphorylation sites, have now been established. For detailed information about different technologies in this phosphoproteomics area, see chapter 4 by de la Fuente van Bentem et al. in this volume.

As mentioned above, we will describe a protein microarray-based in vitro method that allows systematic screening of immobilised proteins for their phosphorylation by specific protein kinases. The main advantage of this intriguing method lies in the large-scale identification of potential kinase substrates. Because the in vivo validation of kinase substrates continues to be laborious and time-consuming, efficient pre-selection of substrate candidates based on protein microarrays before proceeding with in vivo analyses is highly advantageous. However, it should be emphasised that potential substrates thus identified must be verified by in vivo methods, as the kinase and substrate in question may not interact with each other under in vivo conditions.

5.2 Protein Microarray Technology

Protein microarrays containing recombinant proteins have become an important high-throughput tool in the proteomics field because they allow parallel, fast and easy analysis of thousands of addressable immobilised proteins for phosphorylation or molecular interactions (Feilner et al. 2004; Kersten et al. 2004, 2005; Angenendt 2005; Bertone and Snyder 2005; Merkel et al. 2005; Stoll et al. 2005; Hultschig et al. 2006). Due to the small size of such microarrays, only minute amounts of spotted proteins and analytes are needed (Kreutzberger 2006).

5.2.1 History of Microarrays

The development of microarray technology began with DNA microarrays, which led to applications in gene expression profiling and mutation mapping. The basic concept of the miniaturised microspot ligand-binding assay was first introduced by Ekins and colleagues (Ekins 1989; Ekins et al. 1990; Schena et al. 1995), who established that miniaturisation of spot size can increase sensitivity of detection. DNA microarray technology has been powered by the completion of whole genome sequencing projects, which has provided comprehensive nucleotide sequence information. However, DNA microarray technology does not provide information regarding quantitative and qualitative changes in proteins, or the function of proteins. This led to the development of a comparable technology for the analysis of proteins using protein microarrays. In contrast to nucleic acids, proteins have diverse and individual molecular structures, and protein functionality often depends on posttranslational modifications (for example phosphorylation, protein cleavage or multi-protein complex formation). In addition, for functional studies, captured proteins immobilised onto the microarray must remain in Fan active state. Therefore, it is obvious that a major obstacle to the generation of protein microarrays is access to a large number of pure recombinant proteins. From this point of view, it is easy to appreciate why the development of DNA microarrays was much faster and more efficient than the development of protein microarrays.

The first protein arrays were constructed by Buessow and colleagues as high-density arrays to analyse protein expression cDNA libraries (Bussow et al. 1998). These authors spotted 27,648 *Escherichia coli* clones in duplicate onto one 22 cm × 22 cm nylon filter membrane (Hybond-N+; Amersham, Little Chalfont, UK). Subsequently, protein expression and the detection of recombinant expression products were performed directly on these filters without further purification. As this array technology includes high costs regarding the membrane as well as the high consumption of sample material, development in the direction of protein microarrays was indispensable.

A breakthrough in the development of protein microarrays was the universal protein array (UPA) system of Ge (2000). Ge used this system to analyse the interaction of proteins with RNA, DNA, other proteins, ligands and other small chemicals. Nitrocellulose membrane was used as the array surface, and the size was based on a 96-well microtitre plate format. Luecking et al. (1999) were able to miniaturise array size by using PVDF filter strips cut to a microscope slide format. Lysates from 92 expressed cDNA clones were spotted onto these strips and these protein arrays were screened with specific antibodies. Shortly afterwards, two seminal papers were published by MacBeath and Schreiber (2000) and Zhu et al. (2001). MacBeath and Schreiber used protein microarray technology to demonstrate three different applications in a proof-of-principle study: (1) screening for protein–protein interactions, (2) identifying the substrates of protein kinases, and (3) identifying the protein targets of small molecules. Snyder's group reported the yeast proteome array, containing 5,800 unique yeast proteins (Zhu et al. 2001), and demonstrated the usefulness of protein microarray technology to screen for protein–protein interactions by identification of calmodulin- and phospholipid-binding proteins. The resulting protein microarray was later commercialised (Yeast ProtoArray, Protometrix; Invitrogen, Carlsbad, CA) and has since been used in various studies (Zhu et al. 2001; Michaud et al. 2003; Huang et al. 2004; Smith et al. 2005).

The application of protein microarrays also expanded into the plant field when the first *Arabidopsis* protein microarrays were generated (Kersten et al. 2003). These microarrays were used to characterise monoclonal antibodies or polyclonal sera with regard to their specificity and cross-reactivity. Following this initial study, more plant microarrays were generated and a phosphorylation assay allowing candidate kinase substrates to be short-listed from up to thousands of addressable immobilised proteins was developed (see Sect. 5.4).

5.2.2 Types of Protein Microarrays

Protein microarrays are generally divided into three types, according to a recent classification by Hultschig et al. (2006): (1) protein microarrays (PMAs), which contain immobilised purified proteins and may be applied to functional studies or for profiling of antibodies and sera with regard to specificity (Kersten et al. 2003; Lueking et al. 2003; Angenendt 2005; Merkel et al. 2005; Steller et al. 2005). (2) Antibody microarrays (AMAs), which contain antibodies or antibody mimics. This type of microarray is widely used to profile proteins from biological samples (Haab 2001; Robinson et al. 2002; Michaud et al. 2003; Nielsen et al. 2003; Hueber et al. 2005; Kopf et al. 2005). (3) Reverse protein microarrays (RPMAs). In contrast to the two other types, RPMAs contain immobilised cell lysates, tissue lysates or sera, which comprise the whole repertoire of sample proteins. These arrays are being used to determine protein abundance in immobilised biological samples (Nam et al. 2003; Geho et al. 2005; Labaer and Ramachandran 2005; Mircean et al. 2005; Speer et al. 2005; Hultschig et al. 2006; Kreutzberger 2006).

5.2.3 Establishment of Protein Microarrays for Phosphorylation Studies

For the generation of protein microarrays, large numbers of purified, recombinant (preferably native) proteins are required. Proteins are often expressed and purified from expression clones generated using basic technologies (cDNA cloning, recombinant protein expression and purification), which have being applied for a number of years (Bussow et al. 1998; Clark et al. 1999; Heyman et al. 1999; Bussow et al. 2000; Walhout et al. 2000; Kersten et al. 2003; Gong et al. 2004). Another promising method is cell-free expression. Here, proteins are expressed from cDNA templates, which can be easily generated by PCR and simply stored. This actually simplifies the production and purification of proteins, but is unfortunately quite expensive. Additionally, such systems can be easily down-scaled (Angenendt et al. 2004) and can be used for the direct synthesis of proteins on microarrays (Ramachandran et al. 2004; Angenendt et al. 2006).

Protein microarrays used to identify substrate phosphorylation by kinases were first introduced by MacBeath and Schreiber (2000) and Zhu and colleagues (2000). MacBeath and Schreiber used microarrays coated with a BSA-N-hydroxysuccinidimide monolayer. In addition to the discovery of technologies revealing ligand binding and protein–protein interactions on microarrays, they also performed proof-of-principle studies using three different but well known kinase–substrate pairs. They spotted the substrate of each pair in quadruplicate onto three microarrays, and each slide was incubated with one of the three kinases (MacBeath and Schreiber 2000). Zhu and colleagues used polydimethylsiloxane-covered nanowell-microarrays to analyse the activity and specificity of 119 yeast kinases for 17 different substrates. They immobilised one substrate per microarray. To analyse the substrate specificity of each kinase, they applied different kinases to nanowells of a microarray. Using this assay, 32 kinases that preferentially phosphorylated one or two substrates were identified (Zhu et al. 2000).

Another recent application can be found in the expression of mutants of p53 as fusion proteins with a biotin acceptor sequence, and their subsequent spotting and characterisation on streptavidin-coated chips (Boutell et al. 2004). These latter authors used the simultaneous presentation of allelic p53 variants on an array to enable rapid functional characterisation of the p53 protein in relation to its polymorphic forms. The array was also screened for MDM2 interactions as well as for p53 phosphorylation by casein kinase II.

These very successful and promising studies showed that protein phosphorylation by kinases can be detected using protein microarrays. This technology was subsequently used for the screening of potential substrates in large numbers of proteins. We generated plant microarrays and established phosphorylation assays, first for barley (Kramer et al. 2004) and later for *Arabidopsis* (Feilner et al. 2005). For more detailed information about these studies, see section 5.4.

A recent work by the Snyder group (Ptacek et al. 2005, Kung and Snyder 2006) generated the first "phosphorylome map" for yeast by using *Saccharomyces cerevisiae* proteome microarrays containing 4,400 proteins (representing approximately 70% of yeast open reading frames), which they incubated with 87 different yeast protein kinases. Each kinase phosphorylated between 1 and 256 substrates, with an average of 47 substrates per kinase. The authors mentioned that a distinct set of substrates was phosphorylated per kinase indicating that each kinase has a unique substrate recognition profile. The substrates phosphorylated by different kinases were then searched for common sequence motifs. Known consensus motifs were identified for 11 kinases; these were similar to sequence motifs determined for kinase orthologues in other species.

The phosphorylation assays mentioned above were performed using radioactive ATP ([γ-^{33}P]ATP). The first attempts to detect phosphorylation status and kinase activity on microarrays using fluorescence detection were made by Martin et al. (2003) using ProQ-Diamond phosphoprotein strain (Pro-Q DPS). The main advantage of using a phosphostain would be that a common microarray scanner could be used for the analysis of microarrays. Unfortunately, radioactive labelling is still more sensitive. Nevertheless, a number of recent promising studies have used Pro-Q DPS for the detection of phosphorylation events on microarrays (Rupcich et al. 2005; Pal et al. 2006). Depending on the phosphorylation state of the protein, the detection limits of Pro-Q DPS start at the level of picograms (as determined in protein- and peptide-arrays; Martin et al. 2003; Rupcich et al. 2005; Pal et al. 2006) or a few nanograms of protein (as determined on 1- or 2-DE; Steinberg et al. 2003; Agrawal and Thelen 2005), with linear ranges up to 500-fold (Morandell et al. 2006).

Peptide microarrays may also be applied to the identification of novel kinase substrates. Furthermore, this kind of microarray is perfectly suited to the identification of specific phosphorylation sites on the substrates. Different library types may be used for the generation of peptide arrays. Two general types can be defined: knowledge-based libraries, comprising peptides with sequences derived from naturally occurring proteins; and libraries that are designed de novo (Schutkowski et al. 2005).

Other in vitro and in vivo methods applied in the field of plant phosphoproteomics have been recently reviewed in Kersten et al. (2006) as well as in chapter 4 by de la Fuente van Bentem et al. in this volume.

5.3 Using Plant Protein Microarrays for the Identification of Potential Protein Kinase Substrates

Protein microarrays are one of the proteomic tools that can be used to identify potential phosphorylation targets. Our group identified potential substrates for two different plant species: first barley (Kramer et al. 2004) and later *Arabidopsis* (Feilner et al. 2005). For both approaches His-tagged cDNA expression libraries were generated as a source of recombinant proteins. In both studies FAST™ slides, which are coated with a nitrocellulose-derived polymer for non-covalent protein attachment, were used for the generation of the protein microarrays. After protein immobilisation, the microarrays were screened with an anti-RGS-HIS_6 antibody to detect the recombinant proteins.

In the barley study, we successfully established a novel protein microarray-based kinase assay, which was applied to identify potential substrates of a library-derived barley casein kinase2α (CK2α) in the presence of [γ-^{33}P]ATP (Kramer et al. 2004). In total, 768 recombinant barley proteins were arrayed on two microarrays, i.e. 384 proteins were spotted on each microarray in quadruplicate. As positive controls, library-derived barley proteins were used, which share a strong homology with different plant HMG ("high mobility group") proteins. These proteins are well known CK2α substrates (Grasser et al. 1989; Stemmer et al. 2002). Out of the 768 proteins, 21 potential targets were identified, including well-known substrates of CK2α (e.g. different HMG proteins and calreticulin) as well as targets not described in the literature so far. The selection of potential substrates was performed based on qualitative criteria, i.e. the protein was regarded as positive when all four of the corresponding quadruplicates in two independent experiments were detectable.

This medium-throughput assay – with respect to the number of proteins analysed on the microarrays – was followed by an *Arabidopsis* study (Feilner et al. 2005), which, to our knowledge, is the largest to identify potential kinase substrates in plants to date. For this study, we generated *Arabidopsis* proteins microarrays containing approximately 1,700 proteins. Each protein was spotted twice in two identical fields. As well as using a higher number of proteins, a possible signal quantification and threshold-based selection method were introduced to increase the reliability of the method. Microarrays were incubated with the mitogen-activated *Arabidopsis* kinases MPK3 and MPK6 in the presence

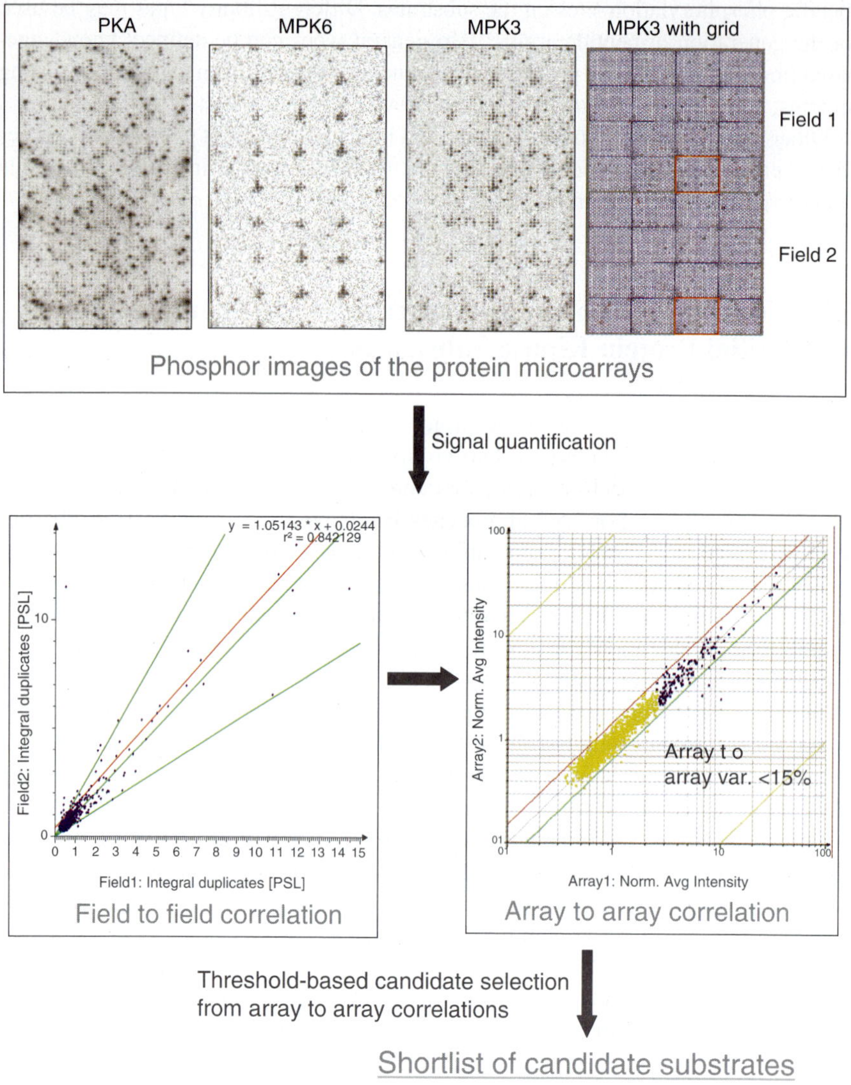

Fig. 5.1 Schematic depiction of identification of substrate candidates for plant kinases using protein microarray-based kinase assays. The design and application of the method for the analysis of Arabidopsis protein kinases was published by Feilner et al. (2005). Briefly, proteins are spotted in two identical fields onto the microarrays. For each kinase two independent experiments with two microarrays each were performed. Detection was performed using phosphorimaging after incubation with [γ-^{33}P]ATP and the corresponding kinase. *Upper panel* Phosphorimages of the representative *Arabidopsis* protein microarrays after incubation with PKA, MPK6 and MPK3, as indicated. The right-most image shows the MPK3 microarray with a superimposed grid for further signal quantification. *Bottom left panel* Field-to-field correlation of the two signals of spotted proteins phosphorylated with MPK6 on one microarray. Signals of the phosphorylated spots in field 2 (*y*-axis) are plotted against the signal intensities of the corresponding proteins in field 1 (*x*-axis). *PSL* Photo-stimulated luminescence, *r* regression coefficient. *Bottom right panel* Slide-to-slide correlation to compare the two slides from one experiment. Signals of the phosphorylated spots in array 2 (*y*-axis) are plotted against the signal of the corresponding proteins in microarray 1 (*x*-axis)

of [γ -^{33}P]ATP. As a control, mouse protein kinase A, which was expected to show a different phosphorylation pattern compared to the two MPKs, was used (Fig. 5.1).

Using the threshold-based quantification method to evaluate the microarray results, 48 potential substrates of MPK3, and 39 of MPK6, were identified from among the 1,700 proteins. As expected, many of them were common to both kinases (26 substrates). Two independent experiments with two microarrays each were performed for these two kinases. A field-to-field correlation of the two signals of spotted proteins that were phosphorylated by each kinase from the two identical fields of a microarray was performed for quantification. First, only duplicates deviating less than 25% from the average intensity of both spots were considered for subsequent analysis (region within both green outer lines in the field-to-field correlation shown in Fig. 5.1). The red line shows the ideal distribution of theoretical 100% reproducible replicates. The regression line of experimental replicates is given by the inner green line in Fig. 5.1. Secondly, this procedure was followed by an array-to-array correlation to compare the two slides from one experiment (see Fig. 5.1). Based on a threshold we then selected positive proteins from each experiment. The threshold was set up in such a way that only spots that showed deviation from the background corrected signals assigned to the same protein of less than a factor of 2 between the two arrays were considered. Additionally, only those spots exceeding the average signal intensity of the background by at least 10 times the standard deviation of the background signals were considered in order to identify potential targets of the kinases (highlighted as blue spots in the array-to-array correlation in Fig. 5.1). Finally, only proteins identified as phosphorylated in both experiments were subsequently defined as substrate candidates or potential substrates of the given kinase.

Nearly all 48 potential MPK3 substrates were confirmed using an in vitro assay with the recombinant candidate proteins, which were refolded on NiNTA-beads prior to a kinase assay in solution (Feilner et al. 2005). Mouse protein kinase A (PKA), which was included in this study as an example of a kinase from a different family to validate the specificity of the method, gave a clearly different phosphorylation pattern compared to the MPKs. For PKA, 35 potential substrates were identified, three of which were in common with MPK3, and four with MAPK6. The relatively high number of mouse PKA substrates identified within the set of Arabidopsis proteins may be due to the fact that several plant protein kinases belong to the same group as PKA, i.e. the AGC group (named after PKA, PKG, and PKC; see www.nih.go.jp/mirror/Kinases/pkr/pk_catalytic/pk_hanks_seq_align_long.html), and they share specific sequence motifs such as the FXXF hydrophobic motif in the C terminus (Bogre et al. 2003).

As an additional control for the quality of the identified substrates, the sequences of the potential substrates of PKA were compared with the sequences of all 1,700 spotted *Arabidopsis* proteins on the microarray. Therefore, all potential substrates were analysed with four well known consensus sequences of PKA target proteins (Kennelly and Krebs 1991; Pearson and Kemp 1991). Significant accumulation of the four consensus sequences within the potential PKA substrates could be seen (data not shown).

5.4 Potential Substrates of Arabidopsis MPK3 and MPK6

5.4.1 Brief Bibliographic Overview of MAPK Pathway

Various types of kinases from plants have been described and classified into different groups (Hardie 1999). One of the largest and most important families of serine/threonine kinases is the mitogen-activated protein kinase (MAPK) family. In yeast, plants and humans, MAPKs and their upstream activators are organised in highly conserved signalling modules that have a major impact on cell growth, differentiation, proliferation, stress response and death (Tena et al. 2001). MAPKs represent the last component of the "three-kinase" module in the MAPK cascade (Tena et al. 2001; Asai et al. 2002; Jonak et al. 2002; see also Fig. 5.2).

This module consists of at least a MAPK kinase kinase (MAPKKK), a MAPKK and finally a MAPK, all representing phosphorylation enzymes. During signal transduction the serine/threonine protein kinase MAPKKK is phosphorylated by upstream activator (another kinase or receptor). The activated MAPKKK then phosphorylates the subsequent dual specific MAPKK, which in turn activates the MAPK by phosphorylation of a threonine as well as a tyrosine residue in the "activation loop" (Keyse 2000; Nuhse et al. 2000). In the fully sequenced Arabidopsis genome, 20 genes encoding MAPKs, 10 genes encoding MAPKKs

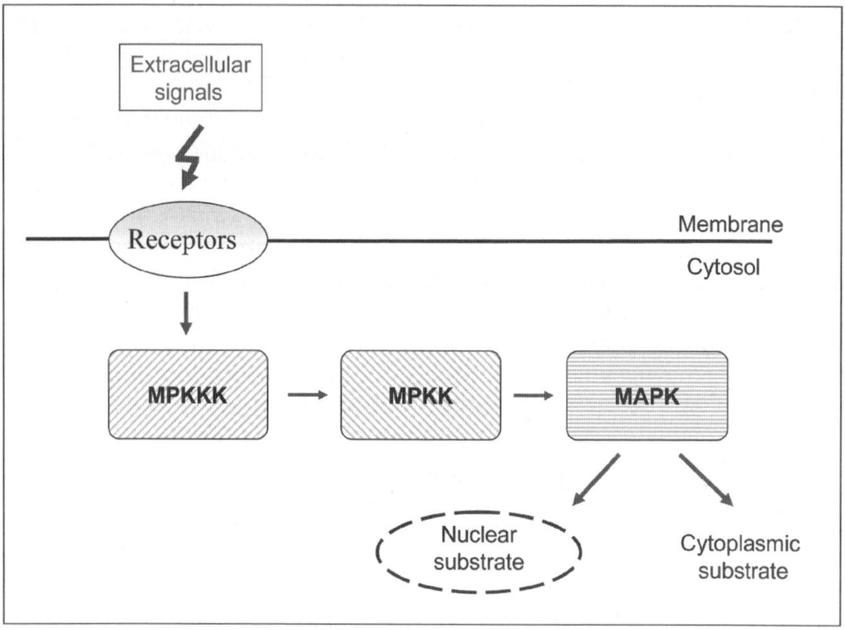

Fig. 5.2 Schematic representation of the mitogen-activated protein kinase (MAPK) pathway

and more than 60 genes encoding MAPKKKs have been identified (MAPK Group 2002).

Plant MAPKs can be classified into four groups: A–D. For more detailed information on the other two kinase families, namely MAPKKs and MAPKKKs, please see the report by the MAPK Group (2002).

The first three MAPK groups (A–C) contain the phosphorylation motif TEY whereas members of group D contains the motif TDY in their loop (Hardie 1999; Zhang and Klessig 2001; Jonak et al. 2002). MAPK group A, which includes tobacco WIPK, alfalfa SIMK as well as Arabidopsis MPK3 and MPK6, is involved mainly in hormonal and environmental responses. The second MAPK group (B) is involved mostly in some environmental responses but also in cell cycle regulation. Little is known so far about the functions of MAPKs in groups C and D.

MAPKs are localised in the cytoplasm as well as in the nucleus (Šamaj et al. 2002), and they link, in multiple ways, upstream intracellular and extracellular stimuli to downstream cellular responses (Jonak et al. 2002; see Fig. 5.2). Plant MAPK cascades are associated with defence responses to abiotic stress, including temperature and drought (Jonak et al. 1996; Mizoguchi et al. 1996; Ichimura et al. 2000), mechanic and osmotic stress (Bogre et al. 1996; Mizoguchi et al. 1996; Tena and Renaudin, 1998; Ichimura et al. 2000) and UV light, as well as the cellular responses to hormones (Tena et al. 2001; Jonak et al. 2002; Lu et al. 2002), pathogens and elicitors (Romeis et al. 1999; Nuhse et al. 2000; Asai et al. 2002). Once the terminal MAPK is activated, it can then phosphorylate downstream substrates including transcription factors, cytoskeleton-associated proteins or other protein kinases (e.g. Šamaj et al. 2004).

Although many plant protein kinases have been shown to be activated in response to numerous stress conditions, the number of identified substrates is still limited. In plants, only a few studies have identified in vivo MAPK substrates to date. Liu and Zhang (2004) found that distinct isoforms of 1-aminocyclopropane-1-carboxylic acid synthase-6 (ACS-6), a rate-limiting enzyme of ethylene biosynthesis, represent MPK6 substrates in Arabidopsis, and that MPK6 is previously activated by MKK4. Phosphorylation by MPK6 leads to the accumulation of ACS proteins and to elevated levels of ACS activity in vivo, consequently increasing ethylene production. In a more recent study, a MPK4 substrate, called MKS1 (MAPK substrate 1), was identified in Arabidopsis (Andreasson et al. 2005). Using the yeast two-hybrid assay, MKS1 was identified to interact both with MPK4 and with different WRKY transcription factors, such as WRKY25 and WRKY33. Another study reported that MPK6 is required to maintain basal resistance to a virulent pathogen and to activate full resistance to avirulent bacteria as well as oomycete pathogens (Menke et al. 2004). Furthermore, these latter authors describe a correlation between VSP1 (pathogen-inducible gene vegetative storage protein 1), which is positively regulated by MPK6, and resistance in MPK6-silenced plants, indicating that VSP1 could be a potential substrate of MPK6.

5.4.2 Large-Scale Identification of Potential MPK3 and MPK6 Substrates

As mentioned above, 48 substrates for MPK3 and 39 for MPK6 (out of 1,700 proteins tested) were identified based on protein microarray studies in combination with a threshold-based selection method (Feilner et al. 2005). The first MPK6 substrate, described here in more detail, is 1-aminocyclopropane-1-carboxylic acid synthase-6 (ACS-6; MPK6: At4g11280). ACS-6 was the first in vivo plant MAPK substrate to be described (Liu and Zhang 2004), and exactly this protein was also detected with the microarray approach. Thus, one can assume that microarray-based detection represents an excellent large-scale screening method to reveal potential substrates, including candidates with physiological relevance.

One of the identified proteins belongs to the LRR family (MPK3 substrate: At4g03260), and may participate as a receptor in elicitor-induced MAPK cascades. As reported previously, the LRR-type receptor-like protein kinase FLS2 (flagellin sensitive), which acts upstream of MPK6 and MPK3, recognises the binding of the flagellin flg22 (Asai et al. 2002). This leads to the activation of transcription factors of the WRKY family (WRKY22 and WRKY29).

The calmodulin-binding family protein (MPK3 substrate: At5g62070) may be a component of the MAPK cascade because a calmodulin-binding protein has been described previously as a negative regulator of stress tolerance due to sodium and osmotic stress (Perruc et al. 2004).

So far no involvement in the MAPK cascade in plants has been shown for the casein kinase (MPK3 and MPK6 substrate: At5g4410). However, in HeLa cells a direct interaction between MAP kinase p38 and CK2, and thereby the stress-induced activation through the MAP kinase, could be observed (Sayed et al. 2000).

Flowering locus T protein (MPK3 and MPK6 substrate: At1g65480) is a putative membrane-associated protein with homology to human phosphatidylethanolamine-binding protein (Kardailsky et al. 1999). This protein is identical to the Raf kinase inhibitor protein involved in regulation of the Raf/MEK/ERK signalling pathway (Yeung et al. 1999).

It is also important to mention further proteins identified as potential kinase substrates, such as several transcription factors (MPK3: At5g58620, At2g02820 and At5g66940, MPK6: At5g58620, At5g54630, At3g60390, and At5g43650) and one transcription regulator (MPK6: At2g46020). In addition to the phosphorylation status of transcription factors, phosphorylation of histones (MPK3: At5g02560, At1g52740 and At5g65360, MPK6: At5g65360) could also be involved in regulating gene transcription. Previously, it has been shown that the tail domain of histones regulates chromatin structure and hence gene transcription (Loury and Sassone-Corsi 2003). Additionally, Yamagata et al. (2001) described the phosphorylation of histones via a MAPK kinase-like protein in soybean cell culture.

Although many core proteins were identified, these results support the assumption that regulation in response to MAPK signalling is very complex and it is not restricted to the transcriptional level (Yang et al. 2003).

5.5 Conclusions and Future Prospects

In conclusion, a very promising and powerful in vitro generic test system was developed by using protein microarrays. Hereby, we were able to identify substrate phosphorylation by kinases followed by an independent in vitro verification of the potential substrates in solution using refolded proteins. We would like to stress that this method is only used to shortlist candidate proteins out of a vast number of proteins and/or even a whole proteome. Follow-up experiments such as in vivo verification and the mapping of phosphorylation sites in substrates are essential to evaluate the physiological relevance of targets in MAPK signalling. As this collection of proteins expands, the demand for more plant and other eukaryotic and human protein microarrays will obviously increase. Furthermore, progress in protein purification methods as well as cell free expression will enhance protein quantity and quality, thereby increasing the functional activity of each protein spot (Kung and Snyder 2006). The establishment of protein microarrays is not without drawbacks and development will obviously not be as fast as in the case of DNA microarrays. Nevertheless, it can be argued that protein microarrays are one of the most promising tools available to identify protein phosphorylation by kinases as well as other protein interactions with various partners in parallel.

Acknowledgements We would like to thank Dr. Claus Hultschig for his helpful comments and his continuous support. The work was supported by a grant from the German Federal Ministry of Education and Research (BMBF, GABI-grant 0313112). Dr. Tanja Feilner is supported by the Irish Research Council for Science Engineering and Technology postdoctoral fellowship scheme. We apologise to our colleagues whose work on plant MAPK signalling and protein microarrays was not included in the reference list due to space limitations.

References

Agrawal GK, Thelen JJ (2005) Development of a simplified, economical polyacrylamide gel staining protocol for phosphoproteins. Proteomics 5:4684–4688

Ahn NG, Resing KA (2001) Toward the phosphoproteome. Nat Biotechnol 19:317–318

Andreasson E, Jenkins T, Brodersen P, Thorgrimsen S, Petersen NH, Zhu S, Qiu JL, Micheelsen P, Rocher A, Petersen M, Newman MA, Bjorn Nielsen H, Hirt H, Somssich I, Mattsson O, Mundy J (2005) The MAP kinase substrate MKS1 is a regulator of plant defense responses. EMBO J 24:2579–2589

Angenendt P (2005) Progress in protein and antibody microarray technology. Drug Discovery Today Targets 10:503–511

Angenendt P, Wilde J, Kijanka G, Baars S, Cahill DJ, Kreutzberger J, Lehrach H, Konthur Z, Glökler J (2004) Seeing better through a MIST: evaluation of monoclonal recombinant antibody fragments on microarrays. Anal Chem 76:2916–2921

Angenendt P, Kreutzberger J, Glokler J, Hoheisel JD (2006) Generation of high density protein microarrays by cell-free in situ expression of unpurified PCR products. Mol Cell Proteomics 5:1658–1666

Arabidopsis Genome Initiative (2000) Analysis of the genome sequence of the flowering plant *Arabidopsis thaliana*. Nature 408:796–815

Asai T, Tena G, Plotnikova J, Willmann MR, Chiu WL, Gomez-Gomez L, Boller T, Ausubel FM, Sheen J (2002) MAP kinase signalling cascade in Arabidopsis innate immunity. Nature 415:977–983

Bertone P, Snyder M (2005) Prospects and challenges in proteomics. Plant Physiol 138:560–562

Bogre L, Ligterink W, Heberle-Bors E, Hirt H (1996) Mechanosensors in plants. Nature 383:489–490

Bogre L, Okresz L, Henriques R, Anthony RG (2003) Growth signalling pathways in Arabidopsis and the AGC protein kinases. Trends Plant Sci 8:424–431

Boutell JM, Hart DJ, Godber BL, Kozlowski RZ, Blackburn JM (2004) Functional protein microarrays for parallel characterisation of p53 mutants. Proteomics 4:1950–1958

Bussow K, Cahill D, Nietfeld W, Bancroft D, Scherzinger E, Lehrach H, Walter G (1998) A method for global protein expression and antibody screening on high-density filters of an arrayed cDNA library. Nucleic Acids Research 26:5007–5008

Bussow K, Nordhoff E, Lubbert C, Lehrach H, Walter G (2000) A human cDNA library for high-throughput protein expression screening. Genomics 65:1–8

Clark MD, Panopoulou GD, Cahill DJ, Büssow K, Lehrach H (1999) Construction and analysis of arrayed cDNA libraries. Methods Enzymol 303:205–233

Cohen P (2000) The regulation of protein function by multisite phosphorylation – a 25 year update. Trends Biochem Sci 25:596–601

Ekins RP (1989) Multi-analyte immunoassay. J Pharm Biomed Anal 7:155–168

Ekins R, Chu F, Biggart E (1990) Multispot, multianalyte, immunoassay. Ann Biol Clin (Paris) 48:655–666

Feilner T, Kreutzberger J, Niemann B, Kramer A, Possling A, Seitz H, Kersten B (2004) Proteomic studies using microarrays. Curr Proteomics 1:283–295

Feilner T, Hultschig C, Lee J, Meyer S, Immink RG, Koenig A, Possling A, Seitz H, Beveridge A, Scheel D, Cahill DJ, Lehrach H, Kreutzberger J, Kersten B (2005) High throughput identification of potential Arabidopsis mitogen-activated protein kinases substrates. Mol Cellular Proteomics 4:1558–1568

Ge H (2000) UPA, a universal protein array system for quantitative detection of protein–protein, protein–DNA, protein–RNA and protein–ligand interactions. Nucleic Acids Res 28:e3

Geho D, Lahar N, Gurnani P, Huebschman M, Herrmann P, Espina V, Shi A, Wulfkuhle J, Garner H, Petricoin E 3rd, Liotta LA, Rosenblatt KP (2005) Pegylated, streptavidin-conjugated quantum dots are effective detection elements for reverse-phase protein microarrays. Bioconjug Chem 16:559–566

Gong W, Shen YP, Ma LG, Pan Y, Du YL, Wang DH, Yang JY, Hu LD, Liu XF, Dong CX, Ma L, Chen YH, Yang XY, Gao Y, Zhu D, Tan X, Mu JY, Zhang DB, Liu YL, Dinesh-Kumar SP, Li Y, Wang XP, Gu HY, Qu LJ, Bai SN, Lu YT, Li JY, Zhao JD, Zuo J, Huang H, Deng XW, Zhu YX (2004) Genome-wide ORFeome cloning and analysis of Arabidopsis transcription factor genes. Plant Physiol 135:773–782

Görg A, Weiss W, Dunn MJ (2004) Current two-dimensional electrophoresis technology for proteomics. Proteomics 4:3665–3685

Grasser KD, Maier UG, Feix G (1989) A nuclear casein type II kinase from maize endosperm phosphorylating HMG proteins. Biochem Biophys Res Commun 162:456–463

Gygi SP, Rochon Y, Franza BR, Aebersold R (1999) Correlation between protein and mRNA abundance in yeast. Mol Cell Biol 19:1720–1730

Haab BB (2001) Advances in protein microarray technology for protein expression and interaction profiling. Curr Opin Drug Discov Dev 4:116–123

Hardie DG (1999) Plant protein serine/threonine kinases: classifications and functions. Annu Rev Plant Physiol Plant Mol Biol 50:97–131

Heyman JA, Cornthwaite J, Foncerrada L, Gilmore JR, Gontang E, Hartman KJ, Hernandez CL, Hood R, Hull HM, Lee WY, Marcil R, Marsh EJ, Mudd KM, Patino MJ, Purcell TJ, Rowland JJ, Sindici ML, Hoeffler JP (1999) Genome-scale cloning and expression of individual open reading frames using topoisomerase I-mediated ligation. Genome Res 9:383–392

Huang J, Zhu H, Haggarty SJ, Spring DR, Hwang H, Jin F, Snyder M, Schreiber SL (2004) Finding new components of the target of rapamycin (TOR) signaling network through chemical genetics and proteome chips. Proc Natl Acad Sci USA 101:16594–16599

Huber SC, Hardin SC (2004) Numerous posttranslational modifications provide opportunities for the intricate regulation of metabolic enzymes at multiple levels. Curr Opin Plant Biol 7:318–322

Hueber W, Kidd BA, Tomooka BH, Lee BJ, Bruce B, Fries JF, Sonderstrup G, Monach P, Drijfhout JW, van Venrooij WJ, Utz PJ, Genovese MC, Robinson WH (2005) Antigen microarray profiling of autoantibodies in rheumatoid arthritis. Arthritis Rheum 52:2645–2655

Hultschig C, Kreutzberger J, Seitz H, Konthur Z, Bussow K, Lehrach H (2006) Recent advances of protein microarrays. Curr Opin Chem Biol 10:4–10

Ichimura K, Mizoguchi T, Yoshida R, Yuasa T, Shinozaki K (2000) Various abiotic stresses rapidly activate Arabidopsis MAP kinases ATMPK4 and ATMPK6. Plant J 24:655–665

Jonak C, Kiegerl S, Ligterink W, Barker PJ, Huskisson NS, Hirt H (1996) Stress signaling in plants: a mitogen-activated protein kinase pathway is activated by cold and drought. Proc Natl Acad Sci USA 93:11274–11279

Jonak C, Okresz L, Bogre L, Hirt H (2002) Complexity, cross talk and integration of plant MAP kinase signalling. Curr Opin Plant Biol 5:415–424

Kardailsky I, Shukla VK, Ahn JH, Dagenais N, Christensen SK, Nguyen JT, Chory J, Harrison MJ, Weigel D (1999) Activation tagging of the floral inducer FT. Science 286:1962–1965

Kennelly PJ, Krebs EG (1991) Consensus sequences as substrate specificity determinants for protein kinases and protein phosphatases. J Biol Chem 266:15555–15558

Kersten B, Feilner T, Kramer A, Wehrmeyer S, Possling A, Witt I, Zanor MI, Stracke R, Lueking A, Kreutzberger J, Lehrach H, Cahill DJ (2003) Generation of Arabidopsis protein chips for antibody and serum screening. Plant Mol Biol 52:999–1010

Kersten B, Possling A, Blaesing F, Mirgorodskaya E, Gobom J, Seitz H (2004) Protein microarray technology and ultraviolet crosslinking combined with mass spectrometry for the analysis of protein-DNA interactions. Anal Biochem 331:303–313

Kersten B, Wanker EE, Hoheisel J, Angenendt P (2005) Multiplex approaches in protein microarray technology. Expert Rev Proteomics 2:499–510

Kersten B, Agrawal GK, Iwahashi H, Rakwal R (2006) Plant phosphoproteomics: a long road ahead. Proteomics 6:5517–5528

Keyse SM (2000) Protein phosphatases and the regulation of mitogen-activated protein kinase signalling. Curr Opin Cell Biol 12:186–192

Knight ZA, Schilling B, Row RH, Kenski DM, Gibson BW, Shokat KM (2003) Phosphospecific proteolysis for mapping sites of protein phosphorylation. Nat Biotechnol 21:1047–1054

Kopf E, Shnitzer D, Zharhary D (2005) Panorama Ab Microarray Cell Signaling kit: a unique tool for protein expression analysis. Proteomics 5:2412–2416

Kramer A, Feilner T, Possling A, Radchuk V, Weschke W, Burkle L, Kersten B (2004) Identification of barley CK2alpha targets by using the protein microarray technology. Phytochemistry 65:1777–1784

Kreutzberger J (2006) Protein microarrays: a chance to study microorganisms? Appl Microbiol Biotechnol 70:383–390

Kung LA, Snyder M (2006) Proteome chips for whole-organism assays. Nat Rev Mol Cell Biol 7:617–622

Labaer J, Ramachandran N (2005) Protein microarrays as tools for functional proteomics. Curr Opin Chem Biol 9:14–19

Liu Y, Zhang S (2004) Phosphorylation of 1-aminocyclopropane-1-carboxylic acid synthase by MPK6, a stress-responsive mitogen-activated protein kinase, induces ethylene biosynthesis in Arabidopsis. Plant Cell 16:3386–3399

Loury R, Sassone-Corsi P (2003) Histone phosphorylation: how to proceed. Methods 31:40–48

Lu C, Han MH, Guevara-Garcia A, Fedoroff NV (2002) Mitogen-activated protein kinase signaling in postgermination arrest of development by abscisic acid. Proc Natl Acad Sci USA 99:15812–15817

Lueking A, Horn M, Eickhoff H, Bussow K, Lehrach H, Walter G (1999) Protein microarrays for gene expression and antibody screening. Anal Biochem 270:103–111

Lueking A, Possling A, Huber O, Beveridge A, Horn M, Eickhoff H, Schuchardt J, Lehrach H, Cahill DJ (2003) A nonredundant human protein chip for antibody screening and serum profiling. Mol Cell Proteomics 2:1342–1349

MacBeath G, Schreiber SL (2000) Printing proteins as microarrays for high-throughput function determination. Science 289:1760–1763

Manning G, Whyte DB, Martinez R, Hunter T, Sudarsanam S (2002) The protein kinase complement of the human genome. Science 298:1912–1934

MAPK Group (2002) Mitogen-activated protein kinase cascades in plants: a new nomenclature. Trends Plant Sci 7:301–308

Martin K, Steinberg TH, Cooley LA, Gee KR, Beechem JM, Patton WF (2003) Quantitative analysis of protein phosphorylation status and protein kinase activity on microarrays using a novel fluorescent phosphorylation sensor dye. Proteomics 3:1244–1255

Menke FL, van Pelt JA, Pieterse CM, Klessig DF (2004) Silencing of the mitogen-activated protein kinase MPK6 compromises disease resistance in Arabidopsis. Plant Cell 16:897–907

Merkel JS, Michaud GA, Salcius M, Schweitzer B, Predki PF (2005) Functional protein microarrays: just how functional are they? Curr Opin Biotechnol 16:447–452

Michaud GA, Salcius M, Zhou F, Bangham R, Bonin J, Guo H, Snyder M, Predki PF, Schweitzer BI (2003) Analyzing antibody specificity with whole proteome microarrays. Nat Biotechnol 21:1509–1512

Mircean C, Shmulevich I, Cogdell D, Choi W, Jia Y, Tabus I, Hamilton SR, Zhang W (2005) Robust estimation of protein expression ratios with lysate microarray technology. Bioinformatics 21:1935–1942

Mizoguchi T, Irie K, Hirayama T, Hayashida N, Yamaguchi-Shinozaki K, Matsumoto K, Shinozaki K (1996) A gene encoding a mitogen-activated protein kinase kinase kinase is induced simultaneously with genes for a mitogen-activated protein kinase and an S6 ribosomal protein kinase by touch, cold, and water stress in *Arabidopsis thaliana*. Proc Natl Acad Sci USA 93:765–769

Morandell S, Stasyk T, Grosstessner-Hain K, Roitinger E, Mechtler K, Bonn GK, Huber LA (2006) Phosphoproteomics strategies for the functional analysis of signal transduction. Proteomics 6:4047–4056

Mukherji M (2005) Phosphoproteomics in analyzing signaling pathways. Expert Rev Proteomics 2:117–128

Nam MJ, Madoz-Gurpide J, Wang H, Lescure P, Schmalbach CE, Zhao R, Misek DE, Kuick R, Brenner DE, Hanash SM (2003) Molecular profiling of the immune response in colon cancer using protein microarrays: occurrence of autoantibodies to ubiquitin C-terminal hydrolase L3. Proteomics 3:2108–2115

Nielsen UB, Cardone MH, Sinskey AJ, MacBeath G, Sorger PK (2003) Profiling receptor tyrosine kinase activation by using Ab microarrays. Proc Natl Acad Sci USA 100:9330–9335

Nuhse TS, Peck SC, Hirt H, Boller T (2000) Microbial elicitors induce activation and dual phosphorylation of the *Arabidopsis thaliana* MAPK 6. J Biol Chem 275:7521–7526

Pal M, Moffa A, Sreekumar A, Ethier SP, Barder TJ, Chinnaiyan A, Lubman DM (2006) Differential phosphoprotein mapping in cancer cells using protein microarrays produced from 2-D liquid fractionation. Anal Chem 78:702–710

Pawson T, Scott JD (2005) Protein phosphorylation in signaling – 50 years and counting. Trends Biochem Sci 30:286–290

Pearson RB, Kemp BE (1991) Protein kinase phosphorylation site sequences and consensus specificity motifs: tabulations. Methods Enzymol 200:62–81

Perruc E, Charpenteau M, Ramirez BC, Jauneau A, Galaud JP, Ranjeva R, Ranty B (2004) A novel calmodulin-binding protein functions as a negative regulator of osmotic stress tolerance in *Arabidopsis thaliana* seedlings. Plant J 38:410–420

Ptacek J, Devgan G, Michaud G, Zhu H, Zhu X, Fasolo J, Guo H, Jona G, Breitkreutz A, Sopko R, McCartney RR, Schmidt MC, Rachidi N, Lee SJ, Mah AS, Meng L, Stark MJ, Stern DF, De

Virgilio C, Tyers M, Andrews B, Gerstein M, Schweitzer B, Predki PF, Snyder M (2005) Global analysis of protein phosphorylation in yeast. Nature 438:679–684
Ramachandran N, Hainsworth E, Bhullar B, Eisenstein S, Rosen B, Lau AY, Walter JC, LaBaer J (2004) Self-assembling protein microarrays. Science 305:86–90
Robinson WH, DiGennaro C, Hueber W, Haab BB, Kamachi M, Dean EJ, Fournel S, Fong D, Genovese MC, de Vegvar HE, Skriner K, Hirschberg DL, Morris RI, Muller S, Pruijn GJ, van Venrooij WJ, Smolen JS, Brown PO, Steinman L, Utz PJ (2002) Autoantigen microarrays for multiplex characterisation of autoantibody responses. Nat Med 8:295–301
Romeis T, Piedras P, Zhang S, Klessig DF, Hirt H, Jones JD (1999) Rapid Avr9- and Cf-9-dependent activation of MAP kinases in tobacco cell cultures and leaves: convergence of resistance gene, elicitor, wound, and salicylate responses. Plant Cell 11:273–287
Rupcich N, Green JR, Brennan JD (2005) Nanovolume kinase inhibition assay using a sol-gel-derived multicomponent microarray. Anal Chem 77:8013–8019
Šamaj J, Ovecka M, Hlavacka A, Lecourieux F, Meskiene I, Lichtscheidl I, Lenart P, Salaj J, Volkmann D, Bögre L, Baluška F, Hirt H (2002) Involvement of the mitogen-activated protein kinase SIMK in regulation of root hair tip-growth. EMBO J 21:3296–3306
Šamaj J, Baluška F, Hirt H (2004) From signal to cell polarity: mitogen-activated protein kinases as sensors and effectors of cytoskeleton dynamicity. J Exp Bot 55:189–198
Sayed M, Kim SO, Salh BS, Issinger OG, Pelech SL (2000) Stress-induced activation of protein kinase CK2 by direct interaction with p38 mitogen-activated protein kinase. J Biol Chem 275:16569–16573
Schena M, Shalon D, Davis RW, Brown PO (1995) Quantitative monitoring of gene expression patterns with a complementary DNA microarray. Science 270:467–470
Schutkowski M, Reineke U, Reimer U (2005) Peptide arrays for kinase profiling. Chembiochem 6:513–521
Smith MG, Jona G, Ptacek J, Devgan G, Zhu H, Zhu X, Snyder M (2005) Global analysis of protein function using protein microarrays. Mech Ageing Dev 126:171–175
Speer R, Wulfkuhle JD, Liotta LA, Petricoin EF 3rd (2005) Reverse-phase protein microarrays for tissue-based analysis. Curr Opin Mol Ther 7:240–245
Steinberg TH, Agnew BJ, Gee KR, Leung WY, Goodman T, Schulenberg B, Hendrickson J, Beechem JM, Haugland RP, Patton WF (2003) Global quantitative phosphoprotein analysis using Multiplexed Proteomics technology. Proteomics 3:1128–1144
Steller S, Angenendt P, Cahill DJ, Heuberger S, Lehrach H, Kreutzberger J (2005) Bacterial protein microarrays for identification of new potential diagnostic markers for *Neisseria meningitidis* infections. Proteomics 5:2048–2055
Stemmer C, Schwander A, Bauw G, Fojan P, Grasser KD (2002) Protein kinase CK2 differentially phosphorylates maize chromosomal high mobility group B (HMGB) proteins modulating their stability and DNA interactions. J Biol Chem 277:1092–1098
Stoll D, Templin MF, Bachmann J, Joos TO (2005) Protein microarrays: applications and future challenges. Curr Opin Drug Discov Dev 8:239–252
Tena G, Renaudin JP (1998) Cytosolic acidification but not auxin at physiological concentration is an activator of MAP kinases in tobacco cells. Plant J 16:173–182
Tena G, Asai T, Chiu WL, Sheen J (2001) Plant mitogen-activated protein kinase signaling cascades. Curr Opin Plant Biol 4:392–400
Venter JC, Adams MD, Myers EW, Li PW, Mural RJ, Sutton GG, Smith HO, Yandell M, Evans CA, Holt RA, Gocayne JD, Amanatides P, Ballew RM, Huson DH, Wortman JR, Zhang Q, Kodira CD, Zheng XH, Chen L, Skupski M, Subramanian G, Thomas PD, Zhang J, Gabor Miklos GL, Nelson C, Broder S, Clark AG, Nadeau J, McKusick VA, Zinder N, Levine AJ, Roberts RJ, Simon M, Slayman C, Hunkapiller M, Bolanos R, Delcher A, Dew I, Fasulo D, Flanigan M, Florea L, Halpern A, Hannenhalli S, Kravitz S, Levy S, Mobarry C, Reinert K, Remington K, Abu–Threideh J, Beasley E, Biddick K, Bonazzi V, Brandon R, Cargill M, Chandramouliswaran I, Charlab R, Chaturvedi K, Deng Z, Di Francesco V, Dunn P, Eilbeck K, Evangelista C, Gabrielian AE, Gan W, Ge W, Gong F, Gu Z, Guan P, Heiman TJ, Higgins ME, Ji RR, Ke Z, Ketchum KA, Lai Z, Lei Y, Li Z, Li J, Liang Y, Lin X, Lu F, Merkulov GV, Milshina

N, Moore HM, Naik AK, Narayan VA, Neelam B, Nusskern D, Rusch DB, Salzberg S, Shao W, Shue B, Sun J, Wang Z, Wang A, Wang X, Wang J, Wei M, Wides R, Xiao C, Yan C, Yao A, Ye J, Zhan M, Zhang W, Zhang H, Zhao Q, Zheng L, Zhong F, Zhong W, Zhu S, Zhao S, Gilbert D, Baumhueter S, Spier G, Carter C, Cravchik A, Woodage T, Ali F, An H, Awe A, Baldwin D, Baden H, Barnstead M, Barrow I, Beeson K, Busam D, Carver A, Center A, Cheng ML, Curry L, Danaher S, Davenport L, Desilets R, Dietz S, Dodson K, Doup L, Ferriera S, Garg N, Gluecksmann A, Hart B, Haynes J, Haynes C, Heiner C, Hladun S, Hostin D, Houck J, Howland T, Ibegwam C, Johnson J, Kalush F, Kline L, Koduru S, Love A, Mann F, May D, McCawley S, McIntosh T, McMullen I, Moy M, Moy L, Murphy B, Nelson K, Pfannkoch C, Pratts E, Puri V, Qureshi H, Reardon M, Rodriguez R, Rogers YH, Romblad D, Ruhfel B, Scott R, Sitter C, Smallwood M, Stewart E, Strong R, Suh E, Thomas R, Tint NN, Tse S, Vech C, Wang G, Wetter J, Williams S, Williams M, Windsor S, Winn-Deen E, Wolfe K, Zaveri J, Zaveri K, Abril JF, Guigo R, Campbell MJ, Sjolander KV, Karlak B, Kejariwal A, Mi H, Lazareva B, Hatton T, Narechania A, Diemer K, Muruganujan A, Guo N, Sato S, Bafna V, Istrail S, Lippert R, Schwartz R, Walenz B, Yooseph S, Allen D, Basu A, Baxendale J, Blick L, Caminha M, Carnes-Stine J, Caulk P, Chiang YH, Coyne M, Dahlke C, Mays A, Dombroski M, Donnelly M, Ely D, Esparham S, Fosler C, Gire H, Glanowski S, Glasser K, Glodek A, Gorokhov M, Graham K, Gropman B, Harris M, Heil J, Henderson S, Hoover J, Jennings D, Jordan C, Jordan J, Kasha J, Kagan L, Kraft C, Levitsky A, Lewis M, Liu X, Lopez J, Ma D, Majoros W, McDaniel J, Murphy S, Newman M, Nguyen T, Nguyen N, Nodell M, Pan S, Peck J, Peterson M, Rowe W, Sanders R, Scott J, Simpson M, Smith T, Sprague A, Stockwell T, Turner R, Venter E, Wang M, Wen M, Wu D, Wu M, Xia A, Zandieh A, Zhu X (2001) The sequence of the human genome. Science 291:1304–1351

Walhout AJ, Temple GF, Brasch MA, Hartley JL, Lorson MA, van den Heuvel S, Vidal M (2000) GATEWAY recombinational cloning: application to the cloning of large numbers of open reading frames or ORFeomes. Methods Enzymol 328:575–592

Yamagata H, Saka K, Tanaka T, Aizono Y (2001) Light activates a 46-kDa MAP kinase-like protein kinase in soybean cell culture. FEBS Lett 494:24–29

Yang SH, Sharrocks AD, Whitmarsh AJ (2003) Transcriptional regulation by the MAP kinase signaling cascades. Gene 320:3–21

Yeung K, Seitz T, Li S, Janosch P, McFerran B, Kaiser C, Fee F, Katsanakis KD, Rose DW, Mischak H, Sedivy JM, Kolch W (1999) Suppression of Raf-1 kinase activity and MAP kinase signalling by RKIP. Nature 401:173–177

Zhang S, Klessig DF (2001) MAPK cascades in plant defense signaling. Trends Plant Sci 6:520–527

Zhu H, Klemic JF, Chang S, Bertone P, Casamayor A, Klemic KG, Smith D, Gerstein M, Reed MA, Snyder M (2000) Analysis of yeast protein kinases using protein chips. Nat Genet 26:283–289

Zhu H, Bilgin M, Bangham R, Hall D, Casamayor A, Bertone P, Lan N, Jansen R, Bidlingmaier S, Houfek T, Mitchell T, Miller P, Dean RA, Gerstein M, Snyder M (2001) Global analysis of protein activities using proteome chips. Science 293:2101–2105

Zolnierowicz S, Bollen M (2000) Protein phosphorylation and protein phosphatases. De Panne, Belgium, 19–24 September 1999. EMBO J 19:483–488

Chapter 6
Discovery via Proteomics of a Novel Cell Signalling Pathway in Plants Involving Extracellular ATP

Stephen Chivasa, William J. Simon, John M. Hamilton, Keith Lindsey, and Antoni R. Slabas

Abstract Proteomic techniques were used to identify plant extracellular matrix (ECM) proteins. Among the many classical cell wall proteins with known biochemical activity that were identified, new proteins predicted from their open reading frame in the database were localised to the ECM. Putative protein kinases and ATP-binding proteins were also identified in these cell wall-enriched protein fractions. Bio-informatic analysis of the primary sequences of the putative kinases and ATP-binding proteins confirmed their ECM localisation, prompting a search for extracellular phosphoproteins. After identification of phosphorylated proteins in the plant ECM, subsequent experiments confirmed the existence of extracellular ATP and led to the discovery that extracellular ATP is a repressor of cell death in plants. Furthermore, external ATP has emerged as a control point of pathogen elicitor-induced programmed cell death. The studies described here constitute an example of how proteomics can be a powerful tool to drive discovery of novel signalling and/or metabolic pathways through integration of systematic protein identification, bio-informatics, and downstream hypothesis testing.

Stephen Chivasa, School of Biological and Biomedical Sciences, University of Durham, Durham, DH1 3LE, UK

William J. Simon, School of Biological and Biomedical Sciences, University of Durham, Durham, DH1 3LE, UK

John M. Hamilton, School of Biological and Biomedical Sciences, University of Durham, Durham, DH1 3LE, UK

Keith Lindsey, School of Biological and Biomedical Sciences, University of Durham, Durham, DH1 3LE, UK

Antoni R. Slabas, School of Biological and Biomedical Sciences, University of Durham, Durham, DH1 3LE, UK, E-mail: a.r.slabas@durham.ac.uk

J. Šamaj and J. Thelen (eds.), *Plant Proteomics*
© Springer 2007

6.1 Introduction

The visionary whole genome sequencing projects of the last two decades are complete or nearing completion and have spawned an exponential growth in the development of sensitive, high-throughput technologies for identification of structural components of living cells and tissues as well as for global analysis of integral components of signalling modules of particular physiological responses. Proteomics has emerged as a powerful tool with great potential to accelerate discovery of new pathways or to make novel links between already known cellular signalling cascades. A commonly used approach for discovery-driven proteomics has been the systematic mapping and identification of structural and non-structural resident proteins of particular cell compartments with the aim of formulating hypotheses based on the data. In this chapter, we discuss the discovery of a novel role for extracellular ATP that emerged from a proteomic study of the plant extracellular matrix (ECM).

6.2 Proteomics of the ECM

6.2.1 The Plant ECM

The plant ECM is a compartment bounded by the external surface of the plasma membrane and the cuticle (for leaves), the bark or cork (for stems), the testa (for seeds), the exine (for pollen grains), or unbounded (for roots). Because of its position in the plant, the ECM is a strategic compartment that can potentially mediate several cellular processes such as cell-to-cell communication, or perception of signals from pathogenic or symbiotic microorganisms invading the intercellular space, as well as coordination of developmental cues within and between various tissues and cell types. As such, the ECM has been the subject of intense research interest, but progress using traditional approaches has been slow. A number of research groups have been quick to take advantage of the opportunities availed by recent developments in proteomics and a substantial amount of data is now emerging.

Proteins within this ECM space include soluble proteins in the interstitial fluid, cell-wall-linked proteins and proteins loosely bound to the external surface of the plasma membrane. The plasma membrane delineates the ECM from the cytosol and contains proteins with both cytosolic and extracellular domains. Although these proteins are classified as integral membrane proteins, physiologically they could participate in or influence events happening in both the ECM and the cytosol. The class of such integral membrane proteins with extracellular domains linked to the cell wall has also been considered to be part of the ECM proteome (Verica and He 2002; Bayer et al. 2006). The goal of proteomic studies is to isolate these proteins in fractions with minimal or no contamination from other cellular compartments and to identify them via mass spectrometry (MS).

6.2.2 Extraction of Plant ECM Proteins

There are three commonly used methods to prepare fractions enriched for plant ECM proteins (reviewed by Lee et al. 2004). First, isolation of the soluble ECM proteins from intact plants has been achieved by vacuum infiltration of buffer or salt solutions into the apoplast and recovering the proteins by centrifugation (Boudart et al. 2005; Dani et al. 2005; Zhu et al 2006), or by harvesting proteins secreted into the growth medium of hydroponic plants (Bardy et al. 1998; Charmont et al. 2005). A second, non-disruptive, method has been the use of cell suspension cultures whereby secreted soluble proteins are recovered from the growth medium after filtering out the cells (Robertson et al. 1997; Oh et al. 2005) and proteins loosely bound to the cell walls and external face of the plasma membrane can be gently removed by washing intact cells in various salt solutions (Robertson et al. 1997; Borderies et al. 2003; Kwon et al. 2005). The third approach involves pulverising the plant tissues or suspension cells and separating the cell wall fragments from the rest of the homogenate. The cell wall fragments are thoroughly washed and protein is then extracted from these fragments once or several times sequentially, using various chelating reagents or salts (Chivasa et al. 2002; Watson et al. 2004; Bayer et al. 2006).

The biggest challenge that has faced researchers attempting to identify plant ECM proteins has been the difficulties associated with collecting ECM proteins without contamination from cytosolic proteins. Indeed, this is a universal problem that impedes any study attempting to isolate pure organelles or organellar protein fractions. Although vacuum infiltration-centrifugation and use of cell suspension cultures are largely non-disruptive methods, there is always the possibility that a few cells can be damaged in the process and lead to a very low level contamination of the resultant protein fractions. In the disruptive procedure, cell wall fragments are in contact with the total cell homogenate and it is possible that some intracellular proteins become ionically attached to the charged cell wall polymers, leading to quite a considerable level of contamination. The purity of ECM protein fractions has been assayed using activity of marker enzymes or antibodies against known proteins (Chivasa et al. 2002; Kwon et al. 2005; Zhu et al. 2006). However, the apparent lack of contamination based on marker enzyme activity or western blotting may only be an indication that contamination is very marginal, but cannot be considered proof of absolute purity. This poses serious problems in the interpretation of proteomic data, especially given that the sensitivity of MS, used for protein identification, is several orders of magnitude higher than both biochemical and immunological tests. In order to extend proteomic studies to the discovery of new metabolic or signalling pathways and new organelle functions, complementary methods to confirm localisation have to be applied.

6.2.3 Identity of ECM Proteins

A pioneering study by the Slabas and Bolwell groups (Robertson et al. 1997) was the first to attempt separation and identification of a large number of ECM proteins. Using N-terminal sequencing, as no plant genome had been completely sequenced at that time, this study analysed ECM proteins from Arabidopsis, carrot, tomato, tobacco and French bean. At the completion of the Arabidopsis genome sequencing, the study was broadened and utilised 2-dimensional gel electrophoresis (2-DE) and MS to identify Arabidopsis cell wall proteins (Chivasa et al. 2002; see also Chap. 12 by Boudart et al. in this volume).

Some of the proteins identified in these two studies had been identified before; the vast majority have putative biological functions assigned on the basis of sequence homology obtained via BLAST searches against the databases. The biological functional classes of the identified proteins include cell wall modifying enzymes, defence proteins, and apoplast metabolism proteins. A significant number of the identified proteins are annotated in the databases as hypothetical proteins; such proteins have insufficient homology to any known proteins and had been predicted from open reading frames in genome databases. Identification of these proteins demonstrates that the open reading frames encoding these hypothetical proteins are expressed and that the proteins are apoplastic, suggesting that their functions are most likely to be within the ECM. Discovery of the functions of these proteins is expected to shed more light on cell wall biosynthesis and restructuring during growth, as well as unravelling new apoplastic metabolic pathways.

There is, however, uncertainty surrounding the identification of some particular proteins in the ECM. These proteins have experimentally determined or putative biological functions in intracellular compartments and their identification in ECM protein fractions has been considered an artefact resulting from contamination. This is because these proteins do not possess an N-terminal signal peptide responsible for targeting to the endoplasmic reticulum (ER), the first organelle of the secretory system. However, a number of studies in animals, yeast, and plants have demonstrated the localisation of such proteins on the cell surface using in situ immunolocalisation (Edwards et al. 1999; Gozalbo et al. 1998; Robertson et al. 2000). These results have led to the proposal that an alternative pathway not requiring recognition of the classical signal peptide exists and accounts for the export of such proteins to the ECM (reviewed by Nickel 2003). It is probable that these proteins serve quite a different function in the ECM to their known biochemical activity required intracellularly since a single protein can have numerous functions depending on its subcellular localisation. For example, although glyceraldehyde-3-phosphate dehydrogenase (GAPDH) is a classical cytosolic enzyme of the glycolytic pathway, it is secreted to the yeast ECM where it mediates adhesion of fungal cells to host tissues (Gozalbo et al. 1998), and also translocates to the nucleus of animal cells where it mediates cell death (Sawa et al. 1997).

Before the non-canonical ECM proteins identified by MS can be pursued in search of novel functions, their extracellular localisation requires confirmation via complementary approaches such as immuno-gold electron microscopy or use of tagged fusion proteins such as green fluorescent protein (GFP)-tagging. This is because the alternative secretory pathway remains uncharacterised and so the proteomic data cannot be safely relied on as absolute purity of ECM protein fractions is almost unachievable and is not guaranteed as explained above. Realising this drawback, the Slabas group decided to focus, in the first instance, on ECM proteins whose extracellular address is predictable from the primary protein sequence.

6.3 Unravelling New Biological Functions of the Apoplast

Although there has been an explosive growth in the number of studies embracing proteomics as a technology to allow high-throughput global analysis of subcellular compartments, relatively few have gone further to mine these data sets to discover new functions of the subcellular compartment in question. The observations made by our group during mapping of plant ECM proteins proved to be fruitful in this vein as described in the following sections.

6.3.1 Phosphorylated and ATP-Binding ECM Proteins

Chivasa et al. (2002) applied a proteomic approach to the identification of *Arabidopsis thaliana* cell wall proteins, using bio-informatic tools to predict the characteristics and fate of the protein in the cell. The primary protein sequences were analysed using Signal-P (Nielsen et al. 1997) for predicting the presence and location of the signal peptide, and the TMHMM algorithm (Sonnhammer et al. 1998) for predicting the location and orientation of alpha helices in membrane-spanning proteins. This study identified a protein designated as a putative protein kinase (At1g53070) in the MIPS database (http://mips.gsf.de). Close examination of the sequence of this protein revealed that, although the protein name is a misnomer in that it does not have a kinase domain, it has the classical ATP/GTP-binding motif known as the P-loop (Walker et al. 1982; Saraste et al. 1990). The characteristic P-loop domain has the pattern GxxxxGK[ST] (x indicates any amino acid residue; alternative residues are shown in brackets). In At1g53070, the P-loop is from positions 47 to 54 (Fig. 6.1). In accordance with a cell wall localisation, it has a predicted cleavable N-terminal signal peptide (Fig. 6.1) and neither a transmembrane domain nor any of the classical C-terminal ER retention motifs (Vitale and Denecke 1999). This suggests that this protein is a bona fide cell wall protein.

In follow-up experiments, ^{32}P-labelled azido-ATP incubated with growth medium proteins harvested from Arabidopsis cell cultures was cross-linked to a number of proteins upon illumination with ultraviolet light (Fig. 6.2). Specificity for ATP was demonstrated by competing out azido-ATP labelling with the addition of excess non-radioactive ATP into the reaction (Fig. 6.2). This result confirms that the plant ECM has ATP-binding proteins. The presence of proteins in the ECM that interact with ATP was intriguing. Neither ATP-binding proteins nor kinases had been localised to the plant ECM before. This finding suggested the possible existence of undiscovered ATP-dependent ECM metabolism or signalling pathways.

a MKIQILCFTT LFLAIFTSQV TTAYKFKFDY FGNGTDPISF HGDAEYGPDT DGKSRSGAIA
LTRDNIPFSH GRAIFTTPIT FKPNASALYP FKTSFTFSIT PKTNPNQGHG LAFIVVPSNQ
NDAGSGLGYL SLLNRTNNGN PNNHLFAVEF DVFQDKSLGD MNDNHVGIDI NSVDSVVSVK
SGYWVMTRSG WLFKDLKLSS GDRYKAWIEY NNNYKVVSVT IGLAHLKKPN RPLIEAKFDL
SKVIHEVMYT GFAGSMGRGV ERHEIWDWTF QN

b MLQGRREAKK SYALFSSTFF FFFICFLSSS SAELTDKGVN FEVVALIGIK SSLTDPHGVL
MNWDDTAVDP CSWNMITCSD GFVIRLEAPS QNLSGTLSSS IGNLTNLQTV LLQNNYITGN
IPHEIGKLMK LKTLDLSTNN FTGQIPFTLS YSKNLQYLRV NNNSLTGTIP SSLANMTQLT
FLDLSYNNLS GPVPRSLAKT FNVMGNSQIC PTGTEKDCNG TQPKPMSITL NSSQNKSSDG
GTKNRKIAVV FGVSLTCVCL LIIGFGFLLW WRRRHNKQVL FFDINEQNKE EMCLGNLRRF
NFKELQSATS NFSSKNLVGK GGFGNVYKGC LHDGSIIAVK RLKDINNGGG EVQFQTELEM
ISLAVHRNLL RLYGFCTTSS ERLLVYPYMS NGSVASRLKA KPVLDWGTRK RIALGAGRGL
LYLHEQCDPK IIHRDVKAAN ILLDDYFEAV VGDFGLAKLL DHEESHVTTA VRGTVGHIAP
EYLSTGQSSE KTDVFGFGIL LLELITGLRA LEFGKAANQR GAILDWVKKL QQEKKLEQIV
DKDLKSNYDR IEVEEMVQVA LLCTQYLPIH RPKMSEVVRM LEGDGLVEKW EASSQRAETN
RSYSKPNEFS SSERYSDLTD DSSVLVQAME LSGPR

c

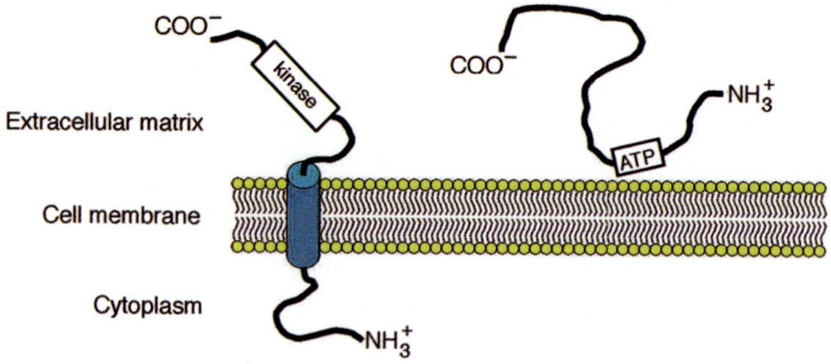

Fig. 6.1 Arabidopsis cell wall proteins with predicted ATP-binding properties. **a** The protein sequence of At1g53070 showing the signal peptide (*red*) and the ATP/GTP-binding motif (*blue*). **b** At3g25560 sequence with the signal peptide (*red*), the transmembrane domain (*underlined*) and the putative kinase domain (*blue*) highlighted. **c** Schematic representation of the predicted location and orientation of the At1g53070 and At3g25560 proteins

The same study identified a putative receptor-like kinase protein (At3g25560), which contains a predicted cleavable N-terminal signal peptide, a transmembrane domain, and no ER retention motif, suggesting that the protein spans the plasma membrane and has cytosolic and extracellular domains (Fig. 6.1). Surprisingly, the TMHMM algorithm predicted the C-terminus and the kinase domain to be extracellular while the N-terminus is cytosolic. The existence of a putative kinase domain in the predicted extracellular region of this protein is similar to the CD4 receptor of the T-helper cells of the immune system, which has kinase activity in its extracellular domain (Yakubov et al. 1998). The implication of this prediction is that the plant ECM contains protein kinase activity.

However, it must be noted that the accuracy of the prediction method is estimated at 77% (Sonnhammer et al. 1998), making it imperative to validate this prediction

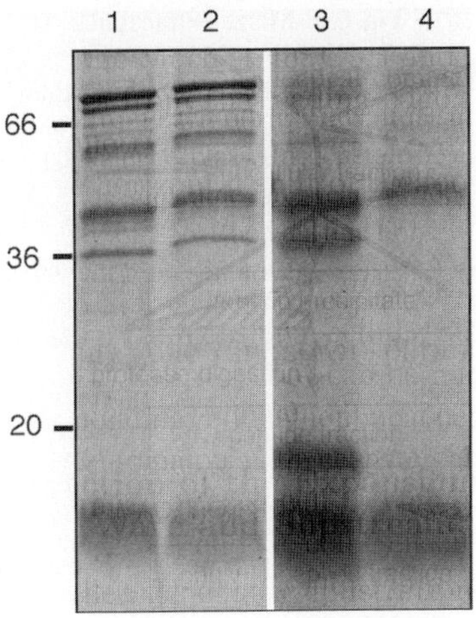

Fig. 6.2 Arabidopsis ATP-binding proteins. Proteins secreted into the growth medium of cell cultures were labelled with 8-azidoadenosine 5'-[α^{32}P]triphosphate by illuminating with ultraviolet light (250 nm) for 5 min in the absence (lane 1) or presence (lane 2) of excess cold ATP. Lanes: *1*, *2* Coomassie blue-stained proteins; *3*, *4* autoradiogram images of *1* and *2*, respectively. It is clear that cold ATP competed out the signal, confirming the specificity for ATP of the labelling reaction. Numbers on the left are positions of molecular weight markers (kDa)

experimentally. Nevertheless, we decided to search for other putative ECM kinases without predicted transmembrane domains so that the presence of kinases in the ECM would not be dependent upon a possibly wrong prediction of the orientation of a membrane-bound kinase, but rather on the presence of putative kinases in the apoplastic fluid or cell wall. Subsequent searches of the Arabidopsis database revealed several putative kinases with signal peptides and neither membrane-spanning helices nor an ER retention motif (S. Chivasa, J. Hamilton, A. Slabas, unpublished data), suggesting that they are ECM residents. Again these findings reinforced the possibility that the ECM is host to ATP-dependent metabolic or signalling events.

These predictions could be verified directly, by cloning the genes and demonstrating kinase activity of the recombinant proteins, or indirectly, by demonstrating the presence of phosphorylated bona fide ECM proteins. The latter option was taken and it was discovered that a number of proteins secreted into the ECM of Arabidopsis cell suspension cultures are phosphorylated and that the phosphorylation status of some of these proteins changes rapidly in response to treatment with pathogen elicitors (Chivasa et al. 2002; Ndimba et al. 2003). A chitinase-like protein (At3g12500) with all the necessary primary sequence information consistent with ECM localisation, was confirmed to be a phosphoprotein (Chivasa et al. 2002). A search for extracellular phosphorylated proteins in the ECM of maize resulted in the identification of phosphorylated ECM peroxidases (Fig. 6.3; Chivasa et al. 2005b).

6.3.2 *Plant Extracellular ATP*

The presence of putative extracellular kinases in the Arabidopsis genome and the identification of phosphorylated proteins and ATP-binding proteins implied the presence of ATP in the ECM. A previous study had demonstrated that Arabidopsis plants in which leaf cuticle development had been suppressed by incubation in high humidity, secreted

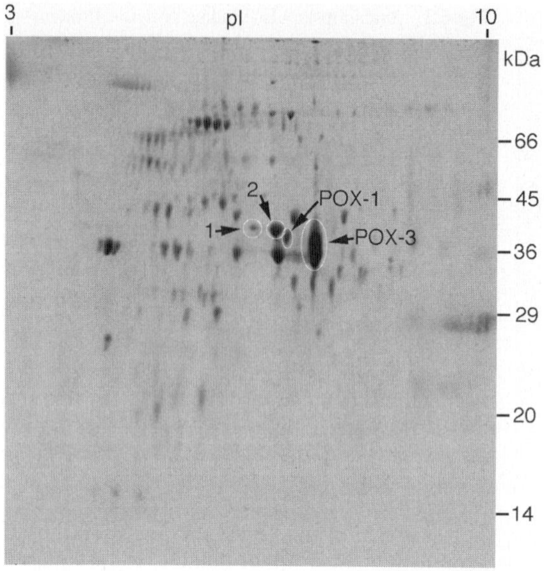

Fig. 6.3 Maize extracellular phosphoproteins. Proteins secreted into the growth medium of maize cell cultures were resolved by 2-dimensional gel electrophoresis (2-DE) and analysed by western blotting using phosphotyrosine antibodies. The four phosphoproteins are annotated; spots *1* and *2* could not be identified; *POX-1* and *POX-3* are two different peroxidases

ATP on the leaf surfaces (Thomas et al. 2000). Chivasa et al. (2005a) used a different approach that entailed feeding Arabidopsis cell suspension cultures with a radio-labelled tracer (^{32}P-phosphate) that was incorporated into ATP; the labelled ATP was subsequently secreted into the growth medium. This demonstrated that live cells take up the phosphate, incorporate it into ATP, and then actively secrete the ATP into the ECM. More importantly, the integrity of the plasma membrane remained intact for the duration of the experiment, confirming that extracellular ATP did not arise from leakage of cells. The recent generation of an ATP reporter fusion construct with a cellulose-binding domain peptide appended to the N-terminus of luciferase allowed in situ visualisation of ATP secretion in transgenic *Medicago truncatula* expressing the protein in the root cell walls (Kim et al. 2006). Thus, the existence of extracellular ATP in plants is not an artefact, but rather a real biological phenomenon.

Realising that determining the role of the putative ECM kinases, ATP-binding proteins, and phosphoproteins was the key to the discovery of novel roles of the plant ECM, we initiated research in this area. Given the redundancy of genes common to many gene families, we realised that a reverse genetic approach would possibly require knockout of multiple genes to obtain a phenotype and that this would be a labour-intensive and circuitous route. Thus, a biochemical approach was chosen, whereby extracellular ATP would be selectively removed from its target sites so that the physiological role of this metabolite and the ECM proteins associated with it could be elucidated.

6.3.3 In Search of the Role of Extracellular ATP

Two approaches for the selective removal of extracellular ATP from its binding sites were selected. The first was destruction of extracellular ATP using membrane-impermeant enzyme systems, while the second was inundation of the ECM with a non-hydrolysable ATP analogue to competitively exclude ATP from its binding sites and inhibit any reactions requiring cleavage of the gamma phosphate. Maize and Arabidopsis cell cultures as well as intact bean, tobacco, and Arabidopsis plants showed a death response under such conditions (Chivasa et al. 2005a). Apyrase and hexokinase were the chosen extracellular ATP sequestering enzymes. Apyrase, an enzyme that hydrolyses ATP and ADP to AMP and inorganic phosphate is commonly used to deplete extracellular ATP without directly affecting pools of internal ATP. Hexokinase phosphorylates glucose to glucose-6-phosphate in a reaction that consumes a molecule of ATP. Application of glucose with hexokinase to the ECM of cell cultures and intact plants is expected to target extracellular ATP since the enzyme is not internalised, even though some of the glucose might be taken up. However, one of the major concerns about this approach is that while the enzymes destroy extracellular ATP, the cells may perceive the deficit and reciprocate by actively pumping out more ATP, thereby causing the intracellular energy levels to plummet and the cells to die as a result. This is why incorporating the non-hydrolysable ATP analogue in the experiments was critical.

The non-hydrolysable analogue of ATP used in experiments by Chivasa et al. (2005a) is β,γ-methyleneadenosine 5’-triphosphate. In this molecule, the phosphodiester bond between the beta and gamma phosphates is replaced with a non-hydrolysable methylene group, rendering the gamma phosphate unavailable for cleavage. Since the molecule resembles ATP, it can still bind to enzyme pockets that usually bind ATP. Addition of excessive amounts of this compound out-competes ATP, and any reactions that require ATP to donate its gamma phosphate are inhibited as a result. Unlike treatments with apyrase or hexokinase, which deplete extracellular ATP, treatments with the non-hydrolysable analogue do not decrease the level of extracellular ATP at all. Thus, cell death induced by these treatments was concluded to be a specific response to the removal of ATP from its binding sites either via its destruction by enzymes or its competitive exclusion by the analogue.

A series of experiments described by Chivasa et al. (2005a) demonstrated that the cell death response was not an artefact, but was triggered by manipulating the ECM using these experimental tools to ablate extracellular ATP or exclude it from its targets. This demonstrated that external ATP is indispensable for supporting plant cell viability. However, the physiological significance of this observation was still unclear.

6.4 Physiological Significance of Extracellular ATP Depletion-Induced Death

There are many cellular processes and structural components that are vital for the growth and development of plants. Depletion of some of these components may compromise some aspects of growth, but the plants still remain viable. However, some of these pathways and structural polymers are critical for viability and their removal is lethal. The finding that availability of external ATP is requisite to maintaining plant cell viability implies that it is intricately involved in some critical biochemical/signalling pathway or maintenance of the integrity of some indispensable structural components. Two important questions are raised by this observation. First, what are the biochemical or signalling pathways that require external ATP? Second, are there instances during growth and development when plants deliberately invoke the requirement for external ATP for viability in order to activate programmed cell death? Addressing these questions would give insight into the physiological significance of the cell death triggered by interference with extracellular ATP utilisation that was achieved under experimental conditions.

The first question is dealt with in section 6.5. The second question can be addressed by investigating if naturally occurring programmed cell death phenomena are associated with depletion of pericellular ATP. Programmed cell death is an important strategy used by plants during development in processes such as the developmental remodelling of leaf shape, differentiation of the water conductive tracheary elements, ablation of tissues with transient functions, senescence, and organ development (Greenberg 1996; Pennel and Lamb 1997; Kuriyama and Fukuda 2002; Van Doorn and Woltering 2005). Plants also activate programmed cell death in response to biotic and abiotic environmental cues such as heat, ozone, hypoxia and pathogen invasion (Pennel and Lamb 1997; Rao et al. 2000; Lam et al. 2001; Swidzinski et al. 2002). We went on to investigate if extracellular ATP is a regulator of pathogen-induced programmed cell death.

6.4.1 Elicitor-Induced Cell Death is Regulated by External ATP

When plants possessing a resistance gene against a specific disease-causing pathogen come into contact with the pathogen, they activate a robust and rapid defence response that is often accompanied by programmed cell death of several cell layers around the infection foci. This defence response is triggered by the interaction of the resistance gene product and the product of the cognate pathogen avirulence gene – the gene-for-gene interaction (Flohr 1971; Dangl and Jones 2001). However, this response can be simulated by treating plants with pathogen-derived elicitors (Montesano et al. 2003). Chivasa et al. (2005a) used fumonisin B1 (FB1), a programmed cell death-eliciting mycotoxin that activates Arabidopsis defence responses (Stone et al. 2000), to simulate pathogen-induced cell death.

The results obtained in this study are represented schematically in Fig. 6.4. Treatment of Arabidopsis cell cultures triggered a heightened destruction of extracellular ATP, which preceded the onset of cell death. Extracellular ATP started to decline ~16h after FB1 addition. Notably, increased hydrolysis of external ATP was initiated before the cells had lost plasma membrane integrity, suggesting that ECM proteins were responsible for the depletion of ATP. Crucially, the study revealed that exogenous ATP could be added to the FB1-treated cell cultures concurrently with FB1 or at any point up to ~40h later to rescue the cells from death. The same response could be demonstrated on roots of hydroponic Arabidopsis plants.

The significance of these findings is that they place extracellular ATP at the centre of the defensive hypersensitive response. It appears as if the external ATP

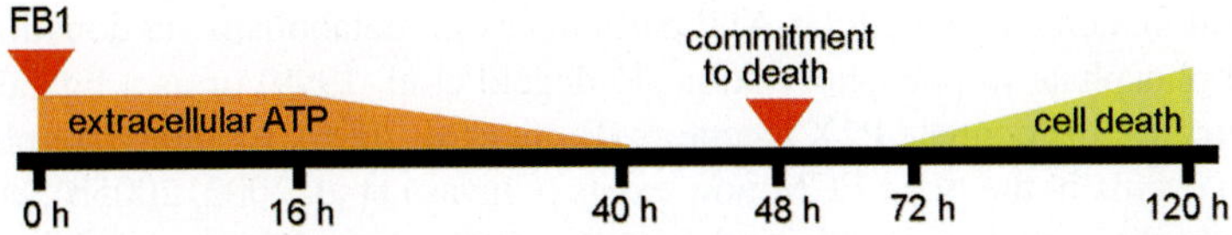

Fig. 6.4 Timing of events during fumonisin B1 (FB1) treatment of cell cultures. After FB1 addition at time 0h, significant depletion of external ATP starts just after 16h and by 40h no external ATP is detectable in the medium. FB1-induced death is abolished by adding exogenous ATP concurrently with FB1 at time 0h or any time after, up to 40h. When added at 48h or later, exogenous ATP fails to rescue the cells from death

level is sensed by the cells and a sustained decline in the level triggers cell suicide. Chivasa et al. (2005a) proposed a hypothesis stating that extracellular ATP is a repressor of a default cell death pathway. According to this hypothesis, pathogen elicitors activate an ATP-degrading activity in the ECM and the removal of external ATP activates programmed cell death. However, the effect of a wider range of pathogen-derived elicitors and real pathogens on external ATP needs to be examined to establish the generality of the role of ATP in this response. Some indirect evidence supporting the notion that pathogen elicitors trigger extracellular ATP hydrolysis was obtained by Kiba et al. (1995, 1999), who observed that pathogen elicitors stimulate membrane- and cell wall-bound nucleoside triphosphatase activity. An alternative possibility is that elicitors repress ATP secretion and the ATP already present in the ECM will eventually decline due to normal turnover. These two mechanisms may function together to bring about external ATP depletion. Ongoing research should clarify what actually happens in this system.

Some of the critical early events in the gene-for-gene pathogen-host interaction include the production of reactive oxygen species in the ECM by membrane-bound and cell wall enzyme systems (Bolwell et al. 2002). The results of Chivasa et al. (2005a) reveal that the determinants of the associated hypersensitive cell death in this interaction include external ATP, implying that the ECM is also central to pathogen-induced programmed cell death.

6.5 How Does Extracellular ATP Function?

The inevitable question arising from these observations is how does extracellular ATP sustain cell viability? It is possible that external ATP represses default cell death or that it participates in an ECM metabolic pathway that is crucial for cell viability. The exact mechanism by which ATP could fulfil this function is still unclear and remains the subject of current research. However, some clues might be gleaned from the animal field, where extracellular ATP has been recognised as a signalling molecule at the centre of many physiological processes (Gordon 1986; Redegeld et al. 1999).

In animal cells, extracellular ATP participates in metabolism via donation of its gamma phosphate in phosphorylation (Redegeld et al. 1999) or as a ligand gating Ca^{2+} entry via the cognate P2X receptors (Baljit et al. 2006). Evidence for phosphorylation events in the plant ECM now exists (Chivasa et al. 2002, 2005b), but cloning of the kinases responsible and experimental proof of kinase activity is still awaited. Though no plant homologues of animal P2X receptors have yet been identified, evidence exists that extracellular ATP depolarises Arabidopsis membranes (Lew and Dearnaley 2000) and gates Ca^{2+} entry into Arabidopsis plants (Demidchik et al. 2003; Jeter et al. 2004). It is quite possible that the role of external ATP in viability could be mediated via external phosphorylation or activation of P2X-type receptors.

A proteomic study of the effects of treatment with glucose-hexokinase on Arabidopsis cell cultures revealed changes in the abundance of several proteins in the total soluble fraction (Chivasa et al. 2005a). Figure 6.5 shows a typical 2-dimensional

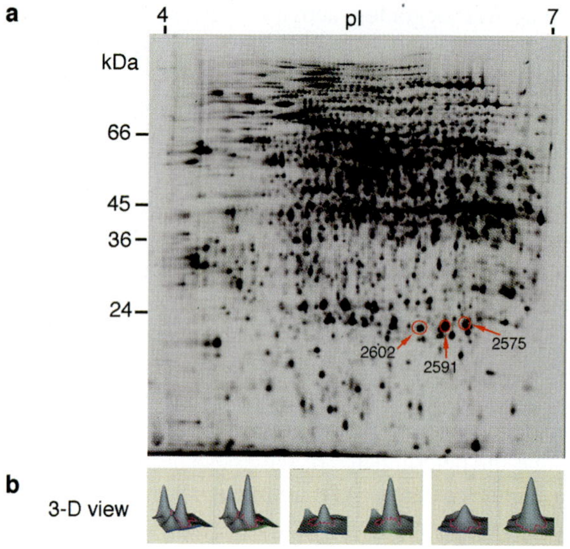

Spot no.	2575	2591	2602
Fold-change:	1.89	2.84	2.83
P-value:	0.027	0.0030	0.0029

Fig. 6.5 Arabidopsis proteins analysed by 2-dimensional difference gel electrophoresis (2-D DiGE). Cell cultures were treated with glucose or glucose-hexokinase and total soluble protein analysed by 2-D DiGE. **a** Typical profile of total soluble protein. The protein spots showing the highest magnitude of change in abundance are annotated with their master gel number. **b** A 3-D view (from DeCyder software; GE Healthcare, Uppsala, Sweden) showing the indicated spot volume in control (*left*) and glucose-hexokinase treated (*right*) samples for each spot. The spot profile is encircled at its base. The spot numbers correspond to annotations in **a.** The fold-change in protein abundance and the probability value showing the statistical significance of the change are also indicated

gel profile of proteins from Arabidopsis cell cultures. The strongest response to external ATP removal by hexokinase was in the abundance of glutathione *S*-transferase (At1g02930) protein spots, which were up-regulated by around twofold or higher (Fig. 6.5). Reverse genetic experiments will be required to investigate the role of this protein in cell viability or response to treatments affecting external ATP utilisation.

6.6 Conclusions and Future Studies

Proteomics was instrumental in bringing to our attention the possibility that the plant ECM could host novel signalling or metabolic pathways. Subsequent biochemical experiments confirmed this, and extracellular ATP has now emerged as an important metabolite central to cell survival and to regulation of the hypersensitive defence response. Current research focuses on the ATP-binding proteins, phosphoproteins, and the associated kinases that utilise extracellular ATP to perform this function in plants. Proteomics will undoubtedly prove to be invaluable in prising open the ECM to dissect the mechanism by which external ATP performs its function.

However, a number of issues need addressing to enable a better understanding of the role of the ECM in cell viability and pathogen-induced cell death. First, it is important to establish the concentration of extracellular ATP in intact plants. Values obtained from cell suspension culture systems would not be useful as these are dependent on the volume of growth medium and the number of cells in the culture (cell density). The apoplastic fluid is very difficult to harvest from intact tissues without vacuum-infiltration of buffer, and so other methods must be developed. This information will be vital, especially in trying to understand the kinetics of extracellular ATP depletion during activation of ECM ATPases by pathogen attack.

The second issue that needs addressing is the mechanism by which plants secrete ATP. In animal neuronal cells and cells of secretory tissues such as the pancreas, ATP is released via exocytosis (Unsworth and Johnson 1990; Sorensen and Novak 2001). In non-neuronal and non-secretory animal cells, the mechanism of ATP release is still unclear, but there is evidence for the involvement of ABC transporters and anion channels (Roman et al. 1997; Dutta et al. 2002, 2004). Transgenic Arabidopsis plants over-expressing the P-glycoprotein have increased ATP release (Thomas et al. 2000), suggesting that plants are capable of releasing ATP via a mechanism requiring ABC transporter activity. A recent study on *Medicago truncatula* demonstrated that ATP secretion from actively growing regions of the roots was sensitive to brefeldin A, implicating vesicular trafficking (Kim et al. 2006).

It should be noted that other functions of plant extracellular ATP are emerging; extracellular ATP is now known to affect gravitropic growth of roots (Tang et al. 2003) and it has been implicated in the activation of mitogen-activated protein kinases and the ethylene biosynthetic pathway (Jeter et al. 2004) as well as in the generation of reactive oxygen species (Kim et al. 2006; Song et al. 2006).

Acknowledgements A.R.S. and S.C. wish to thank the Biotechnology and Biological Sciences Research Council for a grant supporting this work and for continued support in developing proteomic facilities.

References

Baljit S, Khakh R, North A (2006) P2X receptors as cell-surface ATP sensors in health and disease. Nature 442:527–532

Bardy N, Carrasco A, Galaud JP, Pont-Lezica R, Canut H (1998) Free-flow electrophoresis for fractionation of *Arabidopsis thaliana* membranes. Electrophoresis 19:1145–1153

Bayer EM, Bottrill AR, Walshaw J, Vigouroux M, Naldrett MJ, Thomas CL, Maule AJ (2006) *Arabidopsis* cell wall proteome defined using multidimensional protein identification technology. Proteomics 6:301–311

Bolwell GP, Bindschedler LV, Blee KA, Butt VS, Davies DR, Gardner SL, Gerrish C, Minibayeva F (2002) The apoplastic oxidative burst in response to biotic stress in plants: a three component system. J Exp Bot 53:1367–1376

Borderies G, Jamet E, Lafitte C, Rossignol M, Jauneau A, Boudart G, Monsarrat B, Esquerré-Tugayé M-T, Boudet A, Pont-Lezica R (2003) Proteomics of loosely bound cell wall proteins of *Arabidopsis thaliana* cell suspension cultures: a critical analysis. Electrophoresis 24:3421–3432

Boudart G, Jamet E, Rossignol M, Lafitte C, Borderies G, Jauneau A, Esquerré-Tugayé M-T, Pont-Lezica R (2005) Cell wall proteins in apoplastic fluids of *Arabidopsis thaliana* rosettes: identification by mass spectrometry and bioinformatics. Proteomics 5:212–221

Charmont S, Jamet E, Pont-Lezica R, Canut H (2005) Proteomic analysis of secreted proteins from *Arabidopsis thaliana* seedlings: improved recovery following removal of phenolic compounds. Phytochemistry 66:453–461

Chivasa S, Ndimba BK, Simon WJ, Robertson D, Yu XL, Knox PJ, Bolwell P, Slabas AR (2002) Proteomic analysis of the *Arabidopsis thaliana* cell wall. Electrophoresis 23:1754–1765

Chivasa S, Ndimba BK, Simon WJ, Lindsey K, Slabas AR (2005a) Extracellular ATP functions as an endogenous external metabolite regulating plant cell viability. Plant Cell 17:3019–3034

Chivasa S, Simon WJ, Yu XL, Yalpani N, Slabas AR (2005b) Pathogen elicitor-induced changes in the maize extracellular matrix proteome. Proteomics 5:4894–4904

Dangl JL, Jones JDG (2001) Plant pathogens and integrated defense responses to infection. Nature 411:826–833

Dani V, Simon WJ, Duranti M, Croy RRD (2005) Changes in the tobacco leaf apoplast proteome in response to salt stress. Proteomics 5:737–745

Demidchik V, Nichols C, Oliynyk M, Dark A, Glover BJ, Davies JM (2003) Is ATP a signaling agent in plants? Plant Physiol 133:456–461

Dutta AK, Okada Y, Sabirov RZ (2002) Regulation of an ATP-conductive large-conductance anion channel and swelling-induced ATP release by arachidonic acid. J Physiol 542:803–816

Dutta AK, Sabirov RZ, Uramoto H, Okada Y (2004) Role of ATP-conductive anion channel in ATP release from neonatal rat cardiomyocytes in ischaemic or hypoxic conditions. J Physiol 559:799–812

Edwards SR, Braley R, Chaffin WL (1999) Enolase is present in the cell wall of *Saccharomyces cerevisiae*. FEMS Microbiol Lett 177:211–216

Flohr HH (1971) Current status of the gene-for-gene concept. Annu Rev Phytopathol 9:275–296

Gordon JL (1986) Extracellular ATP: effects, source and fate. Biochem J 233:309–319

Gozalbo D, Gil-Navarro I, Azorín I, Renau-Piqueras J, Martínez JP, Gil ML (1998) The cell wall-associated glyceraldehydes-3-phosphate dehydrogenase of *Candida albicans* is also a fibronectin and lamin binding protein. Infect Immun 66:2052–2059

Greenberg JT (1996) Programmed cell death: a way of life for plants. Proc Natl Acad Sci USA 93:12094–12097

Jeter CR, Tang W, Henaff E, Butterfield T, Roux SJ (2004) Evidence of a novel cell signalling role for extracellular adenosine triphosphates and diphosphates in *Arabidopsis*. Plant Cell 16:2652–2664

Kiba A, Toyoda K, Ichinose Y, Yamada T, Shiraishi T (1995) Specific inhibition of cell wall-bound ATPases by fungal suppresser from *Mycosphaerella pinodes*. Plant Cell Physiol 36:809–817

Kiba A, Takeda T, Kanemitsu T, Toyoda K, Ichinose Y, Yamada T, Shiraishi T (1999) Induction of defence responses by synthetic glycopeptides that have a partial structure of the elicitor in the spore germination fluid of *Mycosphaerella pinodes*. Plant Cell Physiol 40:978–985

Kim SY, Sivaguru M, Stacey G (2006) Extracellular ATP in plants. Visualization, localization and analysis of physiological significance in growth and signalling. Plant Physiol 142:984–992

Kuriyama H, Fukuda H (2002) Developmental programmed cell death in plants. Curr Opin Plant Biol 5:568–573

Kwon H-K, Yokoyama R, Nishitani K (2005) A proteomic approach to apoplastic proteins involved in cell wall regeneration in protoplasts of *Arabidopsis* suspension-cultured cells. Plant Cell Physiol 46:843–857

Lam E, Kato N, Lawton M (2001) Programmed cell death, mitochondria and the plant hypersensitive response. Nature 411:848–853

Lee S-J, Saravanan RS, Damasceno CMB, Yamane H, Kim B-D, Rose JKC (2004) Digging deeper into the plant cell wall proteome. Plant Physiol Biochem 42:979–988

Lew RR, Dearnaley JDW (2000) Extracellular nucleotide effects on the electrical properties of growing *Arabidopsis thaliana* root hairs. Plant Sci 153:1–6

Montesano M, Brader G, Palva ET (2003) Pathogen-derived elicitors: searching for receptors in plants. Mol Plant Pathol 4:73–79

Ndimba BK, Chivasa S, Hamilton JM, Simon WJ, Slabas AR (2003) Proteomic analysis of changes in the extracellular matrix of Arabidopsis cell suspension cultures induced by fungal elicitors. Proteomics 3:1047–1059

Nickel W (2003) The mystery of nonclassical protein secretion. Eur J Biochem 270:2109–2119

Nielsen H, Engelbrecht J, Brunak S, von Heijne G (1997) Identification of prokaryotic and eukaryotic signal peptides and prediction of their cleavage sites. Prot Eng 10:1–6

Oh IS, Park AR, Bae MS, Kwon SJ, Kim YS, Lee JE, Kang NY, Lee S, Cheong S, Park OK (2005) Secretome analysis reveals an *Arabidopsis* lipase in defence against *Alternaria brassicicola*. Plant cell 17:2832–2847

Pennel RI, Lamb C (1997) Programmed cell death in plants. Plant Cell 9:1157–1168

Rao MV, Kock JR, Davis KR (2000) Ozone: a tool for probing programmed cell death in plants. Plant Mol Biol 44:345–358

Redegeld FA, Caldwell CC, Sitkovsky MV (1999) Ecto-protein kinases: ectodomain phosphory lation as a novel target for pharmaceutical manipulation? Trends Pharmacol Sci 20:453–459

Robertson D, Mitchell GP, Gilroy JS, Gerrish C, Bolwell GP, Slabas AR (1997) Differential extraction and protein sequencing reveals major differences in patterns of primary cell wall proteins from plants. J Biol Chem 272:15841–15848

Robertson L, Robertson WM, Sobczak M, Helder J, Tetaud E, Ariyanayagam MR, Ferguson MA, Fairlamb A, Jones JT (2000) Cloning, expression and functional characterisation of a peroxi redoxin from the potato cyst nematode *Globodera rostochiensis*. Mol Biochem Parasitol 111:41–49

Roman R, Wang Y, Lidofsky S, Feranchak A, Lomri N, Scharschmidt B, Fitz J (1997) Hepatocellular ATP-binding cassette protein expression enhances ATP release and autocrine regulation of cell volume. J Biol Chem 272:21970–21976

Saraste M, Sibbald PR, Wittinghofer A (1990) The P-loop: a common motif in ATP- and GTP-binding proteins. Trends Biochem Sci 15:430–434

Sawa A, Khan A, Hester L, Snyder S (1997) Glyceraldehyde-3-phosphate dehydrogenase: nuclear translocation participates in neuronal and nonneuronal cell death. Proc Natl Acad Sci USA 94:11669–11674

Song CJ, Steinebrunner I, Wang X, Stout SC, Roux SJ (2006) Extracellular ATP induces the accumulation of superoxide via NADP oxidases in *Arabidopsis*. Plant Physiol 140:1222–1232

Sonnhammer ELL, von Heijne G, Krogh G (1998) A hidden Markov model for predicting transmembrane helices in protein sequences. Intelligent Syst Mol Biol 6:175–182

Sorensen CE, Novak I (2001) Visualization of ATP release in pancreatic acini in response to cholinergic stimulus. J Biol Chem 276:32925–32932

Stone JM, Heard JE, Asai T, Ausubel FM (2000) Simulation of fungal-mediated cell death by fumonisin B1 and selection of fumonisin B1-resistant (fbr) Arabidopsis mutants. Plant Cell 12:1811–1822

Swidzinski JA, Sweetlove LJ, Leaver CJ (2002) A custom microarray analysis of gene expression during programmed cell death in *Arabidopsis thaliana*. Plant J 30:431–446

Tang W, Brady SR, Sun Y, Muday GK, Roux SJ (2003) Extracellular ATP inhibits root gravitropism at concentrations that inhibit polar auxin transport. Plant Physiol 131:147–154

Thomas C, Rajagopal A, Windsor B, Dudler R, Lloyd A, Roux SJ (2000) A role for ectophosphatase in xenobiotic resistance. Plant Cell 12:519–533

Unsworth CD, Johnson RG (1990) Acetylcholine and ATP are coreleased from the electromotor nerve terminals of *Narcine brasiliensis* by an exocytotic mechanism. Proc Natl Acad Sci USA 87:553–557

Van Doorn WG, Woltering EJ (2005) Many ways to exit? Cell death categories in plants. Trends Plant Sci 10:117–122

Verica JA, He Z-H (2002) The cell wall-associate kinase (*WAK*) and *WAK*-like kinase gene family. Plant Physiol 129:455–459

Vitale A, Dinecke J (1999) The endoplasmic reticulum – gateway of the secretory pathway. Plant Cell 11:615–628

Walker JE, Saraste M, Runswick MJ, Gay NJ (1982) Distantly related sequences in the alpha- and beta-subunits of ATP synthase, myosin, kinases and other ATP-requiring enzymes and a common nucleotide binding fold. EMBO J 1:945–951

Watson BS, Lei Z, Dixon RA, Sumner LW (2004) Proteomics of *Medicago sativa* cell walls. Phytochemistry 65:1709–1720

Yakubov LA, Kit YY, Richter VA, Andreeva AY, Karamyshev VN, Stein CA, Valentin Vlassov V (1998) The extracellular domain of CD4 receptor possesses a protein kinase activity. FEBS Lett 431:45–48

Zhu J, Chen S, Alvarez S, Asirvatham VS, Schachtman DP, Wu Y, Sharp RE (2006) Cell wall proteome in the maize primary root elongation zone. I. Extraction and identification of water-soluble and lightly ionically bound proteins. Plant Physiol 140:311–325

Chapter 7
Cereal Proteomics

Hisashi Hirano

Abstract Following the development of genome sequence analysis in cereals such as rice, wheat, barley and maize, attention has focused on proteome analysis. This consists of the large scale separation and identification of proteins, determination of their functions and functional networks, and construction of appropriate databases. Many novel techniques for proteome analysis, such as two-dimensional electrophoresis (2-DE), nano-liquid chromatography and mass spectrometry (MS), have developed rapidly. Such techniques have made possible the efficient separation and identification of cereal proteins, and high-throughput analysis of functions and functional networks of these proteins. This chapter describes the current status of protein expression profiling (2-DE profiling, shotgun profiling, quantitative profiling), analyses of the subcellular localization of proteins, protein post-translational modifications (PTMs; e.g., phosphorylation and glycosylation) and protein–protein interaction in cereals. In addition, the development of software for proteome analysis and the construction of databases for the enormous amount of information obtained from cereal proteome analysis are outlined.

7.1 Introduction: Genomic and Proteomic Analyses

Sequence analysis of genomic DNA, which began full scale in the 1990s, has developed rapidly during the last decade. The entire genomic DNA sequence is available for many organisms, including plants, animals, and humans. Based on such analyses, the number of genes present in each genome has been estimated; however, only 40–50% of the proteins encoded by these genes have been assigned functions, as indicated by the available databases for many organisms, including cereals. This suggests the necessity of determining the functions of these unknown proteins and their functional networks through proteomic analysis.

Although proteomic research was developed after genomic sequence analysis, its growth over the last few years has been remarkable and has resulted in the new

Hisashi Hirano, Yokohama City University, International Graduate School of Arts and Sciences, Graduate School of Medicine, Yokohama, Japan, E-mail: hirano@yokohama-cu.ac.jp

J. Šamaj and J. Thelen (ed.), *Plant Proteomics*
© Springer 2007

field of science we now call proteomics. In proteomic analysis, a large number of proteins are analyzed by high-throughput methods such as two-dimensional electrophoresis (2-DE) and mass spectrometry (MS). From this aspect, proteomic analysis is very different from conventional protein analysis.

Genomic analyses of cereals, which are the most important food crops, began in the 1990s. Table 7.1 shows the genome sizes and the numbers of genes estimated by genomic analysis for various cereals. For example, 37,500 protein-coding genes are believed to be present in the rice genome, suggesting that the rice proteome consists of at least 37,500 proteins. However, this number very likely represents a minimum value. Different structures and functions of proteins are generated by alternative splicing and post-translational modification (PTM); therefore, we currently cannot predict how many proteins are present in a given plant. The proteins that are detected at certain time-points represent only a fraction of the total number of proteins, since not all proteins are constitutively expressed in plants at a particular stage, and neither perfect detection nor perfect identification techniques are available.

On the other hand, there are also polyploid plants among cereals. For example, bread wheat, which is hexaploid, has three different genomes. Its genome size is

Table 7.1 Genome size and number of genes estimated by genome analysis, and number of proteins detected and identified in proteome analysis in cereals

	Rice	Wheat	Barley	Maize
Genome size	420	16,000	5,100	2,400
Number of genes	37,500	–	–	–
Number of proteins detected (identified) in proteome analysis				
Seed	– (877)[a]	612 (301)[e]	1,200 (103)[j]	–
Endosperm	100 (37)[b]	1,300 (177)[f]	–	632 (496)[k]
Embryo	700 (46)[c]	–	–	–
Amyloplast	–	600 (289)[g]	–	–
Anther	1,080 (365)[d]	–	–	–
Lemma	–	300 (56)[h]	–	–
Leaf	– (1,002)[a]	404 (142)[i]	–	900 (149)[l]
Leaf sheath	509 (115)[b]	–	–	–
Stem	567 (186)[b]	–	–	–
Root	– (1,350)[a]	–	–	302 (74)[m]
Organelle				
Chloroplast	252 (89)[b]	–	–	526 (54)[n]
Mitochondria	672 (146)[b]	–	–	300 (38)[o]
Plasma membrane	464 (159)[b]	–	–	–
Golgi membrane	361 (46)[b]	–	–	–
Nucleus	549 (190)[b]	–	–	–
Vacuolar membrane	141 (76)[b]	–	–	–
Cell wall	513 (111)[b]	–	–	225 (95)[p]

[a]Koller et al. 2002, [b]Komatsu 2005, [c]Fukuda et al. 2003, [d]Imin et al. 2001, [e]Mak et al. 2006, [f]Skylas et al. 2001, [g]Balmer et al. 2006, [h]Woo et al. 2002, [i]Donnelly et al. 2005, [j]Ostergaard et al. 2004, [k]Mechin et al. 2004, [l]Porubleva et al. 2001, [m]Hochholdinger et al. 2005, [n]Majeran et al. 2005, [o]Hochholdinger et al. 2004, [p]Zhu et al. 2006

five times larger than that of humans, and 120 times larger than that of Arabidopsis. Bread wheat has many proteins with slightly different amino acid sequences that are controlled by genes in the different genomes.

A number of useful reviews on the current state of plant proteomics are available, including those by Agrawal and Rakwal (2006), Chen and Harmon (2006), Glinski and Weckwerth (2006), Hirano et al. (2004), and Rossignol et al. (2006).

7.2 A Typical Procedure for Proteomic Analysis

Routine proteomic analysis begins with the separation of a number of proteins, often by 2-DE (O'Farrell 1975; Klose 1975). A peptide map called a "peptide mass fingerprint" is then constructed by MS. Alternatively, a partial amino acid sequence is determined by tandem MS/MS ("de novo sequencing"). The appropriate genomic database is then searched using the peptide mass fingerprint and/or amino acid sequence data, and the corresponding gene or gene product is identified. The entries in protein sequence databases show which proteins have been ascribed particular functions. If the sequence of a protein is similar to that of other proteins with known function, the function of the target protein can be inferred. When the function of a protein is not identified using homology, construction of a protein profile (i.e., expression time and level, subcellular localization) and analyses of PTMs and protein–protein interactions are usually performed in order to determine the function of the protein (described below). Finally, the 2-DE pattern, amino acid sequence, and structure and function of each protein are compiled into a database.

Proteomic analysis is a comprehensive analysis in which the functions and functional networks of "all" proteins in an organism are investigated. However, in practice it is difficult to analyze all possible proteins. Therefore, in many cases, particular protein groups, such as phosphoproteins, glycoproteins, organelle proteins, protein complex constituents, and disease-associated proteins, are analyzed. This kind of study is called "focused proteomics." At present, many researchers are interested in focused proteomic analyses of cereals as well as other organisms.

7.3 Mass Spectrometry and Proteomic Analysis

Mass spectrometry for proteins and peptides has developed rapidly since the 1980s. It is now possible to measure the mass of proteins and peptides at the femtomolar level with high accuracy using MS, and to efficiently identify a number of proteins using software developed for proteomic research. MS is also used frequently in the analysis of protein expression, PTMs, and protein–protein interactions.

MS requires an ion source, a mass spectrometer, and an ion detector. Although several methods of protein and peptide ionization exist that differ in terms of their ion source, two "soft" ionization methods, i.e., matrix-assisted laser desorption ionization (MALDI) and electrospray ionization (ESI), are usually used. Generally, time-of-flight (TOF) MS is combined with MALDI, while quadrupole MS (Q MS) and ion trap MS (IT MS) are combined with ESI. Recently, Q/TOF MS, in which TOF is appended to Q MS, was developed. MALDI-TOF/TOF MS and ESI-IT/TOF have also been developed. Q/TOF MS, TOF/TOF MS, and IT/TOF MS can determine amino acid sequences by MS/MS analysis. Each MS is different depending on the type of ionization and the mass spectrometer, and no single MS technique can be applied to all fields of proteomic analysis. Instead, it is necessary to choose the instrument most suited to the purpose of the analysis. MALDI-TOF MS is often used for high-throughput identification of proteins by peptide mass fingerprinting. For the analysis of amino acid sequences and PTMs, MS/MS such as ESI-IT and ESI-Q/TOF MS are used. Recently, a Fourier transform ion cyclotron resonance MS (FT MS) instrument was refined. This instrument uses the phenomenon of ion cyclotron resonance to fragment a protein ionized by ESI or MALDI and determine its amino acid sequence and PTMs. FT MS allows high-resolution analysis. In most MS analyses, such as MALDI-TOF MS, ESI-Q/TOF MS, and ESI-IT/TOF MS, the proteins are analyzed after protease digestion, because high molecular weight proteins cannot be analyzed directly. This approach is called "bottom-up proteomics." In contrast, in FT MS, the entire protein can be applied directly. This approach is called "top-down proteomics" or "top-down MS" (VerBerkmoes et al. 2002).

Recently, peptide fragmentation methods such as electron capture dissociation (ECD) for FT MS and electron transfer dissociation (ETD) for IT MS have been developed. These fragmentation methods can generate *c* and *z* series ions. Since neither ECD nor ETD cleaves amino acid side chains, we can easily analyze PTMs using these MS approaches.

7.4 Expression Profiles of Cereal Proteins

Protein expression varies depending on the species, variety, growth stage, organ, cellular organelle, and environment. Expression profiles are closely related to protein function; therefore, analysis of protein expression dynamics is the first step in the determination of protein function.

7.4.1 2-DE-Based Profiling

Matsudaira (1987) first reported a method in which proteins separated by 2-DE were electroblotted onto polyvinylidene difluoride membranes, with their partial amino acid sequences then being determined by a gas-phase protein sequencer. Using this method, the partial amino acid sequences of a number of plant proteins separated by 2-DE (Hirano 1989) were determined for protein identification. Since that time, this technique has been widely used to identify proteins in rice, Arabidopsis, wheat, barley, and tobacco (Hirano et al. 2004).

Recently, protein identification in cereals and other organisms has been accomplished by comparing experimentally obtained peptide mass fingerprints with theoretical fingerprints based on the information stored in databases. Alternatively, partial amino acid sequences of proteins separated by 2-DE have been determined by MS/MS, allowing protein identification.

Seed endosperm and embryo protein compositions and dynamic changes in seed proteins during development have been extensively characterized. Such information is important for improving the quality and quantity of seed proteins, to determine the functions of seed proteins as food, and to elucidate the mechanisms of seed maturation *and* germination. For example, Vensel et al. (2005) investigated developmental changes in over 250 proteins found in wheat endosperm, and Finnie et al. (2002) analyzed changes in 36 proteins found in barley seeds during grain filling and seed maturation by 2-DE and MS. It is also important to analyze changes in protein expression during germination; however, there are currently only a few reports of this type (Finnie et al. 2004). On the other hand, allergen proteins, wheat germ beer quality-related proteins, and wheat quality-related proteins, among others, were identified by 2-DE (Hirano et al. 2004). These studies are important for understanding the functions of plant proteins in food.

There are many reports on proteins found in plasma membranes, mitochondria, chloroplasts, peroxisomes, amyloplasts, and ribosomes of various plant species, including rice and wheat (Hirano et al. 2004). These proteins were detected by 2-DE and identified mainly by MS. Proteomic analysis of organelles is important for understanding organelle function, biogenesis, and metabolism. Recently, Zabrouskov et al. (2003) identified chloroplast proteins in Arabidopsis by FT MS using top-down proteomics. This was the first application of top-down MS in plant proteomics. This technique will likely be applied to cereal proteomics as well.

To construct expression profiles, proteins are extracted from cells grown under different conditions and protein compositions are compared. This is called "protein differential display" analysis. Such analysis allows interspecies and varietal differences in plant proteins to be studied, and many plant proteins specific to certain organs, tissues, and growth stages have been detected. Proteins that are up- or down-regulated by hormone treatment, disease, and stresses such as low temperature, heat, drought, and ozone, have also been investigated by 2-DE (Hirano et al. 2004). For example, rice lesion-mimic mutants, also known as cell death and resistance (*cdr*) mutants, show significant resistance to rice blast fungus in the form of spontaneous cell death across entire leaves. The *CDR1* and *CDR2* genes negatively regulate certain phosphorylation events. Takahashi et al. (2003) compared the total protein phosphorylation levels between *cdr* mutants and wild-type plants by 2-DE to identify novel factors involved in the phosphorylation events, and showed that the phosphorylation level of four proteins in *cdr1* was increased by treatment with a protein phosphatase inhibitor. One of the four proteins was prohibitin, which is associated with senescence and cell death and functions as a chaperone in the assembly of the mitochondrial respiratory chain complex in yeast and mammals.

Difference gel electrophoresis (DIGE), a recently developed 2-DE technique with high reproducibility (Unlu et al. 1997), is currently gaining in popularity. In this technique, proteins are extracted from different types of cells, labeled with different fluorescent reagents, mixed, and then separated by 2-DE using a single gel. The proteins are detected separately at excitation wavelengths specific for the different fluorescent reagents. The two patterns are analyzed using an image analyzer, and differences in protein expression are easily detected. This technique is useful for reducing gel-to-gel variation; therefore, DIGE can be recommended for use in cereal proteomic analyses.

7.4.2 Shotgun Profiling

Although 2-DE allows simple, rapid, and reproducible separation of a number of proteins, it is difficult to separate high molecular weight and basic proteins. To overcome this problem, a "shotgun" method has been developed (Link et al. 1999; Smith et al. 2002). In this method, extracted cellular proteins are digested with proteases such as trypsin and lysylendopeptidase, and the resultant peptides are analyzed by multidimensional nano liquid chromatography (LC), followed by MS/MS for determination of the sequences. Using the shotgun method, a comprehensive proteomic analysis has been performed for many organisms.

Koller et al. (2002) identified a total of 2,528 unique proteins, including 1,022 proteins from leaves, 1,350 proteins from roots, and 877 proteins from seeds. Most of the proteins (67.2%) had known functions or were homologous with other proteins with known functions. The identified proteins were classified into 16 functional categories. The most abundant class of proteins was involved in metabolic processes (20.8%). There were also many proteins involved in protein synthesis, protein degradation, and signal transduction. Of the 2,528 unique proteins, 189 were expressed in all three organs, but 622, 862, and 512 proteins were expressed only in leaves, roots, and seeds, respectively. Enzymes involved in central metabolic pathways were present in all tissues, while many proteins showed a tissue-specific expression pattern. Similarly, Watson et al. (2003) identified a total of 304 proteins from leaves, stems, roots, flowers, seed pods, and cell suspension cultures of barrel medic by MS. Functions were assigned to 55% of the proteins based on homology with known sequences. Sixty-one percent of the proteins were detected in one or more tissues, while 39% were found in only one tissue and had an assigned function related to that specific tissue.

Around the same time, Gomez et al. (2002) comprehensively analyzed proteins from thylakoid membrane subdomains (grana) of chloroplasts in pea and spinach (see also Chap. 14 by Jarvis, this volume).

7.4.3 *Quantitative Profiling*

Isotope-coded affinity tag (ICAT) labeling was also used in protein differential display analysis using MS. With this technique, proteins from different cells with different molecular masses were labeled by ICAT (Gygi et al. 1999). The proteins in the two samples were then mixed and digested with a protease, such as trypsin, and analyzed by LC-MS/MS. Differences in protein quantity were determined by measuring the peptides labeled with heavy and light ICAT reagents. However, at present, there are only a few reports on protein differential display analysis using ICAT.

Isobaric tag for relative and absolute quantitation (iTRAQ) is an alternative to ICAT. The reagent contains four isobaric tags (145 Da) consisting of a reporter tag (114–117 Da), a balance tag (31–28 Da), and a peptide-reactive group. Peptide counterparts from four different samples, which have identical masses, are labeled with the four different isobaric tags. The signal intensity of the reporter tags shows the abundance of the proteins in each sample. Simultaneously, peptide sequence information is obtained through MS/MS analysis. High peptide coverage and reproducibility can be expected using this method. We compared the quantification data between iTRAQ and DIGE analyses (Fig. 7.1). The data obtained from these analyses were significantly correlated, suggesting that these methods are quantitative. Reports of protein differential display analysis using iTRAQ are currently available for Arabidopsis (see Chap. 8 by Baginsky and Gruissem, this volume), but not for cereals.

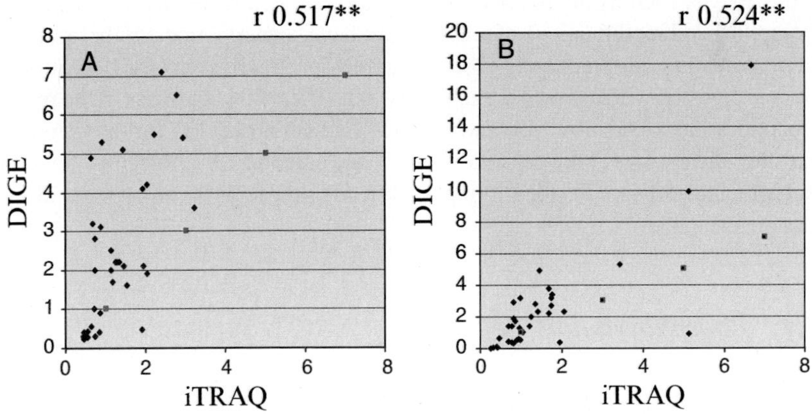

Fig. 7.1 Correlation of difference gel electrophoresis (DIGE) and isobaric tag for relative and absolute quantitation (iTRAQ) data. Proteins differing in amount between two samples (A and B), were detected by DIGE or iTRAQ. The differences in the amounts of these proteins were significantly correlated. Samples, human ovarian clear cell adenocarcinoma and mucinous adenocarcinoma (Y. Yamanaka, H. Hirano unpublished data)

7.4.4 Subcellular Localization of Cereal Proteins

Proteins are spatially organized according to their functions; therefore, analysis of subcellular protein localization is important for understanding protein function. Recently, a high-throughput method using epitope tags or green fluorescent protein (GFP), which can be used to visualize the subcellular localization of proteins, has been developed. In epitope tagging, a DNA chimera, which encodes a target protein and a specific epitope peptide, is introduced into a vector and transfected into a host cell. Expression of the fusion protein can then be detected by microscopy using an antibody. As an alternative to the epitope tag, GFP may be used, allowing the fusion protein to be detected as a fluorescent protein. The localization of almost all proteins in yeast was identified using these methods (Kumar et al. 2002). In plants, a GFP::cDNA fusion library was created and introduced into Arabidopsis and tobacco by *Agrobacterium*-mediated transformation (Cutler et al. 2000; Escobar et al. 2003), and the subcellular localization of the expressed proteins was identified by GFP fluorescence.

7.5 Post-Translational Modifications of Cereal Proteins

To perform their intrinsic functions, most proteins must be post-translationally modified; therefore, the analysis of PTMs is important to determine the in vivo state of proteins. The area of proteomics that deals with PTMs is called "modificomics." Various PTMs, such as the removal of a signal peptide, processing of a precursor polypeptide, and modifications of amino acids, have been reported. Of these, amino acid modifications are frequently observed in MS spectra. In such cases, the protein extract is digested with a protease such as trypsin. The molecular mass of the resultant peptides is analyzed by MS, and the modified amino acid(s) can be identified from the differences between the actual and theoretical molecular masses. Databases and software for identifying modified amino acids based on mass differences are available.

7.5.1 Phosphorylation

Among the various PTMs, phosphorylation and glycosylation have been most widely studied. Protein phosphorylation is closely related to signal transduction. The field of phosphoprotein proteomics is called phosphoproteomics (see Chap. 4 by de la Fuente van Bentem, this volume).

Simple methods have been developed to effectively detect phosphoproteins. For example, phosphorylated proteins can be detected by comparing the 2-DE patterns of proteins before and after phosphatase digestion. Also, phosphorylated proteins separated by gel electrophoresis can easily be detected with the commercially available reagent Pro-Q Diamond (Invitrogen, La Jolla, CA).

Nevertheless, the detection of phosphopeptides and the determination of phosphorylation sites by MS are often difficult without enrichment of the phosphopeptides. A number of techniques have been developed for this purpose (Glinski and Weckwerth 2006; Kersten et al. 2006). Among them, immobilized metal affinity chromatography *(IMAC)* is frequently used in proteomic analysis. IMAC can enrich phosphopeptides based on the affinity between acidic phosphate residues and trivalent metal ions such as Fe^{3+} or Ga^{3+}; however, the detection of phosphopeptides is often hampered by nonspecific binding of peptides rich in glutamic acid and aspartic acid.

A number of studies have reported on the function of protein phosphorylation. In cereals, for example, Tetlow et al. (2004) detected phosphoproteins in the starch-synthesizing amyloplasts of wheat. Three starch branching enzymes (SBEI, SBEIIa, and SBEIIb) and two starch granule-associated SBIIb and starch synthases (SSIIa and SSIIb) were phosphorylated in these structures. The authors found that phosphorylation regulated the activities of SBEIIa and SBEIIb, but not that of SBEI. SBEIIb, starch phosphorylase, and SBEI formed a protein complex, suggesting that protein phosphorylation is involved in the regulation of enzymatic activity and protein–protein interactions.

DNA microarray technology, which was developed in the 1990s, has become a landmark technique in gene expression analysis. Using a similar idea, protein microarrays are used to analyze the expression profiles of proteins and their functions (MacBeath 2002). In the last decade, protein chip technology has developed rapidly and has become a useful tool for the study of protein identity, protein interactions, and biochemical activities. It is also used to analyze post-translationally modified proteins, such as phosphorylated proteins. Two recent reports used this technology to identify novel kinase substrates. Kramer et al. (2004) identified candidate substrates for casein kinase 2α (CK2α) using a protein chip. They immobilized 21,500 barley proteins on a chip, incubated the chip with CK2α and [γ-^{33}P]ATP, and detected the radioactively labeled proteins (Fig. 7.2). As a result, they detected 21 potential target proteins of barley CK2α. In comparison, Ptacek et al. (2005) prepared 87 kinases and added each kinase to yeast proteins immobilized on a chip with [γ-^{33}P]ATP. In this analysis, they identified over 4,000 phosphorylation events involving 1,325 different proteins.

7.5.2 Glycosylation

Protein glycosylation affects cell recognition, membrane binding, enzymatic activity, and protein interactions. Recently, Kaji et al. (2003) developed a novel method for the high-throughput detection of glycoproteins. In their method, proteins are purified using a lectin column, ^{18}O is introduced into the Asn-linked oligosaccharide-binding sites using glycopeptidase, and the ^{18}O-labeled peptides are identified by LC-MS/MS. Using this method, they detected 250 proteins with Asn-linked oligosaccharides in nematodes. This method can also be applied to plant glycoproteins. Gelperin et al.

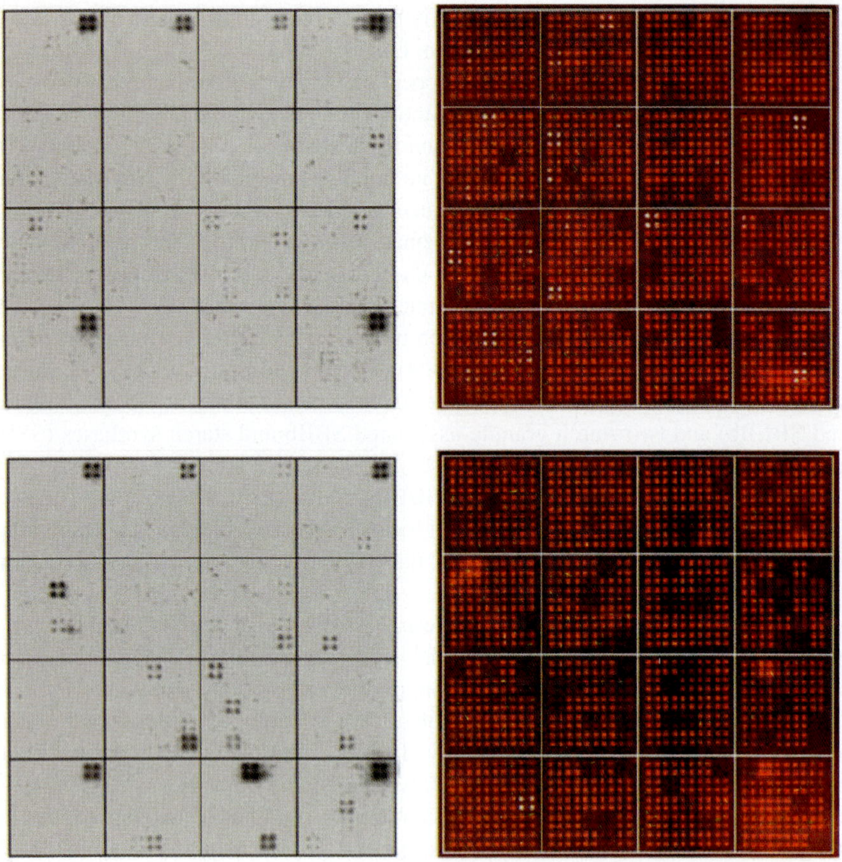

Fig. 7.2 Detection of barley casein kinase 2α (CK2α) targets by a protein microarray technique. A total of 768 recombinant proteins were immobilized on FAST™ slides. *Left* Slides were incubated with CK2α and [γ-^{33}P]ATP, and the CK2α targets were detected by X-ray film. *Right* Targets were detected with an anti-RGS-His$_6$ antibody (Kramer et al. 2004)

(2005) applied the method of Ptacek et al. (2005) to a protein chip with 5,573 over-expressed immobilized proteins and anti-α glycan antibody. In this analysis, they identified 172 known glycoproteins and 217 novel glycoproteins. These techniques can also be applied to cereal proteomic analyses.

Normally, to produce a protein chip, native or recombinant proteins are purified and immobilized on the chip plate; however, the purification of many protein species is not easy, and this produces a bottleneck in the production of high-density protein chips. In response to this problem, Iwafune et al. (2007) developed diamond-like carbon-coated stainless steel (DLC) plates that are modified with an N-hydroxysuccinimide ester. The DLC plate can immobilize gel-resolved proteins with high blotting efficiency (30–70%). A particular protein is then used to probe the proteins on the DLC plate, and interacting proteins are detected by

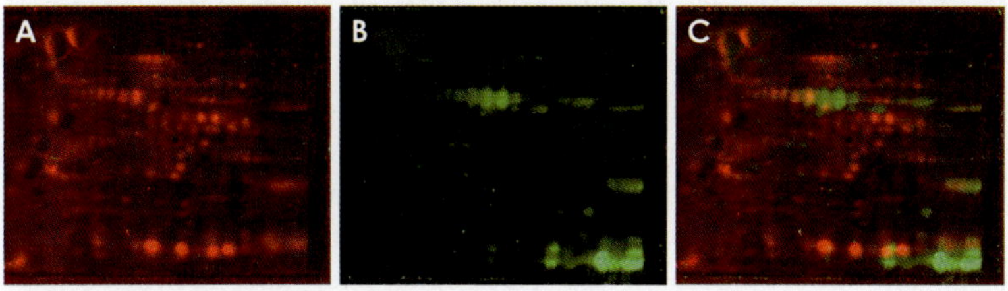

Fig. 7.3 Detection of rice glycoproteins immobilized on a diamond-like carbon coated stainless steel (DLC) plate. Cy3-labeled proteins separated by 2-dimensional gel electrophoresis (2-DE) were electroblotted onto the DLC plate. The plate was then incubated with Cy5-labeled concanavalin A in solution. **a** Cy3-labeled proteins immobilized on the plate. **b** Cy5-labeled proteins interacted with the immobilized glycoproteins. **c** Combined image of **a** and **b**. The detected glycoproteins were identified by matrix-assisted laser desorption ionization time-of-flight (MALDI-TOF) mass spectrometry (MS) analysis

MALDI-TOF MS. Using this technique, proteins on a DLC plate were probed with concanavalin A to detect N-linked high-mannose type glycoproteins (Fig. 7.3). The detected glycoproteins were then digested on the plate and analyzed by MALDI-TOF MS for identification.

7.5.3 Other Post-Translational Modifications

Many kinds of PTMs besides phosphorylation and glycosylation are known in plants; however, little information is available regarding their functions. As discussed above, MS analysis has great potential for the identification of PTMs in a large number of proteins. Nevertheless, even if a PTM is detected, it is not easy to guess the function of a protein based on its modifications, because there is not enough information on the role of PTMs, nor, to our knowledge, is there a database or software that can predict protein function from the PTM information generated by MS analysis. Further developments within this area may significantly contribute to the field of proteomics.

7.6 Protein-Protein Interactions in Cereals

Proteins function by interacting with other proteins and with ligands. Analysis of protein–protein (ligand) interactions is therefore important for determining the function of proteins using proteomic analysis. The study of protein–protein (ligand) interactions is called "interactome" analysis, and the field is referred to as "interaction proteomics" or "interactomics."

7.6.1 Immunoaffinity Purification/MS of Protein Complexes

In proteomic analysis, immunoaffinity purification is an important technique for detecting protein–protein interactions. Epitope tagging is often used; fusion proteins comprised of a target protein and an epitope tag are overexpressed in a cell. Proteins interacting with the target protein are purified by immuno-precipitation using an antibody, separated by electrophoresis, and identified by MS. Alternatively, the native target protein and its interacting proteins may be co-precipitated using an antibody raised against the target protein, then separated by electrophoresis and identified by MS. For example, Alexander and Morris (2006) purified 54 14-3-3 binding proteins, which are eukaryotic regulatory proteins with various functions, by immunoaffinity purification in barley. They found that these binding proteins were involved in carbohydrate metabolism and possibly in plant defense.

7.6.2 Tandem Affinity Purification/MS of Protein Complexes

Tandem affinity purification (TAP) is a powerful technique to purify protein complexes. In this method, a fusion protein that contains the IgG binding domain of protein A and calmodulin binding peptide conjugated to a target protein is overexpressed, and protein complexes consisting of the recombinant target protein and its interactors are purified in two steps using IgG- and calmodulin-beads (Rigaut et al. 1999). This method is frequently used to detect protein–protein interactions in many organisms. Rivas et al. (2002) used this method to purify a 420 kDa hetero-multimeric membrane-associated protein complex that interacted with the tomato Cf-9 protein. The *Cf-9* gene confers race-specific resistance to fungal pathogens expressing the avirulence gene *Arv-9*. On the other hand, Rohila et al. (2006) purified proteins that interacted with 41 TAP-tagged kinases in rice (Fig. 7.4). They fused cDNAs encoding protein kinases to the TAP tag and expressed them in transgenic rice. Proteins that interacted with the kinases were then identified by MS; 23 kinases interacted with other rice proteins. Homologs of some of these interacting proteins have been found in other organisms, suggesting that this method is useful for identifying protein complexes in rice.

7.6.3 Blue Native-/SDS-PAGE/MS of Protein Complexes

Two-dimensional gel electrophoresis using blue native-gel electrophoresis (BN-PAGE) in the first dimension and SDS-PAGE in the second dimension is used to analyze protein complexes. In BN-/SDS-PAGE, protein complexes negatively charged with Coomassie Blue G-250 are separated according to the charge of the complexes in the

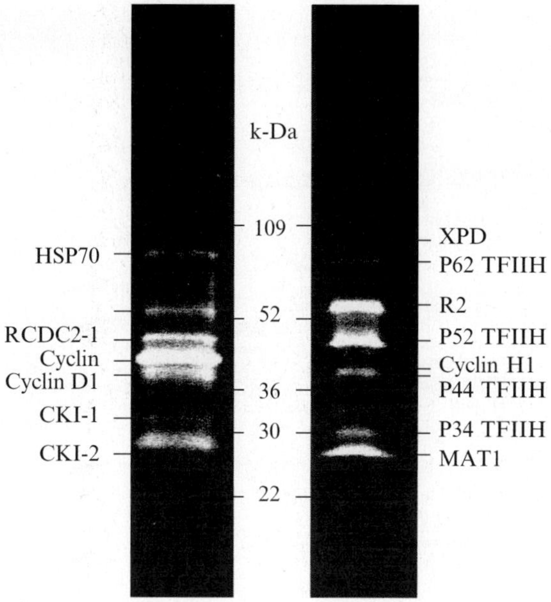

Fig. 7.4 Protein complexes purified from transgenic rice by tandem affinity purification (TAP). TAP-tagged protein kinase complexes were separated by SDS-PAGE and detected with SYPRO Ruby. Proteins were identified by MS. *HSP70* Heat shock protein 70; *RCDC2-1* rice cyclin-dependent kinase; *R2* rice CDK-activating kinase; *MAT1* CDK-activating kinase assembly factor; *CKI-1, -2* cyclin-dependent kinase inhibitor; *TFIIH* general transcription factor IIH; *XPD* nucleotide excision repair protein XP-D homolog (Rohila et al. 2006)

first dimension, and the proteins (subunits) in the complexes are then separated based on their molecular weights in the second dimension under dissociating conditions.

Ciambella et al. (2005) detected thylakoid protein complexes by BN-/SDS-PAGE and identified by MS the proteins contained in the complexes in barley and tomato. They identified about 70% of total integral thylakoid proteins (Fig. 7.5). Heazelwood et al. (2003) analyzed a membrane fraction from rice mitochondria that contained many hydrophobic proteins by BN-/SDS-PAGE and identified by MS a series of large protein complexes with co-migrating subunits that constitute the major electron transport chain.

7.6.4 Protein Interaction Analysis Using Chromosome-Deletion Lines

Islam et al. (2002) studied protein–protein interactions in wheat seed using chromosome-deletion lines. Changes in the protein composition of the wheat seed proteome

B.U.O. Sciences

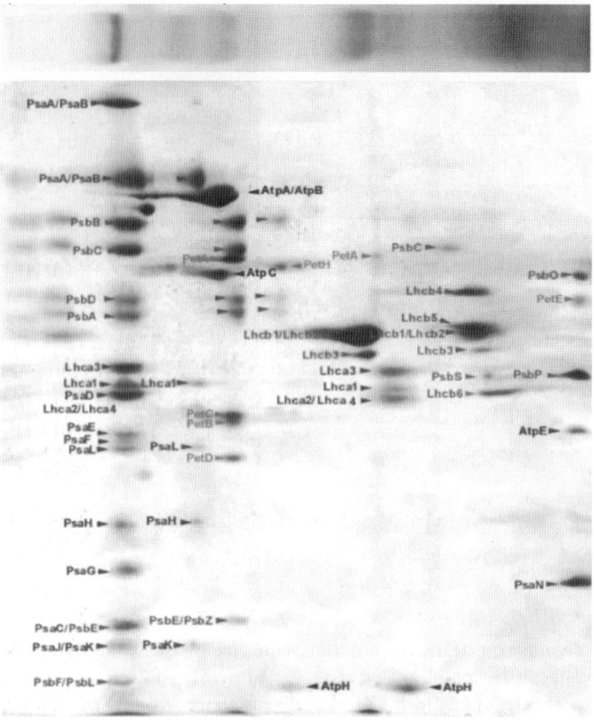

Fig. 7.5 Blue native-polyacrylamide gel electrophoresis (BN)-/SDS-PAGE of barley thylakoid membrane proteins. Proteins were separated by BN-PAGE in the first dimension, and the BN-PAGE lane was excised and incubated in SDS sample buffer containing 5% 2-mercaptoethanol and 6 M urea. The lane was then overlaid on the SDS-PAGE gel for the second dimension. After SDS-PAGE, proteins were detected by Coomassie R-250 staining. The detected proteins were identified by MALDI-TOF MS (Ciambella et al. 2005)

were investigated using 39 ditelocentric lines (Islam et al. 2003b), which carry the normal (euploid) complement of wheat chromosomes except that one arm is missing, and also by using fine deletion lines of chromosome 1B (Islam et al. 2003a). Proteins were separated by 2-DE and visualized by staining with Coomassie Brilliant Blue. Quantitative analysis of the protein spots was performed using PDQuest Software (Bio-Rad Laboratories, UK). Variations in the protein spots between the euploid lines and the 39 ditelocentric lines were evaluated. Out of the 1,755 major spots detected in the ditelocentric lines, 147 (11%) had disappeared, 978 (71%) were up-regulated, and 247 (18%) were down-regulated. Correlation studies of the changes in protein intensity of 24 protein spots from the ditelocentric lines were performed. High correlations between the changes in protein intensities were observed among the proteins encoded by genes located in the homologous arms. The feasibility of a new analytical approach to identify protein–protein interactions based on ICAT labeling of peptides using tryptic digestion followed by ESI Q-TOF MS was also investigated. Down- and up-regulated proteins were detected in the fine chromosome deletion lines. This approach may be useful for the identification of wheat seed proteins and their interactions, which is reportedly difficult by 2-DE due to co-synthesis of proteins by genes from three genomes (A, B, and D).

Although genomic manipulation was successfully used by Islam and Hirano (2003) to predict protein-protein interactions in the hexaploid wheat seed proteome, this technique is not feasible for diploid species where genomic manipulation is not as easy.

7.7 Proteome Informatics

Since the development of proteomics, an enormous amount of information on proteomic analysis has been produced. Unfortunately, in contrast to the multiple DNA sequence databases, no unified proteome database is available for proteomic studies. However, some organizations are attempting to create protein databases. For example, notes (annotations) on the characteristics of proteins and the sequence data of a number of proteins are available on the web at http://www.expasy.ch/alinks.html, and these databases are linked with other databases, such as DNA sequence databases and protein structure databases.

To analyze the functions of a number of proteins, it is important to develop new software for proteomic analysis. Various software programs, which allow 2-DE pattern image analysis, MS analysis, identification of proteins based on peptide mass fingerprints and amino acid sequences, and detection of PTMs are available; however, there is no sophisticated software that can predict the function of proteins based on their amino acid sequences, PTMs, protein–protein interactions, or higher order structures.

7.8 Perspectives

Technology for protein separation and identification has developed rapidly in proteomics. Techniques such as 2-DE, DIGE, ICAT, shotgun analysis, and MS are useful for the separation and identification of proteins in plants, including cereals. Recently, new techniques, such as FT MS for top-down proteomics and TAP for analysis of protein complexes, have been developed for the analysis of protein function and functional networks of proteins. Nevertheless, further development of this technology is required to meet the current demand for high-throughput proteomics. Although a number of software packages are available to identify proteins and to predict their functions, no software is available to analyze PTMs and protein–protein interactions, which are particularly important for the elucidation of protein function.

Efficiency of protein functional analysis in proteomics is strongly "database-dependent." In many cases, protein functions and functional networks between proteins are determined by information retrieved from databases that are based on actual protein analysis. Therefore, improvement of the quality and quantity of proteome databases, and the construction of new software that can analyze protein function, are essential for proteome research.

References

Agrawal GK, Rakwal R (2006) Rice proteomics: a cornerstone for cereal food crop proteomes. Mass Spectrom Rev 25:1–53

Alexander RD, Morris PC (2006) A proteomic analysis of 14-3-3 binding proteins from developing barley grains. Proteomics 6:1886–1896

Chen S, Harmon AC (2006) Advances in plant proteomics. Proteomics 6:5504–5516

Ciambella C, Roepstorff P, Aro EM, Zolla L (2005) A proteomic approach for investigation of photosynthetic apparatus in plants. Proteomics 5:746–757

Cutler SR, Ehrhardt DW, Griffitts JS, Somerville CR (2000) Random GFP::cDNA fusion enable visualization of subcellular structures in cells of *Arabidopsis* at a high frequency. Proc Natl Acad Sci USA 97:3718–3723

Donnelly BE, Madden RD, Ayoubi P, Porter DR, Dillwith JW (2005) The wheat (*Triticum aestivum* L.) leaf proteome. Proteomics 5:1624–1633

Escobar NM, Haupt S, Thow G, Boevink P, Chapman S, Oparka K (2003) High-throughput viral expression of cDNA-green fluorescent protein fusions reveals novel subcellular addresses and identifies unique proteins that interact with plasmodesmata. Plant Cell 15:1507–1523

Finnie C, Melchior S, Roepstorff P, Svensson B (2002) Proteome analysis of grain filling and seed maturation in barley. Plant Physiol 129:1308–1319

Finnie C, Maeda K, Ostergaard O, Bak-Jensen KS, Larsen J, Svensson B(2004) Aspects of the barley seed proteome during development and germination. Biochem Soc Trans 32:517–519

Fukuda M, Islam N, Woo SH, Yamagishi A, Takaoka M, Hirano H (2003) Assessing matrix assisted laser desorption/ ionization-time of flight-mass spectrometry as a means of rapid embryo protein identification in rice. Electrophoresis 24:1319–1329

Gelperin DM, White MA, Wilkinson ML, Kon Y, Kung LA, Wise KJ, Lopez-Hoyo N, Jiang L, Piccirillo S, Yu H, Gerstein M, Dumont ME, Phizicky EM, Snyder M, Grayhack EJ (2005) Biochemical and genetic analysis of the yeast proteome with a movable ORF collection. Genes Dev 19:2816–2826

Glinski M, Weckwerth W (2006) The role of mass spectrometry in plant systems biology. Mass Spectrom Rev 25:173–214

Gomez SM, Nishio JN, Faull KF, Whitelegge JP (2002) The chloroplast grana proteome defined by intact mass measurements from liquid chromatography mass spectrometry. Mol Cell Proteomics 1:46–59

Gygi SP, Rist B, Gerber SA, Turecek F, Gelb MH, Aebersold R (1999) Quantitative analysis of complex protein mixtures using isotope-coded affinity tags. Nat Biotechnol 17:994–999

Heazlewood JL, Howell KA, Whelan J, Millar AH (2003) Towards an analysis of the rice mitochondrial proteome. Plant Physiol 132:230–242

Hirano H (1989) Microsequence analysis of winged bean seed proteins electroblotted from two-dimensional gel. J Protein Chem 8:115–130

Hirano H, Islam N, Kawasaki H (2004) Technical aspects of functional proteomics in plants. Phytochemistry 65:1487–1498

Hochholdinger F, Guo L, Schnable PS (2004) Lateral roots affect the proteome of the primary root of maize (*Zea mays* L.). Plant Mol Biol 56:397–412

Hochholdinger F, Woll K, Guo L, Schnable PS (2005) The accumulation of abundant soluble proteins changes early in the development of the primary roots of maize (*Zea mays* L.). Proteomics 5:4885–4893

Imin N, Kerim T, Weinman JJ, Rolfe BG (2001) Characterisation of rice anther proteins expressed at the young microspore stage. Proteomics 1:1149–1161

Islam N, Hirano H (2003) Exploring protein–protein interactions in wheat proteome by manipulating genome composition: a perspective. J Electrophoresis 47:33–42

Islam N, Woo S-H, Tsujimoto H, Kawasaki H, Hirano H (2002) Proteome approaches to characterize seed storage proteins related to ditelocentric chromosomes in common wheat (*Triticum aestivum* L.). Proteomics 2:1146–1155

Islam N, Tsujimoto H, Hirano H (2003a) Wheat proteomics: relationship between fine chromosome deletion and protein expression. Proteomics 3:307–316

Islam N, Tsujimato H, Hirano H (2003b) Proteome analysis of diploid, tetraploid and hexaploid wheat: towards understanding genome interaction in protein expression. Proteomics 3:549–557

Iwafune Y, Tan JZ, Ino Y, Okayama A, Ishigaki Y, Saito K, Suzuki N, Arima M, Oba M, Kamei S, Tanga M, Okada T, Hirano H (2007) On-chip identification and interaction analysis of gel-resolved proteins using a diamond-like carbon-coated plate. J Proteome Res 6:2315–2322.

Kaji H, Saito H, Yamauchi Y, Shinkawa T, Taoka M, Hirabayashi J, Kasai K, Takahashi N, Isobe T (2003) Lectin affinity capture, isotope-coded tagging and mass spectrometry to identify N-linked glycoproteins. Nat Biotechnol 21:667–672

Kersten B, Agrawal GK, Iwahashi H, Rakwal R (2006) Plant phosphoproteomics: a long road ahead. Proteomics 6:5517–5528

Klose J (1975) Protein mapping by combined isoelectric focusing and electrophoresis of mouse tissues. A novel approach to testing for induced point mutations in mammals. Humangenetik 26:231–243

Koller A, Washburn MP, Lange BM, Andon NL, Deciu C, Haynes PA, Hays L, Schieltz D, Ulaszek R, Wei J, Wolters D, Yates JR 3rd (2002) Proteomic survey of metabolic pathways in rice. Proc Natl Acad Sci USA 99:11969–11974

Komatsu S (2005) Rice proteome database: a step toward functional analysis of the rice genome. Plant Mol Biol 59:179–190

Kramer A, Feilner T, Possling A, Radchuk V, Weschke W, Burkle L, Kersten B (2004) Identification of barley CK2alpha targets by using the protein microarray technology. Phytochemistry 65:1777–1784

Kumar A, Agarwal S, Heyman JA, Matson S, Heidtman M, Piccirillo S, Umansky L, Drawid A, Jansen R, Liu Y, Cheung KH, Miller P, Gerstein M, Roeder GS, Snyder M (2002) Subcellular localization of the yeast proteome. Genes Dev 16:707–719

Link AJ, Eng J, Schieltz DM, Carmack E, Mize GJ, Morris DR, Garvik BM, Yates JR (1999) Direct analysis of protein complexes using mass spectrometry. Nat Biotechnol 17:676–682

MacBeath G (2002) Protein microarrays and proteomics. Nat Genet 32 Suppl:526–532

Majeran W, Cai Y, Sun Q, van Wijk KJ (2005) Functional differentiation of bundle sheath and mesophyll maize chloroplasts determined by comparative proteomics. Plant Cell 17:3111–3140

Mak Y, Skylas DJ, Willows R, Connolly A, Cordwell SJ, Wrigley CW, Sharp PJ, Copeland L (2006) A proteomic approach to the identification and characterisation of protein composition in wheat germ. Funct Integr Genomics 6:322–337

Matsudaira P (1987) Sequence from picomole quantities of proteins electroblotted onto polyvinylidene difluoride membranes. J Biol Chem 262:10035–10038

Mechin V, Balliau T, Chateau-Joubert S, Davanture M, Langella O, Negroni L, Prioul JL, Thevenot C, Zivy M, Damerval C (2004) A two-dimensional proteome map of maize endosperm. Phytochemistry 65:1609–1618

O'Farrell PH (1975) High resolution two-dimensional electrophoresis of proteins. J Biol Chem 250:4007–4021

Ostergaard O, Finnie C, Laugesen S, Roepstorff P, Svennson B (2004) Proteome analysis of barley seeds: identification of major proteins from two-dimensional gels (pI 4–7). Proteomics 4:2437–2447

Porubleva L, Vander Velden K, Kothari S, Oliver DJ, Chitnis PR (2001) The proteome of maize leaves: use of gene sequences and expressed sequence tag data for identification of proteins with peptide mass fingerprints. Electrophoresis 22:1724–1738

Ptacek J, Devgan G, Michaud G, Zhu H, Zhu X, Fasolo J, Guo H, Jona G, Breitkreutz A, Sopko R, McCartney RR, Schmidt MC, Rachidi N, Lee SJ, Mah AS, Meng L, Stark MJ, Stern DF, De Virgilio C, Tyers M, Andrews B, Gerstein M, Schweitzer B, Predki PF, Snyder M (2005) Global analysis of protein phosphorylation in yeast. Nature 438:679–684

Rigaut G, Shevchenko A, Rutz B, Wilm M, Mann M, Seraphin B (1999) A generic protein purification method for protein complex characterization and proteome exploration. Nat Biotechnol 17:1030–1032

Rivas S, Romeis T, Jones JD (2002) The Cf-9 disease resistance protein is present in an approximately 420-kilodalton heteromultimeric membrane-associated complex at one molecule per complex. Plant Cell 14:689–702

Rohila JS, Chen M, Chen S, Chen J, Cerny R, Dardick C, Canlas P, Xu X, Gribskov M, Kanrar S, Zhu JK, Ronald P, Fromm ME (2006) Protein-protein interactions of tandem affinity purification-tagged protein kinases in rice. Plant J 46:1–13

Rossignol M, Peltier JB, Mock HP, Matros A, Maldonado AM, Jorrin JV (2006) Plant proteome analysis: a 2004–2006 update. Proteomics 6:5529–5548

Skylas DJ, Copeland L, Rathmell WG, Wrigley CW (2001) The wheat-grain proteome as a basis for more efficient cultivar identification. Proteomics 1:1542–1546

Smith RD, Anderson GA, Lipton MS, Pasa-Tolic L, Shen Y, Conrads TP, Veenstra TD, Udseth HR (2002) An accurate mass tag strategy for quantitative and high-throughput proteome measurements. Proteomics 2:513–523

Takahashi A, Kawasaki T, Wong HL, Suharsono U, Hirano H, Shimamoto K (2003) Hyperphosphorylation of a mitochondrial protein, prohibitin, is induced by calyculin A in a rice lesion-mimic mutant *cdr1*. Plant Physiol 132:1861–1869

Tetlow IJ, Wait R, Lu Z, Akkasaeng R, Bowsher G, Esposito S, Kosar-Hashemi B, Morell MK, Emes MJ (2004) Protein phosphorylation in amyloplasts regulates starch branching enzyme activity and protein-protein interactions. Plant Cell 16:694–708

Unlu M, Morgan ME, Minden JS (1997) Difference gel electrophoresis: a single gel method for detecting changes in protein extracts. Electrophoresis 18:2071–2077

Vensel WH, Tanaka CK, Cai N, Wong JH, Buchanan BB, Hurkman WJ (2005) Developmental changes in the metabolic protein profiles of wheat endosperm. Proteomics 5:1594–1611

VerBerkmoes NC, Bundy JL, Hauser L, Asano KG, Razumovskaya J, Larimer F, Hettich RL, Stephenson JL Jr (2002) Integrating "top-down" and "bottom-up" mass spectrometric approaches for proteomic analysis of *Shewanella oneidensis*. J Proteome Res 1:239–252

Watson BS, Asirvatham VS, Wang L, Sumner LW (2003) Mapping the proteome of barrel medic (*Medicago truncatula*). Plant Physiol 131:1104–1123

Woo SH, Fukuda M, Islam N, Takaoka M, Kawasaki H, Hirano H (2002) Efficient peptide mapping and its application to identify embryo proteins in rice proteome analysis. Electrophoresis 23:647–654

Zabrouskov V, Giacomelli L, van Wijk KJ, MacLafferty FW (2003) A new approach for plant proteomics: characterization of chloroplast proteins of *Arabidopsis thaliana* by top down mass spectrometry. Mol Cell Proteomics 2:1253–1260

Zhu J, Chen S, Alvarez S, Asirvatham VS, Schachtman DP, Wu Y, Sharp RE (2006) Cell wall proteome in the maize primary root elongation zone. I. Extraction and identification of water-soluble and lightly ionically bound proteins. Plant Physiol 140:311–325

Chapter 8
Current Status of *Arabidopsis Thaliana* Proteomics

Sacha Baginsky and Wilhelm Gruissem

Abstract Proteomics, the systematic identification of proteins, has become an important asset for the study of cellular processes in a systems biology context. During the last few years significant technological improvements have been reported for high-throughput proteomics, both at the level of data analysis software and mass spectrometry hardware. With the maturation of proteomics technology, scientists now aim at proteome-wide protein identification to complement data from genome-wide transcriptional profiling and metabolomics experiments. A complete map of the *Arabidopsis thaliana* proteome is expected to provide important information on genome activities and gene structures. Peptides identified in proteomics experiments are extremely valuable because they manifest the expression of a gene and thus complement the annotation of open reading frames and confirm or correct gene structure prediction. Furthermore, knowledge of repeatedly identified peptides in large-scale proteomics experiments allows peptide arrays with a selected set of proteotypic peptides for absolute protein quantification to be designed. Last but not least, knowledge of protein abundance, posttranslational modification and localisation is the key to a better understanding of the molecular mechanisms of cell functioning and pathway compartmentalisation. In this chapter, we will briefly highlight the current status of Arabidopsis proteomics and discuss existing limitations and anticipated new developments in plant proteomics.

8.1 Current Status of Arabidopsis Proteomics

Arabidopsis thaliana is currently the best understood plant model system that has a large set of tools available for functional genomics (Provart and McCourt 2004). Even with the completion of the Arabidopsis genome sequence, the annotation of gene models remains a major challenge because the structure of several protein

---

Sacha Baginsky, Institute of Plant Sciences, ETH Zurich, Universitätstrasse 2, 8092 Zurich, Switzerland, E-mail: sbaginsky@ethz.ch

Wilhelm Gruissem, Institute of Plant Sciences, ETH Zurich, Universitätstrasse 2, 8092 Zurich, Switzerland

J. Šamaj and J. Thelen (eds.), *Plant Proteomics*
© Springer 2007

coding genes and function of open reading frames is still unclear. Based on available gene prediction algorithms, oligonucleotide microarrays have been developed that allow analysis of the transcriptional activity of most genes. Microarray technology has been extended recently to produce genome tiling arrays that cover nearly the complete genome sequence. RNA profiling using microarrays has revealed the dynamics of transcriptional regulation. In addition, high-throughput quantitative PCR has facilitated analysis of low-abundance RNA transcripts, thus providing a comprehensive view of the dynamic range of the transcriptome. In contrast, the analysis of the proteome and metabolome has proved more difficult. Proteins and metabolites present a significantly larger biochemical heterogeneity than nucleic acids, which requires the application of several different analytical methods (reviewed in Aebersold and Mann 2003; Bino et al. 2004).

Proteomics has a central role in systems biology that complements the analysis of the transcriptome and metabolome. The systematic and quantitative analysis of all proteins is important because concentrations of cellular proteins are difficult to predict from the abundance of mRNAs. Recent reports have shown that positive correlations between transcript and protein abundance can be found for specific subsets of all identified proteins (Gygi et al. 1999b; Griffin et al. 2002; Greenbaum et al. 2003; Washburn et al. 2003; Hack 2004; Kleffmann et al. 2004; Tian et al. 2004). Such positive correlations strongly suggest that the concentration of certain proteins is regulated primarily at the transcriptional level. In most instances, however, protein concentrations are also controlled by posttranslational mechanisms, especially for proteins involved in signal transduction. Such mechanisms may include control of the transcript stability, the rate at which an mRNA is translated, or stability of the protein.

With the development of improved high-throughput techniques, the identification of all proteins that are present in a cell has come within reach, although no comprehensive proteome analysis has been reported to date. Such an approach requires sophisticated protein fractionation techniques and high-sensitivity mass spectrometry (MS) equipment (reviewed in Aebersold and Mann 2003). The standard protein profiling technologies currently have two major shortcomings. First, only the most abundant proteins can be detected in a complex protein mixture – masking the detection of low abundance proteins – and second, high-throughput analysis is restricted to proteins and peptides found in protein databases. As a consequence, low abundance proteins, as well as proteins with unusual or alternative splice sites, with erroneous or false gene prediction, or with posttranslational modifications, will not be identifiable by routine analyses. Despite these drawbacks, protein profiling has been recognised as an important step towards a system-level analysis of a cell, a tissue or an organism, and efforts are currently underway to develop methods to circumvent current difficulties. In particular, the dependence on databases has received much attention over the last few years and several tools for the database-independent detection of peptides from high-throughput MS data have been developed (Fischer et al. 2005; Frank and Pevzner 2005; Grossmann et al. 2005).

With a number of improved MS-data analysis tools in place, an improved coverage of proteome analyses can be expected. So far, however, most reported proteome analyses have reached only very limited coverage of the full proteome (see also below). The depth of a proteome analysis is a critical parameter indicating its comprehensiveness, especially if it aims at a complete description of the full proteome of a cell. Full proteome analyses that do not employ sophisticated protein fractionation strategies prior to MS-analyses usually do not achieve a substantial depth. The largest scale study to date that aimed to describe the complete *A. thaliana* proteome employed 2-dimensional gel electrophoresis (2-DE) prior to matrix-assisted laser desorption ionization–time-of-flight (MALDI-TOF) identification of proteins via peptide mass fingerprinting (Giavalisco et al. 2005). The authors reported the identification of 2,943 spots, which originated from only 663 different genes, even though they used different tissues to increase the diversity among the protein fractions. This is a small number considering the number of predicted proteins in a cell, which is more than 35,000 for Arabidopsis. A slightly better coverage was recently reported from electrospray ionization quadrupole time-of-flight (ESI-Q-TOF) experiments, in which the authors identified 1,032 proteins by a combination of protein fractionation with multi-dimensional monolithic reversed-phase peptide chromatography coupled on-line with MS (Wienkoop et al. 2004). The low proteome coverage achieved in these studies is a result of the repeated detection of abundant proteins and their peptides, while low abundance proteins are not detectable.

More successful large-scale proteome analyses with Arabidopsis have been performed with isolated organelles, membrane systems or subcellular structures (reviewed in Peck 2005). These studies included the analysis of chloroplasts (Peltier et al. 2002; Ferro et al. 2003; Froehlich et al. 2003; Friso et al. 2004; Huber et al. 2004; Kleffmann et al. 2004; Peltier et al. 2004; Baginsky et al. 2005; Vidi et al. 2006; Ytterberg et al. 2006), mitochondria (Brugiere et al. 2004; Heazlewood et al. 2004; Lister et al. 2004; Heazlewood and Millar 2005), peroxisomes (Fukao et al. 2002), the nucleolus (Pendle et al. 2005), vacuoles (Carter et al. 2004; Shimaoka et al. 2004; Endler et al. 2006; Jaquinod et al. 2006), the plasma membrane (Alexandersson et al. 2004; Marmagne et al. 2004; Sazuka et al. 2004; Borner et al. 2005; Jaquinod et al. 2006), the cell wall (Chivasa et al. 2002; Borderies et al. 2003; Boudart et al. 2005), and cytosolic ribosomes (Chang et al. 2005). In addition to cell organelles, initial data on the Arabidopsis pollen proteome were recently published (Holmes-Davis et al. 2005; Noir et al. 2005). Proteome analyses with isolated cell organelles uncover the sub-cellular localisation of proteins that cannot reliably be inferred from genome sequences. The prediction of the sub-cellular localisation of an enzyme from DNA sequences is complicated by sensitivity and specificity constraints of protein targeting prediction software and by potentially unreliable gene prediction. A novel approach to uncover sub-cellular protein localisation was taken recently by Dunkley and colleagues, who used relative protein quantification with mass spectrometry – a technique known as LOPIT – to infer protein localisation by following the enrichment of proteins during organelle isolation (Dunkley et al. 2006). Although promising, their results are currently restricted to proteins from the endo-membrane system, and provided a reliable localisation assessment for only 689 proteins. New insights into protein localisation and targeting make sub-cellular proteome analyses valuable, as a knowledge of protein localisation is important in understanding metabolic pathway compartmentalisation within the cell.

Not surprisingly, unexpected sub-cellular protein targeting routes were recently established using proteomics data for the design of targeted experiments. For example, proteome analyses with isolated chloroplasts uncovered the fact that many proteins without a predictable N-terminal transit peptide enter the chloroplast (Friso et al. 2004; Kleffmann et al. 2004). In one report, the plastid targeting signal is positioned neither at the N- nor at the C-terminus, but rather in stretches of internal amino acid sequences (Miras et al. 2002). More recently, proteins were found to be imported into the chloroplast via the secretory pathway, an exceptionally surprising finding that could explain why some chloroplast proteins are glycosylated (Villarejo et al. 2005). Altogether, existing proteomics data have opened up new avenues in our understanding of intracellular protein targeting and illustrate that some basic cellular concepts are still unknown.

With organelle proteomics as a new source of information about protein localisation, several databases were recently launched that integrate experimental data with targeting prediction. In some instances, novel algorithms with improved performance were designed and benchmarked against different proteomics datasets. Nair and Rost (2005) report a hierarchical prediction system that combines support vector machines with other prediction methods. The novelty of their prediction is its hierarchical nature, i.e. the sub-cellular localisation of a protein is predicted by mimicking the cellular sorting system. The new prediction tool (called LOCtree) was benchmarked against large-scale proteome datasets, and an improved localisation prediction for chloroplast proteins was achieved (Kleffmann et al. 2004; Nair and Rost 2005). In addition to novel prediction algorithms, two sub-cellular protein localisation databases that provide information about Arabidopsis proteins were recently released. A general eukaryotic sub-cellular localisation database (eSLDB) assigns sub-cellular localisation to a protein by integrating three different sources of information: experimental evidence, homology-based localisation and gene structure-based software prediction (Pierleoni et al. 2007). The Arabidopsis-specific database named SUBA (subcellular location database for Arabidopsis proteins) integrates information from high-throughput green fluorescent protein (GFP) localisation assays with other localisation information from the literature such as proteomics data (Heazlewood et al. 2007), and is a further development of the Arabidopsis Mitochondrial Protein Database (AMPDB) (Heazlewood and Millar 2005). Specific plastid protein databases such as PPDB (Friso et al. 2004) and plprot (Kleffmann et al. 2006) assign a subcellular localisation to a protein on the basis of plastid proteomics data.

8.2 Analysis of Post-Translational Modifications

Most large scale proteomics analyses, such as those described above, were designed in a way that does not generally allow the identification of unexpected post-translational modifications or unusual peptides. Although it is possible to identify post-translational modifications by a characteristic mass shift of peptides, such an analysis requires specific database search parameters that are often not used for standard database searches. Furthermore, in many cases the modification decreases the ionisation efficiency of peptides in ESI such that modified peptides will be undetectable in complex mixtures. Exceptions are studies that specifically analysed particular post-translational protein modifications in a targeted analysis. In most instances, the chemical characteristics of the post-translational modifications were used for their enrichment prior to MS/MS analysis. This strategy has been successfully employed for the large-scale detection of phosphoproteins from the plasma membrane of *A. thaliana* cultured cells. Phosphopeptides are enriched by metal ion affinity chromatography (IMAC) and analysed by MS/MS analysis, which reveals information about the exact site of phosphorylation. This comprehensive information was used to assemble a phosphorylation site database (Nuhse et al. 2004). Using the same IMAC method of enrichment of phosphopeptides, de la Fuente and colleagues analysed the phosphoproteome of Arabidopsis leaf extracts (de la Fuente van Bentem et al. 2006). Remarkably, they found that many proteins with a function in RNA metabolism are phosphorylated, suggesting that affinity enrichment of phosphopeptides is an efficient strategy for the enrichment of low abundance proteins from regulatory pathways (de la Fuente van Bentem et al. 2006). Another example for the affinity enrichment of proteins with a specific post-translational modification is the selective isolation of GPI-anchored proteins from the plasma membrane. Here, membrane vesicles were enriched and GPI-anchored proteins specifically released from the membrane by phosphatidylinositol phospholipase C treatment (Borner et al. 2003; Elortza et al. 2006).

In recent years, a remarkable number of redox proteomics studies have been published, most of which have set out to analyse the dithiol/disulfide exchange reactions at cysteines. Redox modulation triggers a number of important signaling cascades that coordinate responses to biotic and abiotic stresses in plants. For example, redox-proteome analyses have focused on the identification of S-nitrosylated proteins (Lindermayr et al. 2005) and the systematic assessment of oxidative stress-induced S-glutathionylation (Dixon et al. 2005). Independently of stress, the reversible dithiol/disulfide exchange at cysteines also plays an important role in the regulation of enzyme activity under normal growth conditions, at both the metabolic and the regulatory level (reviewed in Schurmann, 2003; Gelhaye et al. 2005). For example, thioredoxin (TRX) regulates the activity of enzymes involved in carbohydrate metabolism because its redox status depends on the activity of the photosynthetic electron transport chain (Schurmann 2003). To search for targets of TRX-mediated redox regulation, Marchand and colleagues constructed an affinity column with immobilised thioredoxin to capture soluble proteins from Arabidopsis leaf extracts that specifically interact with TRX. Several proteins were found to bind to the TRX-column, amongst them both known and novel TRX targets from different metabolic and regulatory pathways, including the Calvin cycle, energy metabolism, photosynthesis, protein folding, defence against oxidative stress and amino acid synthesis (Marchand et al. 2004).

8.3 Establishing Protein/Protein Interactions

Cellular metabolism is regulated not only by controlling expression levels and fine-tuning the catalytic activity of individual enzymes, but also by the interaction of cooperating enzymes in macromolecular complexes. Such an organisation is beneficial for the cell, because it allows tight stoichiometric control of all components, and channels metabolites within one enzyme complex by preventing the diffusion of pathway intermediates. Given the advantages of complex assembly for cellular metabolism, it is not surprising that a large number of protein complexes involved in metabolite channeling (so called metabolons) have recently been identified – the reader is referred to the excellent review by Winkel (2004) for an overview. Although a number of such metabolons have been well characterised, it can be expected that many metabolite channeling units remain to be discovered. This is particularly true for those metabolons that form only transiently or whose assembly is regulated in response to prevailing conditions. Analysis of such dynamics is a challenging task, requiring complex biochemical tools. A well established but targeted strategy to identify interaction partners for specific proteins and thus specific metabolite-channeling units, is co-immunoprecipitation and subsequent identification of co-purifying proteins by MS. For this purpose, high quality antibodies are required, which precludes the use of immunoprecipitation in high-throughput protein/protein interaction studies.

So far only a few high-throughput proteomics analyses that attempted to systematically resolve native protein complexes have been reported in Arabidopsis. The method of choice for the native purification of protein complexes is Blue-Native PAGE, which was originally employed to resolve the subunit composition of mitochondrial respiratory complexes in Arabidopsis (reviewed in Millar et al. 2005). Native PAGE was also used to analyse the oligomeric state of proteins in the chloroplast stroma, where many of the identified proteins were found to form homo-oligomers (Peltier et al. 2006). Although native biochemical purification of protein complexes clearly provides some initial insights into complex association, the common presence of contamination with co-purifying proteins that are not associated with the complex in vivo requires that the effective complex-association of the identified proteins be verified using additional tools.

Probably the most promising technology for the detection of true protein/protein interactions is the tandem affinity purification (TAP)-tagging strategy, which has been specifically designed to reduce the rate of co-purification of contaminating proteins with protein complexes, and as such decreases the false positive detection rate compared to other native purification strategies (reviewed in Shevchenko et al. 2002; Ghaemmaghami et al. 2003; Gould et al. 2004). For a TAP-tagging experiment, the protein of interest is tagged with an affinity tag, consisting of two IgG-binding domains and a calmodulin-binding peptide, separated by a cleavage site from a tobacco etch virus (TEV) protein. With this tag the targeted purification of the protein of interest with its cellular interaction partners is possible using two affinity purification steps in tandem. Although the strategy was designed to identify interaction partners with a high degree of purity, a comparison of different datasets obtained from yeast showed that the reliability of the data is still rather limited. In plants, the TAP-tagging strategy has been employed to date only in small-scale experiments to decipher multi-protein complexes involved in fatty acid biosynthesis in *A. thaliana* (Brown et al. 2006). Although high-throughput TAP-tagging strategies have not yet been reported for higher plants, methods that are specifically adapted to the plant system are under development. For example, an improved TAP-tagging protocol, in which the authors replaced the TEV cleavage site between the two tags with the more specific and low-temperature active rhinovirus 3C protease site, was recently reported for Arabidopsis (Rubio et al. 2005). In addition, some improvements on the second purification step have been introduced by modifying the vector construct, which now includes different affinity tags e.g. a hexahistidine tag or c-Myc tags (reviewed in Chang 2006).

8.4 Revealing Proteome Dynamics by Relative and Absolute Protein Quantification

In addition to a description of the static proteome of a cell or an organelle by standard protein profiling techniques, several studies have reported the analysis of Arabidopsis proteome dynamics in response to a perturbation. Such an analysis provides information about changes in protein abundance or post-translational modification in response to a stimulus, thus directly linking specific proteins with a stimulus-triggered response of the cell. Reports of successful analyses of proteome dynamics in Arabidopsis include investigations into changes in the expression of soluble proteins at elevated carbon dioxide levels (Bae et al. 2006), the response of chloroplast proteins to high light stress (Phee et al. 2004), the identity of differentially expressed proteins upon potassium deficiency (Kang et al. 2004), proteome adaptations to cold stress (Amme et al. 2006), the induction of proteins upon osmotic stress (Lee et al. 2004; Ndimba et al. 2005), the response of the chloroplast lumen proteome to cold stress (Goulas et al. 2006), the mechanisms of seed dormancy (Chibani et al. 2006), and resistance mechanisms against biotic and abiotic stresses (Hancock et al. 2005; Oh et al. 2005). Although such proteome analyses provide important data, most studies conducted so far do not provide comprehensive information because the acquisition of quantitative proteome data limits the use of protein fractionation prior to identification. For example, only 49 proteins were found to change in abundance in response to red, far-red and blue light when the full proteome of Arabidopsis was analysed under the different light regimes (Kim et al. 2006). Another study correlated the endogenous auxin content of root cells with dynamic proteome patterns. Altogether, 11 proteins whose abundance changed with changing auxin levels were identified (Sorin et al. 2006).

As mentioned above, the main reason for the low coverage in the above examples is the limited applicability of protein fractionation techniques in quantitative proteome analyses, because every protein fractionation step is a source of error that makes relative protein quantification unreliable. In order to circumvent this constraint, several stable isotope labelling techniques that facilitate the labelling of proteins with different isotope tags have been developed. As a consequence, proteins from different sources can be mixed prior to fractionation because the different isotope tags guarantee that peptides from different sources are distinguishable in later MS analysis. Because fractionation is performed on both samples simultaneously, and handling affects both samples to exactly the same extent, quantification errors are minimised. The best established techniques for relative protein quantification are isotope-coded affinity tag (ICAT) and isobaric tag for relative and absolute quantitation (iTRAQ), both stable isotope-based labelling techniques that allow the labelling of peptides at cysteine residues (ICAT) or at primary amines (iTRAQ) (Gygi et al. 1999a; Ross et al. 2004). The iTRAQ method can be multiplexed and up to four samples can be compared in one run, opening up an opportunity for kinetic or time series experiments.

Despite the great promise of stable isotope labelling techniques for relative quantification of proteins in large-scale proteomics experiments, only two reports that employed iTRAQ quantification for the analysis of Arabidopsis proteome dynamics are available. Jones and colleagues studied the early changes in the phosphoproteome of *A. thaliana* during the defence response to *Pseudomonas syringae* by enriching phosphorylated proteins and subsequently labeling the peptides with iTRAQ reagent (Jones et al. 2006a). Five proteins that showed reproducible differences between a control and three different bacterial challenges were identified, suggesting that they are potentially phosphorylated to constitute the plant basal defence response. A slightly different application of iTRAQ was employed by Rudella and colleagues, who excised gel spots from native 2-D gels containing several subunits of the Clp core complex and labeled the extracted peptides with iTRAQ reagents. The relative quantification suggested a different complex composition in CLPR2 mutants (Rudella et al. 2006).

The range of stable isotope techniques for relative protein quantification was recently expanded by labeling techniques for plant systems. One approach uses $^{15}N\text{-}KNO_3$ for the labeling of proteins, which is an effective strategy because plant cells reduce nitrogen from nitrate and use it for the synthesis of amino acids. When these amino acids are incorporated into proteins, the newly synthesised proteins will differ in mass from unlabelled proteins by a mass shift that depends on the number of incorporated ^{15}N-atoms (Engelsberger et al. 2006). In parallel to metabolic labeling, the first attempts at stable isotope labeling with amino acids in cell culture (SILAC) were recently reported for *A. thaliana* (Gruhler et al. 2005). Such a labeling strategy was previously thought to be unsuitable for plant cells because of their capability to synthesise amino acids, which has been expected to dilute the incorporation of labeled amino acids into proteins at an uncontrolled rate. Gruhler and colleagues have carefully evaluated the SILAC strategy for Arabidopsis cultured cells, and found ^{13}C-arginine labeling to provide consistent results for reliable relative quantification (Gruhler et al. 2005). The authors argue that SILAC can be applied to plant cells because amino acid synthesis is feedback-regulated by the pathway end product. In this way, feeding of amino acids will shut down the de novo synthesis of this particular amino acid, thus resulting in quantitative incorporation of the labeled amino acid into proteins (Gruhler et al. 2005).

Recently, a concept for the absolute quantification of low abundance proteins by high sensitivity mass spectrometry was introduced (known as AQUA; Gerber et al. 2003; Kirkpatrick et al. 2005). This method relies on synthetic internal standard peptides that are spiked at known concentrations into protein fractions of interest. The synthetic peptide has properties identical to a peptide produced during proteolysis of the target protein, except that it has been synthesised in the presence of stable isotopes, making it distinguishable from the native peptide. A targeted analysis of a complex protein sample with scanning mass spectrometers utilising, for example, the "multiple reaction monitoring (MRM)" mode allows detection of both the native peptide and the isotope-labeled standard peptide. The ion count read-back ratio of native and spiked peptide is then used to calculate the absolute quantity of the native peptide. Such an approach has been employed to determine the absolute amount of a low-abundance sucrose synthase isoform in a complex Arabidopsis protein extract (Wienkoop and Weckwerth 2006). Using a standard curve, the authors determined the absolute abundance of sucrose synthase to be 4.5 fmol $g^{-}1$ protein mixture, which equals 2.25 fmol mg^{-1} fresh weight. Although at present in its infancy, absolute quantification by AQUA is a promising approach and probably represents the way ahead for quantitative proteomics.

8.5 Towards a Complete Proteome Map for Arabidopsis

Despite significant advances in MS and protein/peptide fractionation technology, a description of the full proteome of an organism remains a challenging task. One of the great challenges in proteome analyses is the wide dynamic range of protein concentrations in a cell, which by far exceeds the capacity of any mass spectrometer available to date. The most up-to-date mass spectrometers can handle a dynamic range of 3–5 orders of magnitude. The range of protein concentrations in a cell, however, is several orders of magnitude more than this. Therefore, a valid strategy to cope with this challenge is peptide or protein fractionation prior to MS analysis to diminish the dynamic range limitations. New fractionation tools are being developed and it is impossible to highlight them all. MudPit remains the method of choice for peptide fractionation prior to the analysis of complex mixtures by MS/MS (Washburn et al. 2001). Protein fractionation can be achieved by classical high performance liquid chromatography (HPLC) or liquid chromatography (LC) fractionation strategies that may include generic strategies separating proteins by isoelectric point (e.g. free flow electrophoresis), charge, size, hydrophobicity or affinity for specific ligands (e.g. TRX-targets or nucleotide-binding proteins). An additional route towards full proteome coverage is the application of the aforementioned protein fractionation steps to diverse protein fractions from different tissues under different developmental conditions.

It is quite obvious that such a challenge is best tackled as a community effort, and several initiatives for other organisms such as yeast or humans have formed to collect, integrate and disseminate data from and for interested researchers. Several publicly available databases exist for the purpose of capturing and disseminating proteomics MS data. Such databases include the Proteomics Identifications database (PRIDE; Jones et al. 2006b), the Global Proteome Machine Database (gpmDB; Craig et al. 2006), The Institute of Systems Biology's Peptide Atlas (Desiere et al. 2006), and the University of Texas' Open Proteomics Database (opd; Prince et al. 2004). Especially in cases where genome annotation is to be improved by identified and validated peptides, the Peptide Atlas platform has proven very useful for yeast and might also be of use for the Arabidopsis proteomics community (Desiere et al. 2005). An open source platform must ascertain the consistent quality of the data submitted in order to be useful. In the existing Peptide Atlas, this is currently achieved by a standardised data analysis pipeline that includes statistic tools such as PeptideProphet (Keller et al. 2002; Nesvizhskii et al. 2003) to estimate false positive identification rates. In the mid term, such automated pipelines must include tools for the database-independent detection of proteins from MS data in order to increase proteome coverage by those proteins that are not correctly represented in existing gene models and those that carry unpredicted post-translational modifications. We are convinced that this strategy is promising, that it will serve the community by building better tools for functional genomics, and that it will finally increase our understanding of biological systems.

8.6 Conclusions

Plant proteomics is important, because plants are an essential food source for humans and animals, and because a thorough understanding of plant biology is necessary to secure food supply in a changing environment. Studies at the protein level are important to reveal the molecular mechanisms underlying plant growth, development, and interactions with the environment. The analysis of proteins provides several dimensions of information that are inaccessible from the analysis of other cellular components. These include protein abundance, state of modification, subcellular localisation, three-dimensional structure, and their association with each other and/or with biomolecules of different types. Proteomics has furthermore become an indispensable source of information about protein expression, splice variants, and erroneous or incomplete prediction of gene structures in databases. Although currently mostly restricted to *A. thaliana*, other plant species are coming within the focus of proteome research, especially rice and poplar. For these plants, one can expect that proteomics data will be acquired for systems biology approaches that combine and integrate quantitative data from transcriptional profiling, proteomics and metabolomics into robust models of plant metabolism. At the same time, the development of improved tools for MS/MS data analysis as detailed above has made other plant species, especially those relevant to the human diet, amenable to proteomics analyses. An excellent example of such an approach is the recent characterisation of an allergenic protein from a red and a white strawberry variety with a combination of MS and de novo sequencing (Hjerno et al. 2006). This latter study illustrates one future direction of plant proteomics: a diversification of plant species used for proteomics and the increased application of MS-based tools for the characterisation of plant-based foods.

Acknowledgements We would like to apologise to all colleagues whose contributions were not cited in this review due to space constraints. The authors are indebted to Dr. Katja Baerenfaller for critical reading of the manuscript and to Franz Roos and Jonas Grossmann for help with the literature research. The work in the authors' groups was funded by the ETH and grants from the SEP initiative to W.G. and S.B.

References

Aebersold R, Mann M (2003) Mass spectrometry-based proteomics. Nature 422:198–207

Alexandersson E, Saalbach G, Larsson C, Kjellbom P (2004) Arabidopsis plasma membrane proteomics identifies components of transport, signal transduction and membrane trafficking. Plant Cell Physiol 45:1543–1556

Amme S, Matros A, Schlesier B, Mock HP (2006) Proteome analysis of cold stress response in *Arabidopsis thaliana* using DIGE-technology. J Exp Bot 57:1537–1546

Bae H, Kim MS, Sicher RC, Bae HJ, Bailey BA (2006) Necrosis- and ethylene-inducing peptide from *Fusarium oxysporum* induces a complex cascade of transcripts associated with signal transduction and cell death in Arabidopsis. Plant Physiol 141:1056–1067

Baginsky S, Kleffmann T, von Zychlinski A, Gruissem W (2005) Analysis of shotgun proteomics and RNA profiling data from *Arabidopsis thaliana* chloroplasts. J Proteome Res 4:637–640

Bino RJ, Hall RD, Fiehn O, Kopka J, Saito K, Draper J, Nikolau BJ, Mendes P, Roessner-Tunali U, Beale MH, Trethewey RN, Lange BM, Wurtele ES, Sumner LW (2004) Potential of metabolomics as a functional genomics tool. Trends Plant Sci 9:418–425

Borderies G, Jamet E, Lafitte C, Rossignol M, Jauneau A, Boudart G, Monsarrat B, Esquerre-Tugaye MT, Boudet A, Pont-Lezica R (2003) Proteomics of loosely bound cell wall proteins of *Arabidopsis thaliana* cell suspension cultures: a critical analysis. Electrophoresis 24:3421–3432

Borner GH, Lilley KS, Stevens TJ, Dupree P (2003) Identification of glycosylphosphatidylinositol-anchored proteins in Arabidopsis. A proteomic and genomic analysis. Plant Physiol 132:568–577

Borner GH, Sherrier DJ, Weimar T, Michaelson LV, Hawkins ND, Macaskill A, Napier JA, Beale MH, Lilley KS, Dupree P (2005) Analysis of detergent-resistant membranes in Arabidopsis. Evidence for plasma membrane lipid rafts. Plant Physiol 137:104–116

Boudart G, Jamet E, Rossignol M, Lafitte C, Borderies G, Jauneau A, Esquerre-Tugaye MT, Pont-Lezica R (2005) Cell wall proteins in apoplastic fluids of *Arabidopsis thaliana* rosettes: identification by mass spectrometry and bioinformatics. Proteomics 5:212–221

Brown AP, Affleck V, Fawcett T, Slabas AR (2006) Tandem affinity purification tagging of fatty acid biosynthetic enzymes in *Synechocystis* sp. PCC6803 and *Arabidopsis thaliana*. J Exp Bot 57:1563–1571

Brugiere S, Kowalski S, Ferro M, Seigneurin-Berny D, Miras S, Salvi D, Ravanel S, d'Herin P, Garin J, Bourguignon J, Joyard J, Rolland N (2004) The hydrophobic proteome of mitochondrial membranes from Arabidopsis cell suspensions. Phytochemistry 65:1693–1707

Carter C, Pan S, Zouhar J, Avila EL, Girke T, Raikhel NV (2004) The vegetative vacuole proteome of *Arabidopsis thaliana* reveals predicted and unexpected proteins. Plant Cell 16:3285–3303

Chang IF (2006) Mass spectrometry-based proteomic analysis of the epitope-tag affinity purified protein complexes in eukaryotes. Proteomics 6:6158–6166
Chang IF, Szick-Miranda K, Pan S, Bailey-Serres J (2005) Proteomic characterization of evolutionarily conserved and variable proteins of Arabidopsis cytosolic ribosomes. Plant Physiol 137:848–862
Chibani K, Ali-Rachedi S, Job C, Job D, Jullien M, Grappin P (2006) Proteomic analysis of seed dormancy in Arabidopsis. Plant Physiol 142:1493–1510
Chivasa S, Ndimba BK, Simon WJ, Robertson D, Yu XL, Knox JP, Bolwell P, Slabas AR (2002) Proteomic analysis of the *Arabidopsis thaliana* cell wall. Electrophoresis 23:1754–1765
Craig R, Cortens JC, Fenyo D, Beavis RC (2006) Using annotated peptide mass spectrum libraries for protein identification. J Proteome Res 5:1843–1849
De la Fuente van Bentem S, Anrather D, Roitinger E, Djamei A, Hufnagl T, Barta A, Csaszar E, Dohnal I, Lecourieux D, Hirt H (2006) Phosphoproteomics reveals extensive in vivo phosphorylation of Arabidopsis proteins involved in RNA metabolism. Nucleic Acids Res 34:3267–3278
Desiere F, Deutsch EW, Nesvizhskii AI, Mallick P, King NL, Eng JK, Aderem A, Boyle R, Brunner E, Donohoe S, Fausto N, Hafen E, Hood L, Katze MG, Kennedy KA, Kregenow F, Lee H, Lin B, Martin D, Ranish JA, Rawlings DJ, Samelson LE, Shiio Y, Watts JD, Wollscheid B, Wright ME, Yan W, Yang L, Yi EC, Zhang H, Aebersold R (2005) Integration with the human genome of peptide sequences obtained by high-throughput mass spectrometry. Genome Biol 6:R9
Desiere F, Deutsch EW, King NL, Nesvizhskii AI, Mallick P, Eng J, Chen S, Eddes J, Loevenich SN, Aebersold R (2006) The PeptideAtlas project. Nucleic Acids Res 34:D655–658
Dixon DP, Skipsey M, Grundy NM, Edwards R (2005) Stress-induced protein S-glutathionylation in Arabidopsis. Plant Physiol 138:2233–2244
Dunkley TP, Hester S, Shadforth IP, Runions J, Weimar T, Hanton SL, Griffin JL, Bessant C, Brandizzi F, Hawes C, Watson RB, Dupree P, Lilley KS (2006) Mapping the Arabidopsis organelle proteome. Proc Natl Acad Sci USA 103:6518–6523
Elortza F, Mohammed S, Bunkenborg J, Foster LJ, Nuhse TS, Brodbeck U, Peck SC, Jensen ON (2006) Modification-specific proteomics of plasma membrane proteins: identification and characterization of glycosylphosphatidylinositol-anchored proteins released upon phospholipase D treatment. J Proteome Res 5:935–943
Endler A, Meyer S, Schelbert S, Schneider T, Weschke W, Peters SW, Keller F, Baginsky S, Martinoia E, Schmidt UG (2006) Identification of a vacuolar sucrose transporter in barley and Arabidopsis mesophyll cells by a tonoplast proteomic approach. Plant Physiol 141:196–207
Engelsberger WR, Erban A, Kopka J, Schulze WX (2006) Metabolic labeling of plant cell cultures with K15NO3 as a tool for quantitative analysis of proteins and metabolites. Plant Methods 2:14
Ferro M, Salvi D, Brugiere S, Miras S, Kowalski S, Louwagie M, Garin J, Joyard J, Rolland N (2003) Proteomics of the chloroplast envelope membranes from *Arabidopsis thaliana*. Mol Cell Proteomics 2:325–345
Fischer B, Roth V, Roos F, Grossmann J, Baginsky S, Widmayer P, Gruissem W, Buhmann JM (2005) NovoHMM: a hidden Markov model for de novo peptide sequencing. Anal Chem 77:7265–7273
Frank A, Pevzner P (2005) PepNovo: de novo peptide sequencing via probabilistic network modeling. Anal Chem 77:964–973
Friso G, Giacomelli L, Ytterberg AJ, Peltier JB, Rudella A, Sun Q, Wijk KJ (2004) In-depth analysis of the thylakoid membrane proteome of *Arabidopsis thaliana* chloroplasts: new proteins, new functions, and a plastid proteome database. Plant Cell 16:478–499
Froehlich JE, Wilkerson CG, Ray WK, McAndrew RS, Osteryoung KW, Gage DA, Phinney BS (2003) Proteomic study of the *Arabidopsis thaliana* chloroplastic envelope membrane utilizing alternatives to traditional two-dimensional electrophoresis. J Proteome Res 2:413–425
Fukao Y, Hayashi M, Nishimura M (2002) Proteomic analysis of leaf peroxisomal proteins in greening cotyledons of *Arabidopsis thaliana*. Plant Cell Physiol 43:689–696
Gelhaye E, Rouhier N, Navrot N, Jacquot JP (2005) The plant thioredoxin system. Cell Mol Life Sci 62:24–35

Gerber SA, Rush J, Stemman O, Kirschner MW, Gygi SP (2003) Absolute quantification of proteins and phosphoproteins from cell lysates by tandem MS. Proc Natl Acad Sci USA 100:6940–6945

Ghaemmaghami S, Huh WK, Bower K, Howson RW, Belle A, Dephoure N, O'Shea EK, Weissman JS (2003) Global analysis of protein expression in yeast. Nature 425:737–741

Giavalisco P, Nordhoff E, Kreitler T, Kloppel KD, Lehrach H, Klose J, Gobom J (2005) Proteome analysis of *Arabidopsis thaliana* by two-dimensional gel electrophoresis and matrix-assisted laser desorption/ionisation-time of flight mass spectrometry. Proteomics 5:1902–1913

Goulas E, Schubert M, Kieselbach T, Kleczkowski LA, Gardestrom P, Schroder W, Hurry V (2006) The chloroplast lumen and stromal proteomes of *Arabidopsis thaliana* show differential sensitivity to short- and long-term exposure to low temperature. Plant J 47:720–734

Gould KL, Ren L, Feoktistova AS, Jennings JL, Link AJ (2004) Tandem affinity purification and identification of protein complex components. Methods 33:239–244

Greenbaum D, Colangelo C, Williams K, Gerstein M (2003) Comparing protein abundance and mRNA expression levels on a genomic scale. Genome Biol 4:117

Griffin TJ, Gygi SP, Ideker T, Rist B, Eng J, Hood L, Aebersold R (2002) Complementary profiling of gene expression at the transcriptome and proteome levels in *Saccharomyces cerevisiae*. Mol Cell Proteomics 1:323–333

Grossmann J, Roos FF, Cieliebak M, Liptak Z, Mathis LK, Muller M, Gruissem W, Baginsky S (2005) AUDENS: a tool for automated peptide de novo sequencing. J Proteome Res 4:1768–1774

Gruhler A, Schulze WX, Matthiesen R, Mann M, Jensen ON (2005) Stable isotope labeling of *Arabidopsis thaliana* cells and quantitative proteomics by mass spectrometry. Mol Cell Proteomics 4:1697–1709

Gygi SP, Rist B, Gerber SA, Turecek F, Gelb MH, Aebersold R (1999a) Quantitative analysis of complex protein mixtures using isotope-coded affinity tags. Nat Biotechnol 17:994–999

Gygi SP, Rochon Y, Franza BR, Aebersold R (1999b) Correlation between protein and mRNA abundance in yeast. Mol Cell Biol 19:1720–1730

Hack CJ (2004) Integrated transcriptome and proteome data: the challenges ahead. Brief Funct Genomic Proteomic 3:212–219

Hancock JT, Henson D, Nyirenda M, Desikan R, Harrison J, Lewis M, Hughes J, Neill SJ (2005) Proteomic identification of glyceraldehyde 3-phosphate dehydrogenase as an inhibitory target of hydrogen peroxide in Arabidopsis. Plant Physiol Biochem 43:828–835

Heazlewood JL, Tonti-Filippini JS, Gout AM, Day DA, Whelan J, Millar AH (2004) Experimental analysis of the Arabidopsis mitochondrial proteome highlights signaling and regulatory components, provides assessment of targeting prediction programs, and indicates plant-specific mitochondrial proteins. Plant Cell 16:241–256

Heazlewood JL, Millar AH (2005) AMPDB: the Arabidopsis Mitochondrial Protein Database. Nucleic Acids Res 33:D605–610

Heazlewood JL, Verboom RE, Tonti-Filippini J, Small I, Millar AH (2007) SUBA: the Arabidopsis Subcellular Database. Nucleic Acids Res 35:D213–218

Hjerno K, Alm R, Canback B, Matthiesen R, Trajkovski K, Bjork L, Roepstorff P, Emanuelsson C (2006) Down-regulation of the strawberry Bet v 1-homologous allergen in concert with the flavonoid biosynthesis pathway in colorless strawberry mutant. Proteomics 6:1574–1587

Holmes-Davis R, Tanaka CK, Vensel WH, Hurkman WJ, McCormick S (2005) Proteome mapping of mature pollen of *Arabidopsis thaliana*. Proteomics 5:4864–4884

Huber CG, Walcher W, Timperio AM, Troiani S, Porceddu A, Zolla L (2004) Multidimensional proteomic analysis of photosynthetic membrane proteins by liquid extraction–ultracentrifugation–liquid chromatography–mass spectrometry. Proteomics 4:3909–3920

Jaquinod M, Villiers F, Kieffer-Jaquinod S, Hugouvieux V, Bruley C, Garin J, Bourguignon J (2006) A proteomic dissection of *Arabidopsis thaliana* vacuoles isolated from cell culture. Mol Cell Proteomics 6:394–412

Jones AM, Bennett MH, Mansfield JW, Grant M (2006a) Analysis of the defence phosphoproteome of *Arabidopsis thaliana* using differential mass tagging. Proteomics 6:4155–4165

Jones P, Cote RG, Martens L, Quinn AF, Taylor CF, Derache W, Hermjakob H, Apweiler R (2006b) PRIDE: a public repository of protein and peptide identifications for the proteomics community. Nucleic Acids Res 34:D659–663

Kang JG, Pyo YJ, Cho JW, Cho MH (2004) Comparative proteome analysis of differentially expressed proteins induced by K+ deficiency in *Arabidopsis thaliana*. Proteomics 4:3549–3559

Keller A, Nesvizhskii AI, Kolker E, Aebersold R (2002) Empirical statistical model to estimate the accuracy of peptide identifications made by MS/MS and database search. Anal Chem 74:5383–5392

Kim DS, Cho DS, Park WM, Na HJ, Nam HG (2006) Proteomic pattern-based analyses of light responses in *Arabidopsis thaliana* wild-type and photoreceptor mutants. Proteomics 6:3040–3049

Kirkpatrick DS, Gerber SA, Gygi SP (2005) The absolute quantification strategy: a general procedure for the quantification of proteins and post-translational modifications. Methods 35:265–273

Kleffmann T, Russenberger D, von Zychlinski A, Christopher W, Sjolander K, Gruissem W, Baginsky S (2004) The *Arabidopsis thaliana* chloroplast proteome reveals pathway abundance and novel protein functions. Curr Biol 14:354–362

Kleffmann T, Hirsch-Hoffmann M, Gruissem W, Baginsky S (2006) plprot: a comprehensive proteome database for different plastid types. Plant Cell Physiol 47:432–436

Lee S, Lee EJ, Yang EJ, Lee JE, Park AR, Song WH, Park OK (2004) Proteomic identification of annexins, calcium-dependent membrane binding proteins that mediate osmotic stress and abscisic acid signal transduction in Arabidopsis. Plant Cell 16:1378–1391

Lindermayr C, Saalbach G, Durner J (2005) Proteomic identification of S-nitrosylated proteins in Arabidopsis. Plant Physiol 137:921–930

Lister R, Chew O, Lee MN, Heazlewood JL, Clifton R, Parker KL, Millar AH, Whelan J (2004) A transcriptomic and proteomic characterization of the Arabidopsis mitochondrial protein import apparatus and its response to mitochondrial dysfunction. Plant Physiol 134:777–789

Marchand C, Le Marechal P, Meyer Y, Miginiac-Maslow M, Issakidis-Bourguet E, Decottignies P (2004) New targets of Arabidopsis thioredoxins revealed by proteomic analysis. Proteomics 4:2696–2706

Marmagne A, Rouet MA, Ferro M, Rolland N, Alcon C, Joyard J, Garin J, Barbier-Brygoo H, Ephritikhine G (2004) Identification of new intrinsic proteins in Arabidopsis plasma membrane proteome. Mol Cell Proteomics 3:675–691

Millar AH, Heazlewood JL, Kristensen BK, Braun HP, Moller IM (2005) The plant mitochondrial proteome. Trends Plant Sci 10:36–43

Miras S, Salvi D, Ferro M, Grunwald D, Garin J, Joyard J, Rolland N (2002) Non-canonical transit peptide for import into the chloroplast. J Biol Chem 277:47770–47778

Nair R, Rost B (2005) Mimicking cellular sorting improves prediction of subcellular localization. J Mol Biol 348:85–100

Ndimba BK, Chivasa S, Simon WJ, Slabas AR (2005) Identification of Arabidopsis salt and osmotic stress responsive proteins using two-dimensional difference gel electrophoresis and mass spectrometry. Proteomics 5:4185–4196

Nesvizhskii AI, Keller A, Kolker E, Aebersold R (2003) A statistical model for identifying proteins by tandem mass spectrometry. Anal Chem 75:4646–4658

Noir S, Brautigam A, Colby T, Schmidt J, Panstruga R (2005) A reference map of the *Arabidopsis thaliana* mature pollen proteome. Biochem Biophys Res Commun 337:1257–1266

Nuhse TS, Stensballe A, Jensen ON, Peck SC (2004) Phosphoproteomics of the Arabidopsis plasma membrane and a new phosphorylation site database. Plant Cell 16:2394–2405

Oh IS, Park AR, Bae MS, Kwon SJ, Kim YS, Lee JE, Kang NY, Lee S, Cheong H, Park OK (2005) Secretome analysis reveals an Arabidopsis lipase involved in defense against *Alternaria brassicicola*. Plant Cell 17:2832–2847

Peck SC (2005) Update on proteomics in Arabidopsis. Where do we go from here? Plant Physiol 138:591–599

Peltier JB, Emanuelsson O, Kalume DE, Ytterberg J, Friso G, Rudella A, Liberles DA, Soderberg L, Roepstorff P, von Heijne G, van Wijk KJ (2002) Central functions of the lumenal and peripheral thylakoid proteome of Arabidopsis determined by experimentation and genome-wide prediction. Plant Cell 14:211–236

Peltier JB, Ytterberg AJ, Sun Q, van Wijk KJ (2004) New functions of the thylakoid membrane proteome of *Arabidopsis thaliana* revealed by a simple, fast, and versatile fractionation strategy. J Biol Chem 279:49367–49383

Peltier JB, Cai Y, Sun Q, Zabrouskov V, Giacomelli L, Rudella A, Ytterberg AJ, Rutschow H, van Wijk KJ (2006) The oligomeric stromal proteome of *Arabidopsis thaliana* chloroplasts. Mol Cell Proteomics 5:114–133

Pendle AF, Clark GP, Boon R, Lewandowska D, Lam YW, Andersen J, Mann M, Lamond AI, Brown JW, Shaw PJ (2005) Proteomic analysis of the Arabidopsis nucleolus suggests novel nucleolar functions. Mol Biol Cell 16:260–269

Phee BK, Cho JH, Park S, Jung JH, Lee YH, Jeon JS, Bhoo SH, Hahn TR (2004) Proteomic analysis of the response of Arabidopsis chloroplast proteins to high light stress. Proteomics 4:3560–3568

Pierleoni A, Martelli PL, Fariselli P, Casadio R (2007) eSLDB: eukaryotic subcellular localization database. Nucleic Acids Res 35:D208–212

Prince JT, Carlson MW, Wang R, Lu P, Marcotte EM (2004) The need for a public proteomics repository. Nat Biotechnol 22:471–472

Provart NJ, McCourt P (2004) Systems approaches to understanding cell signaling and gene regulation. Curr Opin Plant Biol 7:605–609

Ross PL, Huang YN, Marchese JN, Williamson B, Parker K, Hattan S, Khainovski N, Pillai S, Dey S, Daniels S, Purkayastha S, Juhasz P, Martin S, Bartlet-Jones M, He F, Jacobson A, Pappin DJ (2004) Multiplexed protein quantitation in *Saccharomyces cerevisiae* using amine-reactive isobaric tagging reagents. Mol Cell Proteomics 3:1154–1169

Rubio V, Shen Y, Saijo Y, Liu Y, Gusmaroli G, Dinesh-Kumar SP, Deng XW (2005) An alternative tandem affinity purification strategy applied to Arabidopsis protein complex isolation. Plant J 41:767–778

Rudella A, Friso G, Alonso JM, Ecker JR, van Wijk KJ (2006) Downregulation of ClpR2 leads to reduced accumulation of the ClpPRS protease complex and defects in chloroplast biogenesis in Arabidopsis. Plant Cell 18:1704–1721

Sazuka T, Keta S, Shiratake K, Yamaki S, Shibata D (2004) A proteomic approach to identification of transmembrane proteins and membrane-anchored proteins of *Arabidopsis thaliana* by peptide sequencing. DNA Res 11:101–113

Schurmann P (2003) Redox signaling in the chloroplast: the ferredoxin/thioredoxin system. Antioxid Redox Signal 5:69–78

Shevchenko A, Schaft D, Roguev A, Pijnappel WW, Stewart AF, Shevchenko A (2002) Deciphering protein complexes and protein interaction networks by tandem affinity purification and mass spectrometry: analytical perspective. Mol Cell Proteomics 1:204–212

Shimaoka T, Ohnishi M, Sazuka T, Mitsuhashi N, Hara-Nishimura I, Shimazaki K, Maeshima M, Yokota A, Tomizawa K, Mimura T (2004) Isolation of intact vacuoles and proteomic analysis of tonoplast from suspension-cultured cells of *Arabidopsis thaliana*. Plant Cell Physiol 45:672–683

Sorin C, Negroni L, Balliau T, Corti H, Jacquemot MP, Davanture M, Sandberg G, Zivy M, Bellini C (2006) Proteomic analysis of different mutant genotypes of Arabidopsis led to the identification of 11 proteins correlating with adventitious root development. Plant Physiol 140:349–364

Tian Q, Stepaniants SB, Mao M, Weng L, Feetham MC, Doyle MJ, Yi EC, Dai H, Thorsson V, Eng J, Goodlett D, Berger JP, Gunter B, Linseley PS, Stoughton RB, Aebersold R, Collins SJ, Hanlon WA, Hood LE (2004) Integrated genomic and proteomic analyses of gene expression in mammalian cells. Mol Cell Proteomics 3:960–969

Vidi PA, Kanwischer M, Baginsky S, Austin JR, Csucs G, Dormann P, Kessler F, Brehelin C (2006) Tocopherol cyclase (VTE1) localization and vitamin E accumulation in chloroplast plastoglobule lipoprotein particles. J Biol Chem 281:11225–11234

Villarejo A, Buren S, Larsson S, Dejardin A, Monne M, Rudhe C, Karlsson J, Jansson S, Lerouge P, Rolland N, von Heijne G, Grebe M, Bako L, Samuelsson G (2005) Evidence for a protein transported through the secretory pathway en route to the higher plant chloroplast. Nat Cell Biol 7:1124–1131

Washburn MP, Wolters D, Yates JR 3rd (2001) Large-scale analysis of the yeast proteome by multidimensional protein identification technology. Nat Biotechnol 19:242–247

Washburn MP, Koller A, Oshiro G, Ulaszek RR, Plouffe D, Deciu C, Winzeler E, Yates JR 3rd (2003) Protein pathway and complex clustering of correlated mRNA and protein expression analyses in *Saccharomyces cerevisiae*. Proc Natl Acad Sci USA 100:3107–3112

Wienkoop S, Glinski M, Tanaka N, Tolstikov V, Fiehn O, Weckwerth W (2004) Linking protein fractionation with multidimensional monolithic reversed-phase peptide chromatography/mass spectrometry enhances protein identification from complex mixtures even in the presence of abundant proteins. Rapid Commun Mass Spectrom 18:643–650

Wienkoop S, Weckwerth W (2006) Relative and absolute quantitative shotgun proteomics: targeting low-abundance proteins in *Arabidopsis thaliana*. J Exp Bot 57:1529–1535

Winkel BS (2004) Metabolic channeling in plants. Annu Rev Plant Biol 55:85–107

Ytterberg AJ, Peltier JB, van Wijk KJ (2006) Protein profiling of plastoglobules in chloroplasts and chromoplasts. A surprising site for differential accumulation of metabolic enzymes. Plant Physiol 140:984–997

Chapter 9
Proteomics of *Medicago truncatula*

Zhentian Lei, Satish Nagaraj, Bonnie S. Watson, and Lloyd W. Sumner

Abstract Legumes are unique in their ability to fix atmospheric nitrogen through symbiotic relationships with rhizobia resulting in high protein content in the plants and the portioning of nitrogen in the soil. As a result legumes have become a worldwide staple in both human and animal nutrition. Unfortunately, commercial legumes such as soybean and alfalfa have large complex genomes that make the direct molecular and genetic study of these species more challenging. As a result, *Medicago truncatula* has been adopted as a model species for studying legume biology. These studies now include a large number of detailed proteome analyses, which are reviewed in this chapter. Topics reviewed include proteomic approaches, systematic identification of tissue-specific proteomes, rhizobia interactions, nodulation, arbuscule mycorrhizal interactions, seed development, and embryogenesis. The impact of biotic and abiotic environmental stresses on the *M. truncatula* proteome are also reviewed, including responses to pathogenic interactions, sewage treatment, desiccation tolerance, methyl jasmonate elicitation, and yeast elicitation.

9.1 Introduction

Legumes (Fabaceae), one of the most economically important crop families in the world, consist of over 16,000 species representing about 650 genera (Hymowitz 1990), and are second only to Poaceae (grasses and cereals) in agricultural importance. Common legume commodities include soybean, alfalfa, faba bean, common bean,

Zhentian Lei, Plant Biology Division, The Samuel Roberts Noble Foundation, Ardmore, OK 73401, USA

Satish Nagaraj, Plant Biology Division, The Samuel Roberts Noble Foundation, Ardmore, OK 73401, USA

Bonnie S. Watson, Plant Biology Division, The Samuel Roberts Noble Foundation, Ardmore, OK 73401, USA

Lloyd W. Sumner, Plant Biology Division, The Samuel Roberts Noble Foundation, 2510 Sam Noble Parkway, Ardmore, OK 73401, USA, E-mail: lwsumner@noble.org

J. Šamaj and J. Thelen (eds.), *Plant Proteomics*
© Springer 2007

pea, lentil, and chickpea, which serve as major protein sources for both humans and animals and provide approximately one-half of the world's supply of oilseed. Legumes are planted on about 15% of the world's arable land (270–300 million hectares). In the United States alone, over 73 million acres of soybean (*Glycine max*) and 76 million acres of alfalfa (*Medicago sativa*) were cultivated in 2003, with estimated values of US $17.5 billion and US $7.5 billion, respectively (USDA-NASS 2004). In addition, legumes are also a unique source of natural products such as isoflavonoids, alkaloids, and saponins, many of which have documented antimicrobial, pharmacological, and/or neutraceutical properties (Dixon and Steele 1999; Dixon and Sumner 2003; Sarkar and Li 2003).

Legumes are characterized by their unique capability to fix atmospheric nitrogen through a symbiotic relationship with soil bacteria. These bacteria, collectively termed rhizobia, which include genera such as *Rhizobium*, *Sinorhizobium*, *Bradyrhizobium*, *Mesorhizobium*, and *Azorhizobium*, form specialized organs – nodules – in the roots of plants. Nitrogen fixation occurs in these nodules via the conversion of N_2 to NH_3 by bacterial nitrogenases (see also Chap. 19 by Sarma et al. in this volume). It is estimated that the symbiosis of legumes with soil bacteria fixes nearly 17 million metric tons of atmospheric nitrogen into plants each year, with a value of about US $8 billion. Thus, these beneficial interactions provide a plentiful supply of nitrogen to the plants, which in turn results in very high protein levels in legumes.

The mutualistic interactions between legumes and microbes represent an important aspect of plant biology, which cannot be studied using the model plant *Arabidopsis thaliana* since it is unable to establish root endosymbiosis with rhizobia. The use of agriculturally important legume crops such as soybean and alfalfa to dissect legume biology is complicated by their large and complex polyploid genomes, highly repetitive DNA, and outcrossing nature. Over the last decade, *Medicago truncatula* (common name barrel medic and a close relative to alfalfa) has emerged as a model system for molecular and genetic studies of legume biology (May and Dixon 2004). This is mainly because of its relatively small diploid genome [about 500 megabases (Mb) as compared to about 1,200 Mb for soybean and 3,000 Mb for the human genome], autogamous pollination with a short seed-to-seed generation time, and the ability to regenerate via somatic embryogenesis (Barker et al. 1990; Cook 1999; Trieu et al. 2000; Frugoli and Harris 2001). So far, 226,923 expressed sequence tags (ESTs) have been sequenced for *M. truncatula*. Furthermore, a physical map of the *M. truncatula* genome is now available (Thoquet et al. 2002; Choi et al. 2004), and sequencing of the gene-rich regions of the *M. truncatula* genome is ongoing at the University of Oklahoma and laboratories in the United Kingdom and France (http://www.medicago.org/genome/) (Young et al. 2005; Town 2006). As of December 2006, approximately 190 Mbp of euchromatin sequence has been acquired (http://www.medicago.org/genome/stats.php), representing approximately 25,000 predicted genes. At the same time, the collection of ESTs (http://compbio.dfci.harvard.edu/tgi/cgi-bin/tgi/gimain.pl?gudb=medicago) has increased to 18,238 non-redundant singleton ESTs and 30,000 non-redundant tentative consensus (TC) sequences (MtGI Release 8.0).

The EST sequences, together with the continuously emerging genomic resources, constitute a critical infrastructure that is further propelling *M. truncatula* forward as a model for the study of legume biology.

9.2 Proteomics: an Important Tool in Dissecting Legume Biology

Proteomics using two-dimensional electrophoresis (2-DE) and mass spectrometry (MS) has become an important tool in biological research. The 2-DE/MS-based proteomics approach involves the use of high-resolution 2-DE to separate, visualize, and quantify proteins, followed by the use of MS to qualitatively identify proteins. 2-DE separates proteins in the first dimension based upon isoelectric point (pI) and in the second dimension based upon molecular mass. The use of immobilized pH gradient (IPG) strips in isoelectric focusing has significantly improved 2-DE resolution and reproducibility (Görg et al. 1999). Second dimensional separation is typically performed using polyacrylamide gels (for review, see O'Farrell et al. 1977; Görg et al. 2000; Rabilloud 2002). 2-DE offers semi quantitative measurement of variations in protein expression patterns, including the presence/absence of proteins, migration shifts due to post-translational modifications or the alternative splicing of mRNA, and staining intensity due to differential amounts of the proteins. It has been the method of choice for separation and visualization of complex protein mixtures for several decades owing to continuous improvements in resolution and reproducibility. It has been reported that thousands of proteins can be visualized by a single 2-DE (O'Farrell et al. 1977; Klose 1999) Thus, 2-DE is still a highly competitive technical approach even when compared to more recent methods such as liquid chromatography coupled to tandem mass spectrometry (LC/MS/MS).

The introduction of fluorescent two-dimensional differential gel electrophoresis (2D-DIGE) represents an important breakthrough in 2-DE technology. 2D-DIGE uses two different fluorescent cyanine dyes, Cy3 and Cy5, to covalently label two different protein samples, i.e., control and treatment. The labeled control and treatment protein samples can then be mixed and electrophoresed together. The dyes posses distinct laser excitation and emission spectral properties that can be used to accurately quantify proteins in the two protein samples. DIGE has greatly enhanced 2-DE reproducibility by eliminating inter-gel variations (Unlu et al. 1997).

MS-based methods are commonly utilized for protein identification following excision of protein spots separated by 2-DE. MS-based protein identifications typically involve three different steps: (1) proteolytic digestion of proteins; (2) measurement of mass and/or sequence of peptides generated by the proteolytic digestion; (3) database search to identify proteins by correlating the observed mass and/or fragmentation pattern of peptides to those of *in-silico* digested peptides. Although numerous proteases can be used in protein digestion, the most commonly used protease is trypsin, which hydrolyzes the C-terminal side of lysines and arginines (unless the subsequent amino acid in the sequence is a proline). The advantage of trypsin is that the resulting tryptic digests normally possess two basic sites, i.e., the basic lysine or arginine residual at the C-terminal end and the basic amine group at the N-terminus. This allows peptides to be more efficiently protonated, and hence ionized readily. Multiple basic sites also result in the production of multiple charged species during electrospray ionization (ESI) although not as readily in matrix assisted laser desorption/ionization (MALDI). The multiple charges render one important advantage, i.e., the peptides can easily be distinguished from non-protein impurities, which are typically singly charged. This is especially useful when peptides are selected for MS/MS analysis.

MS-based protein identification relies upon comparative database queries that correlate mass spectrometric data with sequences in the databases. Several algorithms for database searches have been developed, allowing the automation of high-throughput protein identifications. These algorithms use a set of peptide masses for peptide mass fingerprinting (PMF) searches and/or, to a greater extent nowadays, the MS/MS fragmentation spectra of peptides for peptide sequence tag searches. PMF searches are based upon the comparison of experimentally determined peptide masses with the *in-silico*-digested tryptic peptide masses for each entry in the database. The database search results can then be ranked according to the number of peptide matches. More sophisticated scoring algorithms have also been developed. These are normally statistical- or probability-based algorithms that calculate a level of confidence for the match by accounting for mass accuracy, the coverage of the matched protein sequence, the lengths of matched peptides and the size of the database (Berndt et al. 1999; Perkins et al. 1999; Eriksson et al. 2000; Pappin 2003). The mass accuracy of measured peptides was found to strongly influence the specificity of the database search, and therefore, the confidence level of a match. For unambiguous identification by PMF, four to five matched peptides and 15% protein sequence coverage are normally required at a mass accuracy of 100 ppm (parts per million) (Berndt et al. 1999; Mann et al. 2001). Unlike PMF, which uses only a set of peptide masses, MS/MS searches take advantage of the uniqueness of tandem mass spectra that contain highly specific peptide sequence tags. In MS/MS searches, theoretical fragmentation patterns are generated *in silico* from known proteins in the database and matched against the experimental spectra. Probability-based scoring algorithms provide a statistical measure of how well the tandem mass spectrum agrees with the calculated sequence. Several factors are used in calculating the score including cross-correlation, fragment ion frequencies and hypergeometric probability (Perkins et al. 1999; Yates 2004). Intensity models have also been included in correlation analyses to improve matches between sequence and spectra. The sum of scores for all individual matched peptides are reported as the score for the protein match, and scoring cutoffs for a statistical confidence level are also calculated. Many of these search algorithms are publicly available through the Internet as shown in Table 9.1. The development of database search algorithms and MS have converged into a mature, robust, accurate, and rapid technology that has dramatically advanced protein identification and constitutes the basis of the emerging field of proteomics.

Table 9.1 Online protein identification tools using mass spectral data. PMF peptide mass fingerprinting, MS/MS tandem mass spectrometry

Algorithm	Developer	URL	Search Mode
Aldente	Swiss Institute of Bioinformatics	http://www.expasy.org/tools/aldente/	PMF
Popitam	Swiss Institute of Bioinformatics	http://www.expasy.org/tools/popitam/	MS/MS
MS-Fit	UC San Francisco	http://prospector.ucsf.edu/ucsfhtml4.0/msfit.htm	PMF
MS-Tag	UC San Francisco	http://prospector.ucsf.edu/ucsfhtml4.0/mstag.htm	MS/MS
ProFound	The Rockefeller University	http://prowl.rockefeller.edu/prowl-cgi/profound.exe	PMF
PeptideSearch	EMBL-EBI	http://www.narrador.embl-heidelberg.de/GroupPages/Homepage.html	PMF, MS/MS
PepMapper	The University of Manchester	http://wolf.bms.umist.ac.uk/mapper/	PMF
Mascot	Matrix Science Inc.	http://www.matrixscience.com/	PMF, MS/MS
OMSSA	NCBI	http://pubchem.ncbi.nlm.nih.gov/omssa/omssacgi.cgi	MS/MS

9.3 Proteomics of *M. truncatula*

Proteomics has been extensively used in plant biology research (for review, see Rossignol et al. 2006), including the study of legumes such as soybean (Hajduch et al. 2005; Hollung et al. 2005; Natarajan et al. 2006), alfalfa (Watson et al. 2004), *Lotus japonicus* (Wienkoop and Saalbach 2003) and the legume model, *M. truncatula* (Table 9.2). The analytical and biological variances associated with proteomic studies of *M. truncatula* by 2-DE have been assessed using green tissues, i.e., leaves (Asirvatham et al. 2002). Leaf represents the most challenging green tissue in 2-DE-based proteomic studies due to the presence of high levels of the photosynthetic enzyme ribulose 1,5-bisphosphate carboxylase/oxygenase (RuBisCO), which can represent as much as 40–70% of the total soluble protein. The excessive RuBisCO results in poor separation during isoelectric focusing (IEF) and irregularly shaped spots that significantly reduce 2-DE gel quality. Thus, the variability of 2-DE leaf proteomics can be considered the worst case scenario, and the statistical data generated from leaf tissue can be projected onto other plant tissues and cell cultures. Calculation of the average coefficients of variance from 50 spots using ten replicates yielded 16.2% for analytical variance and 24.2% for biological variance. This demonstrated that 2-DE provides a reasonably reproducible quantitative measurement of protein changes in *M. truncatula*. So far, 2-DE has been the primary protein separation technique used in proteomic studies of *M. truncatula* and has yielded significant qualitative and quantitative information regarding the proteome of this organism. Table 9.2 lists the proteomic studies of *M. truncatula* reviewed here.

Table 9.2 Proteomic studies in *Medicago truncatula*

Subjects	References
M. truncatula – microbe interactions	
Sinorhizobium meliloti (rhizobial)	Bestel-Corre et al. 2002, 2004; Catalano et al. 2004
Glomus mosseae (mycorrhizal)	Bestel-Corre et al. 2002, 2004
Glomus intraradices (mycorrhizal)	Valot et al. 2005
Aphanomyces euteiches (oomycete pathogen)	Colditz et al. 2004, 2005
Roots	Mathesius et al. 2001; Watson et al. 2003; Valot et al. 2004
Root suspension cell cultures	Watson et al. 2003; Lei et al. 2005
Leaves	Watson et al. 2003
Stems	Watson et al. 2003
Flowers	Watson et al. 2003
Seed pods	Watson et al. 2003
Seed radicles	Boudet et al. 2006
Seed development	Gallardo et al. 2003
Embryogenic cell cultures	Imin et al. 2005

9.3.1 Proteomics of M. truncatula: *Microbe Interactions*

Medicago truncatula roots can establish symbioses with a wide range of soil microorganisms such as rhizobial bacteria and arbuscular mycorrhizal (AM) fungi. While the rhizobial symbiosis enable legumes to fix atmospheric nitrogen as ammonium ions inside the root nodules, the mycorrhizal symbiosis provides plants with mineral nutrients such as phosphorus. Nodulation is unique to legumes and is of great economic significance. For this reason, nodulation has been the subject of intense research in legume biology. Nodulation that involves specific host–microorganism recognition is mediated by nodulation factors (Nod factors), a class of chitooligosaccharides produced by rhizobial bacteria. Nod factors can mitotically activate the inner cortical cells to grow, leading to the formation of nodules (for review, see Albrecht et al. 1999). Inside the nodules, there are hundreds of symbiosomes where nitrogen is converted into ammonium ions by bacterial nitrogenase. The availability of the genome sequences of two important rhizobial species, *Mesorhizobium loti* and *Sinorhizobium meliloti,* further facilitates the proteomic study of nodulation.

Bestel-Corre and coworkers analyzed the *M. truncatula* root proteome 2, 4, 7, 14, 21, 42 days after inoculation (dai) with *Glomus mosseae* (mycorrhizal) and *S. meliloti* (rhizobial) respectively using pH 3–10 non-linear 13 cm IPG strips (Bestel-Corre et al. 2002). About 500 protein spots were visualized on a single 2-D gel stained with Coomassie Blue at a protein load of 500 μg. Comparative analyses revealed a time-dependent differential protein accumulation in response to mycorrhizal and rhizobial colonization. In both cases, more proteins were induced in the later stages than in the early stages of inoculation. In response to nodulation, 51 proteins were differentially displayed, with one protein being down-regulated, 4 up-regulated and 46 newly induced. In the case of mycorrhizal inoculation, 55 proteins exhibited differential display: 2 were down-regulated, 12 up-regulated and 41 newly induced. These proteins were involved in defense responses (e.g., peroxidase and glutathione transferase), root physiology (e.g., myosin heavy-chain-like protein and fucosidase), and respiratory pathway components (e.g., serine hydroxylmethyl-transferase and cytochrome-*c*-oxidase). Proteins believed to be related to root nodulation included leghemoglobins, *S. meliloti* nitrogenase iron protein, superoxide dismutase and enolase (Bestel-Corre et al. 2002). Although previous genetic work demonstrated that common genes were involved in these two root symbioses (for review, see Albrecht et al. 1999), no common proteins that were differentially regulated were identified in either the mycorrhizal or the rhizobial symbioses. It was suggested that the resolution of the 2-DE used might have been insufficient, and it was recommended that high resolution IPG strips such as ultrazoom or narrow pH range IPG strips of greater resolving power be used together with a parallel study of hydrophobic proteins (Bestel-Corre et al. 2002).

Using a similar approach, Catalano and coworkers investigated the symbiosome membrane proteome from *S. meliloti*-induced nodules (Catalano et al. 2004). The symbiosome membrane inside the root nodule forms an enclosed compartment in which the mature nitrogen-fixing bacteria or the bacteroids localize; there are hundreds of these compartments in the root nodule. The symbiosomes were first isolated and fractionated into symbiosome membrane, bacteroid, and symbiosome space fractions (the symbiosome space is the space between the symbiosome membrane and the bacteroid). Proteins of these three fractions were then profiled using 2-DE. Significant differences were observed among these different protein profiles. Stained with SYPRO Ruby, the symbiosome membrane proteome map revealed more than 100 protein spots, and the symbiosome space and bacteroid maps contained more than 200 proteins spots respectively. Several nodule-specific proteins were first identified from the symbiosome membrane using a proteomics approach, including nodulin 8, 16, and 25 (Catalano et al. 2004). The major groups of proteins identified from the symbiosome membrane were found to be related to protein destination/storage, energy production and transport, e.g., H^+-transporting ATP synthase, multifunctional aquaporin and ABC transporter.

Changes in membrane-associated proteins in response to the AM symbiosis with *Glomus intraradices* have also been investigated (Valot et al. 2005, see also Chap. 20 by Dumas-Gaudot and Recobert, this volume). Comparative analyses between inoculated and non-inoculated root membrane proteins revealed that 36 membrane-associated proteins were differentially displayed in response to the fungal infection, with 15 being newly induced, 3 up-regulated and 18 down-regulated. Of the 36 differentially displayed proteins, 25 were successfully identified by MS. Most of these have not been reported previously as being regulated by AM colonization, demonstrating the utility of the proteomics approach in legume symbiosis research.

Proteomics was also used to investigate the effect of sewage sludges, commonly used as farm fertilizers, on legume-microbe symbioses (Bestel-Corre et al. 2004). Proteomes of *M. truncatula* roots inoculated with the AM fungus *G. mosseae*, or the rhizobial bacteria *S. meliloti*, were analyzed by 2-DE after application of sewage sludges. Although no additional pollutant-responsive proteins were detected in response to the application of sludge enriched with polycyclic aromatics or heavy metals, protein changes in response to colonization were observed. Due to the use of narrow range IPG strips, more differentially displayed proteins were observed than before (Bestel-Corre et al. 2002), demonstrating the higher resolving power of narrow range IPG strips. In addition to previously observed proteins (Bestel-Corre et al. 2002), 23 new proteins were identified in response to fungal colonization and 25 proteins in response to *S. meliloti* colonization (Bestel-Corre et al. 2004). The mycorrhiza-related proteins were found to be of plant origin, e.g., 6-deoxychalcone synthase, while nodule-related proteins were mostly of bacteria origin, e.g., NifK nitrogenase.

Proteomics has also been applied to the field of *M. truncatula*–pathogen interactions (Colditz et al. 2004, 2005; see also Chap. 18 by Colditz et al., this volume). The temporal response of *M. truncatula* roots to *Aphanomyces euteiches*, an oomycete pathogen that causes *M. truncatula* root rot disease, was evaluated at the proteome level (Colditz et al. 2004). Proteins from both infected and non-infected roots were profiled using 2-DE and compared. The differences in the protein profiles were observed as early as 6 h after inoculation. Of the 500 proteins visualized by 2-DE, 16 were found to differentially accumulate in response to the pathogen infection. Most of these proteins were limited to the gel region of 16–18 kDa and pI 4–5 in which pathogen-responsive proteins such as pathogenesis-related (PR)-proteins, abscisic acid (ABA)-responsive proteins and cysteine proteinase were found to dominate (Mathesius et al. 2001; Lei et al. 2005). Most of the induced proteins in this region were identified to be closely associated with members of the PR-10 family (Colditz et al. 2004), which are structurally related to ribonucleases, suggesting that these up-regulated "ribonuclease-like" proteins may be capable of cleaving foreign RNA during pathogen attack (Edreva 2005). Half of these pathogen-induced proteins were not responsive to ABA treatment, indicating the induction of these proteins might be pathogen-specific. The rest of the proteins were also found to be induced by ABA treatment. However, none of the pathogen-induced proteins showed response to drought treatment (Colditz et al. 2004). Other proteins involved in natural product biosynthesis and cell wall modifications were also identified in this study. For example, an isoliquiritigenin-2-O-methyltransferase involved in flavonoid biosynthesis and two putative cell wall structural proteins were induced to a significant degree. Legume flavonoids and isoflavonoids have long been known to play an important role in plant disease and defense response (Dixon 2001). The putative cell wall proteins may be involved in cell wall modifications in response to fungal attack.

Using a similar approach, the same group further compared the root proteomes of two *M. truncatula* lines of increased or decreased resistance to *Aphanomyces euteiches* with that of the widely used A17 line (Colditz et al. 2005). After *A. euteiches* infection, different responses were observed between resistant and susceptible lines. In the susceptible line, proteins involved in flavonoid biosynthesis were typically induced, together with some primary metabolism-related proteins.

Whereas in the resistant line, proteins related to proteasomes were preferentially up-regulated along with N-ethylmaleimide-sensitive fusion (NSF) attachment proteins and membrane polypeptides. The induced flavonoid-related proteins included isoliquiritigenin-2-O-methyltransferase, chalcone reductase and isoflavone reductase. These proteins are involved in the formation of medicarpin, which has been previously demonstrated to be induced by fungal elicitor (Paiva et al. 1991). The induced primary metabolism-related proteins were alcohol dehydrogenase and fructose-bisphosphate aldolase. Proteasomes were also among those proteins significantly induced.

9.3.2 Proteomics of M. truncatula: *Tissues*

Roots of *M. truncatula* have attracted a lot of attention due to their role in symbiosis, and the root proteome has been successfully profiled using 2-DE (IPG strips pH 4–7 and pH 6–11) and proteins identified using PMF (Mathesius et al. 2001). When stained with silver, about 2,500 protein spots could be reproducibly visualized across a pI range of 4 to 7, and 450 across a pI range of 6–11. Approximately 1,500 protein spots were observed between pI 4 and pI 7 and 50 between pI 6 and pI 11 when Coomassie Blue was used to visualize the gels. Similar protein spot numbers (1,661 protein spots) were observed in a suspension cell culture originating from root (Lei et al. 2005). Using PMF, 179 proteins out of the 485 most abundant root proteins visualized on Coomassie Blue-stained gels were identified (Mathesius et al. 2001). Lei and coworkers excised the majority of observable protein spots, i.e. 1,661 protein spots, and analyzed their tryptic digests using nano-LC-MS/MS. Proteins were successfully identified in 1,367 out of the 1,661 protein spots excised, yielding 907 unique protein accession numbers and a identification success rate of 87%. Proteins identified in this work included a complete tricarboxylic acid cycle; a nearly complete glycolytic pathway; a significant portion of the ubiquitin pathway; many enzymes involved in primary and secondary metabolism, and energy production; and proteins implicated in defense, protein destination/storage, and protein synthesis (Lei et al. 2005). This work represents the most extensive proteomic description of *M. truncatula* suspension cells and serves as a reference map for comparative proteomic and functional genomic studies of the response of these cells to biotic and abiotic stress. In both roots and root-derived cell cultures, proteins involved in primary metabolism were found to constitute the major group of proteins (Mathesius et al. 2001; Lei et al. 2005). However, the most abundant proteins in both the root and root suspension cells were found to be plant disease/defense-related proteins such as ABA proteins and PR proteins, suggesting the importance of defense/disease proteins in roots and cell cultures (Mathesius et al. 2001; Lei et al. 2005).

A study of organ-/tissue-specific proteomes from *M. truncatula* successfully identified 304 proteins in a survey of leaves, stems, flowers, seed pods, roots, and suspension cultures (Watson et al. 2003). Distinct proteome patterns were observed in these different tissues. The leaf proteome was dominated by photosynthetic proteins, which accounted for approximately 40% of leaf protein mass as revealed by 2-DE. The highly abundant RuBisCO resulted in poor isoelectric focusing, appearing as horizontal streaks across the 2D gel (Watson et al. 2003). It is believed that removal of RuBisCO prior to 2-DE will significantly enhance the visualization of moderate/low abundance proteins in leaves. Many of the photosynthetic proteins identified in leaves were also observed in stems but at lower levels. The stem proteome also included proteins involved in lignin and phytoalexin biosyntheses, such as cinnamoyl-CoA reductase and isoflavone reductase-like oxidoreductase. Similar to the observations mentioned above (Mathesius et al. 2001; Lei et al. 2005), a significant portion of proteins identified in the root proteome was found to be related to plant defense/disease responses (Watson et al. 2003). These included ABA-responsive proteins, chitinases and peroxidases, and an ABA-responsive protein which was the most abundant protein. The flower proteome was found to have a large complement of energy production-related proteins as well as proteins associated with protein synthesis/targeting. In contrast, the proteome of seed pods, which included seed and pod tissues, was found to consist primarily of storage proteins. These storage proteins serve as energy and nutrient reserves, which are crucial for developing plants.

Temporal changes in the *M. truncatula* seed proteome during seed development were described by Gallardo et al. (2003). 2-DE revealed 120 proteins showing differential accumulation kinetics during seed development. The most abundant proteins were found to be members of the major storage protein families, i.e., the legumins, vicilins, and convicilins. Each of these storage protein families appeared at different developmental stages and continued to accumulate during seed development. The low abundance proteins, in contrast, showed constant as well as transient changes during seed development. Many of these were associated with energy, disease/defense, metabolism, protein targeting/storage, cell growth/division, and cell structure. Although low in abundance, these proteins could be essential to seed development. For example, a low abundance protein found to accumulate constantly during seed development was identified as being related to the biosynthesis of thiamine, and was demonstrated to be indispensable for seed formation. Many proteins involved in methionine (Met) synthesis and S-adenosyl-Met (Ado-Met) formation such as Met synthase and Ado-Met synthetase also changed during seed development. Analysis of the pattern changes associated with the Met-related enzymes led to the hypothesis that Met metabolism may be implicated in the transition from a highly active state to a quiescent state during *M. truncatula* seed development (Gallardo et al. 2003).

Imin and coworkers have generated comprehensive proteome reference maps for *M. truncatula* embryogenic tissue culture generated from single protoplasts. The proteome of the embryogenic tissue culture at the globular stage was profiled using two narrow but overlapping linear IPG strips, i.e., pH 4–7 and 6–11, and visualized with silver staining. Using a protein load of 150 μg and silver staining, approximately 2,700 proteins were resolved reproducibly in the pI range of 4–7 and 500 different proteins in the pI range of 6–11, yielding a total number of more than

3,000 proteins over a pI range of 4–11. For protein identification, a higher protein load (0.75 mg–2 mg) and colloidal Coomassie Blue staining were used. A total of 312 protein spots were excised from Coomassie Blue-stained 2-D gels and analyzed by MALDI-TOF-MS and MS/MS; 169 protein spots representing 128 unique gene products were confidently identified. Comparison of the identified proteins with previously reported proteins in roots (Mathesius et al. 2001) and root-derived suspension cell cultures (Watson et al. 2003) led to the identification of 72 proteins that were unique to the embryogenic cultures. Several of these were thought to be likely candidates for new embryogenesis regulating factors. The same group further investigated the proteome of the *M. truncatula* line 2HA that exhibits a 500-fold greater capacity to regenerate plants in culture through somatic embryogenesis relative to the wild type cultivar, Jemalong (Imin et al. 2005). In medium supplemented with 1-naphthaleneacetic acid and 6-benzylaminopurine, embryos began to form from 2HA tissue 5 weeks after culturing while Jemalong showed only rare embryos even 8 weeks after culturing. The proteomes of Jemalong and 2HA at three different time points of culturing (2, 5 and 8 weeks) were resolved and visualized using 2-DE. Consistent with what others had observed with cell cultures (Mathesius et al. 2001; Watson et al. 2003; Lei et al. 2005), PR proteins were found to be the most abundant proteins in the cultures of both the Jemalong and the 2HA lines. Some of these proteins, such as two of the most abundant PR proteins, PR10-1 and an ABA-responsive protein, remained highly expressed throughout the whole culturing process. In contrast, these proteins were not detectable in the young leaves from which the cultures had been generated. RuBisCO, the most abundant protein in green tissue, was found to gradually decrease during the culturing process. Comparative proteomics revealed that a total of 54 proteins showed differential expression, and 16 of these were confidently identified (Imin et al. 2005). The expression patterns of these proteins varied. Some showed a constant increase or decrease over the culturing process. Some showed increases in the early stage of culturing, but decreased in the late stage of culturing, and others showed the opposite trend. For example, thioredoxin H was found in the early stage of culturing, but became undetectable later. It was not detected in leaf tissue or the isolated protoplasts from which the cultures were generated, suggesting that thioredoxin H is required for the transition from vegetative tissues to cultures. Interestingly, two seed storage-related proteins, i.e., a seed maturation protein and a vicilin, were de novo synthesized only in line 2HA at the late stage of culturing. Another protein showing elevated expression level in 2HA was 1-Cys peroxiredoxin, which was induced slightly at the beginning but dramatically at a later stage of culturing. The expression of 1-Cys peroxiredoxin suggested an important role in the hyperembryogenesis of line 2HA (Imin et al. 2005).

Comparative proteomics was used to investigate desiccation tolerance-related proteins in radicles of *M. truncatula* seeds (Boudet et al. 2006). Desiccation-tolerant and sensitive radicles, i.e., radicles before and after emergence from the seed coat, were used to profile the proteomes. Radicles of different lengths, i.e., 2.8 mm (in which desiccation tolerance can be restored by polyethylene glycol treatment), and 5 mm (in which desiccation tolerance can no longer be re-established), were used to differentiate desiccation tolerance-specific proteins from osmotic-related proteins. Interestingly, more protein spots were visualized by 2-DE in the desiccation-tolerant radicles (328) than in the desiccation-sensitive radicles (252). Protein concentration determination also revealed that the amount of proteins in radicles decreased by more than two-fold during germination. Some proteins remained constant during the germination process while others showed pattern changes. The protein changes could be classified into nine different categories. Using a comparative approach, a total of 23 protein spots were found to be more abundant in both desiccation-tolerant stages (i.e., radicles in nongerminated seeds and protruded radicles of 2.8 mm long treated with polyethylene glycol) compared to the sensitive stage (protruded radicles of 2.8 mm long). Fifteen of them were found to be desiccation tolerance-related proteins, of which 11 were identified by MALDI TOF-MS and MS/MS as late embryogenesis abundant (LEA) proteins (Boudet et al. 2006).

Recently, a National Science Foundation (NSF), Plant Genome Research Program-funded integrated functional genomics project involving *M. truncatula* transcriptomics, proteomics and metabolomics has been conducted at the Noble Foundation (Ardmore, OK). High resolution 2-DE and nano LC-MS/MS were used to dissect the responses of *M. truncatula* cell cultures to external elicitors (yeast and methyl jasmonate) at the proteome level. Preliminary data showed that enzymes associated with isoflavonoid biosynthesis as well as cell wall modifications were differentially up-regulated in response to yeast elicitation. In contrast, proteins involved in protein synthesis and targeting were differentially regulated by methyl jasmonate.

9.4 Future Perspectives

Legume-microbe interactions will continue to be an important area of proteomic studies of *M. truncatula*. Identification and characterization of symbiosis-related proteins and their changes during the nodulation process will remain a primary target of the proteomics of *M. truncatula*. The availability of nodulation-defective or nodulation-altered mutants of *M. truncatula* provides an excellent opportunity for further studies of symbiosis-related proteins (Sagan et al. 1995; Catoira et al. 2000; Wais et al. 2000). It is envisioned that a significant amount of new information related to mutualistic interactions could be gained through proteome studies of nodulation affected mutants such as non-nodulating and non-mycorrhizal, low-nodulating and low-mycorrhizal, nodulating but non-fixing, and supernodulating and mycorrhizal mutants. The successful proteomic studies of symbiosis will most likely require anatomically dissected or sub-cellular fractionation to enrich for specific proteins. For example, fractionation of the symbiosome proteome into bacteroid, symbiosome space, and symbiosome membrane fractions may provide more detailed information about symbiosis. Techniques for isolating proteins in symbiosome membrane, symbiosome space and bacteroid have been reported and have indeed yielded more information (Catalano et al. 2004); however, continued application of these techniques will likely lead to additional novel information and enhanced understanding of mutualism.

Focused studies of *M. truncatula* defense responses are another research area where proteomics offers unique opportunities. Defense responses involve the synthesis of a wide array of secondary metabolites. Many, such as medicarpin, are legume specific. Several groups, including our own, are currently working on the proteomics of the responses of *M. truncatula* to biotic and abiotic stress. We have used a 2-DE/MS approach to dissect the responses of *M. truncatula* suspension cells to yeast and methyl jasmonate elicitors, which showed the differential responses to these two treatments. It is expected that proteomics related to secondary metabolism will continue to attract attention and yield valuable information regarding the synthesis, storage, and transport of secondary metabolites in response to external stimuli.

So far, the majority of proteomic studies of *M. truncatula* have utilized a gel-based approach, mostly 2-DE, to separate and quantify changes in protein accumulation. While this approach is perfectly suitable for analyzing soluble proteins, visualization of changes in membrane proteins by 2-DE is still very challenging (Santoni et al. 2000). Membrane proteins provide many critical cellular functions. They serve as ion and metabolite transporters and signal receptors, which are necessary to the establishment of plant-microbe interactions (Benson et al. 2005). Although the application of 2-DE has been previously reported in the separation of membrane-associated proteins in *M. truncatula* roots (Valot et al. 2005), a significantly larger number of membrane proteins are more likely to be revealed using LC-based multidimensional protein identification technology (MudPIT) (Fischer et al. 2006) and quantified by stable isotope labeling techniques such as isotope-coded affinity tags (ICAT) and isobaric tags for relative and absolute quantitation (iTRAQ) (Gygi et al. 1999; Ross et al. 2004). It is conceivable that, in the near future, membrane-associated proteins will become a focus of proteomics of *M. truncatula* symbiosis and sophisticated technologies like MudPIT and ICAT/iTRAQ will be used.

References

Albrecht C, Geurts R, Bisseling T (1999) Legume nodulation and mycorrhizae formation; two extremes in host specificity meet. EMBO J 18:281–288

Asirvatham VS, Watson BS, Sumner LW (2002) Analytical and biological variances associated with proteomic studies of *Medicago truncatula* by two-dimensional polyacrylamide gel electrophoresis. Proteomics 2:960–968

Barker DG, Bianchi S, Blondon F, Datteé Y, Duc G, Essad S, Flament P, Gallusci P, Génier G, Guy P, Muel X, Tourneur J, Dénarié J, Huguet T (1990) *Medicago truncatula*, a model plant for studying the molecular genetics of the Rhizobium–legume symbiosis. Plant Mol Biol Rep 8:40–49

Benson HP, Boncompagni E, Guerinot ML (2005) An iron uptake operon required for proper nodule development in the *Bradyrhizobium japonicum*–soybean symbiosis. Mol Plant Microbe Interact 18:950–959

Berndt P, Hobohm U, Langen H (1999) Reliable automatic protein identification from matrix-assisted laser desorption/ionization mass spectrometric peptide fingerprints. Electrophoresis 20:3521–3526

Bestel-Corre G, Dumas-Gaudot E, Poinsot V, Dieu M, Dierick JF, van Tuinen D, Remacle J, Gianinazzi-Pearson V, Gianinazzi S (2002) Proteome analysis and identification of symbiosis-related proteins from *Medicago truncatula* Gaertn. by two-dimensional electrophoresis and mass spectrometry. Electrophoresis 23:122–137

Bestel-Corre G, Gianinazzi S, Dumas-Gaudot E (2004) Impact of sewage sludges on *Medicago truncatula* symbiotic proteome. Phytochemistry 65:1651–1659

Boudet J, Buitink J, Hoekstra FA, Rogniaux H, Larre C, Satour P, Leprince O (2006) Comparative analysis of the heat stable proteome of radicles of *Medicago truncatula* seeds during germination identifies late embryogenesis abundant proteins associated with desiccation tolerance. Plant Physiol 140:1418–1436

Catalano CM, Lane WS, Sherrier JD (2004) Biochemical characterization of symbiosome membrane proteins from *Medicago truncatula* root nodules. Electrophoresis 25:519–531

Catoira R, Galera C, de Billy F, Penmetsa RV, Journet EP, Maillet F, Rosenberg C, Cook D, Gough C, Denarie J (2000) Four genes of *Medicago truncatula* controlling components of a nod factor transduction pathway. Plant Cell 12:1647–1666

Choi H-K, Kim D, Uhm T, Limpens E, Lim H, Mun J-H, Kalo P, Penmetsa RV, Seres A, Kulikova O, Roe BA, Bisseling T, Kiss GB, Cook DR (2004) A sequence-based genetic map of *Medicago truncatula* and comparison of marker colinearity with *M. sativa* Genetics 166:1463–1502

Colditz F, Nyamsuren O, Niehaus K, Eubel H, Braun H-P, Krajinski F (2004) Proteomic approach: identification of *Medicago truncatula* proteins induced in roots after infection with the pathogenic oomycete *Aphanomyces euteiches*. Plant Mol Biol 55:109–120

Colditz F, Braun H-P, Jacquet C, Niehaus K, Krajinski F (2005) Proteomic profiling unravels insights into the molecular background underlying increased *Aphanomyces euteiches*-tolerance of *Medicago truncatula*. Plant Mol Biol 59:387–406

Cook DR (1999) *Medicago truncatula* – a model in the making! Curr Opin Plant Biol 2:301–304

Dixon RA (2001) Natural products and disease resistance. Nature 411:843–847

Dixon RA, Steele CL (1999) Flavonoids and isoflavonoids – a gold mine for metabolic engineering. Trends Plant Sci 4:394–400

Dixon RA, Sumner LW (2003) Legume natural products: understanding and manipulating complex pathways for human and animal health. Plant Physiol 131:878–885

Edreva A (2005) Pathogenesis-related proteins: research progress in the last 15 years. Gen Appl Plant Physiol 31:105–124

Eriksson J, Chait BT, Fenyo D (2000) A statistical basis for testing the significance of mass spectrometric protein identification results. Anal Chem 72:999–1005

Fischer F, Wolters D, Rogner M, Poetsch A (2006) Toward the complete membrane proteome: high coverage of integral membrane proteins through transmembrane peptide detection. Mol Cell Proteomics 5:444–453

Frugoli J, Harris J (2001) *Medicago truncatula* on the move! Plant Cell 13:458–463

Gallardo K, Le Signor C, Vandekerckhove J, Thompson RD, Burstin J (2003) Proteomics of *Medicago truncatula* seed development establishes the time frame of diverse metabolic processes related to reserve accumulation. Plant Physiol 133:664–682

Görg A, Obermaier C, Boguth G, Weiss W (1999) Recent developments in two-dimensional gel electrophoresis with immobilized pH gradients: wide pH gradients up to pH 12, longer separation distances and simplified procedures. Electrophoresis 20:712–717

Görg A, Obermaier C, Boguth G, Harder A, Scheibe B, Wildgruber R, Weiss W (2000) The current state of two-dimensional electrophoresis with immobilized pH gradients. Electrophoresis 21:1037–1053

Gygi SP, Rist B, Gerber SA, Turecek F, Gelb MH, Aebersold R (1999) Quantitative analysis of complex protein mixtures using isotope-coded affinity tags. Nat Biotechnol 17:994–999

Hajduch M, Ganapathy A, Stein JW, Thelen JJ (2005) A systematic proteomic study of seed filling in soybean. Establishment of high-resolution two-dimensional reference maps, expression profiles, and an interactive proteome database. Plant Physiol 137:1397–1419

Hollung K, Overland M, Hrustic M, Sekulic P, Miladinovic J, Martens H, Narum B, Sahlstrom S, Sorensen M, Storebakken T, Skrede A (2005) Evaluation of nonstarch polysaccharides and oligosaccharide content of different soybean varieties (*Glycine max*) by near-infrared spectroscopy and proteomics. J Agric Food Chem 53:9112–9121

Hymowitz T (1990) Grain legumes. In: Janick J, Simon JE (eds) Advances in new crops. Timer, Portland, pp 154–158

Imin N, Nizamidin M, Daniher D, Nolan KE, Rose RJ, Rolfe BG (2005) Proteomic analysis of somatic embryogenesis in *Medicago truncatula*. Explant cultures grown under 6-benzylaminopurine and 1-naphthaleneacetic acid treatments. Plant Physiol 137:1250–1260

Klose J (1999) Large-gel 2-D electrophoresis. Methods Mol Biol 112:147–172

Lei Z, Elmer AM, Watson BS, Dixon RA, Mendes PJ, Sumner LW (2005) A two-dimensional electrophoresis proteomic reference map and systematic identification of 1367 proteins from a cell suspension culture of the model legume *Medicago truncatula* Mol Cell Proteomics 4:1812–1825

Mann M, Hendrickson RC, Pandey A (2001) Analysis of proteins and proteomes by mass spectrometry. Annu Rev Biochem 70:437–473

Mathesius U, Keijzers G, Natera SH, Weinman JJ, Djordjevic MA, Rolfe BG (2001) Establishment of a root proteome reference map for the model legume *Medicago truncatula* using the expressed sequence tag database for peptide mass fingerprinting. Proteomics 1:1424–1440

May G, Dixon R (2004) *Medicago truncatula*. Curr Biol 14:R180–181

Natarajan SS, Xu C, Bae H, Caperna TJ, Garrett WM (2006) Characterization of storage proteins in wild (*Glycine soja*) and cultivated (*Glycine max*) soybean seeds using proteomic analysis. J Agric Food Chem 54:3114–3120

O'Farrell PZ, Goodman HM, O'Farrell PH (1977) High resolution two-dimensional electrophoresis of basic as well as acidic proteins. Cell 12:1133–1141

Paiva NL, Edwards R, Sun Y, Hrazdina G, Dixon RA (1991) Stress responses in alfalfa (*Medicago sativa* L.) 11. Molecular cloning and expression of alfalfa isoflavone reductase, a key enzyme of isoflavonoid phytoalexin biosynthesis. Plant Mol Biol 17:653–667

Pappin DJ (2003) Peptide mass fingerprinting using MALDI-TOF mass spectrometry. Methods Mol Biol 211:211–219

Perkins DN, Pappin DJ, Creasy DM, Cottrell JS (1999) Probability-based protein identification by searching sequence databases using mass spectrometry data. Electrophoresis 20:3551–3567

Rabilloud T (2002) Two-dimensional gel electrophoresis in proteomics: old, old fashioned, but it still climbs up the mountains. Proteomics 2:3–10

Ross PL, Huang YN, Marchese JN, Williamson B, Parker K, Hattan S, Khainovski N, Pillai S, Dey S, Daniels S, Purkayastha S, Juhasz P, Martin S, Bartlet-Jones M, He F, Jacobson A, Pappin DJ (2004) Multiplexed protein quantitation in *Saccharomyces cerevisiae* using amine-reactive isobaric tagging reagents. Mol Cell Proteomics 3:1154–1169

Rossignol M, Peltier J-B, Mock H-P, Matros A, Maldonado AM, Jorrín JV (2006) Plant proteome analysis: a 2004–2006 update. Proteomics 6:5529–5548

Sagan M, Morandi D, Tarenghi E, Duc G (1995) Selection of nodulation and mycorrhizal mutants in the model plant *Medicago truncatula* (Gaertn.) after [gamma]-ray mutagenesis. Plant Sci 111:63–71

Santoni V, Molloy M, Rabilloud T (2000) Membrane proteins and proteomics: un amour impossible? Electrophoresis 21:1054–1070

Sarkar FH, Li Y (2003) Soy isoflavones and cancer prevention. Cancer Invest 21:744–757

Thoquet P, Gherardi M, Journet E, Kereszt A, Ane J, Prosperi J, Huguet T (2002) The molecular genetic linkage map of the model legume *Medicago truncatula*: an essential tool for comparative legume genomics and the isolation of agronomically important genes. BMC Plant Biol 2:1–13

Town C (2006) Annotating the genome of *Medicago truncatula*. Curr Opin Plant Biol 9:122–127

Trieu AT, Burleigh SH, Kardailsky IV, Maldonado-Mendoza IE, Versaw WK, Blaylock LA, Shin H, Chiou TJ, Katagi H, Dewbre GR, Weigel D, Harrison MJ (2000) Transformation of *Medicago truncatula* via infiltration of seedlings or flowering plants with Agrobacterium. Plant J 22:531–541

Unlu M, Morgan ME, Minden JS (1997) Difference gel electrophoresis: a single gel method for detecting changes in protein extracts. Electrophoresis 18:2071–2077

USDA-NASS (2004) USDA-NASS Agricultural Statistics 2002, http://www.usda.gov/nass/pubs/agr02/acro02.htm

Valot B, Gianinazzi S, Eliane D-G (2004) Sub-cellular proteomic analysis of a *Medicago truncatula* root microsomal fraction. Phytochemistry 65:1721–1732

Valot B, Dieu M, Recorbet G, Raes M, Gianinazzi S, Dumas-Gaudot E (2005) Identification of membrane-associated proteins regulated by the arbuscular mycorrhizal symbiosis. Plant Mol Biol 59:565–580

Wais RJ, Galera C, Oldroyd G, Catoira R, Penmetsa RV, Cook D, Gough C, Denarie J, Long SR (2000) Genetic analysis of calcium spiking responses in nodulation mutants of *Medicago truncatula*. Proc Natl Acad Sci USA 97:13407–13412

Watson BS, Asirvatham VS, Wang L, Sumner LW (2003) Mapping the proteome of barrel medic (*Medicago truncatula*). Plant Physiol 131:1104–1123

Watson B, Lei Z, Dixon R, Sumner L (2004) Proteomics of *Medicago sativa* cell walls. Phytochemistry 65:1709–1720

Wienkoop S, Saalbach G (2003) Proteome analysis. Novel proteins identified at the peribacteroid membrane from *Lotus japonicus* root nodules. Plant Physiol 131:1080–1090

Yates JR (2004) Mass spectral analysis in proteomics. Annu Rev Biophys Biomol Struct 33:297–316

Young ND, Cannon SB, Sato S, Kim D, Cook DR, Town CD, Roe BA, Tabata S (2005) Sequencing the genespaces of *Medicago truncatula* and *Lotus japonicus*. Plant Physiol 137:1174–1181

Chapter 10
Proteomics of Seed Development in Oilseed Crops

Martin Hajduch, Ganesh Kumar Agrawal, and Jay J. Thelen

Abstract Despite limited genome resources, recent proteomics research on select oilseed crops has shown that global analyses of proteins is currently possible. This review summarizes recent proteomic research efforts on seed development in oilseed crops, most notably soybean and oilseed rape. In-depth, systematic analyses of protein expression during the reserve deposition phase of seed development in soybean and oilseed rape have recently provided the first proteomic perspective of carbon assimilation into storage reserves for any oilseed crop (Hajduch et al. 2005, 2006). This is perhaps the first use of plant proteomics data to predict metabolic pathways with high confidence. In this review we present an updated comparison of intermediary metabolic pathways operating during pod filling in both soybean and oilseed rape. This review also discusses recent systematic analysis of phosphorylated proteins expressed during seed filling in oilseed rape using Pro-Q Diamond phosphoprotein stain in conjunction with high resolution two-dimensional gel electrophoresis. Identification of 103 of these proteins by tandem mass spectrometry revealed approximately 80 novel phosphoproteins, of which 45% were involved in metabolism or energy production. These data suggest that more metabolic enzymes could be regulated by protein phosphorylation than previously thought. Recent large-scale identification of soybean and rapeseed proteins and phosphoproteins by mass spectrometry demonstrates that proteomics research on oilseed crops is an exciting new area of research with tremendous possibilities.

Martin Hajduch, Department of Biochemistry, Life Sciences Center, University of Missouri, Columbia, MO 65211, USA

Ganesh Kumar Agrawal, Department of Biochemistry, Life Sciences Center, University of Missouri, Columbia, MO 65211, USA

Jay J. Thelen, Department of Biochemistry, Life Sciences Center, University of Missouri, 109 Life Sciences Center, Columbia, MO 65211, USA, E-mail: thelenj@missouri.edu

Current address: Martin Hajduch, Institute of Plant Genetics and Biotechnology, Slovak Academy of Sciences, Nitra, Slovakia

J. Šamaj and J. Thelen (eds.), *Plant Proteomics*
© Springer 2007

10.1 Introduction

Oilseeds are generally considered to be non-graminaceous crops that are harvested primarily for their oil and include, but are not limited to, castor, cottonseed, palm kernel, groundnut (peanut), rapeseed, sesame, soybean, and sunflower. Vegetable oils are pressed or extracted primarily from the seed (but also from the fruit) of select plants for edible and industrial applications (Thelen and Ohlrogge 2002). Oilseed production for the past 4 years shows a steady annual increase with a forecasted production in 2006–2007 of 392 million metric tons (Gunstone 2006). In that same time period vegetable oil production increased 21%, with palm (7.7% increase), soybean (5.7%), and rapeseed (4.4%) oils comprising 85% of that growth.

Because of their longstanding economic importance, research on oilseed plants is well established, particularly with respect to seed development. One of earliest published research articles characterizing oil production and accumulation during plant seed development was a report in 1931 by J.V. Eyre (Eyre 1931). More recently, it was noted that changes in the composition of developing rapeseed (oilseed rape, *Brassica napus*) seeds can be correlated with metabolic processes at particular stages of seed development (Norton and Harris 1975). This investigation on developing *B. napus* seeds also pointed out that changes in the composition of developing seeds can be divided into three phases:

- Phase 1 (1–4 weeks after flowering (WAF)) – soluble constituents and starch account for up to 80% of the dry matter, whereas lipids and proteins collectively represent 15%.
- Phase 2 (5–6 WAF) – rise in lipid and protein content and almost equivalent decrease in starch and aqueous fraction.
- Phase 3 (7–12 WAF) – oil and protein reach a maximum, while starch and aqueous soluble constituents decrease to a minimum.

Seed development is a complex, integrated process that requires whole plant coordination, including leaves (sucrose) and roots (micronutrients). Additionally, multiple tissue types are present within the plant seed, notably embryo, endosperm and seed coat. In terms of seed development, the seed filling period (phases 1 and 2) is the period where drastic changes in protein and oil composition occur. It is this period of seed development, rather than embryogenesis and maturation, that has received considerable attention from systems biologists eager to understand metabolic regulation at the transcriptional and post-transcriptional levels. Most oilseed crops have, unfortunately, trailed behind the model plants at the various benchmarks of genomics research. Nevertheless, the lack of complete, annotated genomes and microarray resources does not dramatically hinder proteomics research. And this is perhaps the single greatest advantage proteomics has over most other "omics" technologies – one does not need to wait for extensive technology or resource development (e.g., microarray chips, antibody arrays, or annotated genome of the plant of interest) to embark on proteomics investigations.

Current proteomics technologies, such as high resolution two-dimensional gel electrophoresis (2-DE), are able to elucidate the accumulation of multiple isoforms and post-translational modifications (PTM) of the same protein. The capability of performing such a task on the scale of ca. 1,000 proteins allows for a much greater depth of study than traditional approaches such as western blotting or enzymatic activity analyses, which require specific antibodies or reagents and are comparatively low-throughput. The employment of new post-genomics methodologies promises to provide new insights into already extensively investigated problems. For example, to fully understand carbon assimilation in a developing seed one must consider metabolic mechanisms in the context of both developmental (transcriptional) and biochemical (post-translational) regulation. To perform such in-depth analyses it is necessary to systematically and comprehensively resolve, quantify, and analyze the thousands of proteins expressed throughout seed development. The analysis of whole developing seed is a good "launch point" to provide a system-wide view of the metabolic and structural processes required to make a seed. Proteomic characterization of the organ, tissue, and cellular heterogeneity within a seed on a temporal scale will be a critical step towards understanding the specialized roles of seed coat, embryo, and endosperm. As genomic and microarray resources catch up with proteomics investigations in oilseed crops, it will be particularly interesting to quantitatively compare transcript and protein expression to discover incongruent patterns that might point to regulatory control. These goals will undoubtedly consume the efforts of plant scientists well into the next few decades.

Although the number of oil-bearing crop plants is quite high (Bockisch 1993), research on oilseeds is rather limited. Thus the focus of this review will be on recent proteomics research on soybean, rapeseed, and castor. Perhaps due to its agricultural importance in the United States, soybean has attracted most of the recent proteomics effort among oilseed crops (Table 10.1). Proteomics research in oilseeds has thus far focused primarily on the subject of crop improvement. Many published investigations are centered on the analysis of developing or mature seed in order to understand metabolism or identify and characterize allergenic proteins, with the intent of molecular genetic silencing of these proteins (Table 10.1). Proteomics of seed development in soybean (Hajduch et al. 2005) and rapeseed (Hajduch et al. 2006) are beginning to provide insights into carbon assimilation via the reactions of glycolysis, where relatively little is still known about its regulation and control (Fernie et al. 2004). Besides proteomics research into seed filling in various oilseed crops, this chapter will also discuss the analysis of mature seeds as well as recent seed organelle proteomics.

10.2 Two-Dimensional Electrophoresis to Profile Protein Expression During Seed Filling in Soybean

Proteomics research has transcended from the early stages of protein cataloging or "parts listing" towards a more quantitative, profiling technology useful for making metabolic comparisons and predictions (Mooney et al. 2006). In addition to standard

Table 10.1 Summary of proteomics research on oilseed crop plants

Plant investigated	System investigated	References
	Mature seeds	Herman et al. 2003; Mooney et al. 2004;
		Hollung et al. 2005; Natarajan et al. 2005, 2006
Soybean	Developing seeds	Hajduch et al. 2005
	Root hairs	Wan et al. 2005
	Nodule	Panter et al. 2000
	Nodule – mitochondria	Hoa Le et al. 2004
	Developing seeds	Hajduch et al. 2006
	Phloem	Giavalisco et al. 2006
Brassica and Arabidopsis	Leaf and stem	Albertin et al. 2005; Subramanian et al. 2005
	Mitochondrial compartment, oil body	Mihr et al. 2001; Katavic et al. 2006
Castor	Endoplasmic reticulum	Maltman et al. 2002
Sunflower	Mature seeds	Dong et al. 2006
	Callus lines	Jain et al. 2006
Peanut	Allergenic proteins	Shefcheck and Musser 2004; Magni et al. 2005a, 2005b; Restani et al. 2005; Boldt et al. 2005

pairwise comparisons, quantitative protein profiling across developmental time points or multiple, dissected tissues offers more specific, relational information about a given system. Despite new advancements in the area of gel-free or chemical labeling quantitative proteomics, the time-honored approach of 2-DE is still one of the most reliable options for relative quantification of proteins in a complex sample. The utility of this approach is evident given the extensive quantitative profiling that has been recently demonstrated with developing seed of soybean and rapeseed (Hajduch et al. 2005, 2006). However, for large-scale studies such as these, 2-DE must be performed with the highest level of reproducibility, requiring medium- or narrow-range immobilized pH gradient (IPG) strips to minimize the number of "crowded" or overlapping spot boundaries. Only those spots confirmed in at least three independent, biological replicates should be included for quantification. To achieve this criterion, it is advisable to run a minimum of four biological replicate 2D gels. Although general recommendations, these parameters should be considered before initiating any large scale 2D-gel based study.

Employing these guidelines, a recent investigation on soybean (var. Maverick) seed filling (Hajduch et al. 2005) successfully profiled 679 2-DE protein spots at five sequential developmental stages (2, 3, 4, 5 and 6 WAF). Analysis of each of these protein spot groups by matrix assisted laser desorption ionization-time of flight (MALDI-TOF) mass spectrometry (MS) yielded the identity of 422 of these proteins, representing 216 non-redundant protein activities. These proteins were classified into 14 functional classes according to the classification scheme used for the *Arabidopsis thaliana* genome (Bevan et al. 1998). From this investigation an interactive web database was developed to disseminate both the expression profiles and the peptide mass fingerprint (PMF) information for each of the protein spot groups characterized. Significantly, approximately 22% of the identified proteins were annotated as "unknown" based solely upon the annotation provided with the soybean UniGene database of non-redundant, contiguous expressed sequence tag (EST) sequences. In the absence of genome information, we have previously determined that UniGene or tentative consensus (TC) EST databases are highly useful for protein assignment queries (Mooney et al. 2004). As there are currently over 40 ongoing EST investigations in model and crop plants, annotated assemblies of these sequences will continue to be valuable as database resources for proteomics research.

To characterize expression trends within classes (and subclasses) of proteins (e.g., class: metabolism; subclass: glycolysis) individual expression trends for proteins belonging to these classes were summed to create what is referred to as "composite expression profiles." Composite expression profiling allows some generalizations to be made about soybean seed filling. For example, most metabolic proteins decrease in abundance with seed age, perhaps reflecting metabolic quiescence near the end of the seed filling period and entrance into the seed maturation phase of development. On the other hand, seed storage proteins (SSP; glycinin and conglycinin predominantly) increase spectacularly during this developmental time period, consistent with the well-established function of these proteins as nutrient storage reserves for germination. Three of the more prominent non-SSP proteins from this study include sucrose synthase (SuSy), sucrose-binding proteins (SBP), and lipoxygenases (LOX); 6, 13, and 16 2-DE protein species with these respective activities were observed. The six protein spots identified as SuSy mapped to the same UniGene accession number (Gma.1828). In general, expression of these six protein spots increased until the middle stages of development (3–5 WAF; relative abundance maxima of 0.07, 0.08, 0.13, 0.14, 0.21, and 0.26). This trend agreed with previous transcript expression studies on SuSy during seed development in another legume: *Vicia fabia* (Heim et al. 1993). Unlike mRNA expression data, proteomics provided specific information about individual protein species as some isoforms of SuSy decreased in abundance earlier (or later) than others. The abundance of SuSy, coupled with the multiple isoforms observed, points to this activity as a potentially important step in carbon assimilation during seed filling in soybean.

This investigation also identified 13 protein spots mapping to UniGene accession number Gma.1872 encoding a 62 kDa SBP. This protein has been previously suggested to be involved in sucrose transport (Grimes et al. 1992). In this case, the expression profiles of the 13 isoelectric species were markedly different. Seven protein spots increased throughout the period of seed filling, with relative abundance maxima of 0.05, 0.06, 0.11, 0.19, 0.20, 0.28, and 0.70; in total approximately twice as abundant as SuSy expression. The remaining six spots showed an opposite, decreasing trend during seed filling, with peak maxima of 0.09, 0.10, 0.11, 0.13, 0.21, and 0.42. The overall prominence of SBPs, their dichotomous expression trends, and the unusually high number of 2-DE species indicates that this activity is regulated in a diverse, complex manner. A previously published cDNA microarray investigation of *A. thaliana* seed development showed that SBP transcript abundance increases with seed age (Ruuska et al. 2002), which agrees with the bulk of SBP protein expression in soybean. As most of the phloem-transported carbon is in the form of sucrose it is perhaps not surprising that SuSy and SBP are highly represented proteins in a 2-DE investigation on soybean seed filling. Previous investigations have discussed the importance of sucrose as a signaling molecule for embryo and, more generally, for seed development (reviewed by Borisjuk et al. 2004).

Another prominent activity in developing soybean seed are proteins annotated simply as lipoxygenase. Although the physiological function of seed-expressed LOX is not completely clear, the presence of these proteins in mature soybean seed is highly undesirable due to their ability to modify seed oil with age as well as post-processing. In this investigation, 16 protein spots matching to 8 unique LOX UniGene accessions were identified. The expression trends of these proteins varied widely and the relative abundances at peak of expression ranged from 0.06 to 0.80, with a total relative abundance of 4.16. The diverse expression patterns and high number of unique LOX genes expressed during seed filling suggests highly varied functions of these activities, perhaps including jasmonic acid production (Vick and Zimmerman 1983) and/or defense responses (Keppler and Novacky 1987).

10.3 Proteomic Analysis of Seed Filling in Rapeseed

A parallel 2-DE study of seed filling in rapeseed (*Brassica napus* var. Reston) resulted in expression profiles for 794 protein spots (Hajduch et al. 2006). Each of these 794 spots was subjected to MALDI-TOF MS and nanoLC-MS/MS in parallel, together resulting in the identification of 517 spots representing 289 non-redundant proteins. Hierarchical cluster analysis of the protein expression data (2, 3, 4, 5, and 6 WAF) for each of the 794 protein expression profiles yielded 12 principal expression trends. Like soybean, SSPs (cruciferin, napin) were the most prominent proteins expressed in this developmental study. However, unlike soybean there were no highly expressed non-SSPs such as LOX and SBPs in rapeseed. The reduced dynamic range of protein expression in rapeseed allowed for greater coverage of the seed proteome. Indeed, proteins belonging to energy and metabolism functional classes constituted nearly 50% of the total expressed proteins. These activities were mapped to metabolic pathways to predict the flow of carbon from sucrose to fatty acids, the major storage component of rapeseed. It was concluded that sucrose is cleaved by SuSy, and the products – UDP-glucose and fructose – follow the cytosolic glycolytic pathway to phosphoenolpyruvate (PEP), which is imported into plastids. Following protein expression data, PEP is then converted to pyruvate by a plastid pyruvate kinase (PK), and further to acetyl-CoA by the plastid pyruvate dehydrogenase complex (PDC). Acetyl-CoA is then converted to malonyl-CoA by the heteromeric, plastid acetyl-CoA carboxylase, and malonyl-CoA enters the plastid de novo fatty acid biosynthetic pathway. This conclusion is supported by the expression of proteins for each of the enzymatic steps from sucrose to plastid pyruvate. Although there could be exchange of carbon between the cytosol and plastids at metabolic steps earlier than PEP (notably glucose-6-phosphate and triose phosphate), and the prominence of plastid-targeted triose phosphate isomerase and fructose bisphosphate aldolase (FBA) activities do not discount this, it seems that the principal flow from sucrose to oil in rapeseed involves SuSy and a largely cytosolic glycolytic track. It should be pointed out that, although each of the soluble metabolic activities were identified, no carbon translocators were detected from this 2-DE study. However, this is not surprising given the under-representation of integral membrane proteins in 2-DE experiments. Compared to soybean, metabolic activities in rapeseed appeared to be highly expressed and with a surprising degree of redundancy with respect to the frequency of isoelectric species. For example, nine individual protein spots from 2-DE were identified as cytosolic FBA and an additional eight 2-DE species were mapped to putative plastid-targeted FBAs.

The large subunit of ribulose-1,5-bisphosphate carboxylase/oxygenase (RuBisCO) is encoded by a single gene encoded in the plastid organellar genome. Interestingly, this investigation identified 11 protein spots matching this single gene, suggesting a high level of PTM. It is well known that RuBisCO plays an important role in the Calvin cycle by catalyzing the carboxylation of ribulose 1,5-bisphosphate (R-1,5bisP) to produce two molecules of 3-phosphoglycerate (3-PGA). Recently, it was shown that, in developing rapeseed embryos, RuBisCO functions without a complete Calvin cycle and bypasses glycolytic enzymes glyceraldehyde-3-phosphate dehydrogenase (GAPDH) and phosphoglycerate kinase (PGK) to recycle half of the carbon dioxide released by the plastid PDC (Schwender et al. 2004). The recycling of carbon lost by the plastid PDC is essential to maintain the efficiency of oil production in the embryo (Schwender et al. 2004). The prominence of RuBisCO in this proteomics investigation is therefore not surprising, although the number of isoforms and their expression patterns certainly are. Based on their expression profiles, the 11 2-DE spots identified as RuBisCO in this proteomics study could be divided into two groups. The first group showed maximum abundance at 2 and 3 WAF, and include eight isoforms with peak relative abundances of 0.03, 0.07, 0.09, 0.09, 0.11, 0.13, 0.13, and 0.21. The second group exhibited maximal accumulation at 4 WAF (relative abundance 0.5, 0.13, and 1.4) with typical bell-shaped expression profiles, similar to expression profiles for subunits of the plastid PDC and for enzymes involved in de novo fatty acid biosynthesis.

10.4 Comparison of Carbon Assimilation During Seed Filling in Soybean and Rapeseed

Proteomics investigations on seed filling in soybean and rapeseed are beginning to contribute to the understanding of the molecular and biochemical characteristics of seed development, particularly with respect to metabolic regulation. If proteomic analysis of any single developing seed is helpful to begin predicting metabolic flow, parallel studies on multiple, diverse oilseeds should be invaluable in allowing one to discover and predict conserved, as well as specialized, routes of carbon flow.

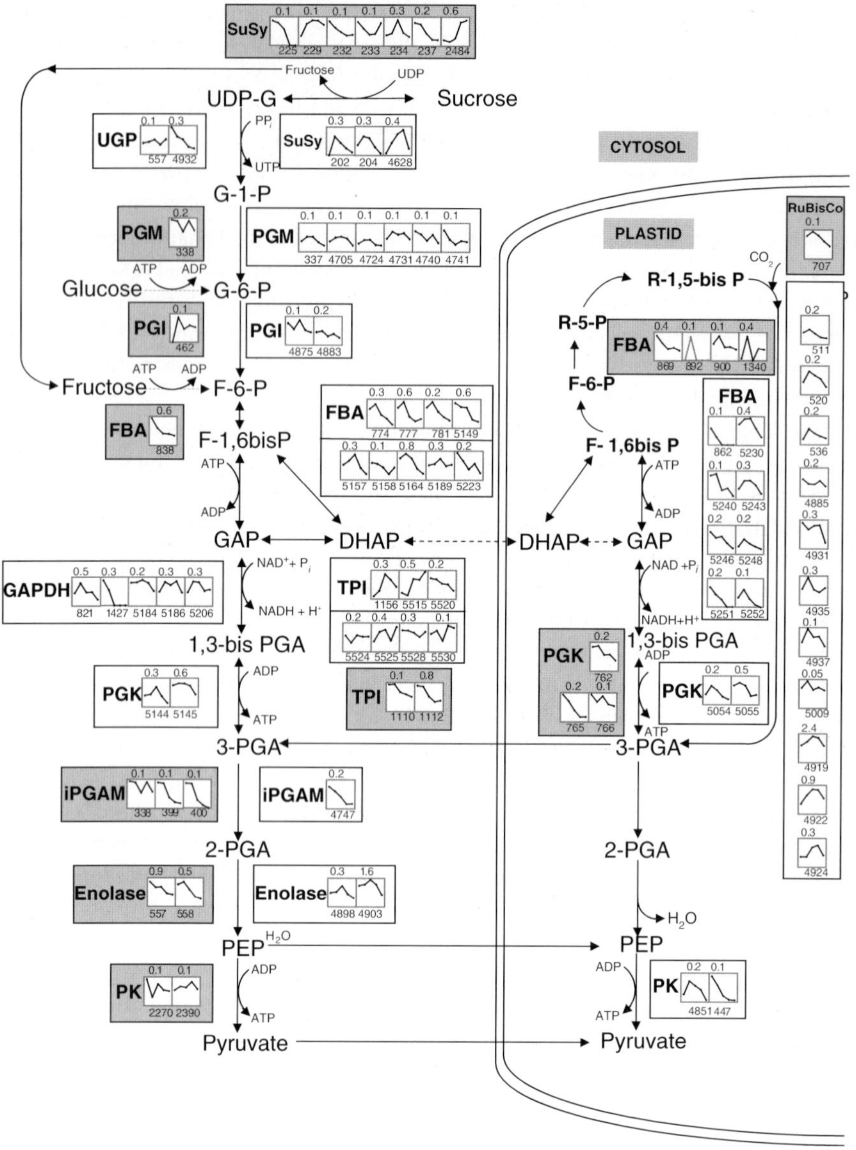

Fig. 10.1 Comparative view of sucrose assimilation during seed filling in soybean and oilseed rape based upon protein expression data (Hajduch et al. 2005, 2006). Proteins (1 mg) isolated from whole seed of soybean (*gray boxes*) and oilseed rape (*white boxes*) at 2, 3, 4, 5, and 6 weeks after flowering (WAF) were resolved by two-dimensional gel electrophoresis (2-DE), minimally in biological triplicate from each of the five stages. Coomassie Brilliant Blue (CBB)-stained 2-DE gels were imaged (16 bit scanning densitometry) and analyzed using ImageMaster Platinum software (GE Healthcare, Piscataway, NJ). Designation of a spot group by ImageMaster required the spot to be detected in a minimum of three replicate gels from at least two developmental stages. Developmental expression data are plotted as a line graph and time points are the mean of three or four biological replicates. Each graph represents a unique 2-DE spot group (number below each

As more investigations such as these are performed on different oilseeds it will therefore be particularly interesting to collect and compare the results to further refine metabolic predictions. As a prelude to this long-term goal, Fig. 10.1 summarizes the combined proteomics results acquired from soybean and rapeseed (Hajduch et al. 2005, 2006) with respect to sucrose assimilation. Figure 10.1 includes updated tandem mass spectrometry (MS/MS)-based protein assignments for previously profiled soybean 2-DE spots (Hajduch et al. 2005).

On a relative abundance basis, SuSy was found to be expressed twofold higher in developing soybean versus oilseed rape seeds. In soybean, seven isoelectric species (1.4 total relative abundance at peak of expression) were detected, whereas in oilseed rape three isoelectric species of SuSy were detected, comprising 0.65 total relative abundance. The prominence of SuSy in soybean and, to a lesser degree, in rapeseed suggests that sucrose degradation into UDP-glucose and fructose initiates phloem-derived disaccharide assimilation in both *Fabaceae* and *Brassicaceae* oilseeds. Cytosolic phosphoglucomutase (PGM) was expressed at 0.1 relative abundance in one detected spot in soybean and 0.31 total abundance in six detected spots in rapeseed. Only one phosphoglucose isomerase (PGI) spot was detected in soybean (0.03 relative abundance) while rapeseed contained two species (0.14 total relative abundance) for PGI. One cytosolic FBA 2-DE spot was detected in soybean (0.6 relative abundance) while nine cytosolic FBA species were detected in rapeseed with a total abundance of 3.1. Multiple putative plastid-targeted FBA protein isoforms were also identified in both soybean and rapeseed (Fig. 10.1). The prominence of FBA expression was undoubtedly the most surprising finding from these investigations of soybean and oilseed rape. It is unclear why both cytosolic and plastidial FBA are so strongly expressed, although it points to the interconversion of 3C and 6C sugars as a potentially important step for carbon homeostasis. Cytosolic triose-phosphate isomerases (TPI) were observed in soybean (two species with total relative abundance of 0.9) although the number of isoforms and their

graph) that was successfully identified by a combination of MALDI-TOF (matrix assisted laser desorption ionization-time of flight) and nanoLC-MS/MS (liquid chromatography-tandem mass spectroscopy). The numbers above each graph show the maximum *y*-axis value expressed in relative volume. Soybean data are the combined profiling and identification data from Hajduch et al. (2005) and a recent re-analysis of each 2-DE spot by nanoLC-MS/MS (G.K. Agrawal, unpublished data). NanoLC-MS/MS acquisitions and database search conditions for soybean proteins were similar to those reported previously for oilseed rape (Hajduch et al. 2006). Metabolites: *UDP-G* UDP-glucose, *G-1-P* glucose 1 phosphate, *G-6-P* glucose 6 phosphate, *F-6-P* fructose 6 phosphate, *F-16-bisP* fructose 16-bisphosphate, *GAP* glyceraldehyde 3-phosphate, *DHAP* dihydroxyacetone phosphate, *13-bis PGA* 13-bisphosphoglyceric acid, *3-PGA* 3 phosphoglyceric acid, *2-PGA* 2 phosphoglyceric acid, *PEP* phosphoenolpyruvate. Enzymes: *SySy* sucrose synthase, *PGM* phosphoglucomutase, *PGI* phosphoglucose isomerase, *FBA* fructose bisphosphate aldolase, *TPI* triose-phosphate isomerase, *PGK* phosphoglycerate kinase, *iPGAM* 2,3-bisphosphoglycerate-independent phosphoglycerate mutase, *PRK* phosphoribulokinase, *RuBisCo* ribulose-15-bisphosphate carboxylase/oxygenase, *PK* pyruvate kinase total abundance was higher in rapeseed (seven at 1.4 total relative abundance). In contrast, the remaining three enzymes of the cytosolic glycolytic pathway, 2,3-bisphosphoglycerate-independent phopsphoglycerate mutase (iPGAM) and enolase, were slightly more abundant in soybean than rapeseed (Fig. 10.1). Surprisingly, PK was detected as a cytosolic form in soybean and as a plastidial form in rapeseed, both as two 2-DE species.

In plastids, a surprisingly big difference in the number of isoforms for ribulose-1,5-bisphosphate carboxylase (RuBisCO) large subunit was observed between rapeseed and soybean (Fig. 10.1). The preponderance of RuBisCO in rapeseed versus soybean is unclear but could reflect different carbon dioxide re-capturing mechanisms in these oilseeds. Alternatively, the stark difference in quantities of oil synthesized in these seed (18% versus 40% seed dry weight in soybean and rapeseed, respectively) could mean reduced emphasis on carbon yield in soybean. Interestingly, plastid FBA, which is likely needed for regenerating R-1,5-bisP, was also reduced in soybean compared to rapeseed, both in quantity and number of isoelectric species. On the other hand, the level of plastid PGK was almost equivalent in these two oilseeds. Overall, it appears that plastid intermediary metabolism may differ substantially between soybean and rapeseed and may therefore warrant further investigation.

10.5 Proteomics of Mature Soybean Seed

Proteomic analysis of seed filling principally reveals the metabolic pathways leading up to seed maturation. As the developing embryo approaches maturation and begins to desiccate, most biosynthetic pathways enter a quiescent state, and it would appear that many proteins are degraded or down-regulated in expression (Hajduch et al. 2005, 2006). In contrast, other proteins such as late embryogenesis (LEA) proteins and albumins are strongly induced. This remodeling of proteins upon embryo maturation is substantial, resulting in the appearance and disappearance of many protein spots on 2-DE gels. Indeed, the first proteomics research on soybean focused on the characterization of proteins present in mature seed. Herman et al. (2003) investigated mature soybean (cv. Jack) seeds for protein composition as compared to transgenic soybean reduced in a low abundance seed allergen, Gly m Bd 30K. This study identified 108 2-DE protein spots (excluding keratin and a yeast protein), revealing beta-conglycinin, glycinin, SBPs, LOXs, LEA, seed maturation proteins, agglutinins, trypsin inhibitor, vacuolar thiol protease, and annexins as the bulk of the expressed proteins. In another investigation, mature soybean (cv. Jefferson) seeds were analyzed by quantitative 2-DE followed by automated PMF resulting in the identification of 44 protein spots (Mooney et al. 2004). In addition to SBPs, LOX and maturation proteins were observed. Apart from the abundant beta-conglycinin, glycinin, and SBP proteins, which accounted for approximately 60% of the seed protein content, two proteasome complex subunits, glutathione S-transferase, and alcohol dehydrogenase were identified. Although these investigations accounted for perhaps 90% of the proteins that accumulate in mature soybean seed, the downside is that another 10% of masked proteins have yet to be identified. To achieve this it might be necessary to employ selective protein subtraction procedures including lectin affinity chromatography to remove the dominant soybean seed proteins (Johnston and Miernyk 2006).

Comparative proteomics utilizing high resolution 2-DE is a powerful strategy for identifying differences in crop cultivars and varieties. In soybean, a combination of near-infrared reflectance spectroscopy with 2-DE resulted in rapid analysis of a total of 832 samples of soybean (Hollung et al. 2005). In this case, near-infrared reflectance spectroscopy was used to screen soybean samples for a lower content of oligosaccharides and nonstarch polysaccharides. Based on these results, 12 samples were selected for further characterization by chromatography and 2-DE. The protein analyses showed varying intensities of several proteins, including the glycinin G1 precursor, between soybean varieties, suggesting a potential relationship between storage protein and complex carbohydrates.

A more in-depth analysis of SSP accumulation in wild vs cultivated soybean seed was performed by Natarajan et al. 2006. Broad and medium range 2-DE was used to resolve and compare total seed proteins from *Glycine soja* and *Glycine max*, with specific emphasis on the two major seed storage proteins glycinin and beta-conglycinin. In total, 44 and 34 2-DE spots were identified as glycinin or beta-conglycinin in wild and cultivated genotypes, respectively. When broad range (pI 3–10) isoelectric focusing (IEF) was used, the 2-DE patterns of cultivated and wild soybean appeared similar. However, medium range IEF (pI 4–7 and pI 6–11) revealed differences between investigated soybean cultivars, emphasizing the importance of spot resolution for comparative 2-DE investigations.

The nutritional value of oilseeds, particularly soybean and peanut, is sometimes compromised by the presence of allergenic proteins, some of which are storage proteins. Identification and characterization of seed allergenic proteins will be a critical step to eventually reduce the expression of these proteins through genetic and biotechnology means. However, reducing the expression of seed allergens could alter the expression of other seed proteins and therefore detrimentally influence other agronomic traits. This question was addressed in an investigation by Herman et al. (2003). Targeted gene silencing of the major soybean allergen Gly m Bd 30 K, which also happens to be a prominent seed protein, revealed strong suppression of this protein but no other significant changes in seed protein composition as determine by 2-DE. This finding suggests that Gly m Bd 30 K does not have a major role in seed protein processing or maturation. This example clearly illustrates the utility of 2-DE (and MS) in applied research aimed at characterizing transgenic crops for potential, unexpected changes in gene/protein expression.

10.6 Organelle and Tissue Specific Proteomics – Towards "In-Depth" Oilseed Proteome Characterization

To date, most proteomics investigations in oilseeds have focused on the whole seed. However, it is evident that low abundance proteins can be difficult to identify from such multi-tissue and mixed organellar protein samples. Tissue dissection or cellular pre-fractionation can greatly improve the detection of low abundance or tissue-/organelle-specific proteins. Indeed, plant "organelle proteomics" is in the advanced stages for organelles that can be readily isolated from prominent organs, particularly leaves. Two chapters in this book are devoted to proteomics of plastids and mitochondria and we point to these chapters for detailed information on the depth of this field in plants [see Chaps. 14 (Jarvis) and 15 (Millar), this volume].

Plant seeds possess a specialized organelle known as an oil body, which functions to store triacylglycerol (TAG) lipid for the germinating embryo. Oil bodies are lipid monolayer organelles that bud from the endoplasmic reticulum (ER) as TAG accumulates within the ER membrane leaflets (Frey-Wyssling et al. 1963). Greater than 50% of the cellular volume of a mature plant embryonic cell can be occupied with oil bodies. Employing both 2-DE and multi-dimensional liquid chromatography-tandem mass spectrometry (also referred to as MudPIT), Katavic et al. (2006) has provided the most comprehensive annotation of the oil body proteome from any plant to date. Oil bodies isolated from two cultivars of oilseed rape (cvs. Westar and Reston) were analyzed in parallel directly by MS or washed with 2 M sodium chloride or 8 M urea to strip off non-specifically associated proteins prior to MS analyses (Katavic et al. 2006). Besides the previously identified oil body proteins oleosin, putative embryo specific protein ATS1 (similar to caleosin), and 11-beta-hydroxysteroid dehydrogenase-like protein (steroleosin), several new proteins were identified from both rapeseed cultivars. For instance, this analysis identified a short-chain dehydrogenase, similar to a TAG-associated factor, tightly associated (resistant to 8 M urea dissociation) with rapeseed oil bodies. Another oil body-associated protein, annotated as myrosinase-associated protein, shows homology to the lipase/hydrolase family of enzymes with GDSL-motifs. It was postulated that these two proteins could be involved in oil body turnover. Another surprising outcome of this study was the strong association between oil bodies and myrosinases, which remain associated with oil bodies even after washing with 8 M urea. Despite this tight association, it is possible that the association of myrosinases with in vitro-isolated oil bodies could be due to their prominence in mature *B. napus* seed. Indeed, cruciferin and napin proteins were also identified in these oil body preparations, suggesting that non-specific association of prominent seed proteins is a possibility. Interestingly, few of these oil body proteins were identified from the whole seed proteome investigation by Hajduch et al. (2006), indicating that organelle proteomics will likely be necessary to fully define the seed proteome.

Upstream of oil body formation, the ER is the site of fatty acid modification and lipid synthesis, most notably TAG packaging. To better understand the protein composition of this organelle, an in vitro isolation procedure was developed and applied to castor seed (Maltman et al. 2002). Analysis of developing castor seed ER proteins by SDS-PAGE and 2-DE (pI 3–10) revealed approximately 30 bands and 300 protein spots, respectively, by Coomassie Brilliant Blue (CBB) or silver staining. Thirteen unique proteins were identified using a combination of N-terminal sequencing, MALDI-TOF MS, and nanoLC-MS/MS (Maltman et al. 2002). Previously known ER proteins, including protein disulfide isomerase, luminal binding protein (BiP), calreticulin, and oleate-12-hydroxylase, were successfully identified, as well as the castor storage lectins ricin and agglutinin. An aspartate protease and N-carbamyl-L-aminohydrolase-like protein (both from Arabidopsis) were also identified in this investigation; the former likely involved in processing of storage proteins. A comparison of ER proteins from germinating and developing castor seed using 2-DE revealed significant changes in protein expression, suggesting that this organellar compartment is dynamically regulated in response to environmental cues (Maltman et al. 2002).

10.7 Presence of Functionally Diverse Phosphoproteins in Developing Rapeseed

As discussed, there has been progress in identifying and profiling proteins expressed in developing or mature oilseed crops, most notably soybean and oilseed rape. These recent large-scale proteomic studies have provided a large volume of information on the metabolic pathways responsible for accumulation of storage reserves such as carbohydrates, oils, and proteins. Despite this information, we know little about the post-translational regulatory mechanisms controlling these pathways in oilseed plants, and plants in general. However, with new proteomics technologies this is likely to change.

For example, a recent report systematically quantified the dynamic changes in phosphorylated proteins in developing rapeseed (Agrawal and Thelen 2006). As this was the first quantitative large-scale analysis of phosphoproteins in plants, the aim of this investigation was to survey the presence and dynamics of phosphoproteins expressed during the seed filling phase using high-resolution 2-DE in combination with a phosphoprotein specific Pro-Q Diamond phosphoprotein fluorescence stain (Pro-Q DPS) and LC-MS/MS.

One of the most important steps in this study was the detection and quantification of phosphoproteins on 2-DE gels using Pro-Q DPS. Pro-Q DPS binds directly to the phosphate moiety of phosphoproteins with high sensitivity and linearity regardless of the phosphoamino acid, and is fully compatible with other staining methods and MS (Steinberg et al. 2003). Recently, a modified protocol for Pro-Q DPS was established in order to overcome some of the inherent limitations of Pro-Q DPS, such as reproducibility, high cost, and the large volume required for large-format 2-DE gels (Agrawal and Thelen 2005). Advantages of the modified protocol are: (1) it is shorter than the existing protocol; (2) it reproducibly detects phosphoproteins in large-format 2-DE gels; (3) it gives lower background and more distinct detection of very low abundance phosphoproteins; (4) it reduces the cost of Pro-Q DPS by threefold (directly by threefold dilution of Pro-Q DPS stock); and (5) it further reduces the overall cost of Pro-Q DPS by decreasing the manufacturer's recommended volume for one 2-DE gel by more than threefold. Overall, the modified protocol is simple, reproducible, and suitable for conducting large-scale phosphoproteomics economically.

A phosphoproteomics study conducted on five sequential stages (2, 3, 4, 5, and 6 WAF) of oilseed rape seed development led to the establishment of a high-resolution 2-DE gel phosphoprotein reference map (Agrawal and Thelen 2006). Approximately 300 phosphoprotein spots were detected on this reference map, and quantitative expression profiles for 234 high-quality spots were determined. Hierarchical cluster analyses of these expression profiles revealed the occurrence of six principal expression trends during seed filling. Out of 234 phosphoprotein spots, the identity of 103 spots was determined using LC-MS/MS, which represented 70 non-redundant phosphoproteins. These non-redundant phosphoproteins were classified into ten major functional categories: energy (25.7%), metabolism (18.6%), protein destination (12.9%), signal transduction (11.4%), disease/defense (8.6%), unclassified (8.6%), protein synthesis (4.3%), storage proteins (4.3%), cell structure (2.9%), and cell growth/division (2.9%). This functional classification revealed that more than 44% of the identified phosphoproteins were enzymes involved in various metabolic pathways. In addition to metabolic enzymes, tubulin beta-8 chain, luminal binding protein, heat shock proteins, proteasome proteins, 14-3-3 proteins, annexins, and cruciferin subunits were also characterized as phosphoproteins.

Surprisingly, 20 2-DE phosphoprotein spots were identified as cruciferin proteins. More than 50% of these proteins were expressed predominantly at 2 WAF followed by a sharp decrease in their expression levels, suggesting that phosphate storage is an unlikely function for cruciferin phosphorylation. A recent study in *Arabidopsis* has also identified storage proteins, including cruciferin, as phosphoproteins (Irar et al. 2006). These two independent studies imply that phosphorylation of seed storage proteins like cruciferin may have an important, evolutionary conserved role in seed development. Further study will help in understanding the regulatory function of seed storage proteins.

Phosphorylation sites were also mapped within 16 non-redundant phosphoproteins, including 14-3-3 proteins, annexin, tubulin beta-8 chain, and 26S proteasome protein. Only one of the identified sites was previously postulated in 14-3-3 protein; the remaining sites were novel. It was proposed that the identified 14-3-3 protein might play a potential role in mediating signaling and metabolic pathways within the developing seed. This study exploited the use of Pro-Q DPS for large-scale analysis of phosphoproteins and provides evidence for the presence of a large number of functionally diverse phosphoproteins within developing rapeseed.

10.8 Conclusions and Future Perspectives

Research into seed development in oilseeds is beginning to take advantage of recent technological developments in the field of proteomics. Compared to six publications leading up to the year 2004, 17 publications pertaining to proteomics research on oilseed crops were reported in 2005 and 2006 (Table 10.1). Despite increased proteomics research on seed development, global systematic investigations into developing seed organelles and specific tissues are relatively rare and remain as opportunities. It would be especially interesting to compare proteomes of the same seed tissue (e.g., seed coat), or perhaps organelles among several diverse oilseed plants, to identify conserved, targeted processes or metabolic pathways. For instance, an in-depth proteomics investigation of organelles might provide empirical evidence for novel metabolic activity within that organelle. Despite the rich history of plant organelle proteomics {particularly mitochondria and plastids, as reviewed elsewhere in this book [see Chaps. 14 (Jarvis) and 15 (Millar), this volume]}, efforts to characterize the diversity of organelle forms (i.e., amyloplasts, leucoplasts, chromoplasts) in a systematic, quantitative manner are non-existent. The dearth of seed organelle proteomics is particularly noticeable; only a few labs have conducted proteomic investigations on plant seed organelles (Maltman et al. 2002; Katavic et al. 2006; Balmer et al. 2006).

As a systematic proteomics strategy, the integration of complementary approaches (2-DE, GelC-MS/MS, MudPIT) appears to be important for in-depth characterization of any given protein sample. The under-representation of membrane proteins and highly basic proteins is perhaps the biggest drawback for 2-DE-based protein profiling. For instance, MudPIT analyses of rapeseed oil bodies resulted in the identification of approximately 15% additional proteins compared to 2-DE (Katavic et al. 2006). However, the implementation of multiparallel approaches for protein identification can be expensive and time-consuming, not to mention technically challenging. On top of all of these concerns is the quantitative capability of the approach. Although comprehensive protein identification is the goal of any approach, if it comes at the cost of quantitative analysis, how useful is that approach beyond the initial "parts-listing" phase? Although MudPIT is arguably the most comprehensive protein identification approach available, the widely publicized lack of reproducibility almost eliminates this technique as a viable approach beyond protein cataloging. Although the field of proteomics is maturing, further method development is certainly necessary to devise facile strategies for protein identification and quantification without bias against certain classes of proteins. This is particularly important for seed biology research. Unbiased, quantitative protein profiling of whole seed has numerous potential applications for both basic and applied gene regulatory or breeding programs.

References

Agrawal GK, Thelen JJ (2005) Development of a simplified, economical polyacrylamide gel staining protocol for phosphoproteins. Proteomics 5:4684–4688

Agrawal GK, Thelen JJ (2006) Large scale identification and quantitative profiling of phosphoproteins expressed during seed filling in oilseed rape. Mol Cell Proteomics 5:2044–2059

Albertin W, Brabant P, Catrice O, Eber F, Jenczewski E, Chevre AM, Thiellement H (2005) Autopolyploidy in cabbage (*Brassica oleracea* L.) does not alter significantly the proteomes of green tissues. Proteomics 5:2131–2139

Balmer Y, Vensel WH, DuPont FM, Buchanan BB, Hurkman WJ (2006) Proteome of amyloplasts isolated from developing wheat endosperm presents evidence of broad metabolic activity. J Exp Bot 57:1591–1602

Bevan M, Bancroft I, Bent E, Love K, Goodman H, Dean C, Bergkamp R, Dirkse W, Van Staveren M, Stiekema W, Drost L, Ridley P, Hudson SA, Patel K, Murphy G, Piffanelli P, Wedler H, Wedler E, Wambutt R, Weitzenegger T, Pohl TM, Terryn N, Gielen J, Villarroel R, De Clerck R, Van Montagu M, Lecharny A, Auborg S, Gy I, Kreis M, Lao N, Kavanagh T, Hempel S, Kotter P, Entian KD, Rieger M, Schaeffer M, Funk B, Mueller-Auer S, Silvey M, James R, Montfort A, Pons A, Puigdomenech P, Douka A, Voukelatou E, Milioni D, Hatzopoulos P, Piravandi E, Obermaier B, Hilbert H, Dusterhoft A, Moores T, Jones JD, Eneva T, Palme K, Benes V, Rechman S, Ansorge W, Cooke R, Berger C, Delseny M, Voet M, Volckaert G, Mewes HW, Klosterman S, Schueller C, Chalwatzis N (1998) Analysis of 1.9 Mb of contiguous sequence from chromosome 4 of *Arabidopsis thaliana*. Nature 391:485–848

Bockish M (1993) Fats and oil handbook. AOCS, Champaign

Boldt A, Fortunato D, Conti A, Petersen A, Ballmer-Weber B, Lepp U, Reese G, Becker WM (2005) Analysis of the composition of an immunoglobulin E reactive high molecular weight protein complex of peanut extract containing Ara h 1 and Ara h 3/4. Proteomics 5:675–686

Borisjuk L, Rolletschek H, Radchuk R, Weschke W, Wobus U, Weber H (2004) Seed development and differentiation: a role for metabolic regulation. Plant Biol 6:375–386

Dong GJ, Pan WD, Liu GS (2006) The analysis of proteome changes in sunflower seeds induced by N^+ implantation. J Biosci 31:247–253

Eyre JV (1931) Notes on oil development in the seed of a growing plant. Biochem J 25:1902–1908

Fernie AR, Carrari F, Sweetlove LJ (2004) Respiratory metabolism: glycolisis, the TCA cycle and mitochondrial electron transport. Curr Opin Plant Biol 7:254–61

Frey-Wyssling A, Grieshaber E, Muhlethaler K (1963) Origin of spherosomes in plant cells. J Ultrastruct Res 8:506–516

Giavalisco P, Kapitza K, Kolasa A, Buhtz A, Kehr J (2006) Towards the proteome of *Brassica napus* phloem sap. Proteomics 6:896–909

Grimes HD, Overvoorde PJ, Ripp K, Franceschi VR, Hitz WD (1992) A 62-kD sucrose binding protein is expressed and localized in tissues actively engaged in sucrose transport. Plant Cell 4:1561–1574

Gunstone FD (2006) Fats and oils forecast. INFORM 17:667–669

Hajduch M, Ganapathy A, Stein JW, Thelen JJ (2005) A systematic proteomic study of seed filling in soybean. Establishment of high-resolution two-dimensional reference maps, expression profiles, and an interactive proteome database. Plant Physiol 137:1397–1419

Hajduch M, Casteel JE, Hurrelmeyer KE, Song Z, Agrawal GK, Thelen JJ (2006) Proteomic analysis of seed filling in *Brassica napus*. Developmental characterization of metabolic isozymes using high-resolution two-dimensional gel electrophoresis. Plant Physiol 141:32–46

Heim U, Weber H, Baumlein H, Wobus U (1993) A sucrose-synthase gene of *Vicia faba* L.: expression pattern in developing seeds in relation to starch synthesis and metabolic regulation. Planta 191:394–401

Herman EM, Helm RM, Jung R, Kinney AJ (2003) Genetic modification removes an immunodominant allergen from soybean. Plant Physiol 132:36–43

Hoa Le TP, Nomura M, Kajiwara H, Day DA, Tajima S (2004) Proteomic analysis on symbiotic differentiation of mitochondria in soybean nodules. Plant Cell Physiol 45:300–308

Hollung K, Overland M, Hrustic M, Sekulic P, Miladinovic J, Martens H, Narum B, Sahlstrom S, Sorensen M, Storebakken T, Skrede A (2005) Evaluation of nonstarch polysaccharides and oligosaccharide content of different soybean varieties (*Glycine max*) by near-infrared spectroscopy and proteomics. J Agric Food Chem 53:9112–9121

Irar S, Oliveira E, Pages M, Goday A (2006) Towards the identification of late-embryogenic-abundant phosphoproteome in Arabidopsis by 2-DE and MS. Proteomics 6:S175–S185

Jain S, Srivastava S, Sarin NB, Kav NN (2006) Proteomics reveals elevated levels of PR 10 proteins in saline-tolerant peanut (*Arachis hypogaea*) calli. Plant Physiol Biochem 44:253–259

Johnston ML, Miernyk JA (2006) Chemical cross-linking immobilized concanavalin A for use in proteomic analyses. Prep Biochem Biotechnol 36:203–214

Katavic V, Agrawal GK, Hajduch M, Harris SL, Thelen JJ (2006) Protein and lipid composition analysis of oil bodies from two *Brassica napus* cultivars. Proteomics 6:4586–4598

Keppler LD, Novacky A (1987) The initiation of membrane lipid peroxidation during bacteria-induced hypersensitive reactions. Physiol Mol Plant Pathol 30:233–245

Magni C, Herndl A, Sironi E, Scarafoni A, Ballabio C, Restani P, Bernardini R, Novembre E, Vierucci A, Duranti M (2005a) One- and two-dimensional electrophoretic identification of IgE-binding polypeptides of *Lupinus albus* and other legume seeds. J Agric Food Chem 53:4567–4571

Magni C, Ballabio C, Restani P, Sironi E, Scarafoni A, Poiesi C, Duranti M (2005b) Two-dimensional electrophoresis and western-blotting analyses with anti Ara h 3 basic subunit IgG evidence the cross-reacting polypeptides of *Arachis hypogaea*, *Glycine max*, and *Lupinus albus* seed proteomes. J Agric Food Chem 53:2275–2281

Maltman DJ, Simon WJ, Wheeler CH, Dunn MJ, Wait R, Slabas AR (2002) Proteomic analysis of the endoplasmic reticulum from developing and germinating seed of castor (*Ricinus communis*). Electrophoresis 23:626–639

Mihr C, Baumgartner M, Dieterich JH, Schmitz UK, Braun H-P (2001) Proteomic approach for investigation of cytoplasmic male sterility (CMS) in Brassica. J Plant Physiol 158:787–794

Mooney BP, Krishnan HB, Thelen JJ (2004) High-throughput peptide mass fingerprinting of soybean seed proteins: automated workflow and utility of UniGene expressed sequence tag databases for protein identification. Phytochemistry. 65:1733–1744

Mooney BP, Miernyk JA, Greenlief CM, Thelen JJ (2006) Using quantitative proteomics of Arabidopsis roots and leaves to predict metabolic activity. Physiol Plant 128:237–250

Natarajan SS, Xu C, Caperna TJ, Garrett WM (2005) Comparison of protein solubilization method suitable for proteomic analysis of soybean seed proteins. Anal Biochem 342:214–220

Natarajan SS, Xu C, Bae H, Caperna TJ, Garrett WM (2006) Characterization of storage proteins in wild (*Glycine soja*) and cultivated (*Glycine max*) soybean seeds using proteomic analysis. J Agric Food Chem 54:3114–3120

Norton G, Harris JF (1975) Composition changes in developing rape seed (*Brassica napus* L.). Planta 123:163–174

Panter S, Thomson R, de Bruxelles G, Laver D, Trevaskis B, Udvardi M (2000). Identification with proteomics of novel proteins associated with the peribacteroid membrane of soybean root nodules. Mol Plant Microbe Interact 13:325–333

Restani P, Ballabio C, Corsini E, Fiocchi A, Isoardi P, Magni C, Poiesi C, Terracciano L, Duranti M (2005) Identification of the basic subunit of Ara h 3 as the major allergen in a group of children allergic to peanuts. Ann Allergy Asthma Immunol 94:262–266

Ruuska SA, Girke T, Benning C, Ohlrogge JB (2002) Contrapuntal networks of gene expression during Arabidopsis seed filling. Plant Cell 14:1191–1206

Schwender J, Goffman F, Ohlrogge JB, Shachar-Hill Y (2004) Rubisco without the Calvin cycle improves the carbon efficiency of developing green seeds. Nature 432:779–782

Shefcheck KJ, Musser SM (2004) Confirmation of the allergenic peanut protein, Ara h 1, in a model food matrix using liquid chromatography/tandem mass spectrometry (LC/MS/MS). J Agric Food Chem 52:2785–2790

Steinberg TH, Agnew BJ, Gee KR, Leung WY, Goodman T, Schulenberg B, Hendrickson J, Beechem JM, Haugland RP, Patton WF (2003) Global quantitative phosphoprotein analysis using Multiplexed Proteomics Technology. Proteomics 3:1128–1144

Subramanian B, Bansal VK, Kav NN (2005) Proteome-level investigation of *Brassica carinata*-derived resistance to *Leptosphaeria maculans*. J Agric Food Chem 53:313–324

Thelen JJ, Ohlrogge JB (2002) Metabolic engineering of fatty acid biosynthesis in plants. Metab Eng 4:12–21

Vick BA, Zimmerman DC (1983) The biosynthesis of jasmonic acid: a physiological role for plant lipoxygenases. Biochem Biophys Res Commun 111:470–477

Wan J, Torres M, Ganapathy A, Thelen J, DaGue BB, Mooney B, Xu D, Stacey G (2005) Proteomic analysis of soybean root hairs after infection by *Bradyrhizobium japonicum*. Mol Plant Microbe Interact 18:458–467

Chapter 11
Proteome Analysis of *Nicotiana tabacum* Suspension Cultures

Kris Laukens, Noor Remmerie, Thomas De Vijlder, Kim Henderickx, and Erwin Witters

Abstract Suspension cultures offer a continuous and homogeneous source of cells growing under strictly controlled conditions, and are ideally suited to study molecular processes at the cellular level. *Nicotiana tabacum* cell suspension cultures are widely used to study cell structure, cell cycle and cell growth in higher plants. Despite the fact that tobacco protein identification is currently hampered by the lack of public genome sequence data, tobacco suspension cells are valuable as a model system to investigate protein dynamics in the context of cellular development. This chapter offers an overview of technical approaches developed for the separation, identification and interaction mapping of the cellular tobacco proteome. A number of proteomics applications used to study molecular regulation of cell cycle, cellular development and response to pathogen attacks are discussed.

11.1 Introduction

A cell suspension culture is a continuous source of homogeneous cells growing under strictly controlled conditions. Experimental variables usually have a strong impact on the outcome and reproducibility of any proteome experiment. Highly standardised systems like cell suspension cultures can reduce biological sample variation and facilitate reproducible proteome analysis. Suspension cells can be

Kris Laukens, Intelligent Systems Laboratory, Department of Mathematics and Computer Science, University of Antwerp, Middelheimlaan 1, 2020 Antwerp, Belgium

Noor Remmerie, Center for Proteome Analysis and Mass Spectrometry, Department of Chemistry, University of Antwerp, Groenenborgerlaan 171, 2020 Antwerp, Belgium

Thomas De Vijlder, Center for Proteome Analysis and Mass Spectrometry, Department of Chemistry, University of Antwerp, Groenenborgerlaan 171, 2020 Antwerp, Belgium

Kim Henderickx, Intelligent Systems Laboratory, Department of Mathematics and Computer Science, University of Antwerp, Middelheimlaan 1, 2020 Antwerp, Belgium

Erwin Witters, Center for Proteome Analysis and Mass Spectrometry, Department of Chemistry, University of Antwerp, Groenenborgerlaan 171, 2020 Antwerp, Belgium, E-mail: Erwin.witters@ua.ac.be

J. Šamaj and J. Thelen (eds.), *Plant Proteomics*
© Springer 2007

sampled within seconds and proteins can be extracted rapidly, making the detection of proteins and complexes with a fast turnover feasible. Since suspension cells can readily be used for feeding experiments they are ideally suited to study short-term responses to environmental stimuli. *Nicotiana tabacum* cell suspension cultures have been widely used for studies on cell structure, cell cycle and cell growth. The major disadvantage facing proteome analysis of *Nicotiana tabacum* – an amphiploid species likely resulting from an interspecific cross between *Nicotiana sylvestris* and *Nicotiana tomentosiformis* – is the lack of public tobacco genome data. With the exception of its fully sequenced plastid and mitochondrion genome (Shinozaki et al. 1986; Sugiyama et al. 2005), only a minor fraction of the tobacco nuclear genome is currently represented in public databases. Tobacco protein identification therefore relies largely on cross-species matching and matching against translated expressed sequence tag (EST)-sequences. In spite of this drawback, tobacco cell suspensions have long been preferred for the study of certain physiological processes thanks to distinct advantageous features as discussed below.

11.1.1 The Tobacco BY-2 Cell Suspension

The most commonly studied tobacco cell suspension system is the *N. tabacum* cv. Bright Yellow-2 (BY-2) cell line. BY-2 was originally derived from tobacco plants for industrial purposes, until Nagata and co-workers discovered its potential to study plant cell cycle regulation (Nagata et al. 1992; Nagata and Kumagai 1999). This model has a number of interesting features for the study of plant cell biology. The cells are relatively large, grow rapidly and can be readily transformed. More important, this cytokinin-autotrophic cell line can be easily synchronised. A highly efficient cell cycle synchronisation protocol has been developed by Nagata and colleagues (Nagata et al. 1992). Synchronisation is based on the use of aphidicolin, a DNA polymerase-inhibiting drug that, upon addition, leads to cell cycle arrest in the DNA-synthesis (S) phase of the cell cycle. When released, all cells simultaneously resume their cell cycle starting from S-phase. Synchrony remains significant until after mitosis. Alternative synchronisation approaches have been described and offer complementary approaches for other intervals of the plant cell cycle (Planchais et al. 2000). As a consequence the use of tobacco BY-2 cells has yielded many key insights into plant cell cycle regulation and has led to the understanding that a large part of the cell cycle machinery has been conserved between plants and animals, but that plants also have unique regulatory mechanisms adapted to their sessile nature (Stals and Inzé 2001; Inzé 2005). The literature is full of descriptions of the dynamics and behaviour of individual cell-cycle-regulating proteins from tobacco BY-2, including reports on their intracellular localisation, abundance, cell-cycle-phase-dependent phosphorylation and interactions. Though this role became somewhat less prevalent with the development of equivalent synchronisation protocols for certain *Arabidopsis thaliana* cell suspension cultures (Menges and Murray 2002), the tobacco BY-2 cell line remains an important model system in specific plant research areas.

11.1.2 Analytical Work Flow Considerations

Any analytical work flow must be adapted to the specific requirements of the organism under investigation. This is particularly true with recalcitrant plant tissues in order to minimise the impact of specific difficulties on both the quality and robustness of the results. The ideal work flow should, if grounded on sufficient standardisation, enable comparative studies and comparisons across different experiments. A number of considerations relevant to working with these systems are outlined in the following paragraphs.

Sampling should be optimised for cell-wall-containing cells, and should be particularly tolerant to phenolic compounds. Sampling should be rapid to permit the analysis of short-term phenomena such as phosphorylation. Our group previously proposed a protocol for sample preparation and subsequent two dimensional gel-electrophoresis (2-DE) separation of the tobacco BY-2 proteome (Laukens et al. 2004) that enables rapid sampling and immediate arrest of protein degradation and modifications, optimised separation and accurate quantification of proteins. These are essential conditions to enable the detection of minor changes in the abundance of individual proteins in fast processes, such as signal transduction or cell cycle regulation.

The most common and straightforward method to identify a protein based on homology searches is bottom-up proteome analysis by means of measuring peptide data. Directed proteolysis of a protein or proteome typically yields a composite mixture that is further analysed to yield a mass spectrum containing either peptide mass data or peptide sequence information. Each biological model and biological question being addressed has its own optimal protein identification protocol. In practice, the pivotal question is whether there is a need to maintain the peptide connectivity of a single protein in order to identify it. For successful identification of proteins from a biological model that has only limited primary sequence information in public databases, one needs to know as much information about the protein as possible when relying on cross-species identification. In general, in the case of *N. tabacum*, sequence information from about four peptides has to be gathered in order to positively identify an orthologous entry in the database (Witters et al. 2003). Another example where peptide connectivity of a protein is desirable is when information about the specific characteristics of a particular protein, such as different isozymes, post-translational modification (PTM) acquired characteristics and others, is under study. If peptide connectivity is required, the proteome should be separated into intact individual proteins by means of liquid chromatography (LC) or gel electrophoresis. The identification strategy should then be based on the combination of all available information for each separated protein in order to maximise identification rates. An additional advantage of keeping the peptide connectivity during mass spectroscopy (MS) analysis is the possibility to acquire peptide mass fingerprint (PMF) spectra. The complementary information provided by the combination of a PMF and peptide sequence information enhances identification confidence.

Two major commercial platforms are available for quantification: difference in gel electrophoresis (DiGE, GE Healthcare, Piscataway, NJ) and isobaric tags for relative and absolute quantification (iTRAQ, Applied Biosystems, Foster City, CA). Both methods allow simultaneous analysis of multiple samples and come with dedicated software support for accurate quantitative analysis. iTRAQ is based on peptide labelling of a proteolysed proteome and hence can be employed for LC-based separation and MS-based quantification. In this work flow all peptides are subjected to fragmentation analysis followed by quantitative analysis and identification. DiGE labelling is optimised to label lysine at the protein level. Gel-based separation of the proteome is followed by qualitative and quantitative detection of the fluorescent dye, and proteins are selected based on their different expression prior to MS analysis.

The outcomes of the proteomic work flow of an unsequenced organism such as tobacco must be integrated in a different manner than in the case of a sequenced model organism. In "pre-annotation proteomics" the conservation of experimental data, and the flexible and continuous reannotation of that data, is essential to turn experimental observations into useful knowledge, and to update resulting interpretations whenever more sequence and annotation information becomes available.

11.2 Two-Dimensional Electrophoretic Analysis of Tobacco Cell Suspension Proteomes

The generic gel-based protocol proposed by our group (Laukens et al. 2004) begins with rapid pipetting of a BY-2 culture sample. The culture medium is then removed by paper filtration and the cell pellet is transferred to liquid nitrogen. The whole process can be carried out in under 30 s. All subsequent actions are performed under denaturing conditions. The protein sample is ground and precipitated using acetone/trichloro-acetic acid, which removes interfering sample matrix compounds that otherwise would have a detrimental effect on the 2-DE protein display. After washing the pellet with acetone, the precipitated proteins are resolubilised in sample buffer containing urea/thiourea prior to 2-DE. Proteins are further separated by conventional methods (Görg et al. 2004) using immobilised pH gradient (IPG) strips in the first dimension and 12% sodium dodecyl sulfate polyacrylamide gel electrophoresis (SDS-PAGE) in the second dimension. Between the two electrophoretic separations, protein disulfide bridges are reduced and the resulting thiol moieties are alkylated.

A 2-DE experiment (pI 3–10, 18 cm first dimension, 12% second dimension) of 50 µg proteins of a tobacco BY-2 cell extract visualised with SYPRO ruby, yields between 1,000 and 1,500 protein spots, depending on the imaging and the geometric and volumetric spot detection parameters used. To visualise a larger number of proteins, the peak capacity can be multiplied by increasing the physical length of the first dimension. Zooming in on the pI gradient, resulting in an increased pI/cm ratio, offers the additional advantage that the sample load can be increased and enhances the detection of low abundant proteins. For example, the number of detected spots exceeds 2,000 when only the isoelectric focusing (IEF) dimension is enlarged from 18 cm to 24 cm, and loaded with a larger amount of proteins (typically 200 µg). Alternatively, certain fractions of the proteome can be enriched using pre-fractionation techniques. Whether this is a suitable option obviously depends on the biological question being addressed.

11.2.1 Tobacco BY-2 Proteome: a Reference Map

The approach described above has lead to the generation of a two-dimensional reference map for the tobacco BY-2 proteome. A large number of proteins in this map have been identified using LC-ESI-Q-TOF MS (liquid chromatography–electrospray ionisation–quadrupole–time-of-flight mass spectrometry) as described by Witters et al. (2003). Basically, the digests of individual spots are introduced to the first dimension of the LC system, where the peptides are focused on a micro-capillary reversed phase column (0.5 mm ID × 5 mm, C18) and desalted. This column is then switched in-line with the analytical dimension (0.2 mm ID × 100 mm, C18). The trapped peptides are co-eluted as a single chromatographic peak to the MS ion source. The chromatography is monitored using the full scan single MS signal. The eluting peptides are screened en masse for charge ratio, abundance and origin (for the latter parameter an extensive list containing *m/z* values of autolytic trypsin peptides and known contaminants such as keratin peptides is checked). At any given time during the chromatography up to eight peptides can be selected for subsequent peptide sequence analysis. The combined information of the individual peptide sequences per sample is then used for protein identification. Likewise, peptides obtained from individual protein spots are analysed using a matrix-assisted laser desorption/ionization (MALDI) tandem MS (MS/MS) set-up. Combined data sets of PMF and MS/MS spectra are then submitted for database querying.

Based on the results, it can be concluded that less than 15% of samples from tobacco can be correctly identified using PMF identification only. Using peptide sequence information, the hit rate for both MALDI MS/MS and ESI MS/MS was higher than 70%. About 10% of the positive hits were identified with protein entries at the species level (*Nicotiana tabacum*), and 20% were identified with protein entries at the family level (*Solanaceae*). All other proteins (69%) were identified with entries of phylogenetically more distant species. It was observed that 80% of the submitted peptide sequences resulted in a significant hit when the protein was present in the database, at the species level so in most cases a single peptide fragmentation spectrum already resulted in protein identification. Only 40% of submitted spectra resulted in an identification with a protein entry at the family level, meaning that, on average, more than two sequence spectra had to be submitted for positive identification. In cases where identification relied on more distantly related species, on average 25% (or only one in four spectra) resulted in unambiguous identification.

Identified proteins were annotated in the on-line accessible database, available at http://www.pdata.ua.ac.be/BY2. The list of identified proteins contains members of many different functional classes, with an abundance distribution covering several orders of magnitude, hydrophilic and hydrophobic proteins and proteins over the full molecular weight and isoelectric point range.

11.2.2 The Tobacco Phosphoproteome: Effects of Bacterial Lipopolysaccharides

Bacterial lipopolysaccharides (LPS) are potent inducers of plant defence reactions, and have been proposed as biological control agents and inducers of innate immunity in plants. It was shown that LPS from an endophytic strain of *Burkholderia cepacia* induced a rapid influx of Ca^{2+} and production of reactive oxygen and nitrogen species during oxidative burst in tobacco suspension cells, as well as K^+/H^+ exchange during alkalinisation of the extracellular culture medium (Gerber and Dubery 2004; Gerber et al. 2004). Moreover, LPS elicitation had specific effects at the level of protein phosphorylation.

Using a gel electrophoresis-based approach, quantitative and qualitative changes in the *N. tabacum* cv. Samsun cell suspension phosphoproteome upon LPS elicitation were studied (Gerber et al. 2006). Treated and control cultures were incubated with ^{32}P-orthophosphate, frozen, and then proteins were extracted. After 2-DE on pH 3–10 gels, autoradiograms revealed differential phosphate incorporation between treated vs non-treated cells, with a peak at about 120 min after elicitation. The phosphoprotein-specific fluorescent dye ProQ Diamond enabled direct visualisation of the phosphoproteome of treated and untreated cells. This approach revealed a number of differences: both up- and down-regulation of a number of spots was observed. The phosphostain was subsequently washed out and, using the general fluorescent stain ruthenium II tris bathophenanthroline disulfonate (RuBP), the global proteome was non-selectively visualised. A number of phosphostain-marked spots appeared to be absent in the RuBP-stained gels, suggesting a high phosphorylation state but low abundance.

In total, 32 significantly up- or down-regulated proteins were analysed using MALDI-TOF MS and nano-ESI-LC MS/MS, of which 27 were positively identified. All identified proteins were known from the literature as phosphoproteins. Among the differentially regulated phosphoproteins were one 14-3-3 protein and two putative 14-3-3 binding proteins as well as calmodulin (presumed to play a role in calcium signalling), suggesting the involvement of 14-3-3-mediated protein / phosphoprotein interactions. The results point towards a role for G-protein-coupled receptor signalling activation of Rho protein, and also activation of vacuolar (H^+)-ATPase, which plays a role in the regulation of the pH of intracellular compartments and extracellular alkalinisation. Both these pathways are known to be involved in the early defence response in plants.

Lipopolysaccharides induce reactive oxygen species (ROS), and a number of phosphoproteome changes seem to be related to this induction, including the activation of nucleoside diphosphate kinases to compensate for ATP-depletion, and thioredoxin-mediated redox regulation. Changes were also observed in the phosphorylation status of a number of chaperones. Other affected proteins were related to stress responses and cytoskeleton reorganisation.

This study demonstrates how a number of intracellular events that occur upon elicitation are accompanied, or even regulated by, targeted phosphorylation, and many targets could directly be related to known elicitor-induced intracellular events. Together, these data provide a comprehensive picture of some elicitation response events in tobacco cells at the phosphoproteome level. The study also demonstrates the suitability of cell suspension cultures for the controlled analysis of signal responses.

11.3 Proteome Analysis of Tobacco BY-2 Plastids

The presence of plastids is one of the features that distinguishes plants from other eukaryotes. Therefore, a better understanding of plastid composition and function is a key interest of plant biologists. The diversity of existing plastids, categorised according to morphology, pigment composition and primary energy metabolism, suggests that they play a central role in differential and crucial metabolic and biosynthetic processes. Unravelling these plastid processes using proteomic technologies has received attention recently (van Wijk 2004; see also Chap. 14 by Jarvis, this volume). Despite the known tobacco plastidial genome, most proteins in the plastids are encoded by the nuclear genome and protein identification still relies on the presence of orthologs in the databases.

Since plastids of tobacco BY-2 cells do not naturally develop into mature chloroplasts but rather remain in a proplastid-like state, they could represent an excellent model system for the study of proplastid and plastid development. Using a shotgun approach, Baginsky and colleagues studied the tobacco BY-2 plastid proteome (Baginsky et al. 2004). Subsequently, proteins from plastids were serially extracted in four fractions according to increasing solubility. Soluble proteins and peripheral membrane proteins were further fractionated by anion exchange chromatography, and proteins from all fractions were separated by SDS-PAGE (10%). These gels were cut into ten slices and subjected to in-gel tryptic digestion. The resulting peptides were separated by reversed-phase LC before analysis on an ion trap MS. Fragmentation data were queried against the NCBI non redundant protein database and the tobacco BY-2-EST database from RIKEN (Matsuoka et al. 2004) using the SEQUEST and MASCOT protein identification tools. An identified protein was considered as plastidic when: (1) the protein or its *Arabidopsis* ortholog has a predicted plastid transit peptide, or (2) the protein already has a reported function in plastids. Nevertheless, one should keep in mind that this approach underestimates the true number of plastidic proteins (Millar et al. 2006).

Many of the identified proteins were found to be part of the nitrogen fixation and amino acid biosynthetic pathways. This is consistent with results obtained at the chloroplast protein and RNA levels (Kleffmann et al. 2004), and suggests a key role for the plastid in providing the amino acids necessary for protein synthesis in rapidly dividing BY-2 cells. Some highly abundant envelope transporters were identified, supporting the heterotrophic nature of plastids in tobacco BY-2. Among them are the ADP/ATP transporter, importing cytosolic ATP to feed heterotrophic metabolic processes, and the glucose 6-phosphate/phosphate translocator, providing glucose 6-phosphate that can be used as a substrate to initiate the pentose phosphate pathway.

Although BY-2 plastids have the ability to differentiate into amyloplast-like organelles upon auxin depletion of the growth medium (Miyazawa et al. 1999), proteome profiles from both plastid types show significant differences (Andon et al. 2002). This suggests that comparisons between different plastid proteomes could provide important insights into plastid-type-specific metabolic functions and pathways. This is one of the reasons that a conglomerate and comparative plastid proteome database was recently developed (Kleffmann et al. 2006).

11.4 Blue-Native/SDS PAGE Analysis of Tobacco BY-2 Protein Complexes

Most physiological processes are not carried out by single proteins, but rather by protein assemblies. The importance of protein complex formation in plants is clearly illustrated by the involvement of such complexes in different physiological processes in plants, such as the cell cycle, cell proliferation, protein synthesis and in the respiratory system. Complex formation and activity is strongly regulated by many parameters, including participating components, cofactors, phosphorylation/dephosphorylation events and interaction with inhibitory or activating proteins. Many techniques to study protein–protein interactions, such as co-immunoprecipitation studies, yeast two-hybrid studies, TAP/FLAG pull down assays and FRET/BRET fluorescence studies in vivo, depend on the use of an antibody or on the presence of a tag to investigate possible interaction partners of a particular protein. Some of these techniques are restricted to binary interactions and none can deliver information about the functional relationship between the different interaction partners of a given protein. Furthermore, attaching a tag to the protein under study may interfere with proper complex assembly and stability, while the genetic engineering needed to express a tagged protein can alter the physiology of the cell. Unlike these techniques, blue native (BN) gel electrophoresis does not rely on any foreknowledge of genomic data or the availability of antibodies. Therefore it can be used to uncover and investigate protein–protein interactions in unsequenced and genetically complicated models like the *N. tabacum* BY-2 cell suspension culture. BN/SDS-PAGE can thus be used as a display technique to visualise a footprint of protein complexes from BY-2 cells.

By using mild detergents during extraction and solubilisation, and the combined use of mild non-ionic detergents and Coomassie Blue during electrophoresis, native protein complexes are separated according to their apparent molecular mass. The combination of BN-PAGE and SDS-PAGE results in a 2D display in which protein complexes are separated in the first dimension, then separated into their respective subunits in the second dimension. With this technique, estimations about the molecular masses of their composite components and their oligomeric state can be obtained.

Although BN-PAGE was originally developed to study mitochondrial protein complexes, today BN-PAGE is used extensively to study protein–protein interactions involved in different metabolic pathways in various organisms. Recently this technique was optimised to analyse whole plant cell lysates (N. Remmerie et al., manuscript submitted). In this study, several well-studied plant protein complexes, such as the 20S proteasome, ribulose 1,5-bisphosphate carboxylase/oxygenase (RuBisCO)-binding protein, and F1F0 ATP synthase, as well as candidate novel protein–protein interactions involved in different pathways, were identified. Putative novel protein–protein interactions include the interplay between glyceraldehyde-3-phosphate dehydrogenase (GAPDH) and phospholipase D, two tobacco enolase isoforms, NADP malic enzyme and NADH glutamate dehydrogenase, and sucrose synthase and glutamine synthase, respectively. Verification by co-immunoprecipitation showed that interactions predicted by BN/SDS-PAGE are genuine. Based upon their migration in first and second dimension, the oligomeric states of different proteins can be derived. Figure 11.1 illustrates that BN/SDS-PAGE is not only useful to study very hydrophobic protein complexes, such as membrane-associated complexes, but can also be used to produce a protein complex footprint of a whole plant cell lysate.

The use of mild non-ionic detergents during BN-PAGE (Eubel et al. 2003) is important to maintain the solubility and stability of the protein complexes from plant species. In general, digitonin seems to be an ideal detergent for protein complex solubilisation and stability (Eubel et al. 2003). A ratio of between 5 and 8.5 g digitonin /g protein seems to stabilise the complexes (Eubel et al. 2003) and facilitate the solubilisation of protein supercomplexes (Heinemeyer et al. 2004). The use of digitonin has already led to the discovery of plant-specific subunits in several protein complexes (Eubel et al. 2003). Several publications (Eubel et al. 2004; Beriault et al. 2005; Sabar et al. 2005) show that the protein complexes remain enzymatically active during BN-PAGE, thereby confirming the presence of all functional subunits.

By combining DiGE and BN-PAGE, multiple samples can be loaded on a single gel (DiGE BN/SDS-PAGE) and comparative studies of complex formation under several physiological conditions can be performed. Using this approach with highly synchronised BY-2 cell suspension cultures is a very promising method for studying the plant cell cycle. After synchronisation by aphidicolin – a reversible DNA-polymerase inhibitor – samples at different phases of the cell cycle can be taken and compared by 2D DIGE-BN-PAGE. This approach allows a dynamic profile of protein complexes in BY-2 cells to be constructed, and reveals activating/inhibiting proteins. By using specific staining methods (e.g. ProQ diamond and ProQ Emerald), additional dynamic information about PTMs such as phosphorylation and glycosylation can be obtained from the same gel.

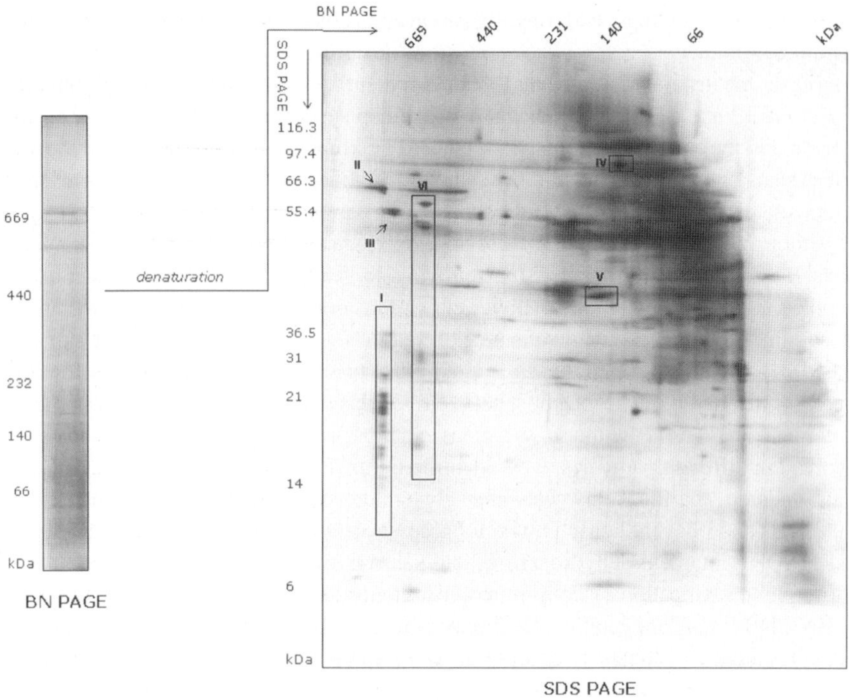

Fig. 11.1 Two dimensional native resolution of protein complexes isolated from *Nicotiana tabacum* cv. Bright Yellow-2 (BY-2) cell suspension cultures. Digitonin-solubilised protein complexes from BY-2 cells were first separated by 1-D blue native (BN)-PAGE. The 1-D separated protein complexes were then dissociated into their subunits by SDS-PAGE in the second dimension. Highlighted spots: *I* 20S proteasome, *II* heat shock protein 60 (HSP60)-complex, *III* ribulose 1,5-bisphosphate carboxylase/oxygenase (RuBisCO)-binding protein complex, *IV* homodimeric transketolase, *V* tetrameric glyceraldehyde-3-phosphate dehydrogenase (GAPDH), *VI* F1F0 ATPase complex

11.5 Identification of Tobacco Proteins

Mass spectrometry-based protein identification relies largely on the comparison of experimentally acquired data with protein databases. The strength of the correlation between theoretical and experimental data gives an indication of the degree of similarity between the given protein and the hit sequence. With the exception of its plastid and mitochondrion genome, the tobacco genome remains poorly represented in public sequence databases. Despite ongoing genome sequencing efforts (Tobacco genome initiative – http://www.tobaccogenome.org/), tobacco protein identification relies heavily on inter-species ortholog matching. Additionally, the EST sequence databases can be queried (in six-frame translation): at the time of writing about 74,000 EST sequences from tobacco are available. Information from

EST sequencing efforts (e.g. Matsuoka et al. 2004) can as such be useful for protein identification.

When relying on cross-species identification it is important to use all available pieces of information to arrive at an optimal assignment for a given protein. This can be illustrated using the case of tobacco BY-2 adenosine kinase (Laukens et al. 2003). This protein was purified based on its affinity for cytokinins, digested with trypsin, and analysed by ESI Q-TOF MS/MS. The combined spectral information was then used to query a number of sequence databases using the Mascot (http://www.matrixscience.com) identification tool. A number of EST hits as well as protein hits from other plants were obtained, each covering some of the peptides. None of the database searches yielded hits for all peptides. All resulting hit sequences could be aligned as their overlapping parts showed high homology. This situation is typical of identification that relies on EST-based and cross-species identification.

Internal evaluation of the accuracy of different identification engines for cross-species identification (unpublished observations) have demonstrated the complementarity of the different algorithms. Querying experimental MS data of human proteins against sequence databases of different closely and less closely related organisms illustrated the sensitivity of different algorithms to increasing sequence dissimilarity. Internal analysis of a tobacco BY-2 MS dataset confirmed the fact that more proteins can be identified effectively by combining the database-dependent identification engine Mascot and a de novo sequencing tool combined with MS blast searches (Shevchenko et al. 2001; Habermann et al. 2004).

11.6 Data Integration

Rather than being considered as "stand-alone" innovations of which only the relevant conclusions are usually absorbed by the scientific community, a shift towards data distribution and integration as a complement to the traditional "printed paper" circuit is now leading to a wealth of data available online.

The first approach towards proteomic data-sharing was the setting up of the federated 2D databases (Appel et al. 1996) that offered researchers a rapid way to locate a certain protein on an experimental "reference" 2D map, and made it possible to derive the identity of an unknown spot by matching it with a corresponding spot on a reference map. Such a reference map has been generated for tobacco BY-2 cells in exponential growth phase (Laukens et al. 2004; http://www.pdata.ua.ac.be/BY2) and can serve as a basis to guide ongoing experiments and as an index for identification data. Federated databases can rapidly provide a presumed identity of a given protein of interest on an experimental gel by matching spots on the experimental gel with annotated spots on a gel in the database. A reference database is most useful in serving as rapid "pre-analysis" information, but another potential application lies in the reverse approach. Given the fact that pI and molecular size of the primary sequence is reasonably well conserved between isoforms and orthologs, these federated repositories can be used to guide various experiments. When a researcher is interested in a certain protein for cloning, sequence analysis, PTM analysis or other purposes, a reference database of the species of interest or a related species could point to the spots on a matched map, which could then be picked for further analyses.

The need for proper integration and electronic sharing of experimental data has further grown with the establishment of minimal reporting requirements imposed by leading proteomic journals and the "Proteomics Standards Initiative". Gel images and identification results are typical features included in proteomic-databases, but the inclusion of other information becomes increasingly important. Processed mass spectra can be stored in libraries, and the possibility to revise them at any time in accordance with the growing sequence databases is of great interest. Other types of biological data that can be deposited in databases include PTMs, protein abundance levels and protein interactions. Besides this biologically relevant information, raw datasets and experimental parameters can be integrated in databases, since these factors are highly relevant for data re-evaluation.

The integration of proteome data in the case of a non-sequenced model species involves some specific concerns. Especially when coming from cross-species protein identification, protein "identity" information cannot be followed back to a single gene locus in a single non-redundant genomic database, but is rather dynamically associated with a number of "most similar" database entries. The typical "identity" field should thus be referred to as "homology", and evaluated together with a statistical parameter describing the significance and completeness of the homology. Regular re-annotation of proteins by re-analysis of their digest MS-spectra against new database versions can lead to further improvement of the existing annotated databases.

An important function of integrated efforts is maintaining, promoting and developing internal relationships between different types of data. Particularly with organisms dependent on cross-species identification, the connectivity between multiple peptides originating from a single protein spot is an important characteristic. With projects such as the tobacco BY-2 proteome analysis in mind, a web-based data integration platform called pProRep was developed (Laukens et al. 2006). Knowledge of a previous analysis of a certain sample or spot can often be of great value for repeated analysis and result interpretation. The pProRep tool enables this type of gel-centric data to be integrated into a relational database. It offers a web-interface to which users can import, analyse, visualise and export experimental datasets. Some advanced query functions allow for novel ways to search the database, for example with experimental peaklists. Query results and their internal relationships can be visualised on, for example, the spot level, and labelled. The application can be used as an "analytical workbench" for experimental proteome data, and will be further developed to offer more advanced data-mining functions. Experimental datasets are growing and could be of great value for the future interpretation of new experiments if the right tools are available. Besides these analytical and integrative functions, pProRep also intends to serve the purpose of online data sharing.

The current possibilities to acquire '–omics'-data impose new challenges to the interpretation, clustering, comparison and functional annotation of biological data. This includes integration with biological knowledge databases such as the Gene Ontology Database (The Gene Ontology Consortium 2000), protein interaction databases, pathway databases, as well as mining the biological literature. A number of automated tools to accomplish this task are now becoming available. The lack of annotated data for tobacco genes and proteins complicates but does not preclude this task; to a certain extent annotation data can be successfully transferred over the species barrier if the (sequence) similarity criteria are carefully selected (Yu et al. 2004). Further maturation of this field can be expected in the near future.

Acknowledgements K.L. is a postdoctoral fellow financed by the Fund for Scientific Research – Flanders (Belgium) (F.W.O. – Vlaanderen).

References

Andon NL, Hollingworth S, Koller A, Greenland AJ, Yates JR, Haynes PA (2002) Proteomic characterization of wheat amyloplasts using identification of proteins by tandem mass spectrometry. Proteomics 2:1156–1168

Appel RD, Bairoch A, Sanchez JC, Vargas JR, Golaz O, Pasquali C, Hochstrasser DF (1996) Federated 2-DE database: a simple means of publishing 2-DE data. Electrophoresis 17:540–546

Baginsky S, Siddique A, Gruissem W (2004) Proteome analysis of tobacco bright yellow-2 (BY-2) cell culture plastids as a model for undifferentiated heterotrophic plastids. J Proteome Res 3:1128–1137

Beriault R, Chenier D, Singh R, Middaugh J, Mailloux R, Appanna V (2005) Detection and purification of glucose 6-phosphate dehydrogenase, malic enzyme, and NADP-dependent isocitrate dehydrogenase by blue native polyacrylamide gel electrophoresis. Electrophoresis 26:2892–2897

Eubel H, Jansch L, Braun HP (2003) New insights into the respiratory chain of plant mitochondria. Supercomplexes and a unique composition of complex II. Plant Physiol 133:274–286

Eubel H, Heinemeyer J, Braun HP (2004) Identification and characterization of respirasomes in potato mitochondria. Plant Physiol 134:1450–1459

Gerber IB, Dubery IA (2004) Protein phosphorylation in *Nicotiana tabacum* cells in response to perception of lipopolysaccharides from *Burkholderia cepacia*. Phytochemistry 65:2957–2966

Gerber IB, Zeidler D, Durner J, Dubery IA (2004) Early perception responses of *Nicotiana tabacum* cells in response to lipopolysaccharides from *Burkholderia cepacia*. Planta 218:647–657

Gerber IB, Laukens K, Witters E, Dubery IA (2006) Lipopolysaccharide-responsive phosphoproteins in *Nicotiana tabacum* cells. Plant Physiol Biochem 44:369–379

Görg A, Weiss W, Dunn MJ (2004) Current two-dimensional electrophoresis technology for proteomics. Proteomics 4:3665–3685

Habermann B, Oegema J, Sunyaev S, Shevchenko A (2004) The power and the limitations of cross-species protein identification by mass spectrometry-driven sequence similarity searches. Mol Cell Proteomics 3:238–249

Heinemeyer J, Eubel H, Wehmhoner D, Jansch L, Braun HP (2004) Proteomic approach to characterize the supramolecular organization of photosystems in higher plants. Phytochemistry 65:1683–1692

Inzé D (2005) Green light for the cell cycle. EMBO J 24:657–662

Kleffmann T, Russenberger D, von Zychlinski A, Christopher W, Sjolander K, Gruissem W, Baginsky S (2004) The *Arabidopsis thaliana* chloroplast proteome reveals pathway abundance and novel protein functions. Curr Biol 14:354–362

Kleffmann T, Hirsch-Hoffmann M, Gruissem W, Baginsky S (2006) plprot: a comprehensive proteome database for different plastid types. Plant Cell Physiol 47:432–436

Laukens K, Lenobel R, Strnad M, Van Onckelen H, Witters E (2003) Cytokinin affinity purification and identification of a tobacco BY-2 adenosine kinase. FEBS Lett 533:63–66

Laukens K, Deckers P, Esmans E, Van Onckelen H, Witters E (2004) Construction of a two-dimensional gel electrophoresis protein database for the *Nicotiana tabacum* cv. Bright Yellow-2 cell suspension culture. Proteomics 4:720–727

Laukens K, Matthiesen R, Lemiere F, Esmans E, Van Onckelen H, Jensen ON, Witters E (2006) Integration of gel-based proteome data with pProRep. Bioinformatics 22:2838–2840

Matsuoka K, Demura T, Galis I, Horiguchi T, Sasaki M, Tashiro G, Fukuda H (2004) A comprehensive gene expression analysis toward the understanding of growth and differentiation of tobacco BY-2 cells. Plant Cell Physiol 45:1280–1289

Menges M, Murray JAH (2002) Synchronous Arabidopsis suspension cultures for analysis of cell-cycle gene activity. Plant J 30:203–212

Millar HA, Whelan J, Small I (2006) Recent surprises in protein targeting to mitochondria and plastids. Curr Opin Plant Biol 9:610–615

Miyazawa Y, Sakai A, Miyagishima S, Takano H, Kawano S, Kuroiwa T (1999) Auxin and cytokinin have opposite effects on amyloplast development and the expression of starch synthesis genes in cultured Bright Yellow-2 tobacco cells. Plant Physiol 121:461–469

Nagata T, Kumagai F (1999) Plant cell biology through the window of the highly synchronized tobacco BY-2 cell line. Methods Cell Sci 21:123–127

Nagata T, Nemoto Y, Hasezawa S (1992) Tobacco BY-2 cell line as the "Hela" cell in molecular biology of higher plants. Int Rev Cytol 132:1–30

Planchais S, Glab N, Inzé D, Bergounioux C (2000) Chemical inhibitors: a tool for plant cell cycle studies. FEBS Lett 476:78–83

Sabar M, Balk J, Leaver CJ (2005) Histochemical staining and quantification of plant mitochondrial respiratory chain complexes using blue-native polyacrylamide gel electrophoresis. Plant J 44:893–901

Shevchenko A, Sunyaev S, Loboda A, Shevchenko A, Bork P, Ens W, Standing KG (2001) Charting the proteomes of organisms with unsequenced genomes by MALDI-quadrupole time-of-flight mass spectrometry and BLAST homology searching. Anal Chem 73:1917–1926

Shinozaki K, Ohme M, Tanaka M, Wakasugi T, Hayashida N, Matsubayashi T, Zaita N, Chunwongse J, Obokata J, Yamaguchi-Shinozaki K, Ohto C, Torazawa K, Meng BY, Sugita M, Deno H, Kamogashira T, Yamada K, Kusuda J, Takaiwa F, Kato A, Tohdoh N, Shimada H, Sugiura M (1986) The complete nucleotide sequence of the tobacco chloroplast genome: its gene organization and expression. EMBO J 5:2043–2049

Stals H, Inzé D (2001) When plant cells decide to divide. Trends Plant Sci 6:359–364

Sugiyama Y, Watase Y, Nagase M, Makita N, Yagura S, Hirai A, Sugiura M (2005) The complete nucleotide sequence and multipartite organization of the tobacco mitochondrial genome: comparative analysis of mitochondrial genomes in higher plants. Mol Genet Genomics 272:603–615

Van Wijk KJ (2004) Plastid proteomics. Plant Physiol Biochem 42:963–977

Witters E, Laukens K, Deckers P, Van Dongen W, Esmans E, Van Onckelen H (2003) Fast liquid chromatography coupled to electrospray tandem mass spectrometry peptide sequencing for cross-species protein identification. Rapid Commun Mass Spectrom 17:2188–2194

Yu H, Luscombe NM, Lu HX, Zhu X, Xia Y, Han JD, Bertin N, Chung S, Vidal M, Gerstein M (2004) Annotation transfer between genomes: protein-protein interologs and protein-DNA regulogs. Genome Res 14:1107–1118

Chapter 12
Cell Wall Proteome

Georges Boudart, Zoran Minic, Cécile Albenne, Hervé Canut, Elisabeth Jamet, and Rafael F. Pont-Lezica

Abstract In this chapter, we will focus on the contribution of proteomics to the identification and determination of the structure and function of CWPs as well as discussing new perspectives in this area. The great variety of proteins found in the plant cell wall is described. Some families, such as glycoside hydrolases, proteases, lectins, and inhibitors of cell wall modifying enzymes, are discussed in detail. Examples of the use of proteomic techniques to elucidate the structure of various cell wall proteins, especially with post-translational modifications such as *N*-glycosylations, proline hydroxylation and *O*-glycosylations, addition of GPI anchors, and phosphorylation, are given. Finally, the emerging understanding of the functions of cell wall proteins is discussed, as well as proposals for future research.

Georges Boudart, Surfaces Cellulaires et Signalisation chez les Végétaux, UMR 5546 CNRS/Université Paul Sabatier Toulouse III, 24 Chemin de Borde Rouge, BP 42617, 31326-Castanet-Tolosan, France

Zoran Minic, Department of Chemistry, University of Saskatchewan, 110 Science Place, Saskatoon, SK S7N 5C9, Canada

Cécile Albenne, Surfaces Cellulaires et Signalisation chez les Végétaux, UMR 5546 CNRS/Université Paul Sabatier Toulouse III, 24 Chemin de Borde Rouge, BP 42617, 31326-Castanet-Tolosan, France

Hervé Canut, Surfaces Cellulaires et Signalisation chez les Végétaux, UMR 5546 CNRS/Université Paul Sabatier Toulouse III, 24 Chemin de Borde Rouge, BP 42617, 31326-Castanet-Tolosan, France

Elisabeth Jamet, Surfaces Cellulaires et Signalisation chez les Végétaux, UMR 5546 CNRS/Université Paul Sabatier Toulouse III, 24 Chemin de Borde Rouge, BP 42617, 31326-Castanet-Tolosan, France

Rafael F. Pont-Lezica, Surfaces Cellulaires et Signalisation chez les Végétaux, UMR 5546 CNRS/Université Paul Sabatier Toulouse III, 24 Chemin de Borde Rouge, BP 42617, 31326-Castanet-Tolosan, France, E-mail: lezica@scsv.ups-tlse.fr

J. Šamaj and J. Thelen (eds.), *Plant Proteomics*
© Springer 2007

12.1 Introduction: Cell Wall Proteins Before Proteomics

Several reports between 1924 and 1960 had presented evidence for the presence of proteins in the cell walls of higher plants. However, with the discovery by Lamport and Northcote (1960) of hydroxyproline (Hyp)-containing proteins in the cell walls of bean and sycamore, the concept of a dynamic structure, containing many more proteins than expected, progressively emerged.

Throughout the decade 1985–1995, sequencing and transcription studies of many genes encoding cell wall proteins (CWPs) in different physiological situations have increased our knowledge of these genes, particularly those encoding structural proteins, e.g. Hyp-rich glycoproteins (HRGPs), Hyp/Pro-rich proteins (H/PRPs), and Gly-rich proteins (GRPs) (Cassab 1998; Ringli et al. 2001). Such proteins are unusually rich in a few specific amino acids, contain highly repetitive sequence domains, and can be highly glycosylated. During the past 5 years, the availability of the complete sequence of the *Arabidopsis thaliana* genome has contributed greatly to the description of other well known CWP families such as endo-1,4-β-D-glucanases (del Campillo 1999), glycoside hydrolase (GH) family 1 (Xu et al. 2004), xyloglucan endotransglucosylase/hydrolases (XTHs) (Fry 2004), pectin methylesterases (PMEs) (Micheli 2001), expansins (Cosgrove et al. 2002), peroxidases (Welinder et al. 2002), leucine-rich repeat-extensins (LRXs) (Baumberger et al. 2003), and arabinogalactan proteins (AGPs) (Gaspar et al. 2001). It was found that most of the known CWPs are encoded by multigene families, which can be very large – up to more than 70 members.

Recently, the emergence of proteomics has been supported by improvements in mass spectrometry (MS) techniques, the complete sequencing of the *A. thaliana* genome, and new bioinformatic tools. A new vision of the cell wall proteome is thus progressively emerging, with the description of new CWP families as well as the characterisation of post-translational modifications (PTMs). In this chapter, we will focus on the contribution of proteomics to the identification and determination of the structure and function of CWPs as well as discussing new perspectives in this area.

12.2 The Surprising Variety of Cell Wall Proteins

Cell wall proteomics is faced with several difficulties, linked mainly to the structure of cell walls as well as to the nature of CWPs (most CWPs have a basic pI, and can be heavily glycosylated). These facts make extraction, separation, and identification procedures challenging (Jamet et al. 2006). Nevertheless, despite this complexity, cell wall proteomics has become a dynamic discipline. Most of the available data has been obtained with Arabidopsis. At present, 365 Arabidopsis CWPs have been identified from various organs and other sources such as rosettes (Boudart et al. 2005), hypocotyls (Feiz et al. 2006), stems at the flowering stage (Minic et al. 2007), cell suspension cultures (Robertson et al. 1997; Chivasa et al. 2002; Borderies et al. 2003; Bayer et al. 2006), and cell wall-regenerating protoplasts (Kwon et al. 2005). Pieces of information are also now beginning to come from other species such as *Zea mays* and *Medicago sativa* (Zhu et al. 2006; Watson et al. 2004).

The great majority of *Arabidopsis* CWPs (88.5%) can be distributed into seven classes based on the presence of functional domains as predicted by bioinformatics (Table 12.1). The remaining 11.5% consists of proteins with unknown function. As expected, proteins acting on carbohydrates constitute the main class (27.4% of identified CWPs), with GHs as the major representatives (20.5%). Carbohydrate esterases/lyases, expansins and glycoside transferases are less abundant. Two functional classes are of equal importance: (1) oxido-reductases (13.7%), include peroxidases, multicopper oxidases, berberine-bridge enzyme (S)-reticulin:oxygen oxido-reductases, laccases and germins; and (2) proteases (12.6%) of several types, i.e. serine proteases (subtilases), aspartic proteases, cysteine proteases and serine carboxypeptidases. Proteins containing interacting domains (8.5%) include carbohydrate-binding proteins (lectins), and protein-binding proteins through leucine-rich repeat (LRR) domains (e.g. polygalacturonase inhibitors), or enzyme inhibitors (e.g. PME and protease inhibitors). Proteins involved in signalling (8.2%) are mainly AGPs and LRR-receptor protein kinases that have been identified through their extracellular LRR domains. Structural proteins are poorly represented (1.9%). Other proteins of various functions (16.2%) are lumped together in a functional class called "miscellaneous". These include proteins awaiting additional experimental data to allow them to be more precisely classified, such as proteins homologous

Table 12.1 Functional classification of cell wall proteins (CWPs) in *Arabidopsis* identified by proteomic approaches

Functional classes (subclasses)[a]	Number of proteins	Identified proteins (%)
Proteins acting on carbohydrates	**100**	**27.4**
Glycoside hydrolases	75	20.5
Esterases and lyases	17	4.7
Expansins	8	2.2
Oxido-reductases	**50**	**13.7**
Peroxidases	23	6.3
Others	27	7.4
Proteases	**46**	**12.6**
Proteins with interacting domains	**31**	**8.5**
Carbohydrate-binding	9	2.5
Protein-binding	22	6.0
Signalling	**30**	**8.2**
Structural proteins	**7**	**1.9**
Miscellaneous	**59**	**16.2**
Unknown function	**42**	**11.5**

[a]Proteins are distributed in functional classes according to the presence of functional domains. Data originate from cell suspension cultures (Robertson et al. 1997; Chivasa et al. 2002; Borderies et al. 2003; Bayer et al. 2006), rosettes (Boudart et al. 2005), hypocotyls (Feiz et al. 2006), stems at the flowering stage (Minic et al., 2007), and cell wall-regenerating protoplasts (Kwon et al. 2005)

to acid phosphatases, blue copper binding proteins and proteins with lipase/acyl hydrolase domains. The final class encompasses proteins of unknown function (11.5%) that contain no previously characterised functional domains; some contain common "domains of unknown function" that can be found in databases, and some are plant-specific proteins.

Bioinformatic analysis of the *Arabidopsis* genome has led to the identification of many gene families encoding CWPs. Table 12.2 presents a comparison between the number of genes predicted to encode CWPs from several families annotated by experts, and the present number of identified proteins. Although this comparison does not give an exhaustive view of the cell wall proteome, it shows that about one-quarter of predicted CWPs have been identified. Based on this figure, we can estimate that about 1,500 genes in the *Arabidopsis* genome code for CWPs. A great disparity can be observed between percentages of identified proteins in each family. Eighty-two percent of the predicted cell wall lectins (legume lectins and monocot mannose-binding lectins) have been identified. In contrast, among structural proteins, not a single extensin has yet been identified by proteomics. There are several likely explanations: extensins are cross-linked into a protein network by covalent bonds (Brady et al. 1996), their high level of glycosylation can prevent accessibility to proteolysis prior to MS analysis, and glycosylated peptides are not identified by the bioinformatic tools used in proteomics. In addition to structural CWPs, low abundant, low molecular mass, short half-life and highly tissue-specific CWPs still escape identification.

12.2.1 Glycoside Hydrolases

A large proportion of proteins identified by cell wall proteomic analyses act on carbohydrates. These includes GHs, which have been classified according to CAZy nomenclature based on sequence homology (http://afmb.cnrs-mrs.fr/CAZY/). The genome of *Arabidopsis* contains 379 genes for GHs from 29 families (Henrissat et al. 2001). Plant GHs are supposed to play a role in various functions in cell wall metabolism, plant defence, signalling, mobilisation of storage reserves (Henrissat et al. 2001) and reorganisation of glycans (Minic et al. 2007). They participate in the regulation of cell wall expansion and alteration during development (Fry 2004; Minic and Jouanin 2006). However, the physiological and biochemical functions of only a small proportion of identified GHs have been demonstrated experimentally.

Plant cell wall polysaccharides are very heterogeneous complex polymers. Consequently, spectra of GH activities are very diverse. The 75 GHs identified by proteomics can be sub-divided into four groups. The first group of GHs comprises enzymes that might be involved in reorganisation of cell wall carbohydrates during growth and development (Minic and Jouanin 2006). Of these, 42 proteins belonging to 10 different families were identified by proteomic analyses (Table 12.3). Their possible *in muro* substrates have also been indicated. For the majority of these enzymes, the substrates are pectins (homogalacturonan and rhamnogalacturonan I)

Table 12.2 Number of predicted genes encoding CWPs in Arabidopsis compared to the number of proteins identified by proteomics

Protein families	Number of predicted genes	Number of identified proteins[a]	Identified proteins (%)	Reference
Glycoside hydrolases family 1	26	6	23.1	Xu et al. 2004
Cellulases	25	2	6.0	del Campillo 1999
Polygalacturonases	69	8	11.6	http://afmb.cnrs-mrs.fr/CAZY/
Pectin methylesterases	67	10	14.9	http://afmb.cnrs-mrs.fr/CAZY/
Pectin acetylesterases	11	3	27.3	http://afmb.cnrs-mrs.fr/CAZY/
Xyloglucan endotransglucosylase/hydrolases	33	7	21.2	http://labs.plantbio.cornell.edu/xth/genes.htm
Expansins	36	8	22.2	http://www.bio.psu.edu/expansins/arabidopsis.htm
Legume and mannose-binding lectins	11	9	81.8	This chapter[b]
Blue copper binding proteins (plantacyanins, stellacyanins, uclacyanins, early nodulins, plastocyanins)	42	7	16.7	Nersissian and Shipp (2002)
Extensins	19	0	0.0	This chapter[b]
Leucine-rich repeat extensins	11	4	36.4	Baumberger et al. 2003
Subtilases	50	18	36.0	Rautengarten et al. 2005
Fasciclin-like arabinogalactan proteins	13	5	38.5	Schultz et al. 2002
Arabinogalactan proteins	26	12	46.2	Schultz et al. 2002
SKS (SKU-similar)	19	8	42.1	Jacobs and Roe 2005
Peroxidases (class III)	73	23	31.5	Welinder et al. 2002
Germin-like proteins	25	7	28.0	This chapter[b]
Total	556	137	24.6	

[a]Origin of data as in Table 12.1, [b]InterProScan searches for motifs or profiles stored in databases (http://www.ebi.ac.uk/InterProScan/)

Table 12.3 Families of Arabidopsis glycoside hydrolases (GHs) identified by cell wall proteomics which are potentially involved in reorganisation of cell wall carbohydrates

CAZy GH family[a]	Predicted enzymatic activity	Cell wall carbohydrates as possible substrates *in muro*	Number of proteins
1	β-D-Glucosidase	Glucan, xyloglucan, cellulose	5
3	β-D-Xylosidase α-L-Arabinofuranosidase	Arabinan, xylan, arabinoxylan	6
9	Endo-1,4-β-glucanase	1,4-β-glucan (cellulose)	2
10	1,4-β-Xylan endohydrolase	Xylan	1
16	Xyloglucan endotransglycosylase (XTH)	Xyloglucan	7
27	α-D-Galactosidase	Galactomannan	3
28	Endo-polygalacturonase Exo-polygalacturonase	Homogalacturonan	8
31	α-D-Xylosidase α-D-glucosidase	Xyloglucan	2
35	β-D-Galactosidase	Galactan	7
51	α-L-Arabinofuranosidase	Arabinoxylan, xylan, arabinan	1

[a]GH have been classified according to CAZy (http://afmb.cnrs-mrs.fr/CAZY/). Predicted enzymatic activities, possible substrates in muro and numbers of identified proteins in each family are indicated. Origin of data as in Table 12.1

and xyloglucans. Interestingly, these two types of polysaccharide are soluble in water, in contrast to cellulose, xylan and galactomannan, which are not. The majority of these enzymes are exo-GHs. A lower number of enzymes are endo-GHs belonging to GH families 9, 10, 16 and 28, which act on cellulose, xylan, xyloglucans and homogalacturonan, respectively. These results suggest that xyloglucans and pectins might undergo important structural changes after their deposition in cell walls. As shown in Table 12.3, some GH families have a broad substrate range. This low specificity has been reported for several purified plant cell wall GHs (Lee et al. 2003; Minic et al. 2006). It was hypothesised that the multifunctionality of plant GHs allows effective modification of complex cell wall carbohydrates without requiring a large number of enzymes.

The second largest group of identified GHs could be involved in defence against pathogens. Chitinases (GH18, GH19) and β-1,3-glucanases (GH17) have been shown to possess antifungal activity (Schlumbaum et al. 1986). Cell walls of various fungi contain chitin and β-1,3- or β-1,6-glucan. Chitinases and β-1,3-glucanases produced by plants can be secreted into intercellular spaces and inhibit fungal growth by degrading their cell wall (Jach et al. 1995; Zhu et al. 1994). Some chitinases (GH18, GH19) also show lysozyme activity hydrolysing β-1,4 linkages between GlcNAc and *N*-acetylmuramic acid (Boller et al. 1983; Brunner et al. 1998). These lysozymes/chitinases show higher activity on bacterial cell wall peptidoglycan.

A third group of GHs could be involved in glycoprotein PTMs. Such enzymes include α-L-arabinofuranosidases (GH3), chitinases (GH18, GH19), β-D-galactosidases (GH35), α-D-mannosidases (GH38), and β-D-glucuronidases (GH79). The role of GHs in hydrolysis of carbohydrate moieties of AGPs has been particularly well studied. Degradation of AGPs seems to occur by the concerted action of several glycosidases. A bifunctional α-L-arabinofuranosidase/β-D-xylosidase (GH3) and a β-D-galactosidase (GH35) from immature seeds of radish (*Raphanus sativus* L.) were shown to be involved in the hydrolysis of carbohydrate moieties of AGP (Kotake et al. 2005). Recently, a β-D-glucuronidase from *A. thaliana* (AtGUS, GH79) was shown to hydrolyse glucuronic acid from carbohydrate chains of AGPs (Minic et al. 2007).

Finally, three GHs identified by proteomics could be involved in the mobilisation of storage reserves. In higher plants, sucrose is one of the predominant initial products of photosynthesis and serves as an intermediate storage system as an alternative to starch. In addition, sucrose is the major carbohydrate translocator, osmoticum, and regulator of gene expression, as well as playing a role in cellular signalling. Use of sucrose depends on its cleavage into glucose and fructose. In plants, this enzymatic reaction is performed by invertases (GH31), which are also called β-D-fructofuranoside-fructohydrolases (Roitsh and Gonzalez 2004). Invertase isozymes are distinguished by their subcellular localisation, such as cell wall, vacuole and cytoplasm. Cell wall invertases are considered to be key enzymes involved in sucrose unloading, cell differentiation, and the response to wounding or pathogen attack.

12.2.2 Proteases

Proteases are necessary for protein turnover, maturation of enzymes, and defence against pathogens (Schaller 2004). Consequently, proteases may play a role in CWP turnover – still a poorly understood process. Previous studies localised protease activities to leaf intercellular fluids (Huangpu and Graham 1995). Results from proteomics have revealed the existence of a great diversity of proteases, such as subtilases, carboxypeptidases, and aspartyl- and cysteine-proteases (Jamet et al. 2006). The subtilase family – the best represented (40%) among the cell wall proteases identified by proteomics – is particularly interesting because in mammals, subtilases are involved in the formation of peptide hormones and growth factors from precursor polypeptides (Rautengarten et al. 2005). Three main functions are proposed for subtilases: control of development, protein turnover, and as components of signalling cascades.

Proteases certainly play key roles in the maturation of CWPs and in the generation of active peptides in the cell wall. For example, using MS techniques, the barley ARAI α-L-arabinofuranosidase/β-D-xylosidase was shown to be processed in vivo by removing about 130 amino acids at its C-terminus (Lee et al. 2003). On the other hand, a role(s) for plant peptides as peptide hormones in intercellular signalling has recently been proposed (Matsubayashi and Sakagami 2006). The group of *CLE* (CLV3/ESR-related) genes encode small, basic, secreted proteins with a conserved stretch of 14 amino acids close to their C-termini that are found only in plants. Interestingly, the 14-amino-acid peptides released by N- and C-terminal processing harbour the biological activity of these proteins (Ito et al. 2006; Kondo et al. 2006).

12.2.3 Lectins

Based on their structure, lectins so far identified in cell walls belong to two out of the seven groups of plant lectins, namely legume lectins and monocot mannose-binding lectins (van Damme et al. 1998). Around 80% of the predicted cell wall lectins from *Arabidopsis* were identified by proteomics (Table 12.2). One mannose-binding lectin has been identified in almost all tissues analyzed. In contrast, a legume-like lectin was only found in the stems of flowering plants. The characteristic feature of lectins is their ability to recognise and bind specific carbohydrates, thus serving as translators of the sugar code. These recognition functions include lectin involvement in interactions with cells or extracellular materials from the same organism (which could be considered as recognition of self, or of endogenous ligands) and interactions with foreign particles or cells (recognition of non-self, or of exogenous ligands). The functions of plant lectins remain elusive, but a role in protection and symbiosis has been proposed (Sharon and Lis 2004). Cell wall lectins are probably involved in recognition and defence against pathogens, but it cannot be ruled out that cell wall lectins are also involved in the assembly of cell wall polysaccharides.

12.2.4 Protein Inhibitors of Cell Wall Modifying Enzymes

Inhibitors of cell wall modifying enzymes (CWMEIs) represent an expanding family of plant proteins that presumably play a role in defence by limiting the rate of degradation of the cell wall by microbial enzymes, thus reducing colonisation of plant tissues by pathogens. Several *Arabidopsis* CWMEIs, such as inhibitors of polygalacturonases (PGIPs), PMEs and xyloglucanases, have been identified by proteomics.

PGIPs are widespread in the plant kingdom (Gomathi and Gnanamanickam 2004). They inhibit only fungal polygalaturonases (PGs), and the degree of susceptibility of the PG depends on the mode of cleavage of the substrate. Many plants possess more than one PGIP with different abilities to inhibit pathogen PG. Overexpression of *PGIP* genes in plants resulted in reduced susceptibility to a fungal pathogen. PGIPs belongs to the superfamily of LRR-proteins. The LRR motif has been shown to be essential for affinity and specificity of PGIP for the PG ligand. Interestingly, Arabidopsis PGIP 2 interacts with polygalacturonic acid via a motif of four clustered bean (Phaseolus vulgaris) Arg and Lys residues located in the vicinity of the PG-binding site, suggesting that PGIPs can be immobilised on the PG substrate (Spadoni et al. 2006).

Pectins are demethylated to polygalacturonic acid by PME, thus favouring further cleavage of the acidic polygalacturonic chains by polygalacturonase. PME can be inhibited by a protein inhibitor (PMEI). Unlike in the case of PGIP, PMEIs exclusively inhibit plant PMEs (Bellincampi et al. 2004). Although *Arabidopsis* PMEI shares strong sequence and structural homology with an inhibitor of tobacco invertase (Nt-CIF), each inhibitor recognises different target enzymes, as recently elucidated by crystallographic analysis (Hothorn et al. 2004).

Inhibitors of hemicellulose-degrading enzymes, such as monocot endoxylanase inhibitors (XIP- and TAXI-type) and dicot xyloglucan endoglucanase inhibiting proteins (XEGIPs), are another interesting family of CWMEIs (Juge 2006). XIP and TAXI are specific inhibitors of microbial endoxylanases but do not inhibit plant endoxylanases. The microbial endoxylanases GH10 and GH11 are inhibited by XIP-I from wheat. The crystal structure of XIP-I in complex with xylanases revealed that the inhibitor possesses two independent enzyme-binding sites, an unusual feature allowing binding to two xylanases that display different folding (Payan et al. 2004). TAXI inhibits only GH11 endoglucanases. Tomato XEGIP inhibits a fungal GH12 xyloglucanase, but not endoxylanases of the GH10 and GH11 families. Based on gene sequence homology and inhibitory activity, XEGIP presumably belongs to an inhibitor protein superfamily that includes the carrot extracellular dermal glycoprotein, the tobacco nectar NEC4 protein, and monocot TAXI-type inhibitors.

The structural similarities of some CWMEI with glucanases (Durand et al. 2005), of wheat XIP-I with chitinases of the GH18 family (Payan et al. 2003), and of TAXI with an aspartyl protease (Sansen et al. 2004) raise the question of their origin. It is probable that gene sequence data associated with elucidation of the biochemical activities of CWMEIs will allow the rapid characterisation of novel families of CWMEIs (Durand et al. 2005).

12.3 Proteomics as an Efficient Way to Depict Cell Wall Protein Structure

MS applied to proteins is being used to explore and elucidate the structure of some CWPs, especially with respect to PTMs. At present, the types of PTMs most studied in plant CWPs are *N*- or *O*-glycosylation, hydroxylation of Pro residues, addition of glycosylphosphatidylinositol (GPI) anchors, and phosphorylation. The following sections present some examples in which PTMs of secreted proteins have been elucidated.

12.3.1 N-*Glycosylation*

Only a few experimental data are available to describe *N*-glycosylations in CWPs. Nevertheless, site-directed mutagenesis preventing correct *N*-glycosylation of a peanut peroxidase showed that *N*-glycosylation is essential for enzymatic activity and appropriate folding of the protein (Lige et al. 2001). MS analysis of a soybean peroxidase showed considerable heterogeneity in its pattern of *N*-glycosylation (Gray and Montgomery 2006). Soybean peroxidase is a glycoprotein with a carbohydrate content of 18% (w/w). It includes seven putative Asn-X-Thr consensus glycosylation sites for *N*-linked glycans on Asn at positions 56, 130, 144, 185, 197, 211 and 216. Matrix-assisted laser desorption/ionisation-time-of-flight (MALDI-TOF) MS analyses of tryptic peptides showed that although all these sites in soybean peroxidase can potentially be glycosylated, not all sites are equally populated. *N*-Glycosylation is complete at Asn^{56}, Asn^{130} and Asn^{144}, since the corresponding non-glycosylated peptides were not detected. The other *N*-glycosylation consensus sites occur in closely spaced pairs: Asn^{185}–Asn^{197} and Asn^{211}–Asn^{216}. The Asn^{185}–Asn^{197} pair was found to be 90% glycosylated. In contrast, Asn^{211}–Asn^{216} is primarily monoglycosylated at Asn^{211}. This can be explained by the fact that these two potential glycosylation sites are separated by only two amino acids. Since translation and glycosylation both occur in the lumen of the endoplasmic reticulum (ER), it is suggested that glycosylation at Asn^{211} will interfere with that of Asn^{216} by steric hindrance. Mapping the glycans onto the crystal structure of soybean peroxidase showed that they are asymmetrically distributed on the molecule, occurring on the substrate-channel face of the enzyme.

12.3.2 Proline Hydroxylation and O-Glycosylation

The 4-hydroxylation of Pro is a PTM found in many plant glycoproteins. However, prolyl hydroxylase does not hydroxylate all Pro residues in HRGPs (extensins and AGPs) or H/PRPs. Modification of Pro residues is thought to occur co-translationally as the protein is inserted in the ER and can be followed by *O*-Hyp glycosylation. A series of experiments using synthetic genes provided the basis for a Hyp-*O*-glycosylation code (Kieliszewski 2001). A range of simple repetitive polypeptides encoding putative glycosylation motifs was designed. These experiments showed that Hyp-*O*-glycosylation is sequence-driven. Hyp-*O*-glycosylation occurs in two ways. Hyp-*O*-arabinosylation is correlated with Hyp contiguity in sequences such as $(Ser\text{-}Hyp_2)_n$ and $(Ser\text{-}Hyp_4)_n$ in *Nicotiana tabacum*. Hyp-*O*-arabinosylation results in short, neutral, linear oligosaccharides such as $Hyp\text{-}(Ara)_{1\text{–}4}$, which are found in extensins. Hyp-*O*-galactosylation is predicted to occur on clustered non-contiguous Hyp residues $(Ser\text{-}Hyp)_n$ typically found in AGP. Hyp-*O*-galactosylation results in addition of large arabinogalactan heteropolysaccharides (Hyp-polysaccharides). In AGPs, these polysaccharides consist of a β-1→3-linked galactan backbone with 1→6-linked side chains containing galactose, arabinose, and often rhamnose and glucuronic acid. Their length is diverse, ranging from 30 sugars/Hyp in gum arabic AGP, up to 150 sugars/Hyp in a radish leaf AGP (Nothnagel 1997).

Two dodecapeptides from the CLE (CLAVATA 3/ESR-related) family were recently shown to contain Hyp by MS: a tracheary differentiation inhibitory factor (TDIF) and the active peptide of CLAVATA 3. TDIF was found in the *Zinnia elegans* xylogenic in vitro culture system (Ito et al. 2006), where it inhibits the formation of tracheids, and promotes cell division. Tandem MS/MS proved that the final product was a dodecapeptide containing two Hyp residues: HEVHypSGHypNPISN. The peptide is produced from a protein of 132 amino acids by removal of 119 amino acids from its N-terminus, and a single Arg residue from its C-terminus. The TDIF sequence from *Z. elegans* was found to be the same as that of Arabidopsis C-terminal 12 amino acids of CLE41 and CLE44.

The Arabidopsis *CLAVATA 3* gene (*CLV3*) encodes a stem cell-specific protein controlling the size of the shoot apical meristem. *CLV3* encodes a 96-amino-acid secreted protein that shares 14 conserved amino acids with other members of the CLE family in its C-terminal region. MALDI-TOF MS performed on calli overexpressing *CL3* detected a typical region corresponding to a 12-amino-acid peptide from Arg^{70} to His^{81} in *CLV3*, in which two of three proline residues were modified to Hyp. Comparison of MS/MS analysis of this natural peptide to the fragmentation patterns of three synthetic peptides containing two Hyp residues covering all possible combinations allowed the two Hyp in the protein sequence to be localised (Kondo et al. 2006).

12.3.3 Addition of GPI Anchors

GPI-anchored proteins are targeted to the plasma membrane. The anchor can be cleaved by specific phospholipases; consequently, the proteins can exist in both soluble and membrane-associated forms. GPI-anchored proteins belong to numerous protein families that have been identified in Arabidopsis (Borner et al. 2003). Integral membrane proteins were prepared by Triton X-114 phase partitioning and assessed for sensitivity to phospholipase C prior to proteomic analysis. Among the identified proteins are classical AGPs and arabinogalactan peptides. The PTMs of arabinogalactan peptides were more precisely characterised, and include addition of a GPI-anchor, cleavage of signal peptide and GPI-anchor, as well as hydroxylation of Gly-Pro motifs (Schultz et al. 2004). The bulk of Arabidopsis AGPs was isolated from seedlings by β-Yariv precipitation, followed by deglycosylation with anhydrous hydrogen fluoride. The resulting AG-peptides were only 10–17 amino acids long, thus allowing MS analysis without tryptic digestion. Peptide sequencing was performed on the deglycosylated peptides and the precise cleavage sites for both N- and C-terminal signals were determined using a combination of MALDI-TOF MS and tandem MS/MS. It was also shown that 8 out of 12 arabinogalactan peptides have all Pro residues fully hydroxylated.

12.3.4 Phosphorylation

Three reports suggest that phospho-Tyr proteins are present in the cell walls of Arabidopsis and maize, and that changes in their level of phosphorylation is induced by pathogen elicitors (Ndimba et al. 2003; Chivasa et al. 2005b). The presence of phosphorylated CWPs was revealed with antibodies raised against synthetic phospho-Tyr. The fact that phosphorylation events can occur outside the cytoplasm was surprising and interesting. The regulation of cell viability via the presence of extracellular ATP in plants suggests that an extracellular phosphorylation mechanism exists (Chivasa et al. 2005a; see Chap. 6 by Chivasa et al., this volume). However, all protein kinases identified in Arabidopsis are predicted to be located in the cytosol (Rudrabhatla et al. 2006). The two extracellular kinases claimed in Arabidopsis (At1g53070 and At1g78850) are actually lectins. This mistake resulted from erroneous annotation since no kinase domain can be found in these proteins by bioinformatic searches for functional motifs (http://www.ebi.ac.uk/InterProScan/). Nevertheless, MS provides an excellent tool to prove the presence of phosphate in peptides derived from phosphoproteins. This evidence is still missing for CWPs.

12.4 Emerging Functions for Cell Wall Proteins

Proteomic data cannot provide information about the roles of the identified proteins. In fact, the functions of most CWPs remain unknown. Bioinformatic searches for functional domains provide only some clues. Additional data (genetic, biochemical and cell biological) are required to fully understand protein function as exemplified in the following.

The functions assigned to GHs are provisional because their biological role, as well as the enzymatic activities of many of them, have not yet been established (Jamet et al. 2006). For example, an enzyme of the GH3 family (XYL3) shares amino acid homology with β-D-xylosidase, but XYL3 was identified as efficiently hydrolysing arabinosyl residues from arabinans, suggesting that it works as an α-L-arabinofuranosidase (Minic et al. 2006). In addition, some GH family members could have multiple physiological functions. For example, β-D-glucosidases (GH1) are involved in lignification, signalling, defence and other functions related to hydrolysis of secondary metabolites. Chitinase-like enzymes are able to degrade chitin in the cell walls of fungal pathogens (Xue et al. 1995). However, the substrates and functional roles of most chitinase-like enzymes are not completely known. For example, a mutation in a chitinase-like gene classified in GH19 (AtCTL1) caused aberrant patterns of lignification with incomplete cell walls in the stem pith (Zhong et al. 2002).

New roles for secreted proteases in the control of plant development and during plant–pathogen interactions are emerging with the description of three Arabidopsis mutants: *sdd1-1* (*stomatal density and distribution1-1*), *ale1* (*abnormal leaf shape1*) and *CDR1-D* (*constitutive disease resistant1*-Dominant). SDD1 and ALE1 belong to a family of 56 members, 46 of which are predicted to be secreted (Plant Subtilase Database, http://csbdb.mpimp-golm.mpg.de/psdb.htm). The establishment of a normal stomatal pattern is disrupted in *sdd1-1*, resulting in stomata clustering and increased stomatal density (von Groll et al.

2002). Overexpression of *SDD1* produced the opposite phenotype. It was proposed that SDD1 may generate an extracellular signal regulating the number of asymmetric divisions in satellite meristemoids. In the *ale1* mutant, epidermal surfaces of embryos and juvenile plants were defective, resulting in excess water loss and organ fusion in plantlets (Tanaka et al. 2001). Remnants of the endosperm remain attached to the developing embryos. It was assumed that ALE1 produces a peptide ligand required for proper differentiation of epidermis. CDR1 encodes an apoplastic aspartic protease (Xia et al. 2004). Overexpression of CDR1 causes dwarfing and resistance to virulent *Pseudomonas syringae*. Antisense *CDR1* plants were less resistant to avirulent *P. syringae* and more susceptible to virulent strains than wild type. Induction of *CDR1* generates a small mobile signal (3–10 kDa) sensitive to heat and protease. CDR1 action is blocked by the protease inhibitor pepstatin and by mutations in the protease active site. CDR1 is assumed to mediate a peptide signal system involved in the activation of inducible resistance mechanisms (Xia et al. 2004). The nature of the peptides involved in all these processes is still unknown.

Several multicopper oxidase-like proteins identified through proteomic analysis are thought to catalyse full, four-electron reduction of dioxygen (O_2) to water (H_2O) using a variety of substrates. However, since they lack a functional copper-binding site, biochemical tests should be done to characterise their enzymatic activities (Jacobs and Roe 2005). These proteins belong to a large gene family with 19 members in Arabidopsis. Functions for only two members of this family have been assessed by studying mutant phenotypes: SKU5 is involved in the control of root growth (Sedbrook et al. 2002) and SKS6 contributes to cotyledon vascular patterning during development (Jacobs and Roe 2005).

12.5 Conclusions

During the last 5 years, the application of proteomic techniques to the cell wall has delivered an overview of CWPs, a description of some CWP structures, and identification of new CWPs, thus opening new horizons for studies on cell wall function. To date, cell wall proteomic studies have focused essentially on CWP identification by peptide mass fingerprinting and characterisation of PTMs by peptide sequencing. Additional cell wall proteomics efforts will permit investigation of CWPs in other plants, organs, and tissues, as well as during biotic or abiotic stresses, allowing the identification of CWPs that are specific to such physiological situations. However, many CWPs might be missing in these analyses due to, e.g., cross-linking to the polysaccharide matrix, fast turnover, low abundance, or large carbohydrate moieties. Progress in MS will facilitate access to these missing proteins. Since proteins, and not genes, are the real actors in the cellular machinery, such future studies are of high importance. They are notably involved in molecular recognition, catalysis, signalling, and protein turnover, critical steps of regulation in many cellular processes.

References

Baumberger N, Doesseger B, Guyot R, Diet A, Parsons RL, Clark MA, Simmons MP, Bedinger P, Goff SA, Ringli C, Keller B (2003) Whole-genome comparison of leucine-rich repeat extensins in *Arabidopsis* and rice. A conserved family of cell wall proteins form a vegetative and a reproductive clade. Plant Physiol 131:1313–1326

Bayer EM, Bottrill AR, Walshaw J, Vigouroux M, Naldrett MJ, Thomas CL, Maule AJ (2006) Arabidopsis cell wall proteome defined using multidimensional protein identification technology. Proteomics 6:301–311

Bellincampi D, Camardella L, Delcour JA, Desseaux V, D'Ovidio R, Durand A, Elliot G, Gebruers K, Giovane A, Juge N, Sorensen JF, Svensson B, Vairo D (2004) Potential physiological role of plant glycosidase inhibitors. Biochim Biophys Acta 1696:265–274

Boller T, Gehri A, Mauch F, Vögeli U (1983) Chitinase in bean leaves: induction by ethylene, purification, properties, and possible function. Planta 157:22–31

Borderies G, Jamet E, Lafitte C, Rossignol M, Jauneau A, Boudart G, Monsarrat B, Esquerré-Tugayé MT, Boudet A, Pont-Lezica R (2003) Proteomics of loosely bound cell wall proteins of *Arabidopsis thaliana* cell suspension cultures: a critical analysis. Electrophoresis 24:3421–3432

Borner GHH, Lilley KS, Stevens TJ, Dupree P (2003) Identification of glycosylphosphatidylinositol-anchored proteins in Arabidopsis. A proteomic and genomic analysis. Plant Physiol 132:568–577

Boudart G, Jamet E, Rossignol M, Lafitte C, Borderies G, Jauneau A, Esquerré-Tugayé M-T, Pont-Lezica R (2005) Cell wall proteins in apoplastic fluids of *Arabidopsis thaliana* rosettes: identification by mass spectrometry and bioinformatics. Proteomics 5:212–221

Brady JD, Sadler IH, Fry SC (1996) Di-isodityrosine, a novel tetrameric derivative of tyrosine in plant cell wall proteins: a new potential cross-link. J Biochem 315:323–327

Brunner F, Stintzi A, Fritig B, Legrand M (1998) Substrate specificities of tobacco chitinases. Plant J 14:225–234

Cassab GI (1998) Plant cell wall proteins. Annu Rev Plant Physiol Plant Mol Biol 49:281–309

Chivasa S, Ndimba B, Simon W, Robertson D, Yu X-L, Knox J, Bolwell P, Slabas A (2002) Proteomic analysis of the *Arabidopsis thaliana* cell wall. Electrophoresis 23:1754–1765

Chivasa S, Ndimba B, Simon WJ, Lindsey K, Slabas AR (2005a) Extracellular ATP functions as an endogenous external metabolite regulating plant cell viability. Plant Cell 17:3019–3034

Chivasa S, Simon WJ, Yu X-L, Yalpan N, Slabas AR (2005b) Pathogen elicitor-induced changes in the maize extracellular matrix proteome. Proteomics 5:4894–4904

Cosgrove DJ, Li LC, Cho HT, Hoffmann-Benning S, Moore RC, Blecker D (2002) The growing world of expansins. Plant Cell Physiol 43:1436–1444

Del Campillo E (1999) Multiple endo-1,4-beta-D-glucanase (cellulase) genes in *Arabidopsis*. Curr Top Dev Biol 46:39–61

Durand A, Hughes R, Roussel A, Flatman R, Henrissat B, Juge N (2005) Emergence of a subfamily of xylanase inhibitors within glycoside hydrolase family 18. FEBS Lett 272:1745–1755

Feiz L, Irshad M, Pont-Lezica RF, Canut H, Jamet E (2006) Evaluation of cell wall preparations for cell wall proteomics. Plant Methods 2:10

Fry SC (2004) Primary cell wall metabolism: tracking the careers of wall polymers in living plant cells. New Phytol 161:641–675

Gaspar Y, Johnson KL, McKenna JA, Bacic A, Schultz CJ (2001) The complex structures of arabinogalactan-proteins and the journey towards understanding function. Plant Mol Biol 47:161–186

Gomathi V, Gnanamanickam SS (2004) Polygalacturonase-inhibiting proteins in plant defence. Curr Sci 87:1211–1217

Gray JS, Montgomery R (2006) Asymmetric glycosylation of soybean seed coat peroxidase. Carbohydr Res 341:198–209

Henrissat B, Coutinho PM, Davies GJ (2001) A census of carbohydrate-active enzymes in the genome of *Arabidopsis thaliana*. Plant Mol Biol 47:55–72

Hothorn M, Wolf S, Aloy P, Greiner S, Scheiffzek K (2004) Structural insights into the target specificity of plant invertase and pectin methylesterase inhibitory proteins. Plant Cell 16:3437–3447

Huangpu J, Graham JS (1995) Purification and developmental analysis of an extracellular proteinase from young leaves of soybean. Plant Physiol 108:969–974

Ito Y, Nakanomyo I, Motose H, Iwamoto K, Sawa S, Dohmae N, Fukuda H (2006) Dodeca-CLE peptides as suppressors of plant stem differentiation. Science 313:842–8845

Jach G, Görnhardt B, Mundy J, Logemann J, Pinsdorf E, Leah R, Schell J, Maas C (1995) Enhanced quantitative resistance against fungal disease by combinatorial expression of different barley antifungal proteins in transgenic tobacco. Plant J 8:97–109

Jacobs J, Roe JL (2005) SKS6, a multicopper oxidase-like gene, participates in cotyledon vascular patterning during *Arabidopsis thaliana* development. Planta 222:652–666

Jamet E, Canut H, Boudart G, Pont-Lezica RF (2006) Cell wall proteins: a new insight through proteomics. Trends Plant Sci 11:33–39

Juge N (2006) Plant protein inhibitors of cell wall degrading enzymes. Trends Plant Sci 11:359–367

Kieliszewski MJ (2001) The latest hype on Hyp-*O*-glycosylation codes. Phytochemistry 57:319–323

Kondo T, Sawa S, Kinoshita A, Mizuno S, Kakimoto T, Fukuda H, Sakagami Y (2006) A plant peptide encoded by CLV3 identified by in situ MALDI-TOF MS analysis. Science 313:845–848

Kotake T, Dina S, Konishi T, Kaneko S, Igarashi K, Samejima M, Watanabe Y, Kimura K, Tsumuraya Y (2005) Molecular cloning of a β-galactosidase from radish that specifically hydrolyzes β-(1→3)- and β -(1→6)-galactosyl residues of Arabinogalactan protein. Plant Physiol 138:1563–1576

Kwon HK, Yokoyama R, Nishitani K (2005) A proteomic approach to apoplastic proteins involved in cell wall regeneration in protoplasts of *Arabidopsis* suspension-cultured cells. Plant Cell Physiol 46:843–857

Lamport DTA, Northcote DH (1960) Hydroxyproline in primary cell walls of higher plants. Nature 18:665–666

Lee RC, Hrmova M, Burton RA, Lahnstein J, Fincher GB (2003) Bifunctional family 3 glycoside hydrolases from barley with α-L-arabinofuranosidase and α-D-xylosidase activity. J Biol Chem 278:5377–5387

Lige B, Ma S, Huystee RB (2001) The effects of the site-directed removal of *N*-glycosylation from cationic peanut peroxidase on its function. Arch Biochem Biophys 386:17–24

Matsubayashi Y, Sakagami Y (2006) Peptide hormones in plants. Annu Rev Plant Biol 57:649–674

Micheli F (2001) Pectin methylesterases: cell wall enzymes with important roles in plant physiology. Trends Plant Sci 6:414–419

Minic Z, Jouanin L (2006) Plant glycoside hydrolases involved in cell wall polysaccharides degradation. Plant Physiol Biochem 44:435–449

Minic Z, Do C-T, Rihouey C, Morin H, Lerouge P, Jouanin L (2006) Purification, functional characterization, cloning and identification of mutants of a seed specific arabinan hydrolase in Arabidopsis. J Exp Bot 57:2339–2351

Minic Z, Jamet E, Negroni L, der Garabedian PA, Zivy M, Jouanin L (2007) A sub-proteome of *Arabidopsis thaliana* trapped on concanavalin A is enriched in cell wall glycoside hydrolases. J Exp Bot (in press)

Ndimba BK, Chivasa S, Hamilton JM, Simon WJ, Slabas AR (2003) Proteomic analysis of changes in the extracellular matrix of *Arabidopsis* cell suspension cultures induced by fungal elicitors. Proteomics 3:1047–1059

Nersissian AM, Shipp EL (2002) Blue copper-binding domains Adv Prot Chem 60:271–340

Nothnagel E (1997) Proteoglycans and related components in plant cells. Int Rev Cytol 174:195–291

Payan F, Flatman R, Porciero S, Williamson G, Roussel A (2003) Structural analysis of xylanase inhibitor protein I (XIP-I), a proteinaceous xylanase inhibitor from wheat (*Triticum aestivum*, var. Soisson). Biochem J 372:399–405

Payan F, Leone P, Porciero S, Furniss C, Tahir T, Williamson G, Durand A, Manzanares P, Gilbert HJ, Juge N, Roussel A (2004) The dual nature of wheat xylanase protein inhibitor XIP-I – structural basis for the inhibition of family 10 and family 11 xylanases. J Biol Chem 279:36029–36037

Rautengarten C, Steinhauser D, Büssis D, Stinzi A, Schaller A, Kopka J, Altmann T (2005) Inferring hypotheses on functional relationships of genes: analysis of the *Arabidopsis thaliana* subtilase gene family. PLoS Comput Biol 1:e40

Ringli C, Keller B, Ryser U (2001) Glycine-rich proteins as structural components of plant cell walls. Cell Mol Life Sci 58:1430–1441

Robertson D, Mitchell GP, Gilroy JS, Gerrish C, Bolwell GP, Slabas AR (1997) Differential extraction and protein sequencing reveals major differences in patterns of primary cell wall proteins from plants. J Biol Chem 272:15841–15848

Roitsch T, Gonzalez MC (2004) Function and regulation of plant invertases: sweet sensations. Trends Plant Sci 9:606–613

Rudrabhatla P, Reddy MM, Rajasekharan R (2006) Genome-wide analysis and experimentation of plant serine/threonine/tyrosine-specific protein kinases. Plant Mol Biol 60:293–319

Sansen S, De Ranter CJ, Gebruers K, Brijs K, Courtin CM, Delcour JA, Rabijns A (2004) Structural basis for inhibition of *Aspergillus niger* xylanase by *Triticum aestivum* xylanase inhibitor I. J Biol Chem 278:36022–36028

Schaller A (2004) A cut above the rest: the regulatory function of plant proteases. Planta 220:183–197

Schlumbaum A, Mauch F, Vogeli U, Boller T (1986) Plant chitinases are potent inhibitors of fungal growth. Nature 324:365–367

Schultz C, Rumsewicz M, Johnson K, Jones B, Gaspar Y, Bacic A (2002) Using genomic resources to guide research directions. The arabinogalactan protein gene family as a test case. Plant Physiol 129:1448–1463

Schultz CJ, Ferguson KL, Lahnstein J, Bacic A (2004) Post-translational modifications of arabinogalactan-peptides of *Arabidopsis thaliana*. J Biol Chem 279:45503–45511

Sedbrook JC, Carroll KL, Hung KF, Masson PH, Sommerville CR (2002) The Arabidopsis SKU5 gene encodes an extracellular glycosyl phosphatidylinositol-anchored glycoprotein involved in directional root growth. Plant Cell 14:1635–1648

Sharon N, Lis H (2004) History of lectins: from hemagglutinins to biological recognition molecules. Glycobiology 14:53R–62R

Spadoni S, Zabotina O, Di Matteo A, Mikkelsen JD, Cervone F, De Lorenzo G, Mattei B, Bellincampi D (2006) Polygalacturonase-inhibiting protein interacts with pectin through a binding site formed of four clustered residues of arginine and lysine. Plant Physiol 141:557–564

Tanaka H, Onouchi H, Kondo M, Hara-Nishimura I, Nishimura M, Machida C, Machida Y (2001) A subtilisin-like serine protease is required for epidermal surface formation in Arabidopsis embryos and juvenile plants. Development 128:4681–4689

Van Damme E, Peumans W, Barre A, Rougé P (1998) Plant lectins: a composite of several distinct families of structurally and evolutionary related proteins with diverse biological roles. Crit Rev Plant Sci 17:575–692

von Groll U, Berger D, Altmann T (2002) The subtilisin-like serine protease SDD1 mediates cell-to-cell signalling during Arabidopsis stomatal development. Plant Cell 14:1527–1539

Watson BS, Lei Z, Dixon RA, Sumner LW (2004) Proteomics of *Medicago sativa* cell walls. Phytochemistry 65:1709–1720

Welinder KG, Justesen AF, Kjaersgard IVH, Jensen RB, Rasmussen SK, Jespersen HM, Duroux L (2002) Structural diversity and transcription of class III peroxidases from *Arabidopsis thaliana*. Eur J Biochem 269:6063–6081

Xia Y, Suzuki H, Borevitz J, Blount J, Guo Z, Patel K, Dixon RA, Lamb C (2004) An extracellular aspartic protease functions in *Arabidopsis* disease resistance signalling. EMBO J 23:980–988
Xu Z, Escamilla-Trevino L, Zeng L, Lalgondar M, Bevan D, Winkel B, Mohamed A, Cheng CL, Shih MC, Poulton J, Esen A (2004) Functional genomic analysis of *Arabidopsis thaliana* glycoside hydrolase family 1. Plant Mol Biol 55:343–367
Xue J, Jorgensen M, Pihlgren U, Rask L (1995) The myrosinase gene family in *Arabidopsis thaliana*: gene organization, expression and evolution. Plant Mol Biol 27:911–922
Zhong R, Kays SJ, Schroeder BP, Ye ZH (2002) Mutation of a chitinase-like gene causes ectopic deposition of lignin, aberrant cell shapes, and overproduction of ethylene. Plant Cell 14:165–179
Zhu J, Chen S, Alvarez S, Asirvatham VS, Schachtman DP, Wu Y, Sharp RE (2006) Cell wall proteome in the maize primary root elongation zone. I. Extraction and identification of water-soluble and lightly ionically bound proteins. Plant Physiol 140:311–325
Zhu Q, Maher EA, Masoud S, Dixon RA, Lamb CJ (1994) Enhanced protection against fungal attack by constitutive co-expression of chitinase and glucanase genes in transgenic tobacco. Biotechnology 12:807–812

Chapter 13
Plasma Membrane Proteomics

Erik Alexandersson, Niklas Gustavsson, Katja Bernfur, Per Kjellbom, and Christer Larsson

Abstract Proteins residing in the plasma membrane have key functions in transport, signal transduction, vesicle trafficking and many other important processes. To better understand these processes it is necessary to reveal the identity of plasma membrane proteins and to monitor modifications and regulation of their expression. This chapter is an overview of the methods used in plant plasma membrane proteomic studies and the results obtained so far. It focuses on studies using mass spectrometry for identification and includes aspects of plasma membrane fractionation, extraction and washing treatments, assessment of purity, separation methods for plasma membrane proteins and choice of techniques for protein cleavage. Finally, the results of plasma membrane proteomic studies are compared and problems with contaminating proteins are discussed.

13.1 Introduction

The plasma membrane is the interface between the cell and its surroundings. This is a position with a central structural role in the cell, and it implies a range of important functions, such as transport of compounds into and out of the cell, communication between the cell exterior and interior, and defence against

Erik Alexandersson, Department of Biochemistry, Center for Chemistry and Chemical Engineering, Lund University, P.O. Box 124, 221 00 Lund, Sweden, E-mail: erik.alexandersson@biochemistry.lu.se

Niklas Gustavsson, Department of Biochemistry, Center for Chemistry and Chemical Engineering, Lund University, P.O. Box 124, 221 00 Lund, Sweden

Katja Bernfur, Department of Biochemistry, Center for Chemistry and Chemical Engineering, Lund University, P.O. Box 124, 221 00 Lund, Sweden

Per Kjellbom, Department of Biochemistry, Center for Chemistry and Chemical Engineering, Lund University, P.O. Box 124, 221 00 Lund, Sweden

Christer Larsson Department of Biochemistry, Center for Chemistry and Chemical Engineering, Lund University, P.O. Box 124, 221 00 Lund, Sweden

J. Šamaj and J. Thelen (eds.), *Plant Proteomics*
© Springer 2007

invading pathogens. These functions are thought to be fulfilled by transport proteins, signal transduction components, and by proteins involved in membrane trafficking.

In Arabidopsis, there are over 26,000 annotated genes and, according to the "ARAMEMNON" database, about one-quarter of the proteins these genes encode are predicted to have transmembrane domains. The proportion of proteins with predicted transmembrane domains is approximately the same in rice – a monocot plant (Schwacke et al. 2003).

The plasma membrane is probably the most diverse membrane of the cell, with a protein composition that varies with cell type, developmental stage and environment; it is likely to harbour thousands of proteins. For instance, receptor-like protein kinases (RLKs) alone, most of which probably reside in the plasma membrane, are represented by more than 600 genes in the Arabidopsis genome (reviewed by Shiu and Bleecker 2001).

Several attempts using mass spectrometry (MS) have been made to identify proteins residing in the plant plasma membrane, primarily in Arabidopsis but also to some extent in rice (Tanaka et al. 2004). According to the Arabidopsis sub-cellular proteomic database, SUBA (www.suba.bcs.uwa.edu.au; Heazlewood et al. 2005), close to 600 unique proteins have been identified by MS in eight major proteomic studies of the Arabidopsis plasma membrane (Borner et al. 2002; Elortza et al. 2003, 2006; Nühse et al. 2003; Santoni et al. 2003; Alexandersson et al. 2004; Dunkley et al. 2006; Nelson et al. 2006). There have also been a few studies on plasma membrane proteomics in plants with as yet un-sequenced genomes, e.g. spinach (Kjell et al. 2004), and on plasma membrane lipid rafts in tobacco (Mongrand et al. 2004). Recently, several integral proteins were identified in the plasma membrane of barley seed aleurone layers (Hynek et al. 2006). Although the rice genome has been sequenced, the number of plasma membrane proteomic studies in rice is still limited in comparison to Arabidopsis.

Three types of proteomic studies on plasma membrane proteins can be distinguished: (1) studies on highly purified plasma membrane fractions aiming at general identification of proteins (Prime et al. 2000; Santoni et al. 2000, 2003; Marmagne et al. 2004; Alexandersson et al. 2004); (2) studies on highly purified plasma membranes targeting certain groups of proteins by taking advantage of their specific structural properties, such as glycosylphosphatidylinositol (GPI)-anchored proteins or phosphoproteins (Borner et al. 2002; Elortza et al. 2003, 2006; Nühse et al. 2003, 2004); and (3) studies that quantify peptides by MS-based methods and compare the amounts of specific proteins between different cell fractions in order to establish their localisations [e.g. LOPIT (localisation of organelle proteins by isotope tagging); Dunkley et al. 2004, 2006; Nelson et al. 2006]. The latter type of study allows for the use of crude membrane fractions. Comparison of proteins in plasma membranes from stressed and non-stressed tissue – sometimes referred to as differential proteomics – has also been reported (Kawamura and Uemura 2003). These and other studies will be further discussed below.

13.2 Plasma Membrane Purification and Fractionation

13.2.1 Plant Material

The first step in plasma membrane fractionation is the choice of plant material. So far, little attention has been devoted to the biological variation of samples. Thus, there exists a great variability in the plant material used in different studies, and in some studies the plant material used is not well defined. Furthermore, plasma membrane preparations originating from different plant organs, such as roots and leaves, are sometimes pooled, which has two obvious drawbacks: (1) the sample becomes unnecessarily complex with regard to the number of proteins (microarray studies show that gene expression in e.g. leaves and roots are quite distinct from each other, and this is most likely reflected in the protein composition), and (2) information about organ-specific protein localisation is lost. Concerning the plant material used for plasma membrane preparation, parallels can be drawn with the development of microarrays, where the initial focus was on technical advancements, only later becoming more centred on biological questions. In addition, more directed studies on proteins expressed in specific cell types or in response to environmental cues will increase the overall number of plasma membrane proteins identified.

13.2.2 Cell Fractionation

The plasma membrane constitutes only a minor part of the total membrane component of plant cells, which are normally dominated by chloroplasts with their extensive thylakoid membrane (green tissue) or by mitochondria with their internal membrane system (non-green tissue). These larger organelles may be removed by low speed centrifugation (typically 10,000 *g* for 10 min) and plasma membranes may then be harvested by subsequent high speed centrifugation of the supernatant. The resulting microsomal fraction will, however, still be dominated by membranes originating from chloroplasts and/or mitochondria depending on the tissue used, and will in addition contain membrane vesicles derived from all other types of organelles present in this tissue. The isolation of reasonably pure plasma membranes from this very complex mixture is thus a demanding task. So far, only two techniques have produced plasma membrane fractions of a purity sufficient for proteomic analyses. These techniques are "partitioning in aqueous two-phase systems" (for latest update, see Larsson et al. 1994) and "preparative free-flow electrophoresis" (Sandelius et al. 1986); the two methods are compared and reviewed in Sandelius and Morré (1990). Notably, both methods separate membrane vesicles according to surface properties, and preparations with a purity of 90% or better may be obtained. Since preparative free-flow electrophoresis requires rather sophisticated equipment whereas partitioning in aqueous two-phase systems can be performed with standard laboratory equipment, the latter technique has been more often employed. Indeed, partitioning in aqueous two-phase systems has been used to obtain plasma membrane preparations from a large number of plant species and tissues ranging from leaves, roots, and wood to cell cultures during the last two decades. Two-phase partitioning has also been the method of choice in Arabidopsis plasma membrane proteomic studies. The use of two-phase partitioning for animal plasma membranes has recently been reviewed by Schindler and Nothwang (2006).

13.2.3 Washing and Extraction Methods

When the plasma membrane fraction has been obtained it is often useful to wash the membranes in order to eliminate soluble contaminating proteins, or to enrich for integral proteins at the expense of peripheral proteins. This also reduces sample complexity and thus enables identification of proteins of lower abundance. Contaminating soluble proteins enclosed in plasma membrane vesicles can be removed by disrupting the vesicles via sonication or freezing and thawing (Palmgren et al. 1990), or by changing the sidedness (polarity) of the vesicles by treatment with the detergent Brij 58 (Johansson et al. 1995). Proteins that are peripherally attached to the plasma membrane can be removed by salt or alkaline treatments, or by organic solvent extraction (for a recent review, see Rolland et al. 2006). An obvious drawback with all membrane washings is the potential loss of water soluble proteins that actually reside in the plasma membrane and consequently are true peripheral plasma membrane proteins. As described in Sect. 13.5.5, quantitative methods such as LOPIT can probably help solve this problem.

13.2.3.1 Sonication/Freezing and Thawing

During sonication or freezing and thawing, membrane vesicles are broken and reformed, which releases proteins enclosed in these vesicles into the soluble phase. These treatments are often combined with salt washing to also remove loosely attached proteins.

13.2.3.1.1 Brij 58

Plasma membrane vesicles obtained by two-phase partitioning have a cytoplasmic-side-in orientation. Treatment with the detergent Brij 58 changes the sidedness of the vesicles and thus effectively removes soluble proteins enclosed in the vesicles (Johansson et al. 1995). Brij 58 treatment was used in Alexandersson et al. (2004) in combination with salt washing.

13.2.3.1.2 Salt Wash

Ions abolish the electrostatic interactions of contaminating soluble proteins and peripheral membrane proteins as well as the polar headgroups of lipids. This method is used regularly to remove proteins attached to membranes. By varying the salt concentration, a more or less stringent removal of proteins is achieved. Low (0.2 M) to high (1 M) concentrations of salts (usually KCl or NaCl) are used. Salt washes were employed by Alexandersson et al. (2004) and Santoni et al. (1999).

13.2.3.1.3 Alkaline Treatment

Soluble, contaminating proteins and peripheral proteins can also be removed by treatment with sodium carbonate (Fujiki et al. 1982) or NaOH. For example, NaOH treatment of Arabidopsis plasma membrane fractions was carried out by Marmagne et al. (2004). Urea and NaOH treatment was combined to strip membranes of soluble proteins in a study focused on the identification of aquaporins residing in the Arabidopsis plasma membrane (Santoni et al. 2003). To maximise the number of identified aquaporins, which are integral proteins with six transmembrane α-helices, both carbonate and urea/NaOH treatments were tested, and it was concluded that urea/NaOH increased the number of identified aquaporins more, probably because it yielded less complex samples by a more stringent removal of peripheral proteins. Ephritikhine et al. (2004) compared the properties of the proteins identified by either NaOH or salt treatments and concluded that no major differences were seen for Arabidopsis plasma membrane fractions between these two methods. This was in contrast to plastid envelope and mitochondrial membranes, where NaOH treatment removed more peripheral proteins than NaCl treatment.

13.2.3.1.4 Organic Solvent Extraction

Alternatively, instead of stripping the plasma membranes of peripheral proteins, integral proteins can be selectively extracted by organic solvents (Seigneurin-Berny et al. 1999). Chloroform/methanol is most commonly used, as reported by Marmagne et al. (2004).

13.2.3.1.5 Triton X-114 Fractionation

Using this technique, integral proteins are enriched in the lower, detergent-rich phase, and peripheral proteins are distributed to the upper, water phase (Bordier 1981). Triton X-114 fractionation was tested by Santoni et al. (2000).

13.2.3.1.6 Other Methods

Additional methods to eliminate contaminating proteins are continuously being developed. For instance, in barley aleurone layers the soluble protein β-amylase was recognised to contaminate plasma membrane preparations in spite of washing by salts and treatment with Triton X-100 or Brij 58. Hynek et al. (2006) therefore developed a method to exclude soluble proteins by employing batch reversed-phase chromatography. In this manner, it was possible to separate soluble and integral proteins, represented by β-amylase and H^+-ATPases, respectively, by elution with different concentrations of 2-propanol. In another attempt to eliminate contaminating proteins, Nielsen et al. (2005) was able to identify a large number of plasma membrane proteins in mouse brain tissue by using a combination of high-salt, carbonate and urea washes, as well as digitonin, which improves the separation of animal plasma membranes from other membranes.

13.2.4 Assessment of Purity

Purity is usually assessed by determining the levels of different markers in the plasma fraction with e.g. the microsomal fraction as a reference fraction. The most commonly used marker for the plasma membrane has been H^+-ATPase, and for the main contaminants in the microsomal fractions, chloroplast thylakoids and mitochondrial inner membranes, chlorophyll and cytochrome *c* oxidase are often used. Suitable marker proteins can be found for most membranes, and the levels can be determined either by measuring enzyme activities or by using antibodies. However, it should be realised that determining enrichment ratios of known plasma membrane proteins and depletion ratios of contaminating proteins in the plasma membrane fraction produces only relative values, and does not indicate whether a fraction is 70% or 90% pure. Absolute purity can be determined by using a specific stain for the plasma membrane: silicotungstic (or phosphotungstic) acid at pH 3 (Roland 1978; Morré 1990). Using this stain, plasma membrane vesicles appear as well stained membranes in electron micrographs, whereas other membranes are only faintly stained. Some simple morphometric procedure may then be used to calculate the proportion of well stained membranes. The purity of plasma membranes obtained by two-phase partitioning and free-flow electrophoresis was determined as 89–97% by this procedure (e.g. Kjellbom and Larsson 1984; Sandelius and Morré 1990; Morré 1990).

MS can also be used to verify purity. As shown in Fig. 13.1, the enrichment or depletion of peptides unique to different marker proteins (or group of proteins) can be determined. This can be done by using stable isotope labelling such as isobaric tags for relative and absolute quantification (iTRAQ) reagent tags (Applied Biosystems, Framingham, MA) as in Fig. 13.1 or by using $H_2{}^{18}O$ or $H_2{}^{16}O$ during trypsinisation to incorporate ^{18}O or ^{16}O into the C-termini of peptides, as in Nelson et al. (2006). Note that the enrichment of plasma membrane proteins in the plasma membrane fraction will vary, as shown in Fig. 13.1 for two peptides representing two aquaporin isoforms, PIP2;1 and PIP2;7, and a peptide

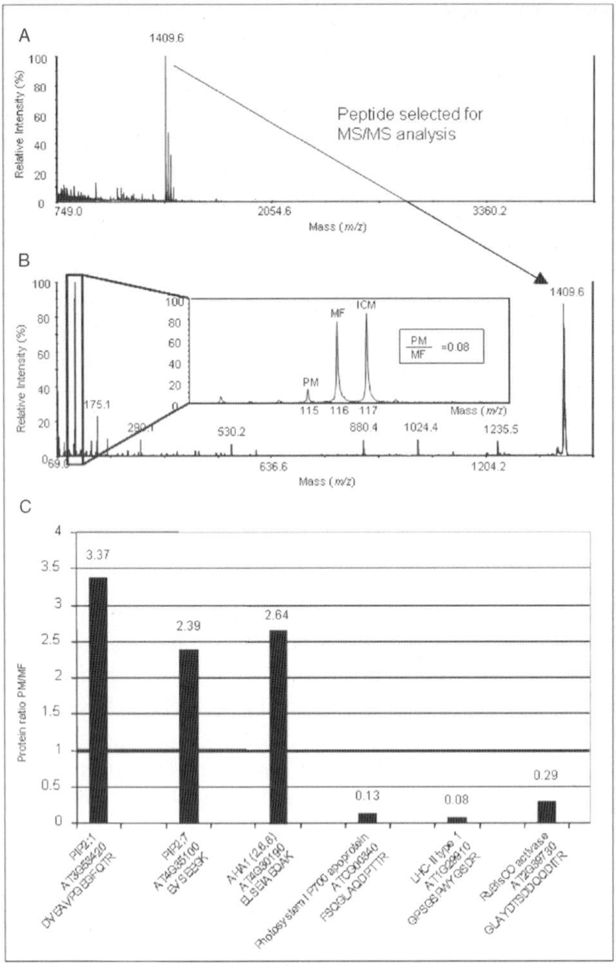

Fig. 13.1 Assessment of plasma membrane purity using isobaric tags for relative and absolute quantification (iTRAQ). Tryptic peptide extracts from the plasma membrane fraction (*PM*), the intracellular membrane fraction (*ICM*), and the corresponding microsomal fraction (*MF*) were labeled with different iTRAQ reagent tags (Applied Biosystems, Framingham, MA) and mixed in a 1:1:1 ratio based on protein concentration. The mixed peptide extract was fractionated by strong cation exchange chromatography followed by reversed phase liquid chromatography, and fractions were collected directly on a matrix assisted laser desorption/ionisation (MALDI) sample plate. Each fraction was analysed by MALDI-time-of-flight (TOF) mass spectrometry (MS) and selected peptides were analysed by tandem MS (MS/MS). Upon MS/MS analysis of a peptide, the three different tags yield different reporter ions of *m/z* 115, 116 and 117, respectively. The intensities of the three reporter ions represent the concentration of the same peptide in each of the three membrane fractions (115–PM, 116–MF, 117–ICM). The PM and ICM fractions were obtained by aqueous two-phase partitioning of an Arabidopsis leaf MF. No washing procedure was used, which may explain the relatively high level of the chloroplast soluble enzyme ribulose 1,5-bisphosphate carboxylase/oxygenase (RuBisCo) activase. **a** MS-spectrum of one LC-fraction in which a peptide with *m/z* 1,409.65 was detected. The peptide was selected for MS/MS-analysis. **b** MS/MS-spectrum of the peptide selected in **a**. *Inset* Mass region in which the iTRAQ reporter ions appear. The peptide was identified as amino acids 44–55 (GPSGSPWYGSDR) from LHC-II type 1, a chloroplast thylakoid protein depleted in the plasma membrane preparation as

(continued)

representing four H^+-ATPase isoforms. The reason may be different rates of turnover, which will result in different proportions of the proteins in the membranes of the endoplasmic reticulum (ER) and in the Golgi stacks. It may also be the result of endocytosis and exocytosis to recycle plasma membrane proteins, which seems to be an important regulatory mechanism also in plants and will result in different proportions of the proteins in endosomes (reviewed by Murphy et al. 2005; Šamaj et al. 2005).

13.3 Protein Separation

13.3.1 Gel-Based Techniques

In early proteomic studies of the plant plasma membrane, two-dimensional gels were used to separate proteins (Prime et al. 2000; Santoni et al. 2000). However, few integral proteins were identified as it proved difficult to resolve these by non-ionic or zwitterionic detergent in the first iso-electrofocusing (IEF) step. Instead, one-dimensional gels were used to increase the number of resolved integral proteins; an approach that was successfully used in several studies on cellular membranes (e.g. Millar and Heazlewood 2003; Alexandersson et al. 2004; Marmagne et al. 2004; Peltier et al. 2004). Further development of two-dimensional electrophoresis (2-DE) to resolve hydrophobic proteins was undertaken; e.g. Navarre et al. (2002) reported the successful identification of integral plasma membrane proteins in yeast by the use of the cationic detergent trimethyl ammonium bromide (CTAB) in the IEF step. Other types of modified two-dimensional systems have also been developed, such as BAC/SDS-PAGE, which involves gel electrophoresis in an acidic buffer system using the cationic detergent benzyldimethyl-*n*-hexadecylammonium chloride (16-BAC) in the first dimension and SDS-PAGE in the second dimension (Hartinger et al. 1996). Lately, different variants of 2-DE, in which SDS-PAGE is used in both the first and second dimensions (termed doubled SDS-PAGE or dSDS-PAGE), have been developed (Rais et al. 2004). In such systems, the acrylamide concentration, urea content and trailing ion used in the gels are altered between the two dimensions. Burré et al. (2006) performed a comparison of identified synaptic vesicle proteins resolved by either one-dimensional SDS-PAGE, BAC/SDS-PAGE or dSDS-PAGE and demonstrated that the three gel-based separation methods were partly complementary to each other regarding resolution and identification of integral proteins.

shown by the low intensity of the 115 reporter ion. **c** Examples of plasma membrane proteins enriched in the plasma membrane fraction (protein ratios >1; aquaporin isoforms PIP2;1 and PIP2;7, and plasma membrane H^+-ATPase isoforms AHA1, 2, 6 and 8), and proteins depleted in the plasma membrane preparation (protein ratios <1; two chloroplast thylakoid proteins and the abundant, soluble chloroplast stroma protein RuBisCo activase). The protein ratios were calculated from MS/MS d ata from one unique peptide for each protein (except for the peptide used for AHA1, which is also present in the isoforms AHA2, 6, and 8)

It is also possible to resolve integral proteins in functional protein complexes by the use of blue-native electrophoresis (BN-PAGE), followed by SDS-PAGE separation in the second dimension. This method was first used by Schägger and von Jagow (1991) and it not only overcomes the difficulties of resolving integral proteins in the first IEF step but also enables the visualisation of multiprotein complexes. In plants, BN-PAGE has been used mostly to resolve protein complexes in chloroplast and mitochondrial membranes, with the exception of one publication on plasma membranes from spinach leaves (Kjell et al. 2004). Lately, several studies in animals have combined BN-PAGE with plasma membrane protein identification (e.g. Brouillard et al. 2005; Yoshinaka et al. 2004).

13.3.2 Non-Gel-Based Techniques

To reduce sample complexity, fractionation of proteins prior to gel electrophoresis and/or of peptides after trypsinisation, is often done by protein/peptide separation by chromatography. To avoid gel-implemented separation problems, entirely non-gel-based techniques have also been developed, e.g. MudPIT (multidimensional protein identification technique), which separates peptides by two-dimensional liquid chromatography (Washburn et al. 2001). However, this technique has, to our knowledge, been used only for plant cell fractions other than the plasma membrane, e.g. the chloroplast envelope membrane (Froehlich et al. 2003) and the cell wall (Bayer et al. 2006).

13.4 Protein Digestion for MS-Analysis

Using standard trypsinisation protocols prior to MS-analysis, it has proved difficult to obtain high sequence coverage of integral proteins. This is likely to be a consequence of the lack of trypsin cleavage sites in membrane-spanning regions and also the limited accessibility to proteases of these regions. Therefore, several alternative methods and proteases have been tested to improve sequence coverage. Van Montfort et al. (2002) were able to double the coverage of the integral domains of a lactose transporter using a combination of in-gel trypsinisation and cyanogen bromide (CNBr) cleavage compared to using either of these methods individually. Han and Schey (2004) reported the 100% sequence coverage of AQP0 in cow eye lens tissue, in which AQP0 is both the only expressed AQP isoform and the most abundant integral protein. For full coverage, digestion with both trypsin and pepsin in optimised solutions were used, and both MALDI (matrix assisted laser desorption/ionisation) and ESI (electrospray ionisation) MS were required for peptide detection, but having taken the effort, one new phosphorylation site, two new oxidation sites and a previously unknown amino acid mutation were detected. Proteinase K has also been successfully used to improve the sequence coverage of integral proteins (Wu et al. 2003; Distler et al. 2006).

13.5 Biological Insights

To date, there are more than ten comprehensive studies identifying plasma membrane proteins in Arabidopsis, which is the only well studied plant species in this field. Consequently, our present knowledge with respect to plasma membrane protein composition has been considerably improved. Not only have close to 600 proteins been predicted to reside in the plasma membrane, about one-fifth of which lack assigned functions, but much knowledge has also been gained about certain protein groups, such as GPI-anchored proteins and phosphoproteins, in the plasma membrane (Elortza et al. 2003, 2006; Nühse et al. 2003, 2004). So far, however, only a few studies have addressed questions regarding changes in the protein composition of the plasma membrane upon abiotic and biotic stresses or during development – so-called differential proteomics.

When evaluating proteomic studies it is easy to become pre-occupied with concerns about contaminating proteins and the detection limits of less abundant proteins. Such concerns, as well as the strikingly low overlap of identified proteins between studies, will be further discussed below. But to begin with, the biological insights gained so far will be highlighted by taking a closer look at the results of proteomic studies on plant plasma membranes.

13.5.1 Proteins Identified in the Plasma Membrane

As might be expected, proteins involved in transport, stress and signal transduction are the major functional groups in the plasma membrane that have been identified by proteomics so far. Indeed, about half of the proteins classified as being involved in signal transduction identified in proteomic studies of *Arabidopsis thaliana* have been found in the plasma membrane (SUBA database). Furthermore, with few exceptions these are found exclusively in the plasma membrane and not in other parts of the cell, and are often identified only by a single plasma membrane proteomic study. This emphasises the importance of the plasma membrane as an interface for signal conductance, but also the divergence of signalling proteins residing in it, e.g. only two RLKs have been identified in more than one plasma membrane proteomic study. This stresses not only the significance of more sensitive detection techniques, but also the importance of studying plasma membrane samples originating from different biological material in order to increase the number of proteins identified. Other important categories of proteins identified in proteomic studies of the plasma membrane are proteins involved in protein metabolism, membrane trafficking and cellular organisation, as well as cell-wall-associated proteins.

The over-representation of proteins involved in signal transduction and transport identified in the plasma membrane is confirmed by looking at the distribution of MS-identified proteins either in the plasma membrane or outside of the plasma membrane categorised by gene ontology (GO) annotations according to TAIR (www.arabidopsis.org; Fig. 13.2). There is also an overrepresentation of proteins identified in the plasma membrane that are involved in protein metabolism. However, many of these are soluble proteins that are highly abundant and may therefore be contaminants of the plasma membrane preparation. In Fig. 13.2, the GO annotations of proteins identified in or outside of the plasma membrane are compared to the distribution of GO annotations of all annotated genes in the Arabidopsis genome. The most obvious difference is the considerably higher level of proteins denoted "biological process unknown" for the annotated genes in the Arabidopsis genome, which probably reflects the fact that MS-identification of proteins so far has been most successful with abundant, and thus characterised, proteins.

The large number of identified proteins with unknown function in specific cellular compartments makes proteomic studies unique in comparison to localisation using other techniques. A localisation not only provides a start to the study of these proteins, but also gives a hint that these previously overlooked proteins might fulfil an important role in the plasma membrane, especially given that the total number of identified plasma membrane proteins is still less than 600 (SUBA database). A localisation to the plasma membrane has also led to reclassification of proteins with previously assigned functions, e.g. remorin was first classified as a DNA-

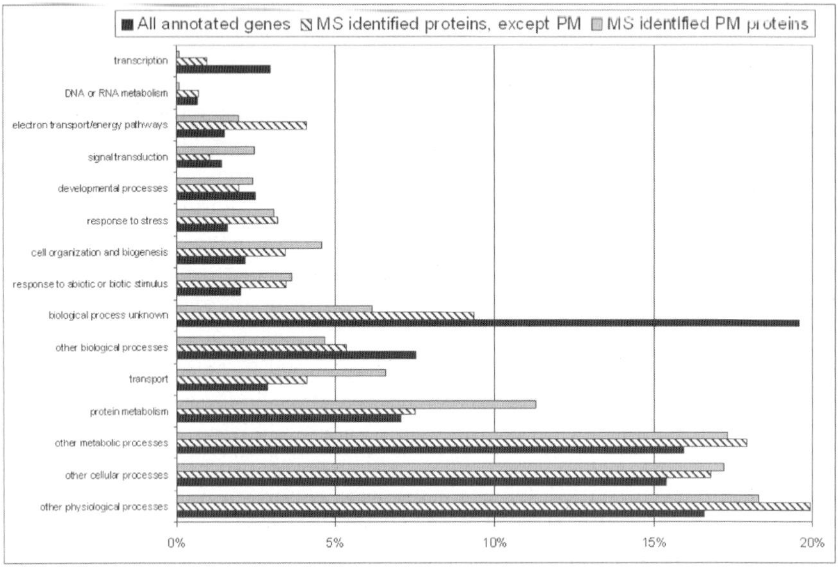

Fig. 13.2 Gene ontology (GO) categorisation of proteins identified by MS in the plasma membrane (PM, *grey bars*) or outside of the plasma membrane (*striped bars*) according to the SUBA database (www.suba.bcs.uwa.edu.au) compared with the GO categorisation of all annotated genes in the Arabidopsis genome retrieved from TAIR (*black bars*, www.arabidopsis.org)

binding protein, but is now believed to reside in the plasma membrane, where it has an unknown function (Alexandersson et al. 2004; Nelson et al. 2006).

13.5.2 Studies Targeted at Plasma Membrane Proteins with Specific Properties

Several proteomic studies of the plasma membrane have targeted certain groups of proteins by taking advantage of their specific structural properties. GPI-anchored proteins have been collected in the soluble phase by treating plasma membranes with phosphatidylinositol phospholipase C (PI-PLC) and subsequent analysis by MS (Borner et al. 2002; Elortza et al. 2003). Later, phospholipase D, which exhibits broader substrate specificity than PI-PLC as it is not hampered by the presence of modifications on the core GPI anchor, was successfully used (Elortza et al. 2006). The Arabidopsis plasma membrane phosphoproteome has been explored by the use of immobilised metal ion affinity chromatography (IMAC), to enrich for negatively charged phosphopeptides (Nühse et al. 2003, 2004). Altogether, more than 300 phosphorylation sites and 200 phosphoproteins have been identified – most of these previously unknown (Nühse et al. 2004). These can be found in the PlantsP database (www.plantsp.sdsc.edu). RLKs form the largest group of identified phosphoproteins. Plasma membrane proteins residing in lipid rafts have also been targetted in tobacco and Arabidopsis by examining membrane fractions insoluble after cold Triton X-100 treatment (Mongrand et al. 2004; Borner et al. 2005; Morel et al. 2006).

13.5.3 Differential Plasma Membrane Proteomics

By comparing different proteomes, changes in protein composition and abundance, e.g. due to different stress conditions, can be detected. This is sometimes referred to as differential proteomics. To date, there are few such proteomic studies focused on plasma membrane proteins. One exception is a study by Kawamura and Uemura (2003), who used one-dimensional SDS-PAGE to examine the profile of purified Arabidopsis plasma membrane proteins upon cold acclimation. In total, 38 proteins that changed in intensity during the treatment were identified by MALDI-TOF MS. However, few of them were integral proteins. There are also relatively few examples of differential plasma membrane proteomics in other organisms. Huang et al. (2006) performed a screening of salt-affected proteins in the plasma membrane of the cyanobacterium *Synechocytis* sp. strain PCC6803 by comparing changes in expression of proteins resolved on two-dimensional IEF/SDS-PAGE gels; 25 plasma membrane proteins were shown to be affected. However, only 11 of the 109 proteins identified had predicted transmembrane domains. More than 200 proteins with predicted transmembrane domains were identified in yeast plasma membrane fractions using a quantitative MudPIT approach (Zybailov et al. 2006). Given that a total of 1,316 proteins were identified, the proportion of predicted transmembrane proteins seems to be somewhat higher with the MudPIT approach compared to 2-DE studies. Another approach to resolve plasma membrane proteins for the purposes of differential proteomics was taken by Delom et al. (2006), who compared 20 protein fractions obtained from yeast either treated or not treated with the anti-fungal compound calcofluor and resolved by one-dimensional lithium dodecyl sulphate (LDS)-PAGE. Changes in expression were monitored by comparing protein band intensities on LDS-PAGE gels. About half of the 90 identified proteins had predicted transmembrane domains. A similar approach to separate and resolve integral proteins has been used for vacuolar proteins in Arabidopsis (Szponarski et al. 2004).

The lack of proteomic studies on differential protein expression is likely to be a consequence of the problems of resolving integral proteins on two-dimensional gels, which is the most established way to visualise up- and down-regulation of protein expression by techniques based on fluorescence dyes such as difference in gel electrophoresis (DIGE). An exception is the study of Lee et al. (2004), who identified and quantified proteins involved in salt stress signalling in Arabidopsis microsomal fractions by the use of two-dimensional gels. As isotope labelling and other labelling techniques become more robust, studies comparing protein abundance also in plasma membranes can be expected to appear. In animals, two studies have used quantitative MS by peptide-labelling in order to identify proteins that are differentially expressed in plasma membrane preparations of the fore- and hindbrain (Olsen et al. 2004; Cheng et al. 2006), and the principle of peptide-labelling has already been used in plant plasma membrane studies (Dunkley et al. 2004, 2006; Nelson et al. 2006).

13.5.4 Overlap of Identified Proteins Among Studies

Although several proteomic studies of the Arabidopsis plasma membrane have been successful in identifying a large number of putative plasma membrane proteins, most notably the comparatively high number of integral proteins, the low level of overlap in identified proteins between different studies is striking. For example, only 34 proteins were identified in both Alexandersson et al. (2004) and Marmagne et al. (2004), which represent the two studies with the highest numbers of proteins identified in the plant plasma membrane. The overlap between proteins identified by MS in Alexandersson et al. (2004) and other plasma membrane proteomic studies, as well as in the further studies focused on cellular compartments other than the plasma membrane, is shown in Fig. 13.3. The comparison was made with the help of the SUBA database, which includes over 40 studies of various

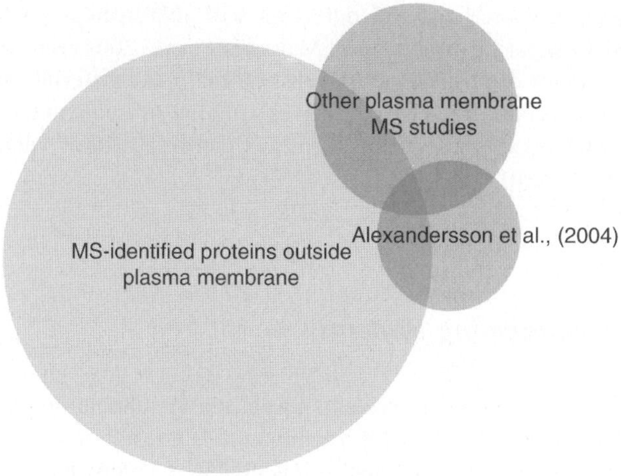

Fig. 13.3 Venn diagram showing the overlap of proteins identified by MS in Alexandersson et al. (2004), and by other plasma membrane and subcellular compartment studies excluding the plasma membrane, all according to the SUBA database. Venn diagram was generated by www.VennDiagram.tk

sub-cellular compartments. Consequently, only studies included in the database were compared and, for unknown reasons, the study of Marmagne et al. (2004) is not included in SUBA. Notably, a relatively small number of proteins overlap between Alexandersson et al. (2004) and other plasma membrane studies. In fact, more than 50% of the total number of proteins identified by plasma membrane proteomic studies included in the SUBA database, have been identified by only one study.

There are several possible reasons for this lack of overlap in identified proteins among different studies. To begin with, the identification of proteins is heavily dependent on the techniques used. Extraction, subcellular fractionation, washing of membranes, amount of sample, protein solubilisation, separation and resolution of proteins, protein cleavage, peptide separation, detection system and in silico identification, including cut-off criteria for a positive match, are all steps that will influence which proteins are identified. For instance, in a comparison of chloroform/methanol extraction, NaOH and NaCl treatments of plastid envelope membranes and mitochondrial membranes, it was found that less than 10% of proteins were identified in all three extraction procedures, while more than 60% of proteins were identified in only one extraction procedure (Ephritikhine et al. 2004). This emphasises the power of using different extraction methods in order to increase protein coverage. The physical properties of proteins and peptides analysed (protein hydrophobicity, size of cleaved peptides depending on protease used, peptide ionisation properties, etc.) together with the type of ionisation source and MS detection will also influence which proteins are identified. Comparing the studies of Marmagne et al. (2004) and Alexandersson et al. (2004), major differences exist between the plant material used (cell suspension cultures versus green rosette leaves) and the subsequent treatment of the isolated plasma membranes (chloroform/methanol extraction versus salt wash and Brij 58 treatment).

13.5.5 *Contaminating Proteins*

The presence of contaminating proteins in plasma membrane preparations represents a challenging problem. Such contamination is a result of sample preparation and handling before protein detection, and is therefore a problem affecting all techniques based on cellular fractionation and localisation of proteins. Unfortunately, determining the purity of the plasma membrane fraction by measuring the presence or enrichment of marker proteins by enzymatic activity or immunodetection, MS or other means is no guarantee that the preparation is free of contaminants, even if a high purity is obtained. In addition, the lack of consensus plasma membrane signal peptides makes bioinformatic predictions of plasma membrane localisation impossible.

MS-based protein identification can be used to identify putative contaminants as recently shown in two studies (Dunkley et al. 2006; Nelson et al. 2006). The LOPIT approach is based on isotope labelling of fractions collected from an iodixanol gradient self-generated by centrifugation (Dunkley et al. 2004). Through a series of comparisons of fractions, the distribution of specific proteins can be determined along the gradient by relative quantification of the isotope tags. Multivariate data analysis is subsequently used to group proteins according to their distributions. Consequently, proteins with unknown localisation can be allocated to an organelle or membrane based on the distribution of proteins with known localisation. By combining iTRAQ-labelling of peptides and two-dimensional liquid chromatography, Dunkley et al. (2006) localised 527 proteins to different parts of the cell, of which 92 were localised to the plasma membrane. Similarly, Nelson et al. (2006) compared the relative enrichment of plasma membrane proteins in the upper phase of the aqueous two-phase system by labelling peptides from the upper and lower phase (the latter is enriched in intracellular membranes, cf. Fig. 13.1) with either $H_2{}^{18}O$ or $H_2{}^{16}O$ (during trypsinisation ^{18}O or ^{16}O is incorporated into the C-termini of peptides). The level of enrichment of canonical plasma membrane proteins, such as H^+-ATPase isoforms, was then used to determine the localisation of proteins. Out of 174 identified and quantified proteins, 70 had a level of enrichment equal to or higher than that of H^+-ATPases. In general, a depletion of canonical chloroplast and mitochondrial proteins was seen in the upper phase. However, some canonical vacuolar proteins, such as V-ATPases, showed less depletion, and in some cases even two- to three-fold enrichment. In the study of Nelson et al. (2006), a relatively high variation in enrichment of peptides originating from the same proteins within the same experiment, and from two repeated experiments, was observed. Furthermore, a high degree of variability in enrichment of canonical plasma membrane proteins was detected in these experiments. Whether such differences between the repetitive experiments are due to technical variability or possible biological variations between samples is not known (see Sect. 13.2.4). Nelson et al. (2006) also found that longer sampling time led to an increased number of detected proteins, and the percentage of proteins considered as contaminants also increased. Their interpretation was that low abundance proteins in the preparation consisted not only of low abundance plasma membrane proteins but also of contaminating proteins that were highly abundant in other parts of the cell (Nelson et al. 2006). However, the use of isotope-labelling per se does not overcome the problem of possible re-distribution of proteins upon plasma membrane preparation. Soluble proteins especially might attach to and contaminate membranes when cells are homogenised. Additionally, it should be considered that virtually all proteins will reside in more than one cellular compartment in vivo due to protein turnover and membrane recycling, as discussed in Section 13.2.4. To evaluate these situations, complementary techniques like immunolocalisation and fluorescence-tagged fusion proteins can be used. This will be particularly important for the identification of soluble proteins as true peripheral proteins of the plasma membrane.

A different indicative technique to recognise putative contaminants is to relate identified proteins to their level of gene expression. This can be done by looking at the number of annotated expressed sequence tags (ESTs), which gives an indication of the expression level. Accordingly, an identified protein with a high number of annotated ESTs indicates an abundant protein and therefore more likely to be a contaminant, whereas the identification of a protein with few annotated ESTs would strengthen the likelihood of a correct localisation. A more refined method would be to compare the gene expression level of the organ studied with the help of gene expression databases e.g. Genevestigator for Arabidopsis (Zimmermann et al. 2004). An obvious drawback of identifying putative contaminants by gene expression is that highly abundant proteins that really reside in the plasma membrane will be assumed to be contaminants. Furthermore, there is not necessarily a correlation between gene expression and protein abundance.

13.6 Conclusions and Future Perspectives

The most obvious advancements in plant plasma membrane proteomics have been fast identification of proteins, confirmation of protein expression, localisation of a large number of proteins, and identification of post-translational modifications. The problems of early proteomic studies, which identified only few integral plasma membrane proteins, have partly been overcome by modified gel-based techniques together with non-gel-based techniques like MudPIT. Still, the difficulties involved in using gel-based two-dimensional separation of integral proteins is reflected in the few reported studies that focus on differential expression of plasma membrane proteins. However, gel-based methods are continuously being developed and in the future appropriate gel-systems for two-dimensional resolution of integral proteins might evolve. Furthermore, labelling of proteins/peptides in order to compare expression by non-gel-based methods has advanced rapidly in the last few years and is a promising way forward for differential proteomics of integral proteins.

The attempts to localise proteins on a sub-cellular level by a proteomic approach have again emphasised the importance of the level of purity of the membrane fractions studied. However, as outlined above, proteomics can itself be a powerful tool to estimate purity and to identify putative contaminants, since various kinds of peptide labelling of cell fractions enable quantitative comparison of proteins and their distribution.

A characteristic feature of MS analysis is that a large fraction of the proteins identified have no assigned functions, as there is no bias against not yet studied proteins in comparison to proteins with known functions in the identification process. The identified proteins with unknown functions are not only confirmed to be expressed, they are also likely to be among the more highly abundant proteins, which itself is an indication that they have important roles. There are also examples of previously characterised proteins that, through their novel localisation, have been assigned new roles.

An important aspect of plant plasma membrane proteomics is to organise the findings of proteomic studies in databases (for a review on plant membrane proteome databases, see Schwacke et al. 2003). In the SUBA database, which focuses on sub-cellular localisation of proteins, the observations of different studies can be combined and compared (Heazlewood et al. 2005). Another example is the rice proteome database, which includes, among other things, spot identifications on two-dimensional gels (gene64.dna.affrc.go.jp/RPD; Komatsu, 2005). The plasma membrane phosphoproteins identified by Nühse et al. (2004) are deposited in the PlantsP database. Future databases and web-based tools are likely to integrate proteomics with other sets of data, e.g. transcriptomics and metabolomic data, placing protein identification and characterisation in a broader context.

References

Alexandersson E, Saalbach G, Larsson C, Kjellbom P (2004) Arabidopsis plasma membrane proteomics identifies components of transport, signal transduction and membrane trafficking. Plant Cell Physiol 45:1543–1556

Bayer EM, Bottrill AR, Walshaw J, Vigouroux M, Naldrett MJ, Thomas CL, Maule AJ (2006) Arabidopsis cell wall proteome defined using multidimensional protein identification technology. Proteomics 6:301–311

Bordier C (1981) Phase separation of integral membrane proteins in Triton X-114 solution. J Biol Chem 256:1604–1607

Borner GH, Sherrier DJ, Stevens TJ, Arkin IT, Dupree P (2002) Prediction of glycosylphosphatidylinositol-anchored proteins in Arabidopsis. A genomic analysis. Plant Physiol 129:486–499

Borner GH, Sherrier DJ, Weimar T, Michaelson LV, Hawkins ND, Macaskill A, Napier JA, Beale MH, Lilley KS, Dupree P (2005) Analysis of detergent-resistant membranes in Arabidopsis. Evidence for plasma membrane lipid rafts. Plant Physiol 137:104–116

Brouillard F, Bensalem N, Hinzpeter A, Tondelier D, Trudel S, Gruber AD, Ollero M, Edelman A (2005) Blue native/SDS-PAGE analysis reveals reduced expression of the mClCA3 protein in cystic fibrosis knock-out mice. Mol Cell Proteomics 4:1762–1775

Burre J, Beckhaus T, Schagger H, Corvey C, Hofmann S, Karas M, Zimmermann H, Volknandt W (2006) Analysis of the synaptic vesicle proteome using three gel-based protein separation techniques. Proteomics 6:6250–6262

Cheng D, Hoogenraad CC, Rush J, Ramm E, Schlager MA, Duong DM, Xu P, Wijayawardana SR, Hanfelt J, Nakagawa T, Sheng M, Peng J (2006) Relative and absolute quantification of postsynaptic density proteome isolated from rat forebrain and cerebellum. Mol Cell Proteomics 5:1158–1170

Delom F, Szponarski W, Sommerer N, Boyer JC, Bruneau JM, Rossignol M, Gibrat R (2006) The plasma membrane proteome of *Saccharomyces cerevisiae* and its response to the antifungal calcofluor. Proteomics 6:3029–3039

Distler AM, Kerner J, Peterman SM, Hoppel CL (2006) A targeted proteomic approach for the analysis of rat liver mitochondrial outer membrane proteins with extensive sequence coverage. Anal Biochem 356:18–29

Dunkley TP, Watson R, Griffin JL, Dupree P, Lilley KS (2004) Localization of organelle proteins by isotope tagging (LOPIT). Mol Cell Proteomics 3:1128–1134

Dunkley TP, Hester S, Shadforth IP, Runions J, Weimar T, Hanton SL, Griffin JL, Bessant C, Brandizzi F, Hawes C, Watson RB, Dupree P, Lilley KS (2006) Mapping the Arabidopsis organelle proteome. Proc Natl Acad Sci USA 103:6518–6523

Elortza F, Nühse TS, Foster LJ, Stensballe A, Peck SC, Jensen ON (2003) Proteomic analysis of glycosylphosphatidylinositol-anchored membrane proteins. Mol Cell Proteomics 2:1261–1270

Elortza F, Mohammed S, Bunkenborg J, Foster LJ, Nühse TS, Brodbeck U, Peck SC, Jensen ON (2006) Modification-specific proteomics of plasma membrane proteins: identification and characterization of glycosylphosphatidylinositol-anchored proteins released upon phospholipase D treatment. J Proteome Res 5:935–943

Ephritikhine G, Ferro M, Rolland N (2004) Plant membrane proteomics. Plant Physiol Biochem 42:943–962

Froehlich JE, Wilkerson CG, Ray WK, McAndrew RS, Osteryoung KW, Gage DA, Phinney BS (2003) Proteomic study of the *Arabidopsis thaliana* chloroplastic envelope membrane utilizing alternatives to traditional two-dimensional electrophoresis. J Proteome Res 2:413–425

Fujiki Y, Hubbard AL, Fowler S, Lazarow PB (1982) Isolation of intracellular membranes by means of sodium carbonate treatment: application to endoplasmic reticulum. J Cell Biol 93:97–102

Han J, Schey KL (2004) Proteolysis and mass spectrometric analysis of an integral membrane: aquaporin 0. J Proteome Res 3:807–812

Hartinger J, Stenius K, Hogemann D, Jahn R (199 6) 16-BAC/SDS-PAGE: a two-dimensional gel electrophoresis system suitable for the separation of integral membrane proteins. Anal Biochem 240:126–133

Heazlewood JL, Tonti-Filippini J, Verboom RE, Millar AH (2005) Combining experimental and predicted datasets for determination of the subcellular location of proteins in Arabidopsis. Plant Physiol 139:598–609

Huang F, Fulda S, Hagemann M, Norling B (2006) Proteomic screening of salt-stress-induced changes in plasma membranes of *Synechocystis* sp. strain PCC 6803. Proteomics 6:910–920

Hynek R, Svensson B, Jensen ON, Barkholt V, Finnie C (2006) Enrichment and identification of integral membrane proteins from barley aleurone layers by reversed-phase chromatography, SDS-PAGE, and LC-MS/MS. J Proteome Res 5:3105–3113

Johansson F, Olbe M, Sommarin M, Larsson C (1995) Brij 58, a polyoxyethylene acyl ether, creates membrane vesicles of uniform sidedness. A new tool to obtain inside-out (cytoplasmic side- out) plasma membrane vesicles. Plant J 7:165–173

Kawamura Y, Uemura M (2003) Mass spectrometric approach for identifying putative plasma membrane proteins of Arabidopsis leaves associated with cold acclimation. Plant J 36:141–154

Kjell J, Rasmusson AG, Larsson H, Widell S (2004) Protein complexes of the plant plasma membrane resolved by Blue Native PAGE. Physiol Plant 121:546–555

Kjellbom P, Larsson C (1984) Preparation and polypeptide composition of chlorophyll-free plasma membranes from leaves of light-grown spinach and barley. Physiol Plant 62:501–509

Komatsu S (2005) Rice proteome database: a step toward functional analysis of the rice genome. Plant Mol Biol 59:179–190

Larsson C, Sommarin M, Widell S (1994) Isolation of highly purified plant plasma membranes and separation of inside-out and right-side-out vesicles. Methods Enzymol 228:451–469

Lee S, Lee EJ, Yang EJ, Lee JE, Park AR, Song WH, Park OK (2004) Proteomic identification of annexins, calcium-dependent membrane binding proteins that mediate osmotic stress and abscisic acid signal transduction in Arabidopsis. Plant Cell 16:1378–1391

Marmagne A, Rouet MA, Ferro M, Rolland N, Alcon C, Joyard J, Garin J, Barbier-Brygoo H, Ephritikhine G (2004) Identification of new intrinsic proteins in arabidopsis plasma membrane proteome. Mol Cell Proteomics 3:675–691

Millar AH, Heazlewood JL (2003) Genomic and proteomic analysis of mitochondrial carrier proteins in Arabidopsis. Plant Physiol 131:443–453

Mongrand S, Morel J, Laroche J, Claverol S, Carde JP, Hartmann MA, Bonneu M, Simon-Plas F, Lessire R, Bessoule JJ (2004) Lipid rafts in higher plant cells: purification and characterization of Triton X-100-insoluble microdomains from tobacco plasma membrane. J Biol Chem 279:36277–36286

Morel J, Claverol S, Mongrand S, Furt F, Fromentin J, Bessoule JJ, Blein JP, Simon Plas F (2006) Proteomics of plant detergent-resistant membranes. Mol Cell Proteomics 5:1396–1411

Morré DJ (1990) Plasma membrane cytochemistry. In: Larsson C, Møller IM (eds) The plant plasma membrane Springer, Berlin, pp 76–92

Murphy AS, Bandyopadhyay A, Holstein SE, Peer WA (2005) Endocytotic cycling of PM proteins. Annu Rev Plant Biol 56:221–251

Navarre C, Degand H, Bennett KL, Crawford JS, Mortz E, Boutry M (2002) Subproteomics: identification of plasma membrane proteins from the yeast *Saccharomyces cerevisiae*. Proteomics 2:1706–1714

Nelson CJ, Hegeman AD, Harms AC, Sussman MR (2006) A quantitative analysis of arabidopsis plasma membrane using trypsin-catalyzed ^{18}O labeling. Mol Cell Proteomics 5:1382–1395

Nielsen PA, Olsen JV, Podtelejnikov AV, Andersen JR, Mann M, Wisniewski JR (2005) Proteomic mapping of brain plasma membrane proteins. Mol Cell Proteomics 4:402–408

Nühse TS, Stensballe A, Jensen ON, Peck SC (2003) Large-scale analysis of in vivo phosphorylated membrane proteins by immobilized metal ion affinity chromatography and mass spectrometry. Mol Cell Proteomics 2:1234–1243

Nühse TS, Stensballe A, Jensen ON, Peck SC (2004) Phosphoproteomics of the Arabidopsis plasma membrane and a new phosphorylation site database. Plant Cell 16:2394–2405

Olsen JV, Andersen JR, Nielsen PA, Nielsen ML, Figeys D, Mann M, Wisniewski JR (2004) HysTag – a novel proteomic quantification tool applied to differential display analysis of membrane proteins from distinct areas of mouse brain. Mol Cell Proteomics 3:82–92

Palmgren MG, Askerlund P, Fredrikson K, Widell S, Sommarin M, Larsson C (1990) Sealed inside-out and right-side-out plasma membrane vesicles—optimal conditions for formation and separation. Plant Physiol 92:871–880

Peltier JB, Ytterberg AJ, Sun Q, van Wijk KJ (2004) New functions of the thylakoid membrane proteome of *Arabidopsis thaliana* revealed by a simple, fast, and versatile fractionation strategy. J Biol Chem 279:49367–49383

Prime TA, Sherrier JD, Mahon P, Packman LC, Dupree P (2000) A proteomic analysis of organelles from *Arabidopsis thaliana*. Electrophoresis 21:3488–3499

Rais I, Karas M, Schagger H (2004) Two-dimensional electrophoresis for the isolation of integral membrane proteins and mass spectrometric identification. Proteomics 4:2567–2571

Roland JC (1978) General preparation and staining of thin sections. In: Hall JL (ed) Electron microscopy and cytochemistry of plant cells. Elsevier, Amsterdam, pp 1–62

Rolland N, Ferro M, Ephritikhine G, Marmagne A, Ramus C, Brugiere S, Salvi D, Seigneurin-Berny D, Bourguignon J, Barbier-Brygoo H, Joyard J, Garin J (2006) A versatile method for deciphering plant membrane proteomes. J Exp Bot 57:1579–1589

Šamaj J, Read ND, Volkmann D, Menzel D, Baluška F (2005) The endocytic network in plants. Trends Cell Biol 15:425–433

Sandelius AS, Morré DJ (1990) Plasma membrane isolation. In: Larsson C, Møller IM (eds) The plant plasma membrane. Springer, Berlin, pp 44–75

Sandelius AS, Penel C, Auderset G, Brightman A, Millard M, Morré DJ (1986) Isolation of highly purified fractions of plasma membrane and tonoplast from the same homogenate of soybean hypocotyls by free-flow electrophoresis. Plant Physiol 81:177–185

Santoni V, Doumas P, Rouquie D, Mansion M, Rabilloud T, Rossignol M (1999) Large scale characterization of plant plasma membrane proteins. Biochimie 81:655–661

Santoni V, Kieffer S, Desclaux D, Masson F, Rabilloud T (2000) Membrane proteomics: use of additive main effects with multiplicative interaction model to classify plasma membrane proteins according to their solubility and electrophoretic properties. Electrophoresis 21:3329–3344

Santoni V, Vinh J, Pflieger D, Sommerer N, Maurel C (2003) A proteomic study reveals novel insights into the diversity of aquaporin forms expressed in the plasma membrane of plant roots. Biochem J 373:289–296

Schägger H, von Jagow G (1991) Blue Native electrophoresis for isolation of membrane protein complexes in enzymatically active form. Anal Biochem 199:223–231

Schindler J, Nothwang HG (2006) Aqueous polymer two-phase systems: effective tools for plasma membrane proteomics. Proteomics 6:5409–5417

Schwacke R, Schneider A, van der Graaff E, Fischer K, Catoni E, Desimone M, Frommer WB, Flugge UI, Kunze R (2003) ARAMEMNON, a novel database for Arabidopsis integral membrane proteins. Plant Physiol 131:16–26

Seigneurin-Berny D, Rolland N, Garin J, Joyard J (1999) Technical advance. Differential extraction of hydrophobic proteins from chloroplast envelope membranes: a subcellular-specific proteomic approach to identify rare intrinsic membrane proteins. Plant J 19:217–228

Shiu SH, Bleecker AB (2001) Plant receptor-like kinase gene family: diversity, function, and signaling. Sci STKE 2001:RE22

Szponarski W, Sommerer N, Boyer JC, Rossignol M, Gibrat R (2004) Large-scale characterization of integral proteins from Arabidopsis vacuolar membrane by two-dimensional liquid chromatography. Proteomics 4:397–406

Tanaka N, Fujita M, Handa H, Murayama S, Uemura M, Kawamura Y, Mitsui T, Mikami S, Tozawa Y, Yoshinaga T, Komatsu S (2004) Proteomics of the rice cell: systematic identification of the protein populations in subcellular compartments. Mol Genet Genomics 271:566–576

Van Montfort BA, Doeven MK, Canas B, Veenhoff LM, Poolman B, Robillard GT (2002) Combined in-gel tryptic digestion and CNBr cleavage for the generation of peptide maps of an integral membrane protein with MALDI-TOF mass spectrometry. Biochim Biophys Acta 1555:111–115

Washburn MP, Wolters D, Yates JR 3rd (2001) Large-scale analysis of the yeast proteome by multidimensional protein identification technology. Nat Biotechnol 19:242–247

Wu CC, MacCoss MJ, Howell KE, Yates JR 3rd (2003) A method for the comprehensive proteomic analysis of membrane proteins. Nat Biotechnol 21:532–538

Yoshinaka K, Kumanogoh H, Nakamura S, Maekawa S (2004) Identification of V-ATPase as a major component in the raft fraction prepared from the synaptic plasma membrane and the synaptic vesicle of rat brain. Neurosci Lett 363:168–172

Zimmermann P, Hirsch-Hoffmann M, Hennig L, Gruissem W (2004) GENEVESTIGATOR. Arabidopsis microarray database and analysis toolbox. Plant Physiol 136:2621–2632

Zybailov B, Mosley AL, Sardiu ME, Coleman MK, Florens L, Washburn MP (2006) Statistical analysis of membrane proteome expression changes in *Saccharomyces cerevisiae*. J Proteome Res 5:2339–2347

Chapter 14
The Proteomes of Chloroplasts and other Plastids

Paul Jarvis

Abstract The post-genomic era of biology has seen a significant shift in focus, from the genes themselves to the proteins they encode. Recent large-scale studies on the proteomes of chloroplasts and others types of plastid have provided significant new insights into the biogenesis, evolution, and functions of these organelles, and have raised some interesting questions. Many of the proteins that define several important sub-organellar compartments (including the envelope and thylakoid membrane systems, the stroma and plastoglobules) have been identified, and this information has been used to make in silico predictions about the entire complement of proteins in each case. Proteomics has revealed that a relatively large number of proteins inside chloroplasts do not possess canonical targeting information (such proteins lack transit peptides for engagement of the general import machinery), and this has led to the elucidation of novel and unusual pathways of chloroplast protein traffic. For example, it is now clear that some Arabidopsis proteins pass through the endoplasmic reticulum and Golgi en route to the chloroplast, and that these proteins may become glycosylated along the way. Comparative studies have been used to characterise organellar proteome changes in response to various environmental cues or genetic perturbations, whilst other approaches have shed light on the oligomerisation and covalent modification of plastidic proteins.

14.1 Introduction

The evolution of the modern plant cell involved the acquisition of mitochondria and chloroplasts through endosymbiosis, and it is generally accepted that these organelles are distant relatives of present-day α-proteobacteria and cyanobacteria, respectively (Margulis 1970). Over the course of evolution, the progenitors of mitochondria and chloroplasts relinquished most of their genes to the nuclear genome, so that now >90% of their constituent proteins are translated

Paul Jarvis, Department of Biology, University of Leicester, Leicester LE1 7RH, UK, E-mail: rpj3@le.ac.uk

J. Šamaj and J. Thelen (eds.), *Plant Proteomics*
© Springer 2007

on cytoplasmic ribosomes, and must engage protein targeting mechanisms that direct them specifically to either organelle (Leister 2003; Bédard and Jarvis 2005; van der Laan et al. 2006). Typically, several thousand different, nucleus-encoded proteins are targeted to these organelles, and it is these proteins that define their distinctive functions.

Chloroplasts are the photosynthetic members of a diverse family of organelles called plastids, which also includes etioplasts (chloroplast precursors that develop in dark-grown, or etiolated, plants), amyloplasts (which are specialised for the storage of starch), and chromoplasts (which accumulate carotenoid pigments) (Whatley 1978; López-Juez and Pyke 2005). Plastids are ubiquitous in plants and algae, and perform numerous essential functions including important steps in the biosynthesis of amino acids, lipids, nucleotides, hormones, vitamins and secondary metabolites, the assimilation of nitrogen and sulphur, and oxygenic photosynthesis (Leister 2003; López-Juez and Pyke 2005). In the latter process, chloroplasts convert energy from sunlight into usable chemical bond energy, and the associated redox reactions lead to the generation of oxygen from water. Chloroplasts are therefore important sites for the production of organic matter and oxygen, and so provide the fuels essential for all higher forms of life (Nelson and Ben-Shem 2004).

Completion of the genome sequencing projects for Arabidopsis, rice and other species, and the development of efficient methods for routine protein identification by mass spectrometry (MS), have together precipitated the onset of the proteomic era; for a discussion of the relevant technologies, see Whitelegge (2003). Due to the overwhelming complexity of cellular proteomes, and the dynamic range limitations associated with analyses on such highly complex mixtures (i.e. the tendency of abundant proteins to mask the presence of less abundant proteins), proteomic studies have tended to focus on isolated subcellular components. In plants, chloroplasts have received considerable attention in this regard (Schröder and Kieselbach 2003; Baginsky and Gruissem 2004; Jarvis 2004; van Wijk 2004; Pan et al. 2005). The value of such proteomic analysis is several-fold: it can confirm the expression and structure of genes predicted based on genome sequence analysis in silico; it can provide information on subcellular and suborganellar protein localisation; it can be used to estimate the abundances of different proteins; and it can even yield information on post-translational modification (PTM) and the composition of multiprotein complexes. Information of this nature will be vital as we move forward into the era of systems biology, especially since it has been estimated that up to 50% of the proteins encoded by the >26,000 genes in the Arabidopsis genome are presently of unknown function (Haas et al. 2005).

The majority of chloroplast proteins bear a cleavable, amino-terminal extension called a transit peptide, which acts as a targeting signal. The transit peptide is recognised by a sophisticated protein import machinery: the so-called TOC and TIC (translocon at the outer / inner envelope membrane of chloroplasts) complexes (Bédard and Jarvis 2005). Since all transit peptides share certain characteristics, it is possible to identify candidate chloroplast proteins through sequence analysis in silico (the TargetP neural network algorithm is a popular method; Emanuelsson et al. 2000). However, transit peptides are not well conserved, and are rather similar to mitochondrial targeting signals, and so such in silico methods are not completely reliable (Richly and Leister 2004). While it is generally accepted that between 2,000 and 4,000 different proteins are targeted to *Arabidopsis* plastids, there are presently less than 900 *Arabidopsis* entries in one database of experimentally determined plastid proteins (Friso et al. 2004; Peltier et al. 2004a). Thus, it is clear that there is a need for further experimentation, and that the proteomic era of chloroplast biology is far from over.

14.2 Proteome Catalogues

Cataloguing aims to identify all of the proteins within a particular cellular or organellar compartment, and so define the functions of that compartment. Although the proteome of a chloroplast is substantially smaller and more manageable than that of an entire cell, it nevertheless comprises several thousand different proteins. Thus, many cataloguing studies have focused on a particular suborganellar compartment: e.g. the internal thylakoid membrane system that bears the photosynthetic complexes, the double-membrane envelope system that surrounds each organelle, the central aqueous matrix, or stroma, and the lipid-containing, thylakoid-associated structures called plastoglobules. Nevertheless, some studies on whole organelles have been done, and these have focused on different plastid types, including chloroplasts, amyloplasts and etioplasts.

14.2.1 The Thylakoid Membrane System

The thylakoids are a complex network of membranous sacks embedded within the stroma of chloroplasts. The membranes themselves harbour the four multiprotein complexes of the photosynthetic light reactions (the photosystems, PSI and PSII, the cytochrome b_6f complex, and the ATP synthase), but also function to form a central aqueous compartment called the thylakoid lumen, which is distinct from the stroma. Difficulties associated with the extraction and analysis of highly hydrophobic membrane proteins have led many thylakoid proteomic studies to focus on lumenal proteins, or proteins peripherally associated with the membranes, which can more easily be resolved by two-dimensional gel electrophoresis (2-DE) (Kieselbach and Schroder 2003). However, the molar ratio between the most abundant and least abundant lumenal proteins has been estimated to be as high as 10^6, which presents a daunting dynamic-range obstacle to comprehensive analysis (Peltier et al. 2002).

While the existence of some lumenal proteins (e.g. plastocyanin and extrinsic components of PSII) has been known for many years, proteomic analysis has revealed an unanticipated level of complexity. Only a relatively small number of different Arabidopsis lumenal proteins have been identified experimentally, but information from these sequences has been used to predict the total lumenal proteome in silico, with estimates ranging from ~80 proteins to ~400 proteins (Peltier et al. 2002; Schubert et al. 2002; Kieselbach and Schroder 2003; Westerlund et al. 2003); the actual size of the proteome likely resides somewhere in between these estimates. The predicted proteome contains many unknown or unexpected proteins, suggesting that the compartment has a much broader spectrum of functions than was previously envisaged. In addition to the expected photosynthetic proteins, and those of unknown function, significant numbers of others involved in protein folding, processing and proteolysis, anti-oxidative defence, and non-photosynthetic redox reactions were present. Some of these proteins likely serve to repair and maintain normal functionality of the photosynthetic complexes, which experience substantial redox stress. The results also revealed that a significant proportion of lumenal proteins (up to 50%, according to one estimate) employ the twin-arginine translocation (TAT) pathway which, unlike the Sec pathway, is able to transport fully folded proteins (Jarvis and Robinson 2004).

The thylakoid membrane itself is dominated by the four photosynthetic complexes, which contain ~100 different proteins in total, and so some early proteomic studies focused on these (Gomez et al. 2002; Whitelegge et al. 2002; Zolla et al. 2002; Whitelegge 2003; Zolla et al. 2003). However, the membrane is also believed to contain many other proteins that are important for the assembly, maintenance and regulation of the photosynthetic apparatus, and which are not themselves components of the complexes. Because strongly hydrophobic proteins are not easily analysed using 2-DE gels, large-scale studies of the thylakoid membrane have employed organic solvent fractionation and other procedures (Friso et al. 2004; Peltier et al. 2004a). These studies achieved near complete coverage of the photosynthetic complexes, but also identified low-abundance components such as those involved in cyclic electron flow around PSI and chlororespiration.

By combining the protein sets derived from the lumenal and membrane proteomic studies mentioned above, and by also including thylakoid proteins identified in classical experiments described in the literature, the total number of experimentally determined thylakoid-associated Arabidopsis proteins is currently about 400 (Peltier et al. 2004a; van Wijk 2004). While some of these may be contaminants from other chloroplast compartments, it is nevertheless interesting to consider the distribution of biological functions: ~30% are involved in photosynthesis, ~25% are of unknown function [including proteins with tetratricopeptide (TPR), pentatricopeptide (PPR), DnaJ and rhodanese domains], ~20% mediate protein translocation, folding, processing and proteolysis, and almost 10% are involved in various aspects of defence against oxidative stress. With respect to protein translocation, it is well documented that nucleus- and chloroplast-encoded proteins are targeted to the thylakoid system through four different targeting pathways, at least three of which are closely related to bacterial transport systems and employ proteinaceous import machineries associated with the thylakoid membrane (Jarvis and Robinson 2004).

14.2.2 The Envelope Membrane System

The envelope is a double-membrane system that forms a semi-permeable barrier between the cytosol and the chloroplast interior. It contains the protein import apparatus responsible for the translocation of nucleus-encoded chloroplast proteins (Bédard and Jarvis 2005), as well as many transporters for the exchange of ions and metabolites (Weber et al. 2005). The envelope also contains a unique biochemical machinery responsible for several important functions, including the synthesis of plastid membrane components and other lipids, and participates in the communication that occurs between the plastid and the nucleus (Jarvis 2003; López-Juez and Pyke 2005).

Like the thylakoid membrane, the envelope proteome is dominated by highly hydrophobic integral membrane proteins, and so proteomic studies of this compartment have employed fractionation procedures other than 2-DE, including differential solvent extraction and multidimensional liquid chromatography (Ferro et al. 2003; Froehlich et al. 2003; Rolland et al. 2003). Up to 50% of the proteins identified as a result of these studies were predicted to have at least one transmembrane domain. Together with the more traditional, single-protein-focused experiments reported in the literature, these proteomic experiments resulted in the identification of more than 400 putative envelope-associated Arabidopsis proteins (Peltier et al. 2004a; van Wijk 2004). Once again, while it is likely that a proportion of these proteins are contaminants from other compartments, it is nonetheless interesting to consider their possible roles: almost 30% of the proteins are of unknown function; 13% mediate protein translocation, folding, processing or degradation; 10% are involved in lipid or fatty acid metabolism; and 9% are transporters of small molecules. Thus, while many of these proteins clearly reflect what are known to be the main functions of the envelope system (functions that are clearly different from those of the thylakoid system discussed earlier), the high proportion of proteins of unknown function indicates that there is still a great deal to be learnt. Interestingly, proteins with similarity to components of the mitochondrial protein import machinery were identified, suggesting the existence of novel, unanticipated protein translocation pathways in the chloroplast envelope (Ferro et al. 2003).

At this point, it should be noted that the intermembrane space, which exists between the two envelope membranes, is considered to be an important suborganellar compartment distinct from the membranes themselves. Unfortunately, very little is known about how proteins are targeted to this space (Kouranov et al. 1999), or indeed about protein targeting to the envelope membranes themselves (Hofmann and Theg 2005; Li and Schnell 2006). While inner envelope membrane proteins possess transit peptides and initially engage the general import machinery (Bédard and Jarvis 2005), much like nucleus-encoded stromal and thylakoidal proteins, little is known about how these proteins reach their final destination. By contrast, the majority of outer envelope membrane proteins do not possess a cleavable targeting signal, and are instead directed to the membrane by virtue of information held in their transmembrane domains. The lack of knowledge concerning envelope targeting mechanisms and signals means that it is presently very difficult to make predictions concerning the composition of the three individual proteomes in silico. Nevertheless, evidence suggests that the outer membrane is characterised by the presence of beta-barrel proteins, and that the inner membrane is dominated by transporters with multiple, helical transmembrane domains (Koo and Ohlrogge 2002; Schleiff et al. 2003).

By analysing carefully collected sets of integral proteins from the inner envelope and thylakoid membranes (identified on the basis of published information), the respective proteomes were found to have quite different characteristics (Sun et al. 2004). On average, thylakoid proteins were smaller and more acidic than envelope proteins, and additionally contained fewer cysteine residues. The larger average size of the envelope set presumably reflects the presence of numerous transporters with multiple membrane spans (Weber et al. 2005), whereas the pI differences may be related to pH differences between the compartments (protons are concentrated in the thylakoid lumen as part of the photosynthetic mechanism). Cysteine residues have unique properties, including the ability to engage directly in redox reactions, and so the reduced cysteine content of thylakoidal proteins might be a consequence of the redox-associated stresses they experience, and may represent a measure to reduce oxidative damage (Sun et al. 2004). These different characteristics should facilitate the formulation of effective methods for the prediction and discrimination of envelope and thylakoidal proteins in the future.

14.2.3 The Stroma

The chloroplast stroma is the aqueous matrix that surrounds the thylakoid membranes and fills the organellar interior. It is the site of the carbon reactions of photosynthesis (the Calvin cycle) and other metabolic pathways, and the location for components of the endogenous genetic system of the plastid. One in silico study estimated that the stroma has the potential to contain over 3,000 different proteins, which is equivalent to ~80% of the total theoretical proteome of plastids (Sun et al. 2004). However, it should be noted that many of these proteins may associate permanently or transiently with the thylakoids or the inner envelope membrane, via protein–protein, electrostatic or hydrophobic interactions, or even lipid anchors, all of which are features that were not screened for in the analysis. Interestingly, evidence suggests that the acetyl-coenzyme A carboxylase complex is envelope-associated, which would place it very close to the site where chloroplast-synthesised fatty acids are used or exported to the cytosol (Rolland et al. 2003). The metabolic channelling advantages of such associations are obvious, and there may be many other similar examples of membrane association that are not easily detected.

Despite its obvious importance, the stroma had not been subjected to large-scale proteomic analysis until recently (Peltier et al. 2006). Using a novel form of 2-DE [involving native gels in the first dimension, instead of isoelectric focusing (IEF) as is more common] a large number of stromal proteins were resolved prior to identification. Well over 200 proteins were identified, and these fall into the following functional categories: 26% mediate protein synthesis, folding, proteolysis and sorting; 12% are involved in primary carbon metabolism, including Calvin cycle and oxidative pentose phosphate pathway enzymes; 11% are of unknown function; while 7%, 6%, 4% and 4% mediate the biosynthesis of amino acids, tetrapyrroles, nucleotides and lipids, respectively. Interestingly, the relative concentrations of the identified proteins were estimated by quantifying spot intensities on the 2-DE gels. As expected, proteins involved in primary carbon metabolism were found to constitute the majority (~75%) of the total stromal mass. Those dedicated to protein synthesis, biogenesis and fate represented nearly 10%, whereas those involved in nitrogen and sulphur assimilation made up ~8%. Proteins involved in other biosynthetic pathways, such as those for fatty acids, amino acids, nucleotides and tetrapyrroles, each represented less than 1% of the total mass.

Other, more focused proteomic studies of the stromal compartment have also been conducted. For example, in a thorough study of the 70S ribosome of spinach chloroplasts, all of the proteins of the 50S and 30S subunits were mapped by 2-DE and identified (Yamaguchi and Subramanian 2000; Yamaguchi et al. 2000). The ribosome was shown to comprise 59 different proteins (33 in the 50S subunit and 25 in the 30S subunit, as well as a ribosome recycling factor found exclusively in the 70S holocomplex), of which 53 are orthologues of bacterial ribosomal proteins and 6 are plastid-specific proteins. It was proposed that the latter components evolved to perform functions unique to plastid translation and its regulation, such as protein targeting to the thylakoid membrane and the mediation of control by nuclear factors.

Interestingly, two 30S ribosomal subunit proteins were identified as targets for regulation by the stromal thioredoxin system (Balmer et al. 2003). This system is composed of ferredoxin, ferredoxin-thioredoxin reductase, and thioredoxin, and it links light to the regulation of photosynthetic enzymes and other plastidic processes such as lipid biosynthesis. Electrons flow from ferredoxin to thioredoxin, which, in its reduced state, regulates the activity of target proteins through the reduction of specific disulphide groups. To gain a more complete picture of the targets of the system, proteomics strategies have been adopted (Motohashi et al. 2001; Balmer et al. 2003; Hisabori et al. 2005). A key feature of the regulatory mechanism is the formation of a transient, intermolecular disulphide bridge between thioredoxin and the target protein. This linkage is then reduced by a second cysteine residue within thioredoxin, releasing the target protein in its reduced and active form. However, if the second thioredoxin cysteine residue is mutated (e.g. to serine or alanine), reduction of the intermolecular linkage is prevented, locking the target in its association with thioredoxin and providing a convenient method for target identification. Using such approaches, ~30 potential targets have been identified, including components acting in established thioredoxin-regulated pathways (e.g. the Calvin cycle, nitrogen and sulphur metabolism, protein synthesis, and the oxidative pentose phosphate pathway) and others not previously recognised as thioredoxin targets (e.g. tetrapyrrole biosynthesis, protein folding, assembly and degradation, starch degradation, DNA replication and transcription, and plastid division). More recently, it has been reported that the reach of this largely stromal regulatory network extends into the thylakoid lumen, where it targets components of the photosynthetic electron transport chain (Motohashi and Hisabori 2006). It should be noted, however, that the biological significance of many of these thioredoxin interactions remains to be established (Hisabori et al. 2005).

14.2.4 *Plastoglobules*

Another plastidic compartment that has been subjected to proteome analysis is the plastoglobule. Plastoglobules are lipid-containing bodies found in different plastid types, including chloroplasts and chromoplasts. Biochemical analyses of these structures revealed that they contain a variety of different lipidic compounds, including galactolipids, fatty acids, carotenoids, tocopherols and plastoquinone, and so they were thought to perform lipid storage functions. Studies using one-dimensional electrophoresis revealed that they contain more than a dozen different proteins, but at the time most were of unknown identity (Kessler et al. 1999). Recently, proteomic analysis led to the identification of about 30 different plastoglobule proteins, and in so doing demonstrated that the role of the plastoglobule is far more complex than was previously envisaged (Vidi et al. 2006; Ytterberg et al. 2006). In addition to the plastid lipid-associated protein (PAP) / fibrillin family of plastoglobule-specific proteins, which are thought to perform a structural role by binding to the surface of the globules and preventing their coalescence, a number of enzymes were identified. The data indicated that the plastoglobules of Arabidopsis chloroplasts play an active role in the synthesis of their lipophilic constituents, including alpha-tocopherol (vitamin E), which has an important anti-oxidative function in the thylakoids. Similarly, the plastoglobules of pepper chromoplasts (which primarily accumulate carotenoids) were found to contain enzymes of carotenoid biosynthesis. In chloroplasts, plastoglobules were shown to be directly and permanently coupled to the thylakoids, which led to the conclusion that their lipidic contents are in a dynamic equilibrium with thylakoid membranes (Austin et al. 2006). This suggests that plastoglobules play an important role in the synthesis of thylakoid constituents.

14.2.5 *Whole Organelles*

An extensive study of the whole chloroplast proteome of Arabidopsis was recently reported (Kleffmann et al. 2004). Using a comprehensive series of fractionation procedures to overcome dynamic range limitations, a total of ~600 different proteins were identified. Interestingly, more than 30% of these proteins are of unknown function. In a recent, similarly comprehensive study of the Arabidopsis mitochondrial proteome, almost 20% of the identified proteins were of unknown function (Heazlewood et al. 2004; see also Chap. 15 by Millar, this volume), and so it seems that we have some way to go before the functions of these two organelles are fully understood. The chloroplast study achieved nearly complete identification coverage for major metabolic pathways, such as the Calvin cycle, but only partial coverage for pathways that are not abundantly expressed (Kleffmann et al. 2004). Parallel RNA profiling experiments revealed a correlation between transcript level and protein abundance for some metabolic pathways, but not others, implying the utilisation of different regulatory mechanisms in different pathways.

Proteome studies have also been conducted on amyloplasts from wheat endosperm (Andon et al. 2002; Balmer et al. 2006a). Amyloplasts are non-photosynthetic plastids specialised for the synthesis and long-term storage of starch. In mature wheat seeds, over 80% of the kernel volume is occupied by starchy endosperm, which is dominated by amyloplasts. By comparing MS data with sequence information from wheat and related cereals, particularly rice, a total of ~400 different proteins were identified (Andon et al. 2002; Balmer et al. 2006a). As expected, most of the enzymes of starch biosynthesis were detected. However, the results revealed a surprisingly broad spectrum of biosynthetic capabilities: like chloroplasts, amyloplasts were found to be endowed with enzymes for the assimilation of nitrogen and sulphur, and for the biosynthesis of amino acids, fatty acids and tetrapyrroles. Nevertheless, when a profile of the functions of the identified amyloplast proteins was compared with a corresponding profile for the chloroplast proteome, some significant differences were observed. For example, the amyloplast proteome contains a substantially higher proportion of proteins involved in carbon, nitrogen and sulphur metabolism, and transport processes. By contrast, the chloroplast proteome contains proportionally more proteins of unknown function, which presumably reflects the more complex set of activities mediated by photosynthetic plastids. Interestingly, while several components of the TOC/TIC import machinery were identified, no proteins of the plastidic ribosome were detected, suggesting that most proteins needed by the developing amyloplast are encoded in the nucleus. This makes sense, since the plastid genome is dominated by genes for components of the photosynthetic apparatus.

It was recently reported that, as in chloroplasts, a fully functional thioredoxin regulatory network is operational in amyloplasts (Balmer et al. 2006b). However, ferredoxin is not reduced by light in this case, but instead by metabolically generated NADPH (via ferredoxin-NADP reductase). As with the chloroplast system, potential targets were identified using a proteomics approach, and were found to act in a range of different processes, including starch metabolism, the biosynthesis of amino acids and lipids, and transport processes. It was suggested that the thioredoxin system might enable amyloplasts to indirectly perceive and respond to light, through the generation of reducing power (leading to reduced thioredoxin) upon the arrival of newly synthesised photosynthate in sink tissues. This would enable the amyloplasts to couple their activities with photosynthesis taking place in leaves.

An analysis of etioplasts in dark-grown rice plants revealed a proteome broadly consistent with what one would expect of plastids in heterotrophic tissue, along with some novel functions (von Zychlinski et al. 2005). Interestingly, whereas the data for amyloplasts and the heterotrophic plastids of tobacco culture cells indicated that these organelles import primarily hexose phosphates (Andon et al. 2002; Baginsky et al. 2004; Balmer et al. 2006a), etioplasts were found to contain triose phosphate translocators, but no hexose phosphate translocator. Thus, while the metabolic functions of etioplasts are similar to those of other heterotrophic plastids, it would appear that etioplasts nevertheless share significant similarities with chloroplasts (the organelles into which they might ultimately develop), since chloroplasts also transport primarily three-carbon sugars (Weber et al. 2005). Consistent with this notion, over 70% of the 240 identified rice etioplast proteins were found to have clear homologues in the Arabidopsis chloroplast proteome (von Zychlinski et al. 2005).

14.3 Protein Targeting Issues

Surprisingly, when the ~600 proteins identified in the whole chloroplast proteome study mentioned above were analysed using the TargetP program (Emanuelsson et al. 2000), only ~60% were predicted to have a chloroplast transit peptide (Kleffmann et al. 2004). Of the remainder, ~40 were predicted to have a mitochondrial presequence, ~50 were predicted to have a signal peptide for translocation into the endoplasmic reticulum (ER), and ~140 were predicted to have no cleavable targeting signal at all. Many of these "misplaced" chloroplast proteins were shown to be of cyanobacterial origin, or are encoded by low abundance transcripts, and so it seems unlikely that they are all simply contaminants from other cellular compartments. In addition, while it is doubtful that the TargetP predictions are accurate for all of these proteins (Emanuelsson et al. 2000; Richly and Leister 2004), it seems equally unlikely that they are wholly incorrect, and so the data suggested that protein targeting to chloroplasts may be more complex than was previously envisaged (Jarvis 2004; Bédard and Jarvis 2005). The existence of mitochondrial proteins with non-canonical targeting signals is well documented (van der Laan et al. 2006), and so it is much less surprising that TargetP predicted only ~50% of the proteins identified in Arabidopsis mitochondria (Heazlewood et al. 2004).

Until recently, all nucleus-encoded chloroplast proteins were thought to arrive in the organelle via one of two post-translational targeting mechanisms: (1) active transport of transit peptide-bearing proteins through the TOC and TIC import complexes (Bédard and Jarvis 2005); (2) direct insertion into the cytosolically exposed outer envelope membrane (Hofmann and Theg 2005). The former mechanism mediates the import of numerous proteins destined for interior locations within chloroplasts (such as the inner envelope membrane, stroma and thylakoids) whereas the latter is exclusively associated with integral proteins of the outer membrane. Thus, it had been assumed that all nucleus-encoded chloroplast interior proteins must have a cleavable, amino-terminal transit peptide.

The first evidence for a more complicated picture of chloroplast protein biogenesis was provided by analyses of the *Arabidopsis* envelope proteome, which identified a protein with homology to quinone oxidoreductases from bacteria, yeast and animals (Miras et al. 2002; Ferro et al. 2003); the protein was termed ceQORH, for chloroplast envelope quinone oxidoreductase homologue. Intriguingly, although alignments of ceQORH with its bacterial, yeast and animal counterparts revealed no amino-terminal extension, or transit peptide, the protein was nevertheless found associated with the inner envelope membrane (inner membrane proteins normally have a transit peptide). Further studies revealed that the extreme amino-terminus of ceQORH is not required for efficient chloroplast targeting, and instead identified an internal sequence of ~40 residues that controls localisation (Miras et al. 2002). Although this internal targeting signal does not bear any obvious resemblance to standard transit peptides, the protein may nevertheless follow the TOC/TIC-mediated import route. Mitochondrial proteins with internal targeting signals also exist, and, while these proteins are recognised by a different primary receptor, they do pass through the same core translocon complex as proteins with cleavable pre-sequences (van der Laan et al. 2006). However, in the absence of any relevant experimental data, it remains to be determined how ceQORH gains access to the chloroplast interior. More recently, another inner membrane protein, IEP32 (inner envelope protein of 32 kDa), was identified and found to lack a canonical transit peptide (Nada and Soll 2004). Import of this protein was shown to proceed at low ATP concentrations and without the assistance of key components of the TOC machinery, suggesting that it follows an import pathway distinct from that used by precursors with transit peptides.

The identification of so many proteins with predicted signal peptides (for ER translocation) in the whole chloroplast proteome was a surprising result (Kleffmann et al. 2004). While some of these proteins may have been contaminants from other compartments, or proteins with amino-terminal transmembrane domains that target them to the outer envelope membrane (such domains are frequently misidentified as signal peptides; Hofmann and Theg 2005), it seems unlikely that they can all be explained away in this fashion. Intriguingly, close physical associations between the ER and the outer envelope membrane have been documented over many years (Crotty and Ledbetter 1973; Whatley et al. 1991), and recent microscopy studies have shown that the envelope exhibits profound structural fluidity (Kwok and Hanson 2004). Furthermore, biochemical interactions between the ER and envelope membrane systems are an essential part of normal lipid metabolism (Awai et al. 2006). Chloroplast protein traffic through the endomembrane system is well documented in algae that have complex plastids; i.e. plastids that are surrounded by three or four membranes, instead of the usual two, and which were derived from algae with simple plastids through secondary endosymbioses (van Dooren et al. 2001). Chloroplast proteins in these species typically have a bipartite targeting signal, comprised of an amino-terminal signal peptide fused to a more-or-less standard chloroplast transit peptide. The signal peptide directs the chloroplast precursor into the ER, where it is removed, and the protein then passes through the endomembrane system until it arrives at the plastid, at which point the transit peptide mediates chloroplast import in the usual fashion (van Dooren et al. 2001). This type of targeting pathway makes sense in these organisms, due to the complex nature of their plastids and the likely autogenous origin of the outer organellar membrane, but would seem unnecessary in higher plants.

Nevertheless, indirect evidence suggesting the existence of a protein transport pathway to chloroplasts through the ER and Golgi in plants has existed for some time: i.e. glycosylated proteins and proteins with predicted signal peptides have been found to localise in plastids (Gaikwad et al. 1999; Chen et al. 2004; Asatsuma et al. 2005). More recently, firm evidence for such a targeting pathway was presented (Villarejo et al. 2005). The Arabidopsis carbonic anhydrase 1 (CAH1) protein, which controls hydration of carbon dioxide, was found to localise in the chloroplast stroma, despite the fact that it was strongly predicted to possess a signal peptide. Intriguingly, CAH1 could not be imported directly by isolated pea chloroplasts, but was instead taken up by pancreatic microsomes and concomitantly processed to its mature size. The protein was predicted to have several acceptor sites for *N*-linked glycosylation, and proteome analysis led to the identification of glycosylated CAH1, as well as several other glycoproteins, in the chloroplast stroma. Because the detected glycans are added only in the Golgi, these data implied the existence of a chloroplast protein transport pathway through the Golgi. Indeed, application of brefeldin A, a commonly used fungal agent that interferes with Golgi-mediated vesicle traffic, was found to obstruct the targeting of CAH1, causing it to arrest within the endomembrane system. The elucidated pathway was proposed to represent an ancestral targeting mechanism that predominated during early evolution (before the development of the now dominant TOC/TIC system), and which for some reason has been retained for a few proteins (Villarejo et al. 2005); the necessity for dual targeting of some proteins, to both plastids and compartments on the secretory pathway, may account for the retention of the ancestral mechanism in some cases (Chen et al. 2004; Asatsuma et al. 2005). It seems likely that many of the other chloroplast proteins with predicted signal peptides follow the same targeting route as CAH1 (Jarvis 2004; Kleffmann et al. 2004).

That some of the proteins identified within the whole chloroplast proteome are predicted to have mitochondrial presequences is much less surprising (Kleffmann et al. 2004). Presequences and chloroplast transit peptides share many similarities and are difficult to distinguish (Bhushan et al. 2006), and it is now becoming increasingly clear that many proteins are dual-targeted to both chloroplasts and mitochondria (Peeters and Small 2001; Duchêne et al. 2005). The identification of chloroplast proteins predicted to have no cleavable targeting signal can be explained in a number of different ways. Some of these may be outer envelope membrane proteins that insert directly into the membrane (Hofmann and Theg 2005), whereas others may have internal targeting signals for chloroplast localisation, like ceQORH or IEP32 (Miras et al. 2002; Nada and Soll 2004). The remainder are presumably contaminants, or proteins that have been classified incorrectly by TargetP (Emanuelsson et al. 2000; Richly and Leister 2004).

The emerging picture of protein targeting to chloroplasts is increasingly complex, demonstrating that transit peptide prediction in silico cannot provide a complete description of the chloroplast proteome. The existence of dual-targeted proteins adds an additional level of complexity (Levitan et al. 2005), further emphasising the need for experimental determination of protein localisation and the value of proteomic analysis.

14.4 Comparative Proteomics

As well as studies that simply catalogue the proteins present in a particular subcellular or suborganellar compartment, comparative proteomics has been employed with considerable success. For example, changes in the organellar proteome were studied during de-etiolation or greening (Lonosky et al. 2004), and the responses of the lumenal, stromal and plastoglobular proteomes to low temperature or light stress have been characterised (Giacomelli et al. 2006; Goulas et al. 2006; Ytterberg et al. 2006). In another example, chloroplasts isolated from Arabidopsis mutants lacking different TOC protein import receptor isoforms were compared with wild-type chloroplasts (Kubis et al. 2003, 2004). Different groups of chloroplast proteins were found to be selectively deficient in different receptor mutants, indicating that the different TOC receptor isoforms likely possess a degree of preprotein recognition specificity (Jarvis and Robinson 2004; Bédard and Jarvis 2005). The data suggested that at least two different import pathways operate in plastids (one for highly abundant components of the photosynthetic apparatus, and another for much less abundant housekeeping proteins). The existence of these separate import pathways may help to prevent deleterious competition effects between preproteins (the housekeeping proteins might otherwise be out-competed), or play a role in the differentiation of different plastid types.

Another interesting study compared the soluble stromal proteomes of mesophyll cell and bundle-sheath cell chloroplasts in maize, a plant that utilises the C_4 photosynthetic mechanism (Majeran et al. 2005). The data not only revealed differential accumulation of photosynthetic carbon metabolism enzymes consistent with our understanding of C_4 photosynthesis, but also shed light on how other plastidic functions are distributed between the two cell types. For example, enzymes involved in nitrogen assimilation and the biosynthesis of lipids and tetrapyrroles were found predominantly in mesophyll cell chloroplasts, whereas those for starch biosynthesis were more abundant in bundle sheath chloroplasts. Many of these differences can be explained by consideration of the fundamental differences that exist between the two types of chloroplast. For instance, nitrogen assimilation has a high requirement for energy (ATP) and reducing power (NADPH), both of which are in short supply in the bundle sheath cell chloroplasts due to the absence of linear photosynthetic electron flow and the high demands of the Calvin cycle; it therefore makes sense to concentrate this process in mesophyll chloroplasts. Similarly, it is quite logical that starch biosynthetic enzymes should be located in the vicinity of the Calvin cycle, which operates exclusively in the bundle sheath cells, since this is the source of new photosynthate for starch synthesis.

14.5 Multiprotein Complexes and Protein Modification

Many proteins do not function alone, but as a part of multiprotein complexes of varying complexity. In order to truly understand the functions of the plastidic proteome, these protein–protein interactions must be identified and characterised.

Furthermore, complexes usually represent functional units, and so hypothetic functions can be assigned to some proteins via their physical association with others of known function. Thus, experimental procedures that maintain the oligomeric status of protein complexes are desirable, since they preserve important interaction information. Several major protein complexes of chloroplasts have been purified to homogeneity and analysed individually; examples include the photosystems (Szabo et al. 2001; Zolla et al. 2002), the cytochrome b_6f complex (Whitelegge et al. 2002), the 70S ribosomal subunits (see Sect. 14.2.3), and a caseinolytic protease (Clp) complex (Peltier et al. 2004b). One approach that has recently come to the fore is native electrophoresis. This provides an effective alternative to IEF as a first dimension in 2-DE analysis, and has the advantage that it preserves oligomeric status. This method has been used to study protein complexes in the thylakoids and the stroma (Granvogl et al. 2006; Peltier et al. 2006).

Protein modification is another important issue that must be taken into consideration. The molecular mass of an intact protein defines its native covalent state, and so its accurate measurement can reveal modifications mediated either post-transcriptionally, through processes such as RNA editing, or post-translationally (Whitelegge 2003; van Wijk 2004). Several studies have revealed covalent modifications of plastidic proteins, including acetylation, glycosylation, palmitoylation, phosphorylation, and N-terminal methionine excision (Yamaguchi et al. 2000; Gomez et al. 2002; Ferro et al. 2003; Giglione et al. 2003; Villarejo et al. 2005). Such modifications may influence the activity, interactions or stability of the protein, or anchor it to a membrane. So far, many proteomic studies have employed a "bottom-up" approach, in which the proteins are first broken up into manageable, characteristic fragments (e.g. using the protease trypsin) prior to MS analysis. The disadvantage of this "peptide mass fingerprinting" strategy is that much of the information inherent in the intact protein (including its overall size and covalent status) may be lost. An alternative, attractive approach is "top-down" proteomics, in which the intact mass of the protein is determined prior to fragmentation (Whitelegge 2003; Zabrouskov et al. 2003; van Wijk 2004). This approach can therefore reveal any modifications that the protein has undergone, as well as its identity.

14.6 Concluding Remarks

The various studies described above have demonstrated the power and utility of plastid proteome analysis, and it is anticipated that proteomics will continue to form an essential component of chloroplast research in the future. In light of the emerging complexity of plastid protein traffic, revealed in part through proteomics, it seems that it will be necessary to develop new prediction tools for the identification of chloroplast proteins, to revise current estimates of the size of the Arabidopsis plastid proteome based on in silico analysis, and to reassess the notion that many plant nuclear genes inherited from the cyanobacterial endosymbiont encode proteins that are not actually targeted back to the chloroplast (Martin et al. 2002). The further and more extensive application of proteomics will play a significant role in achieving these objectives. In conjunction with complementary technologies such as transcriptomics and metabolomics, proteomics will enable the application of unified, systems-based approaches, leading ultimately to a complete and accurate description of the constitution and functioning of the organelles upon which we all depend.

Acknowledgements I thank Enrique López-Juez for helpful comments on the manuscript, and gratefully acknowledge financial support from the Royal Society Rosenheim Research Fellowship and the Biotechnology and Biological Sciences Research Council (BBSRC).

References

Andon NL, Hollingworth S, Koller A, Greenland AJ, Yates JR 3rd, Haynes PA (2002) Proteomic characterization of wheat amyloplasts using identification of proteins by tandem mass spectrometry. Proteomics 2:1156–1168

Asatsuma S, Sawada C, Itoh K, Okito M, Kitajima A, Mitsui T (2005) Involvement of alpha-amylase I-1 in starch degradation in rice chloroplasts. Plant Cell Physiol 46:858–869

Austin JR 2nd, Frost E, Vidi PA, Kessler F, Staehelin LA (2006) Plastoglobules are lipoprotein subcompartments of the chloroplast that are permanently coupled to thylakoid membranes and contain biosynthetic enzymes. Plant Cell 18:1693–1703

Awai K, Xu C, Lu B, Benning C (2006) Lipid trafficking between the endoplasmic reticulum and the chloroplast. Biochem Soc Trans 34:395–398

Baginsky S, Gruissem W (2004) Chloroplast proteomics: potentials and challenges. J Exp Bot 55:1213–1220

Baginsky S, Siddique A, Gruissem W (2004) Proteome analysis of tobacco bright yellow-2 (BY-2) cell culture plastids as a model for undifferentiated heterotrophic plastids. J Proteome Res 3:1128–1137

Balmer Y, Koller A, del Val G, Manieri W, Schurmann P, Buchanan BB (2003) Proteomics gives insight into the regulatory function of chloroplast thioredoxins. Proc Natl Acad Sci USA 100:370–375

Balmer Y, Vensel WH, DuPont FM, Buchanan BB, Hurkman WJ (2006a) Proteome of amyloplasts isolated from developing wheat endosperm presents evidence of broad metabolic capability. J Exp Bot 57:1591–1602

Balmer Y, Vensel WH, Cai N, Manieri W, Schurmann P, Hurkman WJ, Buchanan BB (2006b) A complete ferredoxin/thioredoxin system regulates fundamental processes in amyloplasts. Proc Natl Acad Sci USA 103:2988–2993

Bédard J, Jarvis P (2005) Recognition and envelope translocation of chloroplast preproteins. J Exp Bot 56:2287–2320

Bhushan S, Kuhn C, Berglund AK, Roth C, Glaser E (2006) The role of the N-terminal domain of chloroplast targeting peptides in organellar protein import and mis-sorting. FEBS Lett 580:3966–3972

Chen MH, Huang LF, Li HM, Chen YR, Yu SM (2004) Signal peptide-dependent targeting of a rice alpha-amylase and cargo proteins to plastids and extracellular compartments of plant cells. Plant Physiol 135:1367–1377

Crotty WJ, Ledbetter MC (1973) Membrane continuities involving chloroplasts and other organelles in plant cells. Science 182:839–841

Duchêne AM, Giritch A, Hoffmann B, Cognat V, Lancelin D, Peeters NM, Zaepfel M, Marechal-Drouard L, Small ID (2005) Dual targeting is the rule for organellar aminoacyl-tRNA synthetases in *Arabidopsis thaliana*. Proc Natl Acad Sci USA 102:16484–16489

Emanuelsson O, Nielsen H, Brunak S, von Heijne G (2000) Predicting subcellular localization of proteins based on their N-terminal amino acid sequence. J Mol Biol 300:1005–1016

Ferro M, Salvi D, Brugiere S, Miras S, Kowalski S, Louwagie M, Garin J, Joyard J, Rolland N (2003) Proteomics of the chloroplast envelope membranes from *Arabidopsis thaliana*. Mol Cell Proteomics 2:325–345

Friso G, Giacomelli L, Ytterberg AJ, Peltier JB, Rudella A, Sun Q, Wijk KJ (2004) In-depth analysis of the thylakoid membrane proteome of *Arabidopsis thaliana* chloroplasts: new proteins, new functions, and a plastid proteome database. Plant Cell 16:478–499

Froehlich JE, Wilkerson CG, Ray WK, McAndrew RS, Osteryoung KW, Gage DA, Phinney BS (2003) Proteomic study of the *Arabidopsis thaliana* chloroplastic envelope membrane utilizing alternatives to traditional two-dimensional electrophoresis. J Proteome Res 2:413–425

Gaikwad A, Tewari KK, Kumar D, Chen W, Mukherjee SK (1999) Isolation and characterisation of the cDNA encoding a glycosylated accessory protein of pea chloroplast DNA polymerase. Nucleic Acids Res 27:3120–3129

Giacomelli L, Rudella A, van Wijk KJ (2006) High light response of the thylakoid proteome in arabidopsis wild type and the ascorbate-deficient mutant *vtc2-2*. A comparative proteomics study. Plant Physiol 141:685–701

Giglione C, Vallon O, Meinnel T (2003) Control of protein life-span by N-terminal methionine excision. EMBO J 22:13–23

Gomez SM, Nishio JN, Faull KF, Whitelegge JP (2002) The chloroplast grana proteome defined by intact mass measurements from liquid chromatography mass spectrometry. Mol Cell Proteomics 1:46–59

Goulas E, Schubert M, Kieselbach T, Kleczkowski LA, Gardestrom P, Schröder W, Hurry V (2006) The chloroplast lumen and stromal proteomes of *Arabidopsis thaliana* show differential sensitivity to short- and long-term exposure to low temperature. Plant J 47:720–734

Granvogl B, Reisinger V, Eichacker LA (2006) Mapping the proteome of thylakoid membranes by de novo sequencing of intermembrane peptide domains. Proteomics 6:3681–3695

Haas BJ, Wortman JR, Ronning CM, Hannick LI, Smith RK Jr, Maiti R, Chan AP, Yu C, Farzad M, Wu D, White O, Town CD (2005) Complete reannotation of the *Arabidopsis* genome: methods, tools, protocols and the final release. BMC Biol 3:7

Heazlewood JL, Tonti-Filippini JS, Gout AM, Day DA, Whelan J, Millar AH (2004) Experimental analysis of the Arabidopsis mitochondrial proteome highlights signaling and regulatory components, provides assessment of targeting prediction programs, and indicates plant-specific mitochondrial proteins. Plant Cell 16:241–256

Hisabori T, Hara S, Fujii T, Yamazaki D, Hosoya-Matsuda N, Motohashi K (2005) Thioredoxin affinity chromatography: a useful method for further understanding the thioredoxin network. J Exp Bot 56:1463–1468

Hofmann NR, Theg SM (2005) Chloroplast outer membrane protein targeting and insertion. Trends Plant Sci 10:450–457

Jarvis P (2003) Intracellular signalling: the language of the chloroplast. Curr Biol 13: R314–R316

Jarvis P (2004) Organellar proteomics: chloroplasts in the spotlight. Curr Biol 14:R317–R319

Jarvis P, Robinson C (2004) Mechanisms of protein import and routing in chloroplasts. Curr Biol 14:R1064–R1077

Kessler F, Schnell D, Blobel G (1999) Identification of proteins associated with plastoglobules isolated from pea (*Pisum sativum* L.) chloroplasts. Planta 208:107–113

Kieselbach T, Schroder WP (2003) The proteome of the chloroplast lumen of higher plants. Photosynth Res 78:249–264

Kleffmann T, Russenberger D, von Zychlinski A, Christopher W, Sjolander K, Gruissem W, Baginsky S (2004) The *Arabidopsis thaliana* chloroplast proteome reveals pathway abundance and novel protein functions. Curr Biol 14:354–362

Koo AJ, Ohlrogge JB (2002) The predicted candidates of Arabidopsis plastid inner envelope membrane proteins and their expression profiles. Plant Physiol 130:823–836

Kouranov A, Wang H, Schnell DJ (1999) Tic22 is targeted to the intermembrane space of chloroplasts by a novel pathway. J Biol Chem 274:25181–25186

Kubis S, Baldwin A, Patel R, Razzaq A, Dupree P, Lilley K, Kurth J, Leister D, Jarvis P (2003) The Arabidopsis *ppi1* mutant is specifically defective in the expression, chloroplast import, and accumulation of photosynthetic proteins. Plant Cell 15:1859–1871

Kubis S, Patel R, Combe J, Bédard J, Kovacheva S, Lilley K, Biehl A, Leister D, Ríos G, Koncz C, Jarvis P (2004) Functional specialization amongst the Arabidopsis Toc159 family of chloroplast protein import receptors. Plant Cell 16:2059–2077

Kwok EY, Hanson MR (2004) Stromules and the dynamic nature of plastid morphology. J Microsc 214:124–137

Leister D (2003) Chloroplast research in the genomic age. Trends Genet 19:47–56.

Levitan A, Trebitsh T, Kiss V, Pereg Y, Dangoor I, Danon A (2005) Dual targeting of the protein disulfide isomerase RB60 to the chloroplast and the endoplasmic reticulum. Proc Natl Acad Sci USA 102:6225–6230

Li M, Schnell DJ (2006) Reconstitution of protein targeting to the inner envelope membrane of chloroplasts. J Cell Biol 175:249–259

Lonosky PM, Zhang X, Honavar VG, Dobbs DL, Fu A, Rodermel SR (2004) A proteomic analysis of maize chloroplast biogenesis. Plant Physiol 134:560–574

López-Juez E, Pyke KA (2005) Plastids unleashed: their development and their integration in plant development. Int J Dev Biol 49:557–577

Majeran W, Cai Y, Sun Q, van Wijk KJ (2005) Functional differentiation of bundle sheath and mesophyll maize chloroplasts determined by comparative proteomics. Plant Cell 17:3111–3140

Margulis L (1970) The origin of eukaryotic cells. Yale University Press, New Haven

Martin W, Rujan T, Richly E, Hansen A, Cornelsen S, Lins T, Leister D, Stoebe B, Hasegawa M, Penny D (2002) Evolutionary analysis of *Arabidopsis*, cyanobacterial, and chloroplast genomes reveals plastid phylogeny and thousands of cyanobacterial genes in the nucleus. Proc Natl Acad Sci USA 99:12246–12251

Miras S, Salvi D, Ferro M, Grunwald D, Garin J, Joyard J, Rolland N (2002) Non-canonical transit peptide for import into the chloroplast. J Biol Chem 277:47770–47778

Motohashi K, Hisabori T (2006) HCF164 receives reducing equivalents from stromal thioredoxin across the thylakoid membrane and mediates reduction of target proteins in the thylakoid lumen. J Biol Chem 281:35039–35047

Motohashi K, Kondoh A, Stumpp MT, Hisabori T (2001) Comprehensive survey of proteins targeted by chloroplast thioredoxin. Proc Natl Acad Sci USA 98:11224–11229

Nada A, Soll J (2004) Inner envelope protein 32 is imported into chloroplasts by a novel pathway. J Cell Sci 117:3975–3982

Nelson N, Ben-Shem A (2004) The complex architecture of oxygenic photosynthesis. Nat Rev Mol Cell Biol 5:971–982

Pan S, Carter CJ, Raikhel NV (2005) Understanding protein trafficking in plant cells through proteomics. Expert Rev Proteomics 2:781–792

Peeters N, Small I (2001) Dual targeting to mitochondria and chloroplasts. Biochim Biophys Acta 1541:54–63

Peltier JB, Emanuelsson O, Kalume DE, Ytterberg J, Friso G, Rudella A, Liberles DA, Soderberg L, Roepstorff P, von Heijne G, van Wijk KJ (2002) Central functions of the lumenal and peripheral thylakoid proteome of Arabidopsis determined by experimentation and genome-wide prediction. Plant Cell 14:211–236

Peltier JB, Ytterberg AJ, Sun Q, van Wijk KJ (2004a) New functions of the thylakoid membrane proteome of *Arabidopsis thaliana* revealed by a simple, fast, and versatile fractionation strategy. J Biol Chem 279:49367–49383

Peltier JB, Ripoll DR, Friso G, Rudella A, Cai Y, Ytterberg J, Giacomelli L, Pillardy J, van Wijk KJ (2004b) Clp protease complexes from photosynthetic and non-photosynthetic plastids and mitochondria of plants, their predicted three-dimensional structures, and functional implications. J Biol Chem 279:4768–4781

Peltier JB, Cai Y, Sun Q, Zabrouskov V, Giacomelli L, Rudella A, Ytterberg AJ, Rutschow H, van Wijk KJ (2006) The oligomeric stromal proteome of *Arabidopsis thaliana* chloroplasts. Mol Cell Proteomics 5:114–133

Richly E, Leister D (2004) An improved prediction of chloroplast proteins reveals diversities and commonalities in the chloroplast proteomes of Arabidopsis and rice. Gene 329:11–16

Rolland N, Ferro M, Seigneurin-Berny D, Garin J, Douce R, Joyard J (2003) Proteomics of chloroplast envelope membranes. Photosynth Res 78:205–230

Schleiff E, Eichacker LA, Eckart K, Becker T, Mirus O, Stahl T, Soll J (2003) Prediction of the plant beta-barrel proteome: a case study of the chloroplast outer envelope. Protein Sci 12:748–759

Schröder WP, Kieselbach T (2003) Update on chloroplast proteomics. Photosynth Res 78:181–193

Schubert M, Petersson UA, Haas BJ, Funk C, Schröder WP, Kieselbach T (2002) Proteome map of the chloroplast lumen of *Arabidopsis thaliana*. J Biol Chem 277:8354–8365

Sun Q, Emanuelsson O, van Wijk KJ (2004) Analysis of curated and predicted plastid subproteomes of Arabidopsis. Subcellular compartmentalization leads to distinctive proteome properties. Plant Physiol 135:723–734

Szabo I, Seraglia R, Rigoni F, Traldi P, Giacometti GM (2001) Determination of photosystem II subunits by matrix-assisted laser desorption/ionization mass spectrometry. J Biol Chem 276:13784–13790

Van der Laan M, Rissler M, Rehling P (2006) Mitochondrial preprotein translocases as dynamic molecular machines. FEMS Yeast Res 6:849–861

Van Dooren GG, Schwartzbach SD, Osafune T, McFadden GI (2001) Translocation of proteins across the multiple membranes of complex plastids. Biochim Biophys Acta 1541:34–53

van Wijk KJ (2004) Plastid proteomics. Plant Physiol Biochem 42:963–977

Vidi PA, Kanwischer M, Baginsky S, Austin JR, Csucs G, Dormann P, Kessler F, Brehelin C (2006) Tocopherol cyclase (VTE1) localization and vitamin E accumulation in chloroplast plastoglobule lipoprotein particles. J Biol Chem 281:11225–11234

Villarejo A, Buren S, Larsson S, Dejardin A, Monne M, Rudhe C, Karlsson J, Jansson S, Lerouge P, Rolland N, von Heijne G, Grebe M, Bako L, Samuelsson G (2005) Evidence for a protein transported through the secretory pathway en route to the higher plant chloroplast. Nat Cell Biol 7:1124–1131

Von Zychlinski A, Kleffmann T, Krishnamurthy N, Sjolander K, Baginsky S, Gruissem W (2005) Proteome analysis of the rice etioplast: metabolic and regulatory networks and novel protein functions. Mol Cell Proteomics 4:1072–1084

Weber AP, Schwacke R, Flügge UI (2005) Solute transporters of the plastid envelope membrane. Annu Rev Plant Biol 56:133–164

Westerlund I, Von Heijne G, Emanuelsson O (2003) LumenP – a neural network predictor for protein localization in the thylakoid lumen. Protein Sci 12:2360–2366

Whatley JM (1978) A suggested cycle of plastid developmental interrelationships. New Phytol 80:489–502

Whatley JM, McLean B, Juniper BE (1991) Continuity of chloroplast and endoplasmic reticulum membranes in *Phaseolus vulgaris*. New Phytol 117:209–217

Whitelegge JP (2003) Thylakoid membrane proteomics. Photosynth Res 78:265–277

Whitelegge JP, Zhang H, Aguilera R, Taylor RM, Cramer WA (2002) Full subunit coverage liquid chromatography electrospray ionization mass spectrometry (LCMS+) of an oligomeric membrane protein: cytochrome b(6)f complex from spinach and the cyanobacterium *Mastigocladus laminosus*. Mol Cell Proteomics 1:816–827

Yamaguchi K, Subramanian AR (2000) The plastid ribosomal proteins. Identification of all the proteins in the 50 S subunit of an organelle ribosome (chloroplast). J Biol Chem 275:28466–28482

Yamaguchi K, von Knoblauch K, Subramanian AR (2000) The plastid ribosomal proteins. Identification of all the proteins in the 30 S subunit of an organelle ribosome (chloroplast). J Biol Chem 275:28455–28465

Ytterberg AJ, Peltier JB, van Wijk KJ (2006) Protein profiling of plastoglobules in chloroplasts and chromoplasts. A surprising site for differential accumulation of metabolic enzymes. Plant Physiol 140:984–997

Zabrouskov V, Giacomelli L, van Wijk KJ, McLafferty FW (2003) A new approach for plant proteomics: characterization of chloroplast proteins of *Arabidopsis thaliana* by top-down mass spectrometry. Mol Cell Proteomics 2:1253–1260

Zolla L, Rinalducci S, Timperio AM, Huber CG (2002) Proteomics of light-harvesting proteins in different plant species. Analysis and comparison by liquid chromatography-electrospray ionization mass spectrometry. Photosystem I. Plant Physiol 130:1938–1950

Zolla L, Timperio AM, Walcher W, Huber CG (2003) Proteomics of light-harvesting proteins in different plant species. Analysis and comparison by liquid chromatography-electrospray ionization mass spectrometry. Photosystem II. Plant Physiol 131:198–214

Chapter 15
The Plant Mitochondrial Proteome

A. Harvey Millar

Abstract Plant mitochondria function in respiratory oxidation of organic acids and the transfer of electrons to oxygen via the respiratory electron transport chain, but also participate in wider catabolic and biosynthetic activities vital to plant growth and development. Over 2,000 different proteins are likely to be present in plant mitochondria and many have as yet undefined roles. A fuller systems understanding of the role of mitochondria will require continuing advancements in the isolation of high purity mitochondrial samples and an array of proteomic approaches to categorise their protein composition. In parallel to experimental analysis, prediction programs continue to be useful to search plant genome sequence data to identify putative mitochondrial proteins that are missing due to the physical and chemical limitations of current protein separation and analysis techniques. Studies on post-translational modifications (PTMs) of the primary sequence of mitochondrial proteins are providing important information on putative regulation and damage within the organelle but most such PTMs still need to be investigated in detail to define their biological significance. The interactions between proteins, and between proteins and ligands, define the large-scale structure of the mitochondrial proteome and we are beginning to undercover the dynamic interactions of this protein set, which defines the respiratory apparatus in plants.

15.1 Introduction

The primary role of plant mitochondria is respiratory oxidation of organic acids and the transfer of electrons to O_2 via the respiratory electron transport chain. This catabolic function of mitochondria is then coupled through the membrane potential to the synthesis of ATP. But mitochondria also perform many important secondary functions in plants that are only indirectly related to their primary energetic function; for example they are involved in synthesis of nucleotides, metabolism of

A. Harvey Millar, ARC Centre of Excellence in Plant Energy Biology, The University of Western Australia, 35 Stirling Highway, M316, Crawley 6009, WA, Australia,
E-mail: hmillar@cyllene.uwa.edu.au

J. Šamaj and J. Thelen (eds.), *Plant Proteomics*
© Springer 2007

amino acids and lipids, and synthesis of vitamins and cofactors. Further, mitochondria house particular components of metabolic pathways that begin and end outside the organelle. Key examples include their participation in the photorespiratory pathway, and the export of organic acid intermediates and carbon skeletons for nitrogen assimilation. A range of small molecules are transported to and from the mitochondrion by an array of membrane carriers and channels; these serve as fuel for respiration and as products for cellular biosynthesis. Finally, cellular messages that coordinate cell division, developmental changes and the environmental stress responses of organelle-linked functions have to be propagated to, and transduced throughout, mitochondria via signalling cascades that perceive such factors as phosphorylation status and/or redox poise of critical components. In order to undertake these diverse roles, mitochondria contain many hundreds of distinct proteins. The exact mitochondrial proteome size in plants is not known but is likely to approach 2,000–2,500 distinct gene products based on recent estimates (Millar et al. 2006). The overwhelming majority of mitochondrial proteins are encoded in the nucleus, synthesised in the cytosol and actively imported into mitochondria by protein-selective protein import machinery (Emanuelsson et al. 2000). However, many do not have clearly identifiable targeting sequences that can be recognised in primary sequence data (Heazlewood et al. 2004). Thus direct analysis of the steady state proteome is the key method by which the protein complement of this organelle can be defined. As mitochondrial function and composition will vary across development, between tissue types, in response to the environment and indeed between plant species, a wide variety of proteomic analyses are required to develop a full appreciation of this sub-cellular proteome. Here I have outlined five key components of a historical understanding of research on the mitochondrial proteome and its continuing development towards a systems biology understanding of this organelle. Firstly, the isolation of mitochondria is discussed as it is central to all quality analyses of mitochondrial form, function and composition. Secondly, the experimental analysis of the protein complement is discussed as it represents the heartland of mitochondrial proteomics and our current understanding of it. Thirdly, in parallel to experimental analysis, prediction programs used against the data from plant genome sequencing efforts are seeking to fill the gaps left by experimental limitations. Fourthly within each protein are modifications of primary protein sequence that provide important information on proteome regulation and damage that need to be investigated in detail. Finally, the interactions between proteins, and between proteins and ligands, that define the large scale structure of the proteome are beginning to undercover the key elements in the dynamics of this protein set.

15.2 Isolation of Plant Mitochondria

Techniques for the isolation of plant mitochondria need to avoid dramatically changing the morphological structures observed in planta, and to ensure maintenance of the functional characteristics of the organelles. These prerequisites require methods of cell fractionation and organelle recovery that avoid osmotic rupture of membranes and protect organelles from harmful products released from other cellular compartments. Several extensive methodology reviews on plant mitochondrial purification have been published (Douce 1985; Neuburger 1985; Millar et al. 2001a; Hausmann et al. 2003). The core methods used today are based largely on a series of specific methodology papers (most notably Neuburger et al. 1982; Leaver et al. 1983; Day et al. 1985; Millar et al. 2001b; Keech et al. 2005).

The key to successful mitochondrial isolation is tissue processing as soon as possible following harvesting, and ensuring cooling of solutions and apparatus in order to maintain cell turgor and thus ensure maximal mitochondrial yields following homogenisation. A basic homogenisation medium consists of 0.3–0.4 M of an osmoticum (sucrose or mannitol), 1–5 mM of a divalent cation chelator (EDTA or EGTA), 25–50 mM of a pH buffer (MOPS, TES or Na-pyrophosphate) and 5–10 mM of a reductant (cysteine or ascorbate). The osmoticum maintains the mitochondrial structure and prevents physical swelling and rupture of membranes, the buffer prevents acidification by the contents of ruptured vacuoles, and the EDTA inhibits the function of phospholipases and various proteases requiring Ca^{2+} or Mg^{2+}. The reductant prevents damage from oxidants present in the tissue or produced upon homogenisation. A variety of additions to this basic medium are made to improve yields and protect mitochondria from damage during isolation in some plant tissues. Etiolated seedling tissues and green tissues also often require addition of 0.1–1% (w/v) bovine serum albumin (BSA) to remove free fatty acids, and 1–5% (w/v) polyvinylpyrrolidone (PVP) to remove phenolics that damage organelles in the initial homogenate (Douce 1985). Depending on the tissue, homogenisation of plant samples in this medium can be accomplished by one of several methods: grinding in a pre-cooled mortar and pestle; homogenisation rapidly (seconds) in a square-form beaker with a Polytron or Ultra-Turrax blender; homogenisation more slowly (tens of seconds) in a Waring blender; juicing tuber material directly into x5 homogenisation medium using a commercial vegetable/fruit juice extractor; or grating of tissue by hand using a vegetable grater submerged under homogenisation medium. Alternatively, especially in cell cultures, enzymatic digestion of the cell wall can be used to release protoplasts from which mitochondria are then isolated by a simple Potter homogeniser (Vanemmerik et al. 1992; Zhang et al. 1999; Meyer and Millar 2006). Once homogenisation has been completed, the final suspension must be filtered to remove any starch or cell debris, and kept at 4°C. The filtered homogenate is then poured into centrifuge tubes and centrifuged for 5 min at a maximum of ~1,000 *g* in a fixed angle rotor. The supernatant following centrifugation is then centrifuged for 15–20 min at a maximum of ~12,000 *g* at 4°C and the resulting high speed supernatant is discarded. The tan-, yellow- or green-colored pellet in each tube is re-suspended and can be re-isolated by further ~1,000 *g* and ~12,000 *g* centrifugation steps to maximise purity and wash the crude organelle pellet. This washed organelle preparation was considered adequate for a variety of respiratory measurements in the 1950s–1980s. In mammalian tissues, such washed pellets represent a reasonably enriched mitochondrial preparation (80–90% protein being mitochondrial in origin). However, from plant tissue, such pellets are frequently heavily contaminated by thylakoid membranes, peroxisomes and/or glyoxysomes, and occasionally by bacteria. The protein in this pellet fraction in plants might be only 20–40% mitochondrial in origin (Day et al. 1985; Heazlewood et al. 2004), but it can still represent a 20- to 30-fold enrichment of mitochondrial activities over a whole cell extract.

Further purification of the mitochondrial fraction can be carried out using density gradients, and today this is achieved almost exclusively by the use of the silica sol, Percoll (GE Health Sciences, Piscataway, NJ). Percoll gradients offer more rapid purification, allow separation of mitochondria and thylakoid membranes from green tissues, and also ensure iso-osmotic conditions in the gradient compared to other density media (Neuburger et al. 1982; Day et al. 1985). The most common method used is the sigmoidal, self-generating gradient obtained by centrifugation of a Percoll solution in a fixed-angle rotor. The density gradient is formed during centrifugation at >10,000 *g* due to the sedimentation of the poly-dispersed colloid (average particle size 29 nm diameter, average density $\rho = 2.2$ g/ml). The concentration of Percoll in the starting solution and the time of centrifugation can be varied to optimise a particular separation. Typically, these solutions range from 25% to 35% (v/v) Percoll in a 0.3–0.4 M osmoticum of mannitol or sucrose. Separation of mitochondria from green plant tissues is greatly aided by inclusion of a 0–5% PVP-40 pre-formed gradient in the Percoll solution (Day et al. 1985). Gradients are typically centrifuged at a maximum of 40,000 *g* for 45 min in either a fixed angle rotor or a swing-out rotor in a Sorvall or Beckman preparative centrifuge. Although common fixed angle rotors work, the less common, and significantly more expensive, swing-out rotors give a more defined and condensed mitochondrial band in the gradient. After centrifugation, mitochondria form a buff-colored band below yellow, orange or green plastid membranes. The mitochondria are aspirated from the gradient avoiding collection of the plastid fractions. The Percoll suspension is then diluted with a standard wash medium and centrifuged at a maximum of 15,000 *g* for 15–20 min to recover the mitochondrial fraction as a pellet. Typically, these Percoll-purified preparations from plants contain protein that by mass is 85–95% mitochondrial in origin (Day et al. 1985; Neuburger 1985; Heazlewood et al. 2004). Contaminants in these preparations depend on the tissues used for isolation, but are typically plastidic and peroxisomal in origin.

Improvement in purity can be gained by repeating density gradients of the same type, or by use of high concentration Percoll density gradients that seek to move mitochondria in the sigmoidal gradient, providing a different degree of overlap of mitochondria with the contaminants (Day et al. 1985; Neuburger 1985; Heazlewood et al. 2004). Combining these approaches can yield mitochondrial fractions that are, by protein mass, 95–98% mitochondrial in origin (Millar et al. 2001b; Heazlewood et al. 2004). The key to mitochondrial purity, however, is not simply the steps taken, but the amount of material processed; the higher the amount, the higher the purity of mitochondria that can finally be provided. This is due to several factors: the consolidation of pellets in differential centrifugation improves with amount, the visibility of bands and visibility of contaminant bands improves on gradients with increasing amount, and the core of the banding distribution can be selected when the abundance is higher, without lowering yield to below a useful level. Typically, 45–75% of total mitochondria membrane marker enzyme activity in a tissue is found in the first high-speed pellet, 15–30% is loaded onto the density gradient after washing, and only 5–15% is found in the washed mitochondrial sample at the end of the purification procedure (Millar et al. 2001a). In our hands, plant mitochondrial preparations are at their best when the final yields are about 5–15 mg final mitochondrial protein, which often involves beginning with 500 g–1,500 g fresh weight (FW) of the plant tissue of interest.

Marker enzymes for contaminants commonly found in plant mitochondrial samples can be used to assess the purity of preparations. Peroxisomes can be identified by catalase, hydroxy-pyruvate reductase or glycolate oxidase activities. Chloroplasts can be identified by chlorophyll content, etioplasts by carotenoid content and/or alkaline pyrophosphatase activity, and glyoxysomes by isocitrate lyase activity. Endoplasmic reticulum can be identified by antimycin-A-insenstive cytochrome *c* reductase activity and plasma membranes by K^+-ATPase activity. Cytosolic contamination is rare if density centrifugation is performed properly and care is taken in removal of mitochondrial fractions, but such contamination can be easily measured as alcohol dehydrogenase or lactate dehydrogenase activities. Alternatively, marker antibodies for particular compartments can be used; a recent kit for compartment markers in plants is commercially available from AgriSera (www.agrisera.com). Analysis of mitochondrial integrity is very useful to ensure that mitochondria are intact and thus contain their full proteome. For example, the classic latency test for cytochrome *c* oxidase activity measures outer mitochondrial membrane integrity (Neuburger 1985). Alternatively, the ability of mitochondria samples to maintain a proton-motive force across the inner membrane for ATP synthesis can be assessed by the ratio of respiratory rates in the presence and absence of added ADP (respiratory control ratio) and/or by measurement of the ADP consumed/oxygen consumed ratio (ADP:O ratio).

15.3 Experimental Analysis of the Proteome and Future Plans

Polyacrylamide gel electrophoresis (PAGE) has long been the method of choice for the separation of complex protein mixtures. Plant mitochondrial researchers have used this method extensively to unravel different aspects of the organelle proteome. Classical isoelectric focusing/sodium dodecyl sulfate (IEF/SDS)-PAGE was first used by Rötig and Chauveau (1987) to look at the potato mitochondrial proteome and sub-organellar fractions. This was followed by the work of Colas des Francs-Small et al. (1992, 1993) on mitochondria from different potato tissues. Early 2D gel separations of mitochondrial proteins have also been reported from work on pea (Remy et al. 1987; Humphrey-Smith et al. 1992), wheat (Remy et al. 1987), maize (Barent and Elthon 1992; Dunbar et al. 1997; Lund et al. 2001), and Arabidopsis (Davy de Virville et al. 1998). However, few proteins spots that were visualised on these gels were actually identified in these early studies. In the two proteome mapping studies by Millar et al. (2001a) and Kruft et al. (2001) in Arabidopsis, IEF/SDS-PAGE 2D gels were presented on which 650–800 separable proteins were displayed, and over 100 proteins were identified by a combination of Edman sequencing and mass spectrometry (MS). Subsequent studies in pea, rice and maize have identified ~100 mitochondrial proteins by IEF/SDS-PAGE separation and MS (Bardel et al. 2002; Taylor et al. 2002, 2004b, 2005; Heazlewood et al. 2003b).

Alternative strategies were needed to analyse beyond this set of 100 identifications in order to characterise the electron transport chain, membrane carriers and regulatory components that were missing from the identified sets. The blue-native (BN) PAGE separation of mitochondrial electron transport chain complexes (Schägger and von Jagow 1991; Jansch et al. 1996) has been an important tool for analysing the mitochondrial proteome. Clear separation of OX PHOS chain complex components from Complex I (Heazlewood et al. 2003a), Complexes II and IV (Eubel et al. 2003; Millar et al. 2004) and Complex V (Heazlewood et al. 2003c) have been analysed in Arabidopsis, while Complex III was already well defined in plants (Jansch et al. 1995; Brumme et al. 1998). The assembly of Complex I of mitochondria from maize (Karpova and Newton 1999) as well as Chlamydomonas (Cardol et al. 2004) has also been studied by BN-PAGE. The translocase of the outer membrane (TOM) was analysed by separation and analysis of outer mitochondrial membrane protein complexes by BN/SDS-PAGE (Werhahn et al. 2003). Digitonin stabilisation of protein complexes on BN gels revealed the presence of respiratory supercomplexes, formed by different components of the electron transfer chain (Eubel et al. 2004). These studies resulted in a substantial increase in the identified mitochondrial proteome, especially the membrane components. Small membrane carrier proteins do not resolve well on BN gels, but carbonate stripped membranes were found to be a good source for direct analysis of one-dimensional SDS-PAGE gels to identify a series of these proteins (Millar and Heazlewood 2003). There followed a systematic analysis of hydrophobic proteins from Arabidopsis mitochondria using a mixture of carbonate stripped membranes and organic solvent extractions from mitochondrial membranes (Brugiere et al. 2004).

Diagonal PAGE gels have also been used to some effect to select proteins with particular properties from mitochondrial extracts for analysis. In diagonal PAGE, the same separation is performed in two dimensions leading to a diagonal line of proteins. Subtle changes in the running conditions or sample treatment are then used to move proteins or protein complexes off this diagonal to provide specific sets of significantly purified proteins for analysis. Eubel et al. (2003, 2004) and Sunderhaus et al. (2006) used this BN/BN PAGE approach to provide an intricate means of gel-based separation of protein complexes. However, this approach still awaits full exploitation from a protein identification point of view as many protein features revealed in these gels have not yet been analysed. Diagonal SDS/SDS-PAGE experiments have also been used to separate disulfide-linked dimers such as the mitochondrial alternative oxidase (Holtzapffel et al. 2003) and to purify divalent metal-binding proteins (Herald et al. 2003) in plant mitochondria.

Advances in MS and micro- and nano-flow liquid chromatography have provided further methods for the direct analysis of mitochondrial proteomes without gel-based separation. Instead, whole organelle lysates are digested to peptides, separated by one- or two-dimensional liquid chromatography and on-line analysed by electrospray tandem mass spectrometry to identify peptides. This approach has been applied to both Arabidopsis and rice mitochondrial samples (Heazlewood et al. 2003a, 2004; Brugiere et al. 2004; Kristensen et al. 2004).

Combining all the available published data in Arabidopsis, 547 non-redundant proteins have been located in mitochondrial samples by MS (www.suba.bcs.uwa.edu.au); in rice the number approaches only 150, while in other plant species the lists typically contain only 10–40 entries. In the latter cases the matches are often cross-species, so the specific genes involved have not been defined.

Analysis of the proteins identified using broad categorisation based on functional grouping (Table 15.1) and physical parameters (Fig. 15.1) gives an overview of the currently defined proteome. Notably, over 20% of the defined proteome is still without known function (Table 15.1), while most of the known function proteome is centered on the categories energy, metabolism and protein fate. The physical parameters show that the use of various separation techniques has provided a broad coverage of the proteome in terms of molecular mass and isoelectric point, and has even identified a range of proteins with grand average of hydrophobicity (GRAVY) scores over +0.3 (highly hydrophobic).

Much work remains to be done for the completion of the mass spectrometric proteome mapping of plant mitochondria. Firstly, the issue of contamination of mitochondrial preparations by non-mitochondrial proteins requires resolution. Several pathways are currently being explored. Firstly, higher purity mitochondrial extracts can be prepared by the use of non-density-gradient-based separation techniques. Exciting possibilities include immuno-capture of mitochondria and sub-mitochondrial structures, and also free flow electrophoresis that can provide purification of organelles based on differences in surface charge (Zischka et al.

Table 15.1 Functional categorisation of the 547 proteins identified by mass spectrometry (MS) in plant mitochondria according to gene ontology annotations

Functional category	Number of proteins	% of total
Cellular communication / signal transduction	20	4
Cellular structural organisation	10	2
Cellular transport and transport mechanisms	21	4
Defence stress and detoxification	19	3
DNA synthesis and processing	10	2
Energy	146	27
Metabolism	77	14
Protein fate	70	13
Protein synthesis	29	5
RNA processing	17	3
Transcription	11	2
Miscellaneous	4	1
Unknown function	113	21

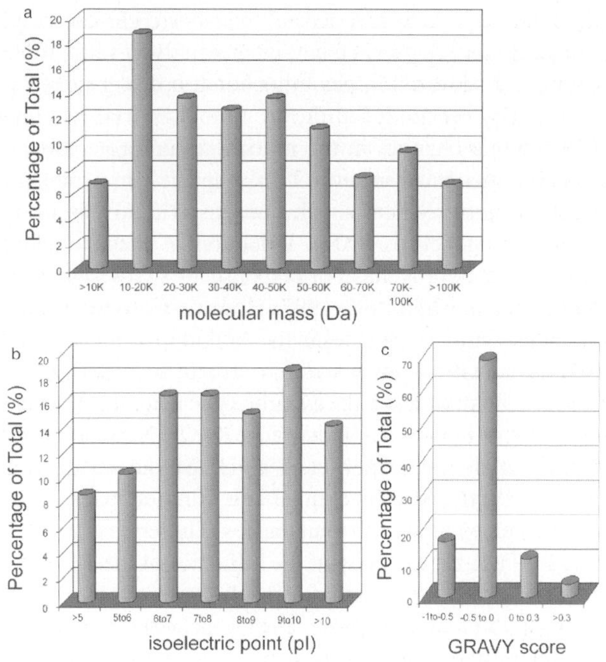

Fig. 15.1 Histograms of the physical properties of the proteins identified in the Arabidopsis mitochondrial proteome of 547 members. **a** Molecular mass, **b** isoelectric point pI, **c** grand average of hydropathicity (GRAVY). Further details of the 547 proteins can be found by searching SUBA (www.suba.bcs.uwa.edu.au)

2003, 2006). Secondly, when suspected contaminant proteins are found by independent research groups looking at different subcellular structures, then these cases of conflict can be targets to be resolved by fluorescent protein targeting studies and/or in vitro import studies of radio-labelled precursors. Thirdly, new tools are needed to ensure that protein subsets are not being missed due to the physical properties of the proteins, e.g., very small proteins, extremely basic or acidic proteins or very hydrophobic proteins.

In addition to the general exploration of the proteome, a range of studies looking at the dynamics of the mitochondrial proteome in response to development or growth conditions has been published. These have sometimes revealed new proteins that are found only in mitochondrial preparations under specific circumstances. Differences between mitochondria from different plant organs were clearly highlighted in the work of Colas des Francs-Small et al. (1992, 1993) looking at potato mitochondria, but it was not until 2002 that Bardel et al. (2002) attempted systematic protein identification of the complexity of these tissue-specific changes. This latter study identified 37 different proteins in the soluble fraction of pea mitochondria by MS or Edman degradation. The protein compositions of mitochondria from green leaves, etiolated leaves, roots and seeds were characterised. Changes in photorespiratory machinery dominated but, interestingly, several different aldehyde dehydrogenases were discovered in pea mitochondria that may be involved in the detoxification of aldehydes produced during mitochondrial dysfunction (Bardel et al. 2002). More subtle changes in the mitochondrial proteome within one cell type have also been investigated in several plant species. In maize and tomato, the heat stress induction of small heat shock proteins (HSPs) has been investigated (Banzet et al. 1998; Lund et al. 2001). Induction of the alternative respiratory bypasses and formate dehydrogenase in response to stress signals has also been reported (Vanlerberghe and McIntosh 1997; Hourton-Cabassa et al. 1998; Møller 2001). Oxidative stress-induced changes in the Arabidopsis mitochondrial proteome reveal a pattern of protein breakdown and new protein accumulation that reflect the loss of major susceptible proteins in the tricarboxylic acid (TCA) cycle and respiratory apparatus, and the induction of thioredoxin (TRX)- and glutathione-based defences (Sweetlove et al. 2002; Chew et al. 2003). Studies to identify the cause of cytoplasmic male sterility (CMS) by comparison of mitochondrial proteomes have highlighted subtle changes in protein abundances (Gutierres et al. 1997; Witt et al. 1997; Mihr et al. 2001). A recent study of the influence of the T-cytoplasm in maize on the mitochondrial proteome revealed a much larger range of nuclear-encoded proteins differing in abundance between the mitochondria isolated from T- and NA-cytoplasm backgrounds, suggesting a role for cytoplasmic factors, such as the expression of the mitochondrial genome, on the transcription/translation of nuclear-encoded genes for mitochondria-targeted proteins, ultimately influencing the mitochondrial proteome (Hochholdinger et al. 2004). Finally, in rice embryo germination, surprisingly large changes in mitochondrial protein content were observed during 48 h in the transition from promitochondrial structures in dry seeds to mature mitochondrial in imbibed, germinating seedlings (Howell et al. 2006). This maturation was facilitated by high levels of protein import components already present in promitochondria and driven by oxidation of external NADH, allowing for the rapid initiation of respiratory and metabolic functions to support early seedling establishment before TCA cycle enzymes are assembled.

15.4 Advances in Predicting the Remaining Proteome

It has long been anticipated that the full mitochondrial proteome that is encoded in the nuclear genome might be revealed by the prediction of targeting sequences in primary amino acid sequences. The targeting information of a range of well studied mitochondrial proteins from a range of organisms is known to be contained in a cleavable N-terminal presequence. It has been shown experimentally that this sequence alone is often sufficient to confer mitochondrial targeting on unrelated proteins. As the import machinery can clearly recognise this information, a range of attempts have been made to develop algorithms to achieve this recognition in silico. Early rule-based predictors such as Psort and MitoProt II (Claros and

Vincens 1996; Nakai and Horton 1999) provide defined outcomes with a clear logical framework for each decision made. However, these predictors have a notorious false-positive and false-negative rate in prediction in plants. Most success in plants has been achieved with machine learning algorithms such as neural networks. The most popular prediction program of this type for use in plants has clearly been Target P (Emanuelsson et al. 2000). It has been very difficult to define a rigorously unbiased test to fully evaluate the performance of the various proposed predictors as the list of known, well researched examples of any given class of proteins is small. Several recent programs claim better results than TargetP, with some justification. Predotar (Small et al. 2004) uses neural networks, while LOCtree (Nair and Rost 2005) and MultiLoc (Hoglund et al. 2006) use primarily support vector machines. Interestingly, the authors of both Predotar and LOCtree cite the use of expanded training sets, rather than better algorithms, as the major determinant of improved prediction performance, given the substantial increase in experimentally determined localisations over the period 2001–2004. The results of nine programs that claim to predict mitochondrial localisation against the MS-derived set of 547 is shown in Table 15.2. On average, each program predicts only ~40% of the experimental set to be located in mitochondria, while nearly 80% of the experimental set is predicted by at least one of the programs; however, all nine programs agree on a set of only 11 proteins, less than 2% of the total experimental set. It seems probable that the next round of predictors will be significantly better than the current ones.

Despite advances in experimental analysis, it is probable that if the full mitochondrial proteome encoded in the nuclear genome of model plants approaches 2,000–2,500 (Millar et al. 2006), then a large proportion will still need to be predicted rather than being identified experimentally by MS. Further, as a sizable proportion of mitochondrial proteins are not targeted by N-terminal sequences but by other cryptic means (Heazlewood et al. 2004), new kinds of algorithms built on

Table 15.2 Prediction of targeting to mitochondria based on 9 currently available tools. *At least one* Number of protein predicted from the set of 547 by at least one of the nine prediction tools, *All nine* number of proteins predicted from the set of 547 by all nine prediction tools. Further details of the 547 proteins and the prediction programs can be found by searching SUBA (www.suba.bcs.uwa.edu.au)

Prediction of targeting	Number of proteins	% of total
Target P	202	37
MitoProt 2	247	45
Predotar	203	37
MitoPred	259	47
Ipsort	242	44
WolfPSort	117	21
MultiLoc	179	33
LocTree	133	24
SubLoc	192	35
At least one	427	78
All nine	11	2

different criteria will be required. A subset of those that cannot be found by MS but are predicted in other ways can then be experimentally targeted by fluorescent protein tagging. Overall, this patchwork approach is the method most likely to define the full mitochondrial proteome within the next decade.

15.5 Beyond the List of the Proteome to its Regulation and Function

To list the gene products that find their way into the membranes and aqueous spaces of mitochondria is not in itself the end goal of mitochondrial proteomics, but it provides the set of cards that dictate the boundaries of the 'mitochondrial game'. Beyond the list itself comes investigation into the details of the function and regulation of each member, followed by attempts to gain a broader understanding of what members share in terms of functional redundancy, functional cooperation and regulatory linkages. These latter activities are many times larger in terms of both time commitment and expertise required than the initial list-generating activity itself.

15.5.1 Post-Translational Modifications

Proteins can undergo modification following translation by proteolytic cleavage of amino acid residues, as well as chemical derivatisation of their side chains, which can include acetylation, glycosylation, hydroxylation, methylation, acylation, phosphorylation, ubiquitination, and sulfation (Jensen 2004). During oxidative stress, a further range of modifications occur, including carbonyl formation, disulfide formation, S-nitrosylation and attachment of lipid aldehydes. This wide range of post-translational modifications (PTMs) can have importance as reversible means of functional regulation; PTM can tag proteins for rapid turnover or can damage protein function either reversibly or irreversibly, leading to a lack of correlation between protein abundance and protein function. Many PTMs have been reported within the mitochondrial proteomes of plants and animals. Understanding such PTMs is thus fundamental to understanding the mitochondrial proteome. Two areas have received particular attention from plant mitochondrial researchers to date: phosphorylation and oxidative modification.

15.5.2 Phosphorylation

The first report of protein phosphorylation in plant mitochondria was the inactivation and phosphorylation of the cauliflower pyruvate dehydrogenase complex (PDC) by a pyruvate dehydrogenase kinase (Randall et al. 1977). Characterisation in pea and Arabidopsis revealed reversible multi-site seryl-phosphorylation of the E1-alpha subunit of PDC (Thelen et al. 2000; Tovar-Mendez et al. 2003). Larger 'proteomic' investigations into plant mitochondrial protein phosphorylation emerged in the 1990s, with initial reports on one-dimensional polyacrylamide gels of 20–30 unidentified proteins labelled with γ-^{32}P in purified potato mitochondrial extracts (Sommarin et al. 1990; Pical et al. 1993) and of 10–15 unidentified γ-^{32}P-labelled proteins in pea mitochondria (Hakansson and Allen 1995). Phosphorylation of three of these latter proteins in pea mitochondria was shown to be on tyrosine residues and to be influenced by the phosphorylation of another protein on histidine residues (Hakansson and Allen 1995). Histidine phosphorylation in the inner membrane of potato mitochondria was also reported, as well as evidence for two autophosphorylated putative kinases that differed in their requirement for divalent cations (Struglics et al. 1999). Evidence for an inner membrane phosphoprotein phosphatase was revealed by the use of a protein phosphatase inhibitor, sodium fluoride, to inhibit γ-^{32}P-labelling of inner mitochondrial proteins in potato (Struglics et al. 2000). The first systematic identification of mitochondrial γ-^{32}P-labelled proteins was reported in 2003 by Bykova et al., who used two-dimensional BN- and IEF-PAGE together with MS to identify peptides from a set of γ-^{32}P-labelled potato mitochondrial gel spots. Further analysis of these proteins revealed that threonine residues 76 and 333 of formate dehydrogenase, and serine residue 294 of the PDC E1-alpha subunit, were phosphorylated, with the remaining 16 proteins still awaiting confirmation as phosphorylated proteins (Bykova et al. 2003). Reports of individual plant mitochondrial phosphoproteins have also appeared, including a 70 kDa HSP of bean mitochondria (Vidal et al. 1993), the β and δ subunits of ATP synthase in potato (Struglics et al. 1998), mitochondrial nucleoside diphosphate kinase from pea (Struglics et al. 1999), the small 22 kDa mt sHSP in maize mitochondria (Lund et al. 2001), and prohibitin in rice (Takahashi et al. 2003). Taken together, the list comprise proteins involved in major mitochondrial processes including the TCA cycle and associated enzymes, components of the electron transport chain and oxidative phosphorylation, HSPs, and defence against oxidative stress. These proteins reside predominantly in the mitochondrial matrix or the inner membrane. This is interesting because the phospholipid bilayer of the inner membrane is seen as an impermeable barrier even for protons, let alone signalling molecules. Hence, any signals that influence phosphorylation of this proteome will need to be actively transported into the matrix to influence this system or else it will respond primarily to intra-organelle generated stimuli. Although the Arabidopsis genome encodes approximately 1,000 protein kinases and hundreds of protein phosphatases, only the pyruvate dehydrogenase kinase/phosphopyruvate dehydrogenase phosphatase system has been clearly elucidated in plant mitochondria. At least 25 protein kinases and 8 phosphatases localise to the mitochondria in mammalian systems (Pagliarini and Dixon 2006), with many of them also performing roles unrelated to the mitochondria and residing primarily on the outside the organelle attached to the outer mitochondrial membrane. Grouping of these kinases and phosphatases into subfamilies shows that they cover most mammalian kinase and phosphatase subgroups, which suggests that mitochondria are subject to a diverse range of signalling pathways. This may also hold true for plant mitochondria, but the low abundance of eukaryotic protein kinases and a dearth of reports has prevented a systematic comparison with their mammalian counterparts at this point in time. The set of ten protein kinases reported in the LC-MS/MS study of Arabidopsis mitochondria (Heazlewood et al. 2004) remain of interest, but the unambiguous identification of protein kinases and phosphatases that operate inside mitochondria and the demonstration that their roles affect mitochondrial function will be paramount in establishing protein phosphorylation as a major regulatory mechanism in plant mitochondria.

15.5.3 Oxidation

The first reported example of direct inhibition of a plant mitochondrial protein by active oxygen species (AOS) was aconitase (Verniquet et al. 1991). These authors demonstrated changes in electron paramagnetic resonance spectra of aconitase, indicating modification of the 4Fe-4S cluster. In a proteomic study of the impact of oxidative stress on Arabidopsis mitochondria (Sweetlove et al. 2002), decreases in the abundance of aconitase, Fe-S centres of the NADH dehydrogenase (complex I), and core subunits of ATP synthase were reported. These losses from the proteome induced by oxidative stress are likely underscored by oxidative modification of proteins. Such damage can be broadly split into two groups: direct effects of reactive oxygen or nitrogen species on amino acid side chains (e.g., carbonyl group formation, S-nitrosylation, aberrant disulphide bond formation), and indirect effects by the modification of amino acid sides by reactive oxygen species (ROS)-generated cytotoxins (e.g., attachment of lipid peroxidation products).

Carbonyl groups formed by oxidative modification of arginine, lysine, proline and threonine amino acid residues can be detected in plant mitochondrial samples. Comparison of control rice mitochondrial samples and a mild oxidative treatment of this material revealed 20 constitutively oxidised proteins and a number of newly oxidised proteins observed only following oxidation treatment. This reveals the existence of a basal oxidation status in vivo and highlights a highly susceptible set from the proteome for further investigation (Kristensen et al. 2004). This set included aconitase, ATP synthase subunits, a range of TCA cycle enzymes and HSPs. The product of the dioxidation of tryptophan residues, known as *N*-formylkynurenine, has been known for decades, but only recently has this oxidation product been shown to be induced in vivo and linked to oxidative damage in the human heart mitochondrial proteome (Taylor et al. 2003). A recent study in plant mitochondria has also revealed a small set of rice and potato proteins that contain *N*-formylkynurenine, with a distribution among dehydrogenases and complexes I and III of the respiratory chain (Møller and Kristensen 2006).

Probably the most cytotoxic and best studied of the lipid peroxidation end products that modify proteins are 4-hydroxy-2-alkenals such as 4-hydroxy-2-nonenal (HNE). Application of this compound is known to inhibit a range of plant mitochondrial enzymes, including glycine decarboxylase, pyruvate dehydrogenase, 2-oxoglutarate dehydrogenase, malic enzyme and alternative oxidase (Millar and Leaver 2000, Winger et al 2005; Taylor et al. 2004a). For the first three of these enzymes, the primary site of action of HNE has been determined to be modification of the cofactor lipoic acid (Taylor et al. 2002). Lipoic acid is a critical cofactor for decarboxylating dehydrogenases with central roles in the TCA cycle, photorespiration and amino acid catabolism. Proteomic analysis of the lipoic acid binding proteins from plant mitochondria has defined differential susceptibility to lipid peroxidation modification (Taylor et al. 2002, 2005). Using HNE antibodies, a range of other mitochondrial protein bands that change in immunoreactivity following oxidative stress has been reported (Winger et al. 2005), but the proteins have not been identified to date.

The regulation of protein function by oxidation and/or reduction in plant mitochondria can also be linked to inter-molecular or intra-molecular disulfide bonds; during stress, aberrant disulfide formation can alter protein functions. Key enzymatic players in this system are likely to be the mitochondrial TRX system and protein disulfide isomerases. The importance of disulfide for respiratory function was first highlighted by the inter-molecular disulfide of the alternative oxidase dimer, which is most likely mediated by TRX action (Vanlerberghe and McIntosh 1997). A broad screen for TRX targets in plant mitochondria has recently been conducted (Balmer et al. 2004). Affinity chromatography in which a mutated TRX traps protein targets by forming a stable heterodisulfide was used. Fluorescent labeling of the potential TRX target proteins was also analysed by blocking free cysteines with *N*-ethylmaleimide, incubating samples (and a water-treated control sample) with the *Escherichia coli* TRX system and then treating both samples with monobromobimane, as a fluorescent thiol-specific probe to expose the new SH groups. A wide range of major mitochondrial proteins was identified by Balmer et al. (2004) as putative TRX-regulated proteins. Interestingly, a significant overlap exists with the set of putatively phosphorylated proteins listed in plant mitochondrial samples (Millar et al. 2005). The interplay of oxidative damage, redox control of disulfides, and detection and repair of incapacitated proteins remains an important field for proteomic analysis in plant mitochondria.

15.6 Non-Covalent Interactions Within the Proteome

Non-covalent interactions include protein–protein interactions and protein–ligand interactions. Both represent core ways in which the proteome operates to provide functional products and both also represent key pathways for discovering the role of proteins of unknown function within mitochondria.

Protein–protein interactions can be transient (typically very hard to study) or stable. Stable interactions represent protein complexes that can be probed by classical purification techniques, by immuno-capture and by in vivo protein–protein interaction techniques such the yeast two-hybrid technique (Y2H), fluorescence resonance energy transfer (FRET) or bioluminescence resonance energy transfer (BRET). The Y2H system has been used extensively to study protein–protein interactions in *Saccharomyces cerevisae* (Uetz et al. 2000; Ito et al. 2001) but is known to produce a relatively high number of false positive results (Ito et al. 2001). In one of the largest uses of immuno-capture and MS to date, 493 bait proteins in *S. cerevisae* were found to be involved in 3,617 interactions after correction for false positives; 74% of these interactions could be confirmed by immuno-precipitaton-immunoblot experiments (Ho et al. 2002). Similar approaches could be undertaken in plant mitochondria, which are clearly less complex than yeast. Currently, Y2H technology is able to handle hundred of baits and thousands of interactions, which is the level that would be required for a serious attempt at defining interactions in plant mitochondria. With the genetic manipulation tools available in Arabidopsis this project is currently possible, albeit substantial. Most of the work on protein interactions to date in plant mitochondria has centered around the use of BN-PAGE (Eubel et al. 2005), as outlined earlier, to define the composition of protein complexes. But while this is very useful in defining large complexes with numerous components, it fails to provide information on the precise protein–protein partners that build these complex sets of interactions. Key examples of the linkage of proteins of unknown function to functional complexes (but not to other discrete proteins in these complexes) can be seen in the BN-PAGE analysis of the electron transport chain. A series of proteins of unknown function observed in IEF-SDS-PAGE gels of mitochondria have been subsequently linked to particular protein complexes by BN-PAGE. For example, the carbonic anhydrase-like proteins attached to complex I (Sunderhaus et al. 2006), the four unknown function proteins associated with complex II (Eubel et al. 2003) and the five unknown function proteins associated with complex IV (Millar et al. 2004).

Protein–ligand interactions represent the myriad interactions that are only beginning to be explored in plant mitochondria. Substrates, products and small effector molecules bind transiently to proteins during catalysis of reactions and to regulate protein binding and activation state. Knowing the affinity of proteins for particular small molecules is thus a key component in developing a broad picture of proteome function. Traditionally, many of these links are made through perusal of protein name annotation or protein sequence motif databases, followed up by enzymatic analysis of purified protein samples to confirm function. However, a proteomic style of investigation is possible through the use of affinity-capture of proteins on media with bound ligands. This can currently be undertaken using an array of commercial media with bound nucleotides, organic acids, cofactors, metal ions, etc. A detailed study of the ATP affinity of mitochondrial soluble proteins has been undertaken recently (Ito et al. 2006) revealing a selective set of ATP-binding proteins. This study highlights the potential for ligand affinity studies in plant mitochondria and identifies over 50 proteins with high or moderate ATP binding. This included proteins not previously experimentally identified in plant mitochondria due to low abundance, but which are selectively enriched by this ligand affinity approach.

15.7 Conclusion

The plant mitochondrial proteome represents a relatively discrete proteome within the cell. Its proteome size is conducive to study and its high protein:lipid ratio makes it relatively amenable to many of the protocols of proteomic analysis. At least in Arabidopsis, mitochondria represent one of the best-studied organelles using proteomic techniques. However, its study by researchers is stimulated not only by these purely technical considerations but by a variety of overarching motivations. Mitochondria contain the core functions associated with energy metabolism. The proteins responsible for these functions can be compared and contrasted with those in other organisms to build a valuable picture of the robustness and plasticity of the respiratory mechanism and its regulation from yeast, to plants, to insects to humans (Richly et al. 2003). Building such a picture can bring broad insights ranging from the mechanisms of respiratory disease to the pathway of evolutionary development of this fundamental pathway in aerobic cell function. The plant mitochondrial proteome also contains proteins with myriad functions peripheral to the central role of energy generation. These are linked to mitochondrial maintenance as a genome-containing organelle, but also to the co-option of cellular functionality to reside near the site of ATP, NADH and organic acid generation in order to bleed off mitochondrial intermediate products into biosynthetic pathways. The peripheral proteins appear significantly plant-specific (Richly et al. 2003; Heazlewood et al. 2004) and analysis of these in plants links to the wider aim of building a systems biology model of plant cell function to drive predictive modelling and the directed engineering of plant production.

References

Balmer Y, Vensel WH, Tanaka CK, Hurkman WJ, Gelhaye E, Rouhier N, Jacquot JP, Manieri W, Schurmann P, Droux M, Buchanan BB (2004) Thioredoxin links redox to the regulation of fundamental processes of plant mitochondria. Proc Natl Acad Sci USA 101:2642–2647

Banzet N, Richaud C, Deveaux Y, Kazmaier M, Gagnon J, Triantaphylidès C (1998) Accumulation of small heat shock proteins, including mitochondrial HSP22, induced by oxidative stress and adaptive response in tomato cells. Plant J 13:519–527

Bardel J, Louwagie M, Jaquinod M, Jourdain A, Luche S, Rabilloud T, Macherel D, Garin J, Bourguignon J (2002) A survey of the plant mitochondrial proteome in relation to development. Proteomics 2:880–898

Barent RL, Elthon TE (1992) Two-dimensional gels: an easy method for large quantities of proteins. Plant Mol Biol Rep 10:338–344

Brugiere S, Kowalski S, Ferro M, Seigneurin-Berny D, Miras S, Salvi D, Ravanel S, d'Herin P, Garin J, Bourguignon J, Joyard J, Rolland N (2004) The hydrophobic proteome of mitochondrial membranes from Arabidopsis cell suspensions. Phytochemistry 65:1693–1707

Brumme S, Kruft V, Schmitz UK, Braun HP (1998) New insights into the co-evolution of cytochrome *c* reductase and the mitochondrial processing peptidase. J Biol Chem 273:13143–13149

Bykova NV, Stensballe A, Egsgaard H, Jensen ON, Møller IM (2003) Phosphorylation of formate dehydrogenase in potato tuber mitochondria. J Biol Chem 278:26021–26030

Cardol P, Vanrobaeys F, Devreese B, Van Beeumen J, Matagne RF, Remacle C (2004) Higher plant-like subunit composition of mitochondrial complex I from *Chlamydomonas reinhardtii*: 31 conserved components among eukaryotes. Biochim Biophys Acta 1658:212–224

Chew O, Whelan J, Millar AH (2003) Molecular definition of the ascorbate-glutathione cycle in Arabidopsis mitochondria reveals dual targeting of antioxidant defenses in plants. J Biol Chem 278:46869–46877

Claros MG, Vincens P (1996) Computational method to predict mitochondrially imported proteins and their targeting sequences. Eur J Biochem 241:779–786

Colas des Francs-Small C, Ambard-Bretteville F, Darpas A, Sallantin M, Huet JC, Pernollet JC, Rémy R (1992) Variation of the polypeptide composition of mitochondria isolated from different potato tissues. Plant Physiol 98:273–278

Colas des Francs-Small C, Ambard-Bretteville F, Small ID, Rémy R (1993) Identification of a major soluble protein in mitochondria from nonphotosynthetic tissues as NAD-dependant formate dehydrogenase. Plant Physiol 102:1171–1177

Davy de Virville J, Alin M-F, Aaron Y, Remy R, Guillot-Salomon T, Cantrel C (1998) Changes in functional properties of mitochondria during growth cycle of *Arabidopsis thaliana* cell suspension cultures. Plant Physiol Biochem 36:347–356

Day DA, Neuburger M, Douce R (1985) Biochemical characterization of chlorophyll-free mitochondria from pea leaves. Aust J Plant Physiol 12:219–228

Douce R (1985) Mitochondria in higher plants: structure, function and biogenesis. Academic Press, London

Dunbar B, Elthon TE, Osterman JC, Whitaker BA, Wilson SB (1997) Identification of plant mitochondrial proteins: a procedure linking two-dimensional gel electrophoresis to protein sequencing from PVDF membranes using a fastblot cycle. Plant Mol Biol Rep 15:46–61

Emanuelsson O, Nielsen H, Brunak S, von Heijne G (2000) Predicting subcellular localization of proteins based on their N-terminal amino acid sequence. J Mol Biol 300:1005–1016

Eubel H, Jansch L, Braun HP (2003) New insights into the respiratory chain of plant mitochondria. Supercomplexes and a unique composition of complex II. Plant Physiol 133:274–286

Eubel H, Heinemeyer J, Braun HP (2004) Identification and characterization of respirasomes in potato mitochondria. Plant Physiol 134:1450–1459

Eubel H, Braun HP, Millar AH (2005) Blue-native PAGE in plants: a tool in analysis of protein–protein interactions. Plant Methods 1:11

Gutierres S, Sabar M, Lelandais C, Chetrit P, Diolez P, Degand H, Boutry M, Vedel F, de Kouchkovsky Y, De Paepe R (1997) Lack of mitochondrial and nuclear-encoded subunits of complex I and alteration of the respiratory chain in *Nicotiana sylvestris* mitochondrial deletion mutants. Proc Natl Acad Sci USA 94:3436–3441

Hakansson G, Allen JF (1995) Histidine and tyrosine phosphorylation in pea mitochondria: evidence for protein phosphorylation in respiratory redox signalling. FEBS Lett 372:238–242

Hausmann N, Werhahn W, Huchzermeyer B, Braun HP, Papenbrock J (2003) How to document the purity of mitochondria prepared from green tissue of pea, tobacco and *Arabidopsis thaliana*. Phyton Ann Rei Bot 43:215–229

Heazlewood JL, Howell KA, Millar AH (2003a) Mitochondrial complex I from Arabidopsis and rice: orthologs of mammalian and fungal components coupled with plant-specific subunits. Biochim Biophys Acta 1604:159–169

Heazlewood JL, Howell KA, Whelan J, Millar AH (2003b) Towards an analysis of the rice mitochondrial proteome. Plant Physiol 132:230–242

Heazlewood JL, Whelan J, Millar AH (2003c) The products of the mitochondrial orf25 and orfB genes are FO components in the plant F1FO ATP synthase. FEBS Lett 540:201–205

Heazlewood JL, Tonti-Filippini JS, Gout AM, Day DA, Whelan J, Millar AH (2004) Experimental analysis of the Arabidopsis mitochondrial proteome highlights signaling and regulatory components, provides assessment of targeting prediction programs, and indicates plant-specific mitochondrial proteins. Plant Cell 16:241–256

Herald VL, Heazlewood JL, Day DA, Millar AH (2003) Proteomic identification of divalent metal cation binding proteins in plant mitochondria. FEBS Lett 537:96–100

Ho Y, Gruhler A, Heilbut A, Bader GD, Moore L, Adams SL, Millar A, Taylor P, Bennett K, Boutilier K, Yang LY, Wolting C, Donaldson I, Schandorff S, Shewnarane J, Vo M, Taggart J, Goudreault M, Muskat B, Alfarano C, Dewar D, Lin Z, Michalickova K, Willems AR, Sassi H, Nielsen PA, Rasmussen KJ, Andersen JR, Johansen LE, Hansen LH, Jespersen H, Podtelejnikov A, Nielsen E, Crawford J, Poulsen V, Sorensen BD, Matthiesen J, Hendrickson RC, Gleeson F, Pawson T, Moran MF, Durocher D, Mann M, Hogue CWV, Figeys D, Tyers M (2002) Systematic identification of protein complexes in *Saccharomyces cerevisiae* by mass spectrometry. Nature 415:180–183

Hochholdinger F, Guo L, Schnable PS (2004) Cytoplasmic regulation of the accumulation of nuclear-encoded proteins in the mitochondrial proteome of maize. Plant J 37:199–208

Hoglund A, Donnes P, Blum T, Adolph HW, Kohlbacher O (2006) MultiLoc: prediction of protein subcellular localization using N-terminal targeting sequences, sequence motifs and amino acid composition. Bioinformatics 22:1158–1165

Holtzapffel RC, Castelli J, Finnegan PM, Millar AH, Whelan J, Day DA (2003) A tomato alternative oxidase protein with altered regulatory properties. Biochim Biophys Acta 1606:153–162

Hourton-Cabassa C, Ambard-Bretteville F, Moreau F, Davy de Virville J, Remy R, Francs-Small CC (1998) Stress Induction of Mitochondrial Formate Dehydrogenase in Potato Leaves. Plant Physiol 116:627–635

Howell KA, Millar AH, Whelan J (2006) Ordered assembly of mitochondria during rice germination begins with pro-mitochondrial structures rich in components of the protein import apparatus. Plant Mol Biol 60:201–223

Humphrey-Smith I, Colas des Francs-Small C, Ambart-Bretteville F, Rémy R (1992) Tissue-specific variation of pea mitochondrial polypeptides detected by computerized image analysis of two-dimensional electrophoresis gels. Electrophoresis 13:168–172

Ito J, Heazlewood JL, Millar AH (2006) Anaylsis of the soluble ATP-binding proteome of plant mitochondria identifies new proteins and nucleotide triphosphate interactions within the matrix. J Proteome Res 5:3459–3469

Ito T, Chiba T, Ozawa R, Yoshida M, Hattori M, Sakaki Y (2001) A comprehensive two-hybrid analysis to explore the yeast protein interactome. Proc Natl Acad Sci USA 98:4569–4574

Jansch L, Kruft V, Schmitz UK, Braun HP (1995) Cytochrome *c* reductase from potato does not comprise three core proteins but contains an additional low-molecular-mass subunit. Eur J Biochem 228:878–885

Jansch L, Kruft V, Schmitz UK, Braun HP (1996) New insights into the composition, molecular mass and stoichiometry of the protein complexes of plant mitochondria. Plant J 9:357–368

Jensen ON (2004) Modification-specific proteomics: characterization of post-translational modifications by mass spectrometry. Curr Opin Chem Biol 8:33–41

Karpova OV, Newton KJ (1999) A partially assembled complex I in NAD4-deficient mitochondria of maize. Plant J 17:511–521

Keech O, Dizengremel P, Gardestrom P (2005) Preparation of leaf mitochondria from *Arabidopsis thaliana*. Physiol Planta 124:403–409

Kristensen BK, Askerlund P, Bykova NV, Egsgaard H, Møller IM (2004) Identification of oxidised proteins in the matrix of rice leaf mitochondria by immunoprecipitation and two-dimensional liquid chromatography–tandem mass spectrometry. Phytochemistry 65:1839–1851

Kruft V, Eubel H, Jansch L, Werhahn W, Braun HP (2001) Proteomic approach to identify novel mitochondrial proteins in Arabidopsis. Plant Physiol 127:1694–1710

Leaver CJ, Hack E, Forde BG (1983) Protein synthesis by isolated plant mitochondria. Methods Enzymol 97:476–484

Lund AA, Rhoads DM, Lund AL, Cerny RL, Elthon TE (2001) In vivo modifications of the maize mitochondrial small heat stress protein, HSP22. J Biol Chem 276:29924–29929

Meyer EH, Millar AH (2006) Isolation of mitochondria from plant cell culture. In: Posch A (ed) Sample preparation and fractionation for 2-D PAGE/proteomics. Humana, Totowa, NJ

Mihr C, Baumgärtner M, Dieterich JH, Schmitz UK, Braun HP (2001) Proteomic approach for investigation of cytoplasmic male sterility (CMS) in Brassica. J Plant Physiol 158:787–794

Millar AH, Leaver CJ (2000) The cytotoxic lipid peroxidation product, 4-hydroxy-2-nonenal, specifically inhibits decarboxylating dehydrogenases in the matrix of plant mitochondria. FEBS Lett 481:117–121
Millar AH, Heazlewood JL (2003) Genomic and proteomic analysis of mitochondrial carrier proteins in Arabidopsis. Plant Physiol 131:443–453
Millar AH, Liddell A, Leaver CJ (2001a) Isolation and subfractionation of mitochondria from plants. Methods Cell Biol 65:53–74
Millar AH, Sweetlove LJ, Giege P, Leaver CJ (2001b) Analysis of the Arabidopsis mitochondrial proteome. Plant Physiol 127:1711–1727
Millar AH, Eubel H, Jansch L, Kruft V, Heazlewood JL, Braun HP (2004) Mitochondrial cytochrome *c* oxidase and succinate dehydrogenase complexes contain plant specific subunits. Plant Mol Biol 56:77–90
Millar AH, Heazlewood JL, Kristensen BK, Braun HP, Møller IM (2005) The plant mitochondrial proteome. Trends Plant Sci 10:36–43
Millar AH, Whelan J, Small I (2006) Recent surprises in protein targeting to mitochondria and plastids. Curr Opin Plant Biol 9:610–615
Møller IM (2001) Plant mitochondria and oxidative stress: electron transport, NADPH turnover, and metabolism of reactive oxygen species. Annu Rev Plant Physiol Plant Mol Biol 52:561–591
Møller IM, Kristensen BK (2006) Protein oxidation in plant mitochondria detected as oxidized tryptophan. Free Radic Biol Med 40:430–435
Nair R, Rost B (2005) Mimicking cellular sorting improves prediction of subcellular localization. J Mol Biol 348:85–100
Nakai K, Horton P (1999) PSORT: a program for detecting sorting signals in proteins and predicting their subcellular localization. Trends Biochem Sci 24:34–36
Neuburger M (1985) Preparation of plant mitochondria, criteria for assessment of mitochondrial integrity, and purity, survival in vitro. In: Douce R, Day DA (eds) Higher plant cell respiration, vol 18. Springer, Berlin, pp 7–24
Neuburger M, Journet E, Bligny R, Carde J, Douce R (1982) Purification of plant mitochondria by isopycnic centrifugation in density gradients of Percoll. Arch Biochem Biophys 217:312–323
Pagliarini DJ, Dixon JE (2006) Mitochondrial modulation: reversible phosphorylation takes center stage? Trends Biochem Sci 31:26–34
Pical C, Fredlund KM, Petit PX, Sommarin M, Møller IM (1993) The outer membrane of plant mitochondria contains a calcium-dependent protein kinase and multiple phosphoproteins. FEBS Lett 336:347–351
Randall DD, Rubin PM, Fenko M (1977) Plant pyruvate dehydrogenase complex purification, characterization and regulation by metabolites and phosphorylation. Biochim Biophys Acta 485:336–349
Remy R, Ambard-Bretteville F, Colas des Frances C (1987) Analysis of two-dimensional gel electrophoresis of the polypeptide composition of pea mitochondria isolated from different tissues. Electrophoresis 8:528–532
Richly E, Chinnery PF, Leister D (2003) Evolutionary diversification of mitochondrial proteomes: implications for human disease. Trends Genet 19:356–362
Rötig A, Chauveau M (1987) Polypeptide composition of potato tuber mitochondria and submitochondrial fractions analysed by one- and two-dimensional gel electrophoresis. Plant Physiol Biochem 25:11–20
Schägger H, von Jagow G (1991) Blue native electrophoresis for isolation of membrane protein complexes in enzymatically active form. Anal Biochem 199:223–231
Small I, Peeters N, Legeai F, Lurin C (2004) Predotar: a tool for rapidly screening proteomes for N-terminal targeting sequences. Proteomics 4:1581–1590
Sommarin M, Petit PX, Møller IM (1990) Endogenous protein phosphorylation in purified plant mitochondria. Biochim Biophys Acta 1052:195–203

Struglics A, Fredlund KM, Møller IM, Allen JF (1998) Two subunits of the F0F1-ATPase are phosphorylated in the inner mitochondrial membrane. Biochem Biophys Res Commun 243:664–668

Struglics A, Fredlund KM, Møller IM, Allen JF (1999) Purification of a serine and histidine phosphorylated mitochondrial nucleoside diphosphate kinase from *Pisum sativum*. Eur J Biochem 262:765–773

Struglics A, Fredlund KM, Konstantinov YM, Allen JF, Møller IM (2000) Protein phosphorylation/dephosphorylation in the inner membrane of potato tuber mitochondria. FEBS Lett 475:213–217

Sunderhaus S, Dudkina NV, Jansch L, Klodmann J, Heinemeyer J, Perales M, Zabaleta E, Boekema EJ, Braun HP (2006) Carbonic anhydrase subunits form a matrix-exposed domain attached to the membrane arm of mitochondrial complex I in plants. J Biol Chem 281:6482–6488

Sweetlove LJ, Heazlewood JL, Herald V, Holtzapffel R, Day DA, Leaver CJ, Millar AH (2002) The impact of oxidative stress on Arabidopsis mitochondria. Plant J 32:891–904

Takahashi A, Kawasaki T, Wong HL, Suharsono U, Hirano H, Shimamoto K (2003) Hyperphosphorylation of a mitochondrial protein, prohibitin, is induced by calyculin A in a rice lesion-mimic mutant cdr1. Plant Physiol 132:1861–1869

Taylor NL, Day DA, Millar AH (2002) Environmental stress causes oxidative damage to plant mitochondria leading to inhibition of glycine decarboxylase. J Biol Chem 277:42663–42668

Taylor NL, Rudhe C, Hulett JM, Lithgow T, Glaser E, Day DA, Millar AH, Whelan J (2003) Environmental stresses inhibit and stimulate different protein import pathways in plant mitochondria. FEBS Lett 547:125–130

Taylor NL, Day DA, Millar AH (2004a) Targets of stress-induced oxidative damage in plant mitochondria and their impact on cell carbon/nitrogen metabolism. J Exp Bot 55:1–10

Taylor NL, Heazlewood JL, Day DA, Millar AH (2004b) Lipoic acid-dependent oxidative catabolism of alpha-keto acids in mitochondria provides evidence for branched-chain amino acid catabolism in Arabidopsis. Plant Physiol 134:838–848

Taylor NL, Heazlewood JL, Day DA, Millar AH (2005) Differential impact of environmental stresses on the pea mitochondrial proteome. Mol Cell Proteomics 4:1122–1133

Thelen JJ, Miernyk JA, Randall DD (2000) Pyruvate dehydrogenase kinase from *Arabidopsis thaliana*: a protein histidine kinase that phosphorylates serine residues. Biochem J 349:195–201

Tovar-Mendez A, Miernyk JA, Randall DD (2003) Regulation of pyruvate dehydrogenase complex activity in plant cells. Eur J Biochem 270:1043–1049

Uetz P, Giot L, Cagney G, Mansfield TA, Judson RS, Knight JR, Lockshon D, Narayan V, Srinivasan M, Pochart P, Qureshi-Emili A, Li Y, Godwin B, Conover D, Kalbfleisch T, Vijayadamodar G, Yang MJ, Johnston M, Fields S, Rothberg JM (2000) A comprehensive analysis of protein-protein interactions in *Saccharomyces cerevisiae*. Nature 403:623–627

Vanemmerik WAM, Wagner AM, Vanderplas LHW (1992) A quantitative comparison of respiration in cells and isolated mitochondria from *Petunia hybrida* suspension cultures – a high yield isolation procedure. J Plant Physiol 139:390–396

Vanlerberghe GC, McIntosh L (1997) Alternative oxidase: from gene to function. Annu Rev Plant Physiol Plant Mol Biol 48:703–734

Verniquet F, Gaillard J, Neuburger M, Douce R (1991) Rapid inactivation of plant aconitase by hydrogen peroxide. Biochem J 276:643–648

Vidal V, Ranty B, Dillenschneide M, Charpenteau M, Ranjeva R (1993) Molecular characterization of a 70 kDa heat-shock protein of bean mitochondria. Plant J 3:143–150

Werhahn W, Jänsch L, Braun H-P (2003) Identification of novel subunits of the TOM complex from *Arabidopsis thaliana*. Plant Physiol Biochem 41:407–416

Winger AM, Millar AH, Day DA (2005) Sensitivity of plant mitochondrial terminal oxidases to the lipid peroxidation product 4-hydroxy-2-nonenal (HNE). Biochem J 387:865–870

Witt U, Lührs R, Buck F, Lembke K, Grünegerg-Seiler M, Abel W (1997) Mitochondrial malate dehydrogenase in *Brassica napus*: altered protein patterns in different nuclear mitochondrial combinations. Plant Mol Biol 35:1015–1021

Zhang Q, Wiskich JT, Soole KL (1999) Respiratory activities in chloramphenicol-treated tobacco cells. Physiol Planta 105:224–232

Zischka H, Weber G, Weber PJ, Posch A, Braun RJ, Buhringer D, Schneider U, Nissum M, Meitinger T, Ueffing M, Eckerskorn C (2003) Improved proteome analysis of *Saccharomyces cerevisiae* mitochondria by free-flow electrophoresis. Proteomics 3:906–916

Zischka H, Braun RJ, Marantidis EP, Buringer D, Bornhovd C, Hauck SM, Demmer O, Gloeckner CJ, Reichert AS, Madeo F, Ueffing M (2006) Differential analysis of *Saccharomyces cerevisiae* mitochondria by free-flow electrophoresis. Mol Cell Proteomics 5:2185–2200

Chapter 16
Proteomic Analysis of the Plant Nucleolus

Olga Koroleva, Peter McKeown, Alison Pendle, and Peter Shaw

Abstract The nucleolus is a prominent sub-nuclear structure found in all eukaryotes. It is where the ribosomal RNA genes are transcribed and ribosomes are synthesised. However, much evidence has now accumulated that the nucleolus is involved in many other nuclear processes. Nucleoli are of moderate protein complexity, comprising a few hundred proteins, and can be isolated for proteomic analysis. In this chapter we describe the purification and analysis of plant nucleoli by proteomic methods and summarise the current results. We also discuss more specific tagging methods that have been used to analyse individual protein complexes, as well as methods for analysing post-translational modifications of nucleolar proteins. Finally we discuss the assessment of the reliability of such proteomic data, and the presentation and curation of this type of data.

16.1 Introduction

The nucleolus is a prominent structure found in the nuclei of all eukaryotes. It is where ribosomal RNA genes are transcribed by RNA polymerase I and where these transcripts are processed to form pre-ribosomes. Like all such nuclear sub-structures, the nucleolus lacks any bounding membrane, and must therefore be held together by intermolecular interactions as well as by the continuity of the rDNA

Olga Koroleva, Department of Cell and Developmental Biology, John Innes Centre, Colney, Norwich NR4 7LX, UK

Peter McKeown, Department of Cell and Developmental Biology, John Innes Centre, Colney, Norwich NR4 7LX, UK

Alison Pendle, Department of Cell and Developmental Biology, John Innes Centre, Colney, Norwich NR4 7LX, UK

Peter Shaw, Department of Cell and Developmental Biology, John Innes Centre, Colney, Norwich NR4 7LX, UK, E-mail: peter.shaw@bbsrc.ac.uk

J. Šamaj and J. Thelen (eds.), *Plant Proteomics*
© Springer 2007

within it with the rest of the nuclear DNA. It has been suggested that the biochemistry of ribosome biogenesis itself is responsible for the existence of the nucleolus as a distinct compartment (Melese and Xue 1995; Hernandez-Verdun et al. 2002) and nucleolar volume correlates with transcription levels. However rRNA transcription occurs without nucleolus formation in Archaea (Omer et al. 2000), whereas nucleoli are formed in organisms lacking RNA polymerase I (Conrad-Webb and Butow 1995) or multiple rRNA repeats (Matsuzaki et al. 2004), suggesting that the formation of a nucleolus is not essential to rRNA transcription, but is an adaptation to create a compartment in which pre-ribosomes may be formed with high efficiency (Raska et al. 2006).

However, much evidence now suggests that there must be more to nucleolar structure and function than making ribosomes. In the first place, nucleolar structure varies considerably between species, between cell types and between individual cells. Second, a number of unexpected, unconventional activities have been localised, at least in part, to the nucleolus (Raska et al. 2006). In most cases these unconventional activities have been investigated in only one or two species, so that it is not yet clear whether they are general nucleolar functions or specific adaptations. For example the nucleolus has been implicated in many other aspects of RNA biology, including biogenesis of snRNAs and snoRNAs, of the signal recognition particle, of tRNAs and RNAse P, and of telomerase, as well as in mRNA surveillance. The nucleolus has also been linked to viral infections, to cell cycle regulation, to cancer and to stress responses (Hiscox 2002; Rubbi and Milner 2003; Maggi and Weber 2005; Mayer and Grummt 2005; Yuan et al. 2005). In many cases the involvement of the nucleolus was discovered by localising factors involved in the various processes to the nucleolus. However, recent approaches using high throughput localisation studies and proteomic analysis of purified nucleoli have put these unconventional activities on a more systematic basis (Scherl et al. 2002; Andersen et al. 2005; Pendle et al. 2005).

For the most part, proteins are recruited to the nucleolus as a consequence of their intermolecular interactions rather than by recognisable nucleolar targeting sequences. This has frustrated attempts to predict nucleolar localisation in silico. Nucleolar localisation sequences (NoLS) exist in viral proteins, such as the plant umbravirus ORF3 protein (Kim et al. 2004) and both potato leaf-roll virus (PLRV) capsid proteins (Haupt et al. 2005). Indeed, the nucleolar location of one PLRV capsid protein was predicted through NoLS identification (Haupt et al. 2005), and comparison of coronavirus N-protein sequences allowed similar predictions for components of the SARS virus (Reed et al. 2006). However, these reflect binding to existing nucleolar proteins (nucleolin, in the latter case) and the interacting sequences are not well conserved. Similarly, a native nucleolar protein may include sequences important for localisation, as in the basic C-terminal domain of yeast Nop25 (Fujiwara et al. 2006) but even this is not widely conserved. Hence, nucleolar localisation remains difficult to predict.

Without reliable prediction, direct determination of the composition of nucleoli is the best option, and it has recently become possible to use high-throughput proteomic techniques to identify large numbers of nucleolar proteins from complex mixtures derived from purified nucleoli. Direct visualisation of subcellular localisation has also been used in mice to identify nucleolar proteins amongst other nuclear proteins via an enhancer-trap system, although this study took as a starting point a broad range of predicted nuclear proteins of which only 10% were exclusively nucleolar (Sutherland et al. 2001). In pilot screens aimed at systematically localising proteins in mammalian (Simpson et al. 2000) and plant cells (Cutler et al. 2000; Escobar et al. 2003; Tian et al. 2004; Koroleva et al. 2005) by systematic green fluorescent protein (GFP) open reading frame fusion expression, 2–3% of GFP fusions were highly enriched within the nucleolus. As expected, these included many RNA processing functions as well as protein kinases and phosphatases that may regulate various aspects of rRNA metabolism (Koroleva et al. 2005). Perhaps more surprising was the finding that certain transcription factor-like proteins were preferentially located in the nucleolus, whereas related family members were found mainly in the nucleoplasm (Tian et al. 2004; Koroleva et al. 2005).

16.2 Proteomics Techniques used for Plants

The two techniques commonly used for proteomic purposes are 2D polyacrylamide gel electrophoresis (2-DE) and liquid chromatography / mass spectrometry (LC/MS) methods. Although 2D gel approaches do have some advantages (see Chap. 2 by Hurkman and Tanaka, this volume), such as quick visualisation of the distribution of major protein groups, accessibility and relatively low basic costs, and ease of sample and data storage, they are applicable only to the major protein components of the proteome, and not suitable for proteins with extreme values of Mr (relative molecular mass) or pI (isoelectric point), hydrophobic membrane components or low-abundance proteins. More advanced uses of this methodology, such as difference in gel electrophoresis (DiGE) (Unlu et al. 1997; Lilley and Dupree 2006) significantly increase the costs – fluorescent dyes and image analysis software are expensive – but the dynamic resolution is still not very high. Furthermore, since many plant proteins are extensively modified, plant protein mixtures do not generally resolve very well on 2-DE gels. It has been estimated that single gel-based analyses allow the identification of approximately 5% of expressed cellular proteins (Heazlewood and Millar 2003). Only an estimated 120 nucleolar proteins had been identified by such techniques prior to the use of high throughput MS methods (Coute et al. 2006).

Therefore, there seems be more future in non-gel-based approaches for analysing the plant proteome, such as recently developed techniques like LC-MALDI MS; LC ESI MS/MS and MuDPIT (Aebersold and Mann 2003). Nevertheless, there are several examples where a combination of 2D gel separation with subsequent identification of proteins from excised spots have allowed analysis of the plant nuclear proteome. In Arabidopsis, 500–700 spots were detected on 2D gels, and analysis using MALDI-TOF (matrix assisted laser desorption / ionisation–time of flight) MS led to the identification of 184 spots corresponding to 158 different proteins (Bae et al. 2003). In rice nuclei isolated from suspension culture cells, from a total of 549 proteins resolved on 2-DE, 190 proteins were identified by MALDI-TOF MS from 257 major protein spots (Khan and Komatsu 2004).

MALDI-TOF is the most widespread type of MS analysis. This method simply measures the mass of each ionised peptide, providing a "fingerprint" of protein composition, with protein identification based on matches between measured masses of peptides and predicted masses of proteolytic cleavage products of the proteins present in databases. The caveat is that with the increasing length of the peptide chain, many peptide fragments with different sequences will have very similar masses. The same problem sometimes results from protein modifications. Besides MALDI, another common ionisation method is electro-spray ionisation (ESI). The multiply charged ions generated by ESI often produce MS/MS spectra that are cleaner and potentially easier to interpret (Bodnar et al. 2003). In general, thorough protein separation/purification is required for ESI, since this technique is very intolerant of any contaminants.

A more complex analytical approach is tandem mass spectrometry (MS/MS), which has two MS stages, often a combination of quadrupole and time-of-flight mass detectors (Q-TOF). In this technique, peptide fragments of a given mass are resolved by the first MS stage, then fragmented in a collision chamber and the mass fragmentation patterns are recorded by the second MS stage. The advantage of tandem MS over single MS fingerprinting is that the precise sequence of amino acids in each peptide can be determined, which allows much more reliable identification.

An important principle for the preparation of samples for proteomic analysis is to reduce sample complexity by protein fractionation, therefore increasing the possibility of detecting proteins with lower abundance in the complex protein mixture.

Very commonly, LC separation precedes the ionisation and MS stages. MALDI tandem mass spectrometers allow a preliminary LC separation "off-line", in which the eluted fractions are spotted onto a MALDI sample plate for later MS analysis. Therefore, LC-MALDI techniques do not suffer from the time constraints imposed by the transient presence of peptides eluting from a column; if necessary, each sample can be analysed more than once. On the other hand, with LC-ESI techniques the sample eluted from the LC is sprayed directly into the ESI source and data is acquired immediately. Because in LC-MALDI the LC is decoupled from the MS analysis, the data is not available immediately, as it is with ESI analyses, and chromatography problems can be present but undetected until too late. MuDPIT (multi-dimensional protein identification technology) is a development of an online LC approach, in which two sequential microcapillary columns (an ion exchange column followed by a reverse phase column) are used to separate very complex mixtures of peptides for MS/MS analysis. This type of analysis allows the identification of very large numbers of proteins/peptides from complex mixtures with the minimum of pre-treatment.

16.3 Purification of Nucleoli

Nucleoli are not bounded by membranes and, since they contain rDNA, they are connected by these DNA strands to the rest of the nuclear chromatin. For these reasons, all methods for preparing nucleoli have relied on mechanical fragmentation procedures. In the case of plant cells, the cell wall must be digested by degrading enzymes, such as cellulases and pectinases, to produce protoplasts. This is necessary to avoid the nuclei and nucleoli being trapped inside the cell wall residues after cellular fragmentation. Pendle et al (2005) used Arabidopsis suspension cultures as a convenient, reproducible and abundant source of cells. The cells were protoplasted, and then the protoplasts were carefully disrupted by a small number of strokes of a stainless steel homogeniser, with a clearance between the piston and chamber of 25 μm, observing the state of the preparation after every few strokes. In general the first few strokes released mostly intact nuclei, while further homogenisation then disrupted the nuclei to release nucleoli.

Nucleoli constitute the densest cellular component with the exception of starch granules, which may be an unavoidable contaminant, and can be quickly and efficiently purified from the other cellular debris by differential centrifugation. One of the most critical factors is the Mg^{2+} concentration. Magnesium ions causes chromatin to cross-link into an unworkable network, and centrifugation steps then bring down the chromatin with the nucleoli enmeshed in it. To solve this problem, Pendle et al (2005) simply left Mg^{2+} out of the homogenisation buffer. The nucleoli could then be separated from the chromatin. However, a small concentration of Mg^{2+} was added immediately after centrifugation, since lack of Mg^{2+} caused a gradual disintegration of the nucleoli.

Similar methods have been used for the purification of nucleoli from human cell culture (Andersen et al. 2002; Scherl et al. 2002), although in this case nuclei were purified first, with sonication being necessary to release the nucleoli. This may indicate that human nucleoli are more tightly associated with the chromatin and the rest of the nucleoplasm. We have extracted nucleoli from Arabidopsis seedlings on a pilot scale, although not so far on a preparative scale, by chopping tissue with a razor blade to release nuclei, which can them be homogenised in the same way as culture cells and nuclei.

Studies during the past few years using live cell imaging of fluorescently tagged proteins have caused a reappraisal of the dynamic nature of the nucleolus, and indeed all other nuclear compartments. It has become clear that virtually all nuclear and nucleolar proteins are rapidly exchanged with the nucleoplasm, and that what distinguishes 'nucleolar' components is a mean nucleolar residence time of the order of seconds rather than of fractions of a second (Phair and Misteli 2000; Raska et al. 2006). This raises the interesting questions of how nucleoli can be isolated as distinct structures over the timescale of minutes or hours necessary for the purification, and how the structures isolated relate to nucleoli seen in living cells. It would be expected that most of the associated protein and other mobile factors would be lost to the medium during extraction and purification, leaving only the most strongly attached factors. The fact that most of the expected proteins are present in purified nucleolar fractions suggests that the situation is more complicated. One possibility is that, along with the rapidly exchanging population of proteins, there is a more stable population. Another possibility is that the changes in medium, ionic strength, metabolite concentrations, etc., that occur during nuclear fragmentation and nucleolar purification cause the dynamic exchange of many of the proteins to stop, either because active processes necessary are halted or because changes in the medium cause an effective precipitation of nucleolar contents into a less soluble state. It is important to bear in mind that these processes may affect different nucleolar constituents to different extents, and that 'purified' nucleoli are unlikely to have exactly the same composition as their in vivo counterparts. However, extracted Arabidopsis nucleoli remain transcriptionally competent (P. McKeown and P. Shaw, unpublished data), which gives some confidence in the validity of the purified fractions.

16.4 Isolation and Analysis of Arabidopsis Nucleolar Proteins

These approaches were successfully applied in the two first published proteomic analyses of human nucleoli purified from HeLa cells, in which 257 proteins (Andersen et al. 2002) and 210 proteins (Scherl et al. 2002), respectively, were identified. Recently, a new study identified 667 proteins within nucleoli prepared from HeLa cells (Andersen et al. 2005). So in total, the results obtained from the three independent analyses of HeLa nucleoli provided a list of 713 individual proteins (Coute et al. 2006). The latest study by Andersen et al (2005) employed both an LC MS/MS Q-ToF instrument and a linear ion trap Fourier-transform ion-cyclotron resonance mass spectrometer (FT-ICR-MS), which provided very high resolution and mass accuracy. The development of hybrid mass spectrometers employing FT-ICR potentially presents new opportunities for proteomics analysis of the nucleolus (Coute et al. 2006).

The Arabidopsis nucleolar preparation obtained as described above from suspension cell cultures was subjected to high throughput proteomic analysis as described by Pendle et al (2005), which identified 217 proteins. This has been subsequently increased to over 500 proteins by a MuDPIT approach using two-stage microcapillary LC linked to a Q-TOF instrument (P. McKeown, P. Shaw, A. Bottrill, unpublished data). Many of the proteins that were identified were expected: known nucleolar proteins, ribosomal proteins, proteins involved in rDNA transcription, and other RNA-interacting proteins involved in ribosome biogenesis. However, many unexpected proteins were also found in the nucleolus, including for example, spliceosomal proteins, small nuclear RNP (snRNP) proteins and translation factors (see Fig. 16.1). These results reinforce the results of several previous studies, implicating the nucleolus in a variety of functions in addition to ribosome biogenesis, including the biogenesis or transport of a range of RNAs and RNPs, and roles in mRNA maturation, cell cycle control and stress

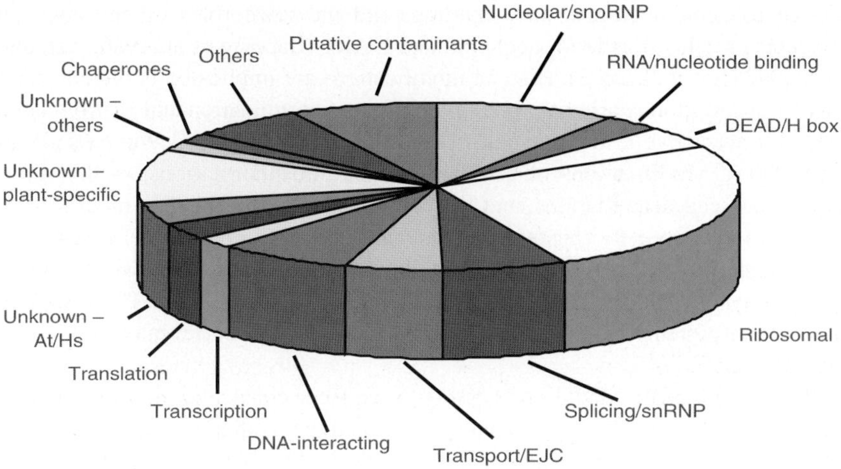

Fig. 16.1 Relative proportions of different protein classes found in Arabidopsis nucleolar proteome

responses (Rubbi and Milner 2003; Andersen et al. 2005; Olson and Dundr 2005; Pendle et al. 2005; Pontes et al. 2006).

It could be argued that many of the proteins identified were contaminants – a large and poorly defined structure such as the nucleolus is likely to be impossible to fully separate from other nuclear and cellular components. Nevertheless, the proteomic analysis has been confirmed by systematic protein expression studies using GFP-tagged proteins. For example, Pendle et al (2005) showed that the vast majority (87%) of the proteins identified by proteomic analysis were also found to be nucleolar-located by the structural criterion of expressing GFP fusions of the identified proteins. Many of the unexpected human proteins have also subsequently been confirmed as being in the nucleolus by structural methods such as GFP tagging and immunofluorescence (Fox et al. 2002). It is also notable that a large proportion of the nucleolar proteins identified in these proteomic analyses are currently completely uncharacterised, further supporting the view that we have as yet a very incomplete picture of the biochemistry that is taking place in the nucleolus. The following websites host current databases of identified nucleolar proteins: http://lamondlab.com/NOPdb/; http://bioinf.scri.sari.ac.uk/cgi-bin/atnopdb/home; http://www.expasy.org/ch2d/.

16.5 Analysis of Post-Translational Protein Modifications

Post-translational modifications (PTMs) are very common in nucleolar proteins, and several hundred specific modifications, many of which are likely to be functionally important in nucleoli, are known. For example, arginine dimethylation has defined roles in nucleoplasmic shuttling, and unknown roles on the abundant nucleolar proteins fibrillarin, nucleolin and GAR1 (Lapeyre et al. 1986; Najbauer et al. 1993; Xu et al. 2003). Histone modifications are implicated in the control of rDNA transcription (Earley et al. 2006) and both ubiquitination and SUMOylation pathways are active in the nucleolus (Mo et al. 2002; Song and Wu 2005; Panse et al. 2006). Modifications of histones include phosphorylations, methylations, acetylations and deaminations, and the complex patterns of histone modifications have been proposed to constitute a 'histone code' (Nightingale et al. 2006). A detailed analysis of nucleolar histone modifications could in principal explore the potential histone code as it applies to the rDNA. It is also likely that PTMs of other non-histone proteins have been underestimated, and will be revealed by detailed MS analyses.

The use of tandem MS/MS processing to sequence covalently modified peptides was first reported by Olsen and Mann (2004), and the method is amenable to high-throughput techniques by using a quadrupole as the first MS stage. As with identification of unmodified peptides, it is possible to study either a pre-selected group of proteins of interest, or to sample the entire proteome. More focussed approaches have been used in several organisms for different modifications, and may allow a greater level of detail. Nevertheless, full determination of all modification sites is still technically difficult. For example, analysis of phosphorylation sites of the nucleolar RENT complex in yeast failed to detect all known sites, even when several different techniques were focussed on just these proteins (Chen et al. 2002).

When extracted Arabidopsis nucleoli were studied with Q-ToF MS/MS, a number of modifications were detected, including a ribosomal protein found in an acetylated form (P. McKeown and P. Shaw, unpublished work). In the same study, eEF1a (1g07930) was found acetylated at two sites, and the mammalian equivalent has also been detected in an acetylated form (Kim et al. 2006). The total number of acetylations found was fewer than in mouse, partly because the majority of acetylated proteins in mouse are mitochondrial or cytoplasmic, and many acetylated nuclear proteins are probably not enriched in the nucleolus (Kim et al. 2006). Kim et al. made use of affinity purification with an anti-acetyl lysine antibody, but were still unable to detect many known acetylations, especially those associated with rare transcription factors. Combining fractionation with affinity tagging may increase the number of nucleolar covalent modifications detected, but restricts the number of modifications that can be searched. For example, our survey detected seven sites of methionine acetylation that would not have been detected had an acetyl lysine-specific antibody been used.

Care is needed, however, in the interpretation of this type of result. Some small modifications have very similar masses: methylation and formylation are close enough in Mr for methylated peptides to also be detected in a screen for formylations, although with lower confidence ratings, which may allow them to be rejected. Dimethylation and citrullination are isobaric and hence cannot be distinguished by this method. Additionally, it may not be possible to distinguish between the modifications of different residues within peptides, a particular problem with highly modified proteins such as histones. Modifications have to be actively sought in search engines such as MASCOT, precluding the detection of novel modifications, although known modifications can be found on novel substrates or at positions not previously identified.

In studies of large modifications such as ubiquitination or SUMOylation, which themselves fragment, it has been found to be necessary to isolate the modified proteins before carrying out MS analysis. Plasma membrane-bound proteins modified with glycosylphosphatidylinositol (GPI)-anchor proteins were identified by a 'shave and conquer' technique: isolated membranes were treated with phospholipase D to liberate any GPI-anchored protein, and this protein fraction was analysed (Elortza et al. 2003). Covalent modifications have also been determined by SILAC (stable isotope labelling by amino acids in cell culture), with one population of HeLa cells fed with labelled tyrosine and the other fed with labelled arginine and lysine. Again, this allowed identification of unknown gamma-phosphorylation sites, and could be applicable to nucleoli (Amanchy et al. 2005).

16.6 Quantification Techniques

A typical proteomic analysis of a protein mixture produces a list of identified components but without any quantification of the relative amounts of the different proteins. The detection and successful identification of a particular peptide will depend on a several other factors apart from the relative abundance of the original protein in the mixture: first, on effective digestion by the proteolytic enzyme used to produce the peptide; second, on how easily the peptide is volatilised and ionised; third, on the way the peptide fragments in the collision cell; fourth, on the presence of modifications, which may not have been predicted. Finally, it will also depend on a presence of the peptide sequence in the database or databases to be searched subsequently. The last factor is a particular problem for species without completely sequenced genomes, since MASCOT and similar search engines search for a sequence match that is based on accurate mass data and therefore cannot rely on homology.

For many proteomic applications, for example, cataloguing a particular tissue/organellar proteome or getting a list of binding partners in a protein complex, simple identification of proteins present in the sample is all that is required. On the other hand, quantitative analysis is essential for many other proteomic applications, where the research involves comparison between different tissues and/or developmental stages of an organism, specific conditions or treatments, mutants or transgenic organisms with over-expressed or silenced genes. Therefore, methods have been developed for both relative and absolute quantification of protein amounts in proteomic analysis. Most biological applications require relative analysis, to compare two or more datasets, and in this review we will consider current proteomics techniques for relative quantification.

One possibility for quantification is 2-DE comparison. Although significant progress has been achieved recently with the development of fluorescent stains and the DiGE technique (Unlu et al. 1997; Lilley and Dupree 2006), 2-DE still allows only a subset of the proteome to be analysed, as certain groups of proteins, such as membrane proteins, co-migrating, low abundance proteins and those with extreme values of molecular mass and pI, cannot be clearly separated and/or visualised. In our experience this means that the majority of plant nuclear and nucleolar proteins either do not run cleanly in 2D gels or do not enter the gel at all. The low dynamic range of DiGE is also a major limitation of this technique. Protein identification from excised gel spots also has limitations because the spots are often contaminated with other proteins. Alternative approaches, non-gel-based quantitative proteomic methods, are based on comparison of peptide abundance, determining the isotopic composition of one or more elements in a compound after stable isotope labelling of a sample, either in vivo or in vitro. We will describe here a selection of isotopic labelling techniques that have been used on plant material. The labels can either be incorporated in vivo as substituted metabolites such as amino acids, or after purification and digestion by reaction with the resulting peptides.

16.6.1 In Vivo Isotope Labelling: SILAC

SILAC uses the in vivo incorporation of isotopically substituted amino acids to shift the molecular masses of the resulting peptides. Two or more cell cultures are grown in parallel using different substituted amino acid mixtures, and then the cultures are subjected to different conditions or treatments, such as drug treatments or stress. After treatment, the cultures are pooled and used for biochemical purification. At the MS stage the relative amounts of each peptide can be quantified, since the origin of the peptide can be determined by its isotopic composition. In general, isotopically substituted lysine and arginine are used together, since all tryptic peptides should then be labelled. It is usually necessary to grow the cell culture in the substituted amino acids for a few days to get a good level of incorporation. This approach was used by Andersen et al (2005) to determine the effects of different drug treatments on human nucleoli. Other typical applications are analysing dynamics events by adding isotopically substituted amino acids to a cell culture at a given time, and analysing samples during a time course (Ong et al. 2002; Blagoev et al. 2004). A large scale proteomic study tested the feasibility and technical challenges associated with SILAC to uncover quantitative changes during apoptosis in the nuclear proteome, resulting in the identification and quantification of 1,174 putative nuclear proteins (Hwang et al. 2006). Another recent example of the application of this technique is selective isotope labelling of proteins from Arabidopsis cell cultures by growing cells in the presence of a single stable isotopically labelled amino acid (Gruhler et al. 2005). A potential problem in applying this method to plants is that plant cells are better at inter-converting amino acids than animal cells, and so there is a risk that other amino acids may eventually be labelled.

16.6.2 In Vitro Isotope Labelling

16.6.2.1 Isotope Coded Affinity Tagging

In isotope coded affinity tagging (ICAT), free cysteines in a protein are reacted with a special affinity tag. Labeled proteins are enzymatically digested and labelled peptides are separated from the bulk mixture, first using affinity chromatography and then, in a second round, using ion-exchange chromatography prior to MS. The tag has three functional elements: a biotin tag, used during affinity capture (avidin chromatography); the isotopically encoded linker chain (with either eight hydrogens or eight deuteriums); and the reactive group, which will bind to and modify cysteine residues of the protein (Gygi et al. 1999; Li et al. 2003). The tag is marketed by Applied Biosystems (Foster City, CA). However, ICAT has limitations, because it only labels cysteine-containing peptides; 10–15% of every genome codes for proteins with no cysteine, and many proteins contain only a single cysteine, providing only a single peptide that can be potentially quantified.

16.6.2.2 Localisation of Organelle Proteins by Isotope Tagging

Localisation of organelle proteins by isotope tagging (LOPIT), a variation of the ICAT technique, was developed for discovering novel proteins in endomembrane organelles (Dunkley et al. 2004b; Lilley and Dupree 2006), and uses analytical centrifugation in combination with differential isotope labelling. The method involves partial separation of organelles by density gradient centrifugation, thus producing over-lapping fractions, followed by the analysis of protein distributions in the gradient by ICAT labelling (Dunkley et al. 2004a, 2004b) and MS. Multivariate data analysis techniques are then used to group proteins. A good correlation was observed between identification lists of proteins clusters in LOPIT and previous experimental evidence of protein locations within sub-cellular structures, and it has been shown that the LOPIT technique can be used to discriminate Golgi, endoplasmic reticulum (ER), plasma membrane, and mitochondrial/plastid proteins (Lilley and Dupree 2006). In the latter paper the more versatile ITRAQ method (see below) was used instead of ICAT.

16.6.2.3 Isobaric Tag for Relative and Absolute Quantification

Isobaric tag for relative and absolute quantification (ITRAQ) involves chemical tagging of the N-terminus of peptides generated from protein digests; the reagent used reacts with primary amines (Zieske 2006). Fragmentation of the tag attached to the peptides generates a low mass unique reporter ion. There are four tags available, all with an identical mass of 145 Da. The advantages of this approach are that all tryptic peptides are labelled, resulting in higher quality data; up to four labels can be used for multiple experiments; improved MS/MS fragmentation results in better confidence identification; and post translational modifications can be detected. The subsequent data analysis requires specialised software, and apart from the ProQuant software supplied by Applied Biosciences, non-commercial i-Tracker software is also available for quantitative proteomics using iTRAQ (Shadforth et al. 2005). Disadvantages of iTRAQ include the increased MS time required because of the increased number of peptides, and the fact that samples must be prepared according to strict guidelines.

16.6.3 Comparison of Quantification Methods

What is the best MS quantification technique to use? Recently a comparative study was carried out on three proteomic quantitative methods, DiGE, ICAT, and iTRAQ, using either 2-DE or LC-MALDI TOF/TOF (Wu et al. 2006). All three approaches yielded quantitative results with reasonable accuracy when the same protein mixture was used. In DiGE, accurate quantification was sometimes compromised due to the full or partial co-migration of proteins. The iTRAQ method was more susceptible to errors in precursor ion isolation, especially with increasing sample complexity. The global-tagging iTRAQ technique was more sensitive than the cysteine-specific ICAT method, which in turn was as sensitive as, if not more sensitive than, the DiGE technique.

16.7 Proteomic Analysis of Nuclear Protein Complexes

16.7.1 TAP Tag Strategy and Variations

The tandem affinity purification (TAP) method (Rigaut et al. 1999) was developed to improve the purification of protein complexes, particularly for subsequent proteomic analysis. The most common previous purification methods for protein complexes used antibody pull-down, and this caused a high background of peptides arising from the antibodies used. The original TAP tag consists of tandem protein A domains, linked via a tobacco etch-virus (TEV) protease site to a calmodulin-binding protein (CBP) domain. In practice, the protein of interest is expressed as a TAP-tagged fusion protein, and a cell extract is made after expression of the fusion. The tagged protein, along with complexed proteins, is absorbed on to IgG beads via the protein A domains, washed, and then released by TEV protease treatment. A second round of purification involves binding to calmodulin beads in the presence of Ca^{2+}, followed by washing and release by a buffer containing EGTA to remove the Ca^{2+}, and dissociate the calmodulin complex.

Subsequently, the TAP tag has been made more general by removing a nuclear localisation signal, and has been adapted for use in plant cells by removing a cryptic splice site, polyadenylation sites and AT- or GC-rich regions, and by the inclusion of castor bean catalase intron 1 for improved expression in plants (Rohila et al. 2004, 2006). Two GATEWAY-compatible binary vectors, NTAPi and CTAPi, were constructed and initially tested using *Agrobacterium*-mediated transformation for transient expression in *Nicotiana benthamiana* leaves. Recently, in a project studying protein kinase signaling networks in cereal leaves, transgenic stable transformants expressing 41 TAP-tagged rice protein kinases were produced and used for subsequent analysis of interacting proteins in rice plants (Rohila et al. 2006). In total, all 41 rice kinases were purified, and 23 of these were isolated as complexes with one or more interacting proteins (Rohila et al. 2006).

However, the NTAPi and CTAPi tags also have some disadvantages. First, many plant proteins have CBDs and therefore will be co-purified with the tagged protein (Reddy et al. 2002). We regularly observe co-purification of elongation factor (EF) 1-alpha proteins, from a small EF-Tu/EF-1A subfamily of identical genes At1g07920, At1g07930, At1g07940, At5g60390 (an example present in Table 16.1), which have a CBD, in protein mixtures pulled down by a number of different NTAPi- and CTAPi-tagged proteins. Two rice EF-homologous proteins were reported to be among the recurring cellular proteins in the TAP procedure by Rohila et al. (2006). Second, the proteolytic treatment by TEV protease involves incubation at 16°C, which may lead to proteolysis by endogenous proteases, because at this stage general protease inhibitors cannot be added to the reaction buffer because they would inhibit the TEV protease. The exception is E-64, but this inhibits only cysteine proteases.

These problems have been addressed recently by development of an alternative tandem affinity purification tag for the isolation of plant protein complexes, called pC-TAPa (Rubio et al. 2005). This construct is also a Gateway-compatible vector, allowing convenient recombination of ORFs from pre-existing Gateway entry clones. The pC-TAPa tag consists of two protein A binding domains, a 3C HRV protease site, six histidine repeats (6-His) and nine myc epitope repeats (myc tag). An advantage of 3C HRV protease is that this enzyme is active at a wide range of temperatures from as low as 4°C (although this requires an increase of either the amount of the enzyme in the reaction mixture, or in incubation time, compared to the relatively fast cleavage at 16°C). The second purification step can use either the 6-His tag or the myc tag, but 6-His tag Ni affinity chromatography has been reported to be more efficient than myc-tag epitope immunoprecipitation (Rubio et al. 2005).

16.7.2 TAP Tag Purification of Nuclear Transport Factor

We have used NTAPi-fusion for purification and proteomic analysis of proteins interacting with nuclear transport factor p15h2 (At1g27970) using transient expression in Arabidopsis cell culture, as described previously (Koroleva et al. 2005). This produced relatively high levels of expression of this particular protein. Another significant change made to the original protocol was using TCA

Table 16.1 List of interacting proteins identified by proteomic analysis of p15h2-NTAPi pulldown

Gene ID	Function	Mass	Peptides (n; unique)	Score	Other match	Comment
At3g04120	GAPDH	37,005	23; 6	1,504		
At1g27970	p15h2	14,107	16; 3	867		Bait
At5g09510	40S Ribosomal protein S1-54	17,061	20; 3	733	At1g04270	Ribosomal
At2g01250	60S Ribosomal protein L7-1	28,210	26; 2	611		Ribosomal
At1g22780	40S Ribosomal protein S18	17,591	5; 1	591	At1g34030; At4g09800	Ribosomal
At5g17330	Glutamate decarboxylase	57,429	20; 6	584		Contaminant?
At2g44120	60S Ribosomal protein L7	27,979	25; 2	497	At2g44120; At3g13580	Ribosomal
At5g23860	Tubulin beta-8 chain	51,258	10; 3	345	At5g62690; At5g62700; Q93Y15	Tubulin
At1g51060	Histone H2A	13,926	8; 2	332	At4g27230; At5g54640	Histone
At1g07920	Elongation factor 1-alpha	49,813	6; 2	281	At1g07930; At1g07940; At5g60390; etc	–
At1g07790	Histone H2B	16,392	4; 1	218	At2g28720; At2g37470; At3g45980; etc	Histone
At4g34670	40S Ribosomal protein S3A	29,956	3; 1	191		Ribosomal
At2g29550	Tubulin beta-7 chain	51,341	7; 3	190		Tubulin
At2g37270	40S Ribosomal protein S5-1	23,090	3; 1	183	At3g11940	Ribosomal
At1g09590	60S Ribosomal protein L21	18,698	6; 2	150	At1g09690; At1g57660; At1g57860	Ribosomal
At5g10860	CBS domain; cAMP-binding	22,829	4; 3	139		–
At2g19740	60S Ribosomal protein L31	13,750	4; 2	127	At4g26230; At5g56710	Ribosomal

At3g02080	40S Ribosomal protein S19-1	15,875	4; 2	125		Ribosomal
At2g09990	40S Ribosomal protein S16	16,849	6; 2	124		Ribosomal
At1g27310	p15h3 (NTF2)	13,633	2; 1	106		Heterodimer?
At1g23290	60S Ribosomal protein L27a-2	16,339	4; 1	104	At1g70600	Ribosomal
At3g25520	60S Ribosomal protein L5	34,394	2; 1	101	At5g39740	Ribosomal
At2g47610	60S Ribosomal protein L7a	29,339	3; 1	99	At3g62870	Ribosomal
At3g48170	Betaine aldehyde dehydrogenase	55,681	5; 2	94		Contaminant?
At5g11110	Sucrose-phosphate synthase	100,178	4; 1	94		Contaminant?
At5g17920	Methionine synthase	84,646	2; 2	90		Contaminant?
At1g48830	408 Ribosomal protein S7	21,909	3; 1	86		Ribosomal
At1g58380	40S Ribosomal protein S2	30,950	6; 1	81	At1g58684; At1g58983; At1g59359; At3g57490; At2g41840	Ribosomal
At5g20980	Methionine synthase	90,993	2; 2	68		Contaminant
At1g79330	Metacaspase 7	45,217	1; 1	68	At1g79340	Contaminant?
At4g31700	Ribosomal protein S6	28,578	3; 1	68	At5g10360	Ribosomal
At1g65960	Glutamate decarboxylase	56,504	9; 5	68		Contaminant?
At2g360 70	Mitochondrial translocator TIM44-2	53,784	4; 1	66		–
At1g67430	60S Ribosomal protein L17-2	20,013	2; 1	63	At1g27400	Ribosomal
At1g22360	UDP-Glucose glucosyltransferase	55,014	1; 1	60		Contaminant?
At3g09680	40S Ribosomal protein S23-1	15,931	1; 1	52	At5g02960	Ribosomal
At5g15520	40S Ribosomal protein S19-2	15,864	9; 1	51	At5g61170	Ribosomal
At1g72150	SEC 14 Phosphogltransfer	64,121	1; 1	44		Contaminant?

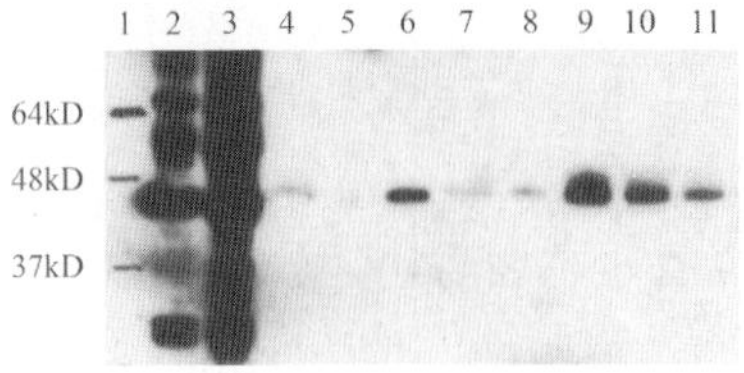

Fig. 16.2 Purification steps of p15h2-NTAPi and interacting proteins for proteomic analysis. Western blot with rabbit anti-CBP primary antibody and anti-rabbit secondary antibody. Lanes: *1* Protein relative molecular mass (Mr) markers, *2* initial cell extract, *3* 10x extract after IgG bead adsorption, *4* wash from IgG beads, *5* tobacco etch-virus (TEV) wash, *6* TEV eluate, *7* wash through from calmodulin column, *8–11* fractions eluted from calmodulin column

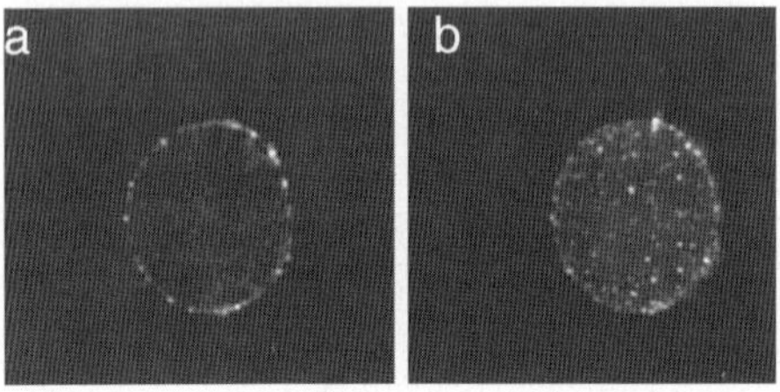

Fig. 16.3 Localisation of p15h2 by green fluorescent protein (GFP) fusion. Numerous small foci at the nuclear periphery are labelled, consistent with localisation to the nuclear pores and with the function of p15 in nuclear export. **a** A single optical section showing the foci labelled at the nuclear periphery. **b** 3D projection from the entire focal section stack through the nucleus shown in **a**

protein precipitation and tryptic digestion of the pellet, instead of separating proteins on a gel and cutting out bands with subsequent digestion. This modification of the procedure avoided losses of yield at the stage of gel separation and extraction of peptides from the gel, which can be 50% or more. It also minimised the time required for MS analysis, as 39 proteins (Table 16.1) were identified simultaneously during a single 1 h LC run followed by nano ESI Q-TOF analysis. Several fractions were set aside from the affinity purification steps for subsequent Western blot analysis, which showed the molecular mass for the fusion protein band to be much higher than expected for this small 14 kDa protein (Fig 16.2). This suggested that the protein was isolated in its homo- or hetero-dimeric form. Another nuclear transport factor, p15h3 (At1g27310), was present among the copurified proteins (Table 16.1), and so it is likely that these proteins form a stable hetero-dimer or higher order complex. Half of the identified proteins (Table 16.1) were components of the large and small subunits of the ribosome, which would be expected for a nuclear transport factor. We expressed a GFP fusion of p15h2 ORF in Arabidopsis suspension culture and observed a striking pattern of localisation on the surface of nuclear envelope, which probably corresponds to nuclear pores with which p15h2 is associated (Fig 16.3).

From our experience with several other TAP-tagged proteins, the most essential factors for the successful isolation of protein complexes are the abundance and stability of the complex, a conclusion also drawn by Rohila et al (2006).

16.8 Reliability of Proteomic Data

Proteomic analysis is an exciting new technique that can provide large data sets in short periods of time. However, when data from different large-scale projects are compared, certain groups of proteins appear to be ubiquitous despite pre-proteomic sample fractionation techniques. So an important challenge is how to set up criteria to distinguish between true positive identifications and false positives. A consortium has been established to develop general criteria for the publication and exchange of MS data and database search results (Kaiser 2002), and guidelines have been developed to assist researchers in the publication of protein identification from MS data (Carr et al. 2004). A method to estimate the rate of false positive identifications using a reverse database (Peng et al. 2003) has provided a technique for reporting the stringency of the search parameters (Cargile et al. 2004). This approach was applied to the analysis of nucleolar proteins in the most recent publication of the nucleolar proteome (Andersen et al. 2005). The criterion used in this study was set up as at least two matching peptides per protein, a mass accuracy within 3 ppm, a MASCOT score for individual peptides of more than 20, and a delta score of more than 5. The threshold of statistical significance in MASCOT searches is a generally accepted criterion for protein identifications. However, researchers working with unsequenced or incompletely sequenced genomes have to perform searches against expressed sequence tag (EST) databases and sometimes have to consider single peptide matches.

Although some aberrant proteins can be distinguished by the use of replicates, this is insufficient to deal with 'persistent contaminants' – high abundance proteins, or those with a tendency to aggregate with nucleoli during extraction (cytoplasmic metabolic enzymes seem especially prone to this). In fact, as far as proteomic analysis is concerned, these proteins are genuine components of the starting preparation and it is only from the cell biological point of view that they are contaminants. Proteomic analysis is unlikely to resolve this problem, although comparison with other studies and databases of common contaminants can help to identify suspect proteins.

The best solution is to use a completely different technique to confirm the sub-cellular location of proteins identified. The most direct technique is visualisation of cellular location by fluorescent tagging or immunofluorescence, either by sampling a small sample of proteins whose localisation is unconfirmed (Andersen et al. 2002) or by the use of high-throughput approaches such as the Gateway system to analyse a large proportion of the proteins identified (Pendle et al. 2005). This also allows a distinction to be drawn between proteins that are largely nucleolar, and those that are also found in other cellular compartments.

16.9 Data Analysis and Databases

The great advantage of high-throughput techniques is that single sets of experiments produce large volumes of data, but this can be a mixed blessing, especially in the absence of generally agreed strategies for ensuring data quality. Hence, the utility of proteomic data to the scientific community depends on curation of data into easily navigable databases, and the use of regularly updated formats that allow an assessment of the biological relevance of any given protein or proteins within a proteome.

Following the identification of the proteins from Arabidopsis nucleoli, the results were classified by probable protein function and placed online at the Arabidopsis Nucleolar Protein Database (http://bioinf.scri.sari.ac.uk/cgi-bin/atnopdb/home; Brown et al. 2005). Potential human and yeast orthologues of the Arabidopsis nucleolar proteins were identified by reciprocal BLAST search, and comparisons with the human nucleolar proteome were performed. A database of the human nucleolar proteome is also available, and is sufficiently well characterised for use as a test system for novel MS techniques (Vollmer et al. 2006). Over 700 proteins are listed, 500 dynamically characterised, with orthologues annotated from yeast, Drosophila and *Caenorhabditis elegans* (http://www.lamondlab.com/NOPdb/; Leung et al. 2003). Both databases are cross-referenced to the PubMed systems and the Lamond NOPdb database also includes raw data, i.e. sequenced peptides from tandem MS/MS. As Coute et al. (2006) note, the data on the basis of which such identifications were made may need to be provided to ensure valid comparisons between different experiments or systems (such as the full list of peptides made available following recent analysis of the human acetylproteome; Kim et al. 2006) in a manner analogous to the 'MIAME' standards (http://www.mged.org/Workgroups/MIAME/miame_1.1.html), which allow direct comparisons between microarray data.

Proteomic analysis of isolated organelles cannot determine whether a protein is specific to the organelle in question, or whether it is also located in other parts of the cell. Neither can it determine the distribution of that protein within the organelle. All of this information provides important guides to function. For these reasons, and also to assess the rate of false positives, about half of the proteins identified in our original Arabidopsis nucleolar proteome (those for which we could obtain ORFs at the time) were transiently expressed as GFP-fusions (Pendle et al. 2005). These data are also provided in the Atnopdb database, including specificity for sub-nucleolar compartments and presence in other parts of the nucleus and cytoplasm. Attempts to understand the complexities of protein localisation could be aided by the use of databases that compile and compare localisation data from different sources. Unfortunately, neither the cross-species Organelle DB (http://organelledb.lsi.umich.edu; Wiwatwattana and Kumar 2005) nor the Arabidopsis-specific GFP/MS-based "sub-cellular location database for Arabidopsis proteins (SUBA)" at www.suba.bcs.uwa.edu.au (Heazlewood and Millar 2003) or GFP-based database at http://aztec.standford.edu/gfp/ (Li et al. 2006)) specify sub-nuclear regions.

Even after reciprocal BLAST search, 37 Arabidopsis nucleolar proteins are of unknown function due to the lack of characterised orthologues. The presence of protein domains of known function within such proteins can help both in suggesting potential functions and in the broader study of the nucleolus in evolution. For example, an analysis of protein domains of the human nucleolar proteome concluded that the core functions of the nucleolus (i.e. those connected to ribosome biosynthesis) were of archaeal origin, but that the many eukaryote-specific domains suggested that the nucleolus had undergone a massive subsequent enlargement, driving the evolution of, for example, RNA helicases (Staub et al. 2004). Accordingly, the human proteome database is searchable by protein domain, as well as by gene ontology.

In the examples cited above, genome sequences can be used to identify the proteins detected by MS. This is not possible in organisms with as yet unsequenced genomes, such as wheat, and alternative strategies will be needed to present MS data in a useable manner. Full use will have to be made of EST databases, and it will be necessary to ensure that such libraries are as amenable to automated searching by MASCOT or related software as are genome databases. BLAST search against the genomes of sequenced relatives such as rice will also be important. Although considered the 'model grass', rice is only distantly related to wheat (diverging 130–240 million years ago; Crane et al. 1995), and the proposed sequencing of the Brachypodium genome may be important for searching for homologues of wheat proteins within a more closely related species.

16.10 Future Perspectives

The presumed goal of proteomics is to catalogue all the different proteins that can be expressed in a given organism along with all the PTMs of each protein, to quantify the expression and modification levels and to determine the way these levels change in different cell types and cell stages. Cell biologists would add to this the determination of the sub-cellular and sub-organellar location and the dynamics of all the cell's proteins. We are a long way from achieving these ambitious goals, if indeed they are achievable or if they are still considered worth achieving when they become technically possible. One major problem is the sheer number of possibilities; most mammalian genes are subject to alternative splicing, and alternative splicing is likely to be much more widespread in plants than currently expected. Each resulting polypeptide is then potentially subject to hundreds of PTMs. In addition to the enormous number of possible molecular species arising from each gene, the dynamic range in concentrations of proteins is also enormous, ranging from as low as one or two molecules to many billions per cell. This is currently one of the most significant limitations in proteomics technology; low abundance species are simply swamped by the overpowering concentrations of common proteins.

Nevertheless, huge improvements have been made recently both in the experimental methods and in the instrumental flexibility and sensitivity for proteomics. Each new generation of MS machines has greater sensitivity, speed and throughput. The result is that current technologies are proving highly productive for samples of medium complexity, such as nucleoli, which contain a few hundred different proteins. In many cases, these 'partial proteomes' seem more accessible than total cellular proteomes and, at least in the short term, the information that this type of analysis can provide seems more useful. Many practical problems of displaying, interpreting and curating the information remain, as well as issues of reproducibility and control of artefacts, but these technical problems are likely to be alleviated as the technology matures. As with other types of large-scale bioinformatics data, accessibility and interoperability of different types of data is still a problem. In our experience, the power of proteomics analysis has to be seen in the context of the complementary cell biology techniques for localisation of the proteins identified. Proteomics will clearly be most powerful when closely integrated with other techniques in multi-disciplinary approaches to biology.

Acknowledgements We are grateful to Andrew Bottrill and Mike Naldrett (John Innes Centre) for their help with proteomic analysis. Financial support for this research was provided by the UK Biotechnology and Biological Sciences Research Council (BBSRC).

References

Aebersold R, Mann M (2003) Mass spectrometry-based proteomics. Nature 422:198–207

Amanchy R, Kalume DE, Iwahori A, Zhong J, Pandey A (2005) Phosphoproteome analysis of HeLa cells using stable isotope labeling with amino acids in cell culture (SILAC). J Proteome Res 4:1661–1671

Andersen JS, Lyon CE, Fox AH, Leung AK, Lam YW, Steen H, Mann M, Lamond AI (2002) Directed proteomic analysis of the human nucleolus. Curr Biol 12:1–11

Andersen JS, Lam YW, Leung AK, Ong SE, Lyon CE, Lamond AI, Mann M (2005) Nucleolar proteome dynamics. Nature 433:77–83

Bae MS, Cho EJ, Choi EY, Park OK (2003) Analysis of the Arabidopsis nuclear proteome and its response to cold stress. Plant J 36:652–663

Blagoev B, Ong SE, Kratchmarova I, Mann M (2004) Temporal analysis of phosphotyrosine-dependent signaling networks by quantitative proteomics. Nat Biotechnol 22:1139–1145

Bodnar WM, Blackburn RK, Krise JM, Moseley MA (2003) Exploiting the complementary nature of LC/MALDI/MS/MS and LC/ESI/MS/MS for increased proteome coverage. J Am Soc Mass Spectrom 14:971–979

Brown JW, Shaw PJ, Shaw P, Marshall DF (2005) Arabidopsis nucleolar protein database (AtNoPDB). Nucleic Acids Res 33:D633–636

Cargile BJ, Bundy JL, Stephenson JL Jr (2004) Potential for false positive identifications from large databases through tandem mass spectrometry. J Proteome Res 3:1082–1085

Carr S, Aebersold R, Baldwin M, Burlingame A, Clauser K, Nesvizhskii A (2004) The need for guidelines in publication of peptide and protein identification data: Working Group on Publication Guidelines for Peptide and Protein Identification Data. Mol Cell Proteomics 3:531–533

Chen SL, Huddleston MJ, Shou WY, Deshaies RJ, Annan RS, Carr SA (2002) Mass spectrometry-based methods for phosphorylation site mapping of hyperphosphorylated proteins applied to Net1, a regulator of exit from mitosis in yeast. Mol Cell Proteomics 1:186–196

Conrad-Webb H, Butow RA (1995) A polymerase switch in the synthesis of ribosomal-RNA in *Saccharomyces cerevisiae*. Mol Cell Biol 15:2420–2428

Coute Y, Burgess JA, Diaz JJ, Chichester C, Lisacek F, Greco A, Sanchez JC (2006) Deciphering the human nucleolar proteome. Mass Spectrom Rev 25:215–234

Crane PR, Friis EM, Pedersen KR (1995) The origin and early diversification of angiosperms. Nature 374:27–33

Cutler SR, Ehrhardt DW, Griffitts JS, Somerville CR (2000) Random GFP::cDNA fusions enable visualization of subcellular structures in cells of Arabidopsis at a high frequency. Proc Natl Acad Sci USA 97:3718–3723

Dunkley TP, Dupree P, Watson RB, Lilley KS (2004a) The use of isotope-coded affinity tags (ICAT) to study organelle proteomes in *Arabidopsis thaliana*. Biochem Soc Trans 32:520–523

Dunkley TP, Watson R, Griffin JL, Dupree P, Lilley KS (2004b) Localization of organelle proteins by isotope tagging (LOPIT). Mol Cell Proteomics 3:1128–1134

Earley K, Lawrence RJ, Pontes O, Reuther R, Enciso AJ, Silva M, Neves N, Gross M, Viegas W, Pikaard CS (2006) Erasure of histone acetylation by Arabidopsis HDA6 mediates large-scale gene silencing in nucleolar dominance. Genes Dev 20:1283–1293

Elortza F, Nuhse TS, Foster LJ, Stensballe A, Peck SC, Jensen ON (2003) Proteomic analysis of glycosylphosphatidylinositol-anchored membrane proteins. Mol Cell Proteomics 2:1261–1270

Escobar NM, Haupt S, Thow G, Boevink P, Chapman S, Oparka K (2003) High-throughput viral expression of cDNA-green fluorescent protein fusions reveals novel subcellular addresses and identifies unique proteins that interact with plasmodesmata. Plant Cell 15:1507–1523

Fox AH, Lam YW, Leung AKL, Lyon CE, Andersen J, Mann M, Lamond AI (2002) Paraspeckles: a novel nuclear domain. Curr Biol 12:13–25

Fujiwara T, Suzuki S, Kanno M, Sugiyama H, Takahashi H, Tanaka J (2006) Mapping a nucleolar targeting sequence of an RNA binding nucleolar protein, Nop25. Exp Cell Res 312:1703–1712

Gruhler A, Schulze WX, Matthiesen R, Mann M, Jensen ON (2005) Stable isotope labeling of *Arabidopsis thaliana* cells and quantitative proteomics by mass spectrometry. Mol Cell Proteomics 4:1697–1709

Gygi SP, Rist B, Gerber SA, Turecek F, Gelb MH, Aebersold R (1999) Quantitative analysis of complex protein mixtures using isotope-coded affinity tags. Nat Biotechnol 17:994–999

Haupt S, Stroganova T, Ryabov E, Kim SH, Fraser G, Duncan G, Mayo MA, Barker H, Taliansky M (2005) Nucleolar localization of potato leafroll virus capsid proteins. J Gen Virol 86:2891–2896

Heazlewood JL, Millar AH (2003) Integrated plant proteomics – putting the green genomes to work. Funct Plant Biol 30:471–482

Hernandez-Verdun D, Roussel P, Gebrane-Younes J (2002) Emerging concepts of nucleolar assembly. J Cell Sci 115:2265–2270

Hiscox JA (2002) The nucleolus – a gateway to viral infection? Arch Virol 147:1077–1089

Hwang SI, Lundgren DH, Mayya V, Rezaul K, Cowan AE, Eng JK, Han DK (2006) Systematic characterization of nuclear proteome during apoptosis: a quantitative proteomic study by differential extraction and stable isotope labeling. Mol Cell Proteomics 5:1131–1145

Kaiser J (2002) Proteomics. Public-private group maps out initiatives. Science 296:827

Khan MM, Komatsu S (2004) Rice proteomics: recent developments and analysis of nuclear proteins. Phytochemistry 65:1671–1681

Kim SC, Sprun R, Chen Y, Xu YD, Ball H, Pei JM, Cheng TL, Kho Y, Xiao H, Xiao L, Grishin NV, White M, Yang XJ, Zhao YM (2006) Substrate and functional diversity of lysine acetylation revealed by a proteomics survey. Mol Cell 23:607–618

Kim SH, Ryabov EV, Brown JWS, Taliansky M (2004) Involvement of the nucleolus in plant virus systemic infection. Biochem Soc Trans 32:557–560

Koroleva OA, Tomlinson ML, Leader D, Shaw P, Doonan JH (2005) High-throughput protein localization in Arabidopsis using Agrobacterium-mediated transient expression of GFP-ORF fusions. Plant J 41:162–174

Lapeyre B, Amalric F, Ghaffari SH, Rao SVV, Dumbar TS, Olson MOJ (1986) Protein and cDNA sequence of a glycine-rich, dimethylarginine-containing region located near the corboxyl-terminal end of nucleolin (C23 and 100 kDa). J Biol Chem 261:9167–9173

Leung AKL, Andersen JS, Mann M, Lamond AI (2003) Bioinformatic analysis of the nucleolus. Biochem J 376:553–569

Li J, Steen H, Gygi SP (2003) Protein profiling with cleavable isotope-coded affinity tag (cICAT) reagents: the yeast salinity stress response. Mol Cell Proteomics 2:1198–1204

Li SJ, Ehrhardt DW, Rhee SY (2006) Systematic analysis of Arabidopsis organelles and a protein localization database for facilitating fluorescent tagging of full-length Arabidopsis proteins. Plant Physiol 141:527–539

Lilley KS, Dupree P (2006) Methods of quantitative proteomics and their application to plant organelle characterization. J Exp Bot 57:1493–1499

Maggi LB, Weber JD (2005) Nucleolar adaptation in human cancer. Cancer Invest 23:599–608

Matsuzaki M, Misumi O, Shin-I T, Maruyama S, Takahara M, Miyagishima SY, Mori T, Nishida K, Yagisawa F, Yoshida Y, Nishimura Y, Nakao S, Kobayashi T, Momoyama Y, Higashiyama T, Minoda A, Sano M, Nomoto H, Oishi K, Hayashi H, Ohta F, Nishizaka S, Haga S, Miura S, Morishita T, Kabeya Y, Terasawa K, Suzuki Y, Ishii Y, Asakawa S, Takano H, Ohta N, Kuroiwa H, Tanaka K, Shimizu N, Sugano S, Sato N, Nozaki H, Ogasawara N, Kohara Y, Kuroiwa T (2004) Genome sequence of the ultrasmall unicellular red alga *Cyanidioschyzon merolae* 10D. Nature 428:653–657

Mayer C, Grummt I (2005) Cellular stress and nucleolar function. Cell Cycle 4:1036–1038

Melese T, Xue Z (1995) The nucleolus – an organelle formed by the act of building a ribosome. Curr Opin Cell Biol 7:319–324

Mo YY, Yu YN, Shen ZY, Beck WT (2002) Nucleolar delocalization of human topoisomerase I in response to topotecan correlates with sumoylation of the protein. J Biol Chem 277:2958–2964

Najbauer J, Johnson BA, Young AL, Aswad DW (1993) Peptides with sequences similar to glycine, arginine-rich motifs in proteins interacting with RNA are efficiently recognized by methyltransferase(s) modifying arginine in numerous proteins. J Biol Chem 268:10501–10509

Nightingale KP, O'Neill LP, Turner BM (2006) Histone modifications: signalling receptors and potential elements of a heritable genetic code. Curr Opin Genet Dev 16:125–136

Olsen JV, Mann M (2004) Improved peptide identification in proteomics by two consecutive stages of mass spectrometric fragmentation. Proc Natl Acad Sci USA 101:13417–13422

Olson MOJ, Dundr M (2005) The moving parts of the nucleolus. Histochem Cell Biol 123:203–216

Omer AD, Lowe TM, Russell AG, Ebhardt H, Eddy SR, Dennis PP (2000) Homologs of small nucleolar RNAs in archaea. Science 288:517–522

Ong SE, Blagoev B, Kratchmarova I, Kristensen DB, Steen H, Pandey A, Mann M (2002) Stable isotope labeling by amino acids in cell culture, SILAC, as a simple and accurate approach to expression proteomics. Mol Cell Proteomics 1:376–386

Panse VG, Kressler D, Pauli A, Petfalski E, Gnadig M, Tollervey D, Hurt E (2006) Formation and nuclear export of preribosomes are functionally linker to the small-ubiquitin-related modifier pathway. Traffic 7:1311–1321

Pendle AF, Clark GP, Boon R, Lewandowska D, Lam YW, Andersen J, Mann M, Lamond AI, Brown JW, Shaw PJ (2005) Proteomic analysis of the Arabidopsis nucleolus suggests novel nucleolar functions. Mol Biol Cell 16:260–269

Peng J, Elias JE, Thoreen CC, Licklider LJ, Gygi SP (2003) Evaluation of multidimensional chromatography coupled with tandem mass spectrometry (LC/LC-MS/MS) for large-scale protein analysis: the yeast proteome. J Proteome Res 2:43–50

Phair RD, Misteli T (2000) High mobility of proteins in the mammalian cell nucleus. Nature 404:604–609

Pontes O, Li CF, Nunes PC, Haag J, Ream T, Vitins A, Jacobsen SE, Pikaard CS (2006) The Arabidopsis chromatin-modifying nuclear siRNA pathway involves a nucleolar RNA processing center. Cell 126:79–92

Raska I, Shaw PJ, Cmarko D (2006) Structure and function of the nucleolus in the spotlight. Curr Opin Cell Biol 18:325–334

Reddy VS, Ali GS, Reddy AS (2002) Genes encoding calmodulin-binding proteins in the Arabidopsis genome. J Biol Chem 277:9840–9852

Reed ML, Dove BK, Jackson RM, Collins R, Brooks G, Hiscox JA (2006) Delineation and modelling of a nucleolar retention signal in the coronavirus nucleocapsid protein. Traffic 7:833–848

Rigaut G, Shevchenko A, Rutz B, Wilm M, Mann M, Seraphin B (1999) A generic protein purification method for protein complex characterization and proteome exploration. Nat Biotechnol 17:1030–1032

Rohila JS, Chen M, Cerny R, Fromm ME (2004) Improved tandem affinity purification tag and methods for isolation of protein heterocomplexes from plants. Plant J 38:172–181

Rohila JS, Chen M, Chen S, Chen J, Cerny R, Dardick C, Canlas P, Xu X, Gribskov M, Kanrar S, Zhu JK, Ronald P, Fromm ME (2006) Protein–protein interactions of tandem affinity purification-tagged protein kinases in rice. Plant J 46:1–13

Rubbi CP, Milner J (2003) Disruption of the nucleolus mediates stabilization of p53 in response to DNA damage and other stresses. EMBO J 22:6068–6077

Rubio V, Shen Y, Saijo Y, Liu Y, Gusmaroli G, Dinesh-Kumar SP, Deng XW (2005) An alternative tandem affinity purification strategy applied to Arabidopsis protein complex isolation. Plant J 41:767–778

Scherl A, Coute Y, Deon C, Calle A, Kindbeiter K, Sanchez JC, Greco A, Hochstrasser D, Diaz JJ (2002) Functional proteomic analysis of human nucleolus. Mol Biol Cell 13:4100–4109

Shadforth IP, Dunkley TP, Lilley KS, Bessant C (2005) i-Tracker: for quantitative proteomics using iTRAQ. BMC Genomics 6:145

Simpson JC, Wellenreuther R, Poustka A, Pepperkok R, Wiemann S (2000) Systematic subcellular localization of novel proteins identified by large-scale cDNA sequencing. EMBO Rep 1:287–292

Song ZY, Wu M (2005) Identification of a novel nucleolar localization signal and degradation signal in Survivin-deltaEx3: a potential link between nucleolus and degradation. Oncogene 24:2723–2734

Staub E, Fiziev P, Rosenthal A, Hinzmann B (2004) Insights into the evolution of the nucleolus by an analysis of its protein domain repertoire. Bioessays 26:567–581

Sutherland HGE, Mumford GK, Newton K, Ford LV, Farrall R, Dellaire G, Caceres JF, Bickmore WA (2001) Large-scale identification of mammalian proteins localized to nuclear sub-compartments. Human Mol Genet 10:1995–2011

Tian GW, Mohanty A, Chary SN, Li S, Paap B, Drakakaki G, Kopec CD, Li J, Ehrhardt D, Jackson D, Rhee SY, Raikhel NV, Citovsky V (2004) High-throughput fluorescent tagging of full-length Arabidopsis gene products in planta. Plant Physiol 135:25–38

Unlu M, Morgan ME, Minden JS (1997) Difference gel electrophoresis: a single gel method for detecting changes in protein extracts. Electrophoresis 18:2071–2077

Vollmer M, Horth P, Rozing G, Coute Y, Grimm R, Hochstrasser D, Sanchez JC (2006) Multi-dimensional HPLUMS of the nucleolar proteome using HPLC-chip/MS. J Sep Sci 29:499–509

Wiwatwattana N, Kumar A (2005) Organelle DB: a cross-species database of protein localization and function. Nucleic Acids Res 33:D598–D604

Wu WW, Wang G, Baek SJ, Shen RF (2006) Comparative study of three proteomic quantitative methods, DIGE, cICAT, and iTRAQ, using 2D gel- or LC-MALDI TOF/TOF. J Proteome Res 5:651–658

Xu C, Henry PA, Setya A, Henry MF (2003) In vivo analysis of nucleolar proteins modified by the yeast arginine methyltransferase Hmt1/Rmt1p. RNA 9:746–759

Yuan XJ, Zhou YG, Casanova E, Chai MQ, Kiss E, Grone HJ, Schutz G, Grummt I (2005) Genetic inactivation of the transcription factor TIF-I! leads to nucleolar disruption, cell cycle arrest, and p53-mediated apoptosis. Mol Cell 19:77–87

Zieske LR (2006) A perspective on the use of iTRAQ reagent technology for protein complex and profiling studies. J Exp Bot 57:1501–1508

Chapter 17
Pollen and Pollen Tube Proteomics

Tong Chen, Xiaoqin Wu, Yanmei Chen, Nils Böhm, Jinxing Lin, and Jozef Šamaj

Abstract Germinated pollen grains form pollen tubes that accomplish rapid polarized growth within female gametophytic tissues in order to deliver the sperm cell to the ovule. This process is essential for successful plant fertilization in vivo. Pollen tubes are considered to be an ideal model system to study tip growth. A more comprehensive understanding of proteins involved in pollen development, germination and tube elongation is important not only for unraveling the intricate machinery of sexual reproduction in flowering plants, but has considerable potential for the improvement of crop plant production. Integrated proteomic, genomic and cell biological studies offer powerful tools for dissection of the fascinating mechanisms of pollen development and tip-growth regulation in higher plants. Here, we provide a brief overview of current advances in pollen and pollen tube proteomics.

Tong Chen, Key Laboratory of Photosynthesis and Molecular Environmental Physiology, Institute of Botany, The Chinese Academy of Sciences, Beijing 100093, China

Xiaoqin Wu, Key Laboratory of Photosynthesis and Molecular Environmental Physiology, Institute of Botany, The Chinese Academy of Sciences, Beijing 100093, China

Yanmei Chen, Key Laboratory of Photosynthesis and Molecular Environmental Physiology, Institute of Botany, The Chinese Academy of Sciences, Beijing 100093, China

Nils Böhm, Institute of Cellular and Molecular Botany, University of Bonn, Kirschallee 1, 53115 Bonn, Germany

Correspondence author: Jinxing Lin, Key Laboratory of Photosynthesis and Molecular Environmental Physiology, Institute of Botany, The Chinese Academy of Sciences, Xiangshan, Beijing 100093, China, E-mail: linjx@ibcas.ac.cn

Jozef Šamaj, Institute of Cellular and Molecular Botany, University of Bonn, Kirschallee 1, 53115 Bonn, Germany

Jozef Šamaj, Institute of Plant Genetics and Biotechnology, Slovak Academy of Sciences, Akademicka 2, 95007, Nitra, Slovak Republic

J. Šamaj and J. Thelen (eds.), *Plant Proteomics*
© Springer 2007

17.1 Introduction

The pollen tube, formed from the germinated pollen grain, delivers male gametes to the egg apparatus in flowering plants. Therefore, pollen and pollen tubes are essential for sexual reproduction in higher plants (Feijo et al. 2004; Malho et al. 2006). Moreover, pollen tubes represent an ideal model system for the investigation of polarized tip growth in plant cells (Fig. 17.1; Hepler et al. 2001; Šamaj et al. 2006). A comprehensive understanding of pollen germination and plant pollination is not only important in our efforts to unravel the intricate machinery of sexual reproduction in flowering plants, but also has immense potential for the improvement of crop plant production. Pollen biology offers a number of features making pollen an attractive system for the analysis of molecular events involved in pollen development, tube growth and plant pollination. For reproductive biology, it is also important to study pollen as the male gametophyte of flowering plants.

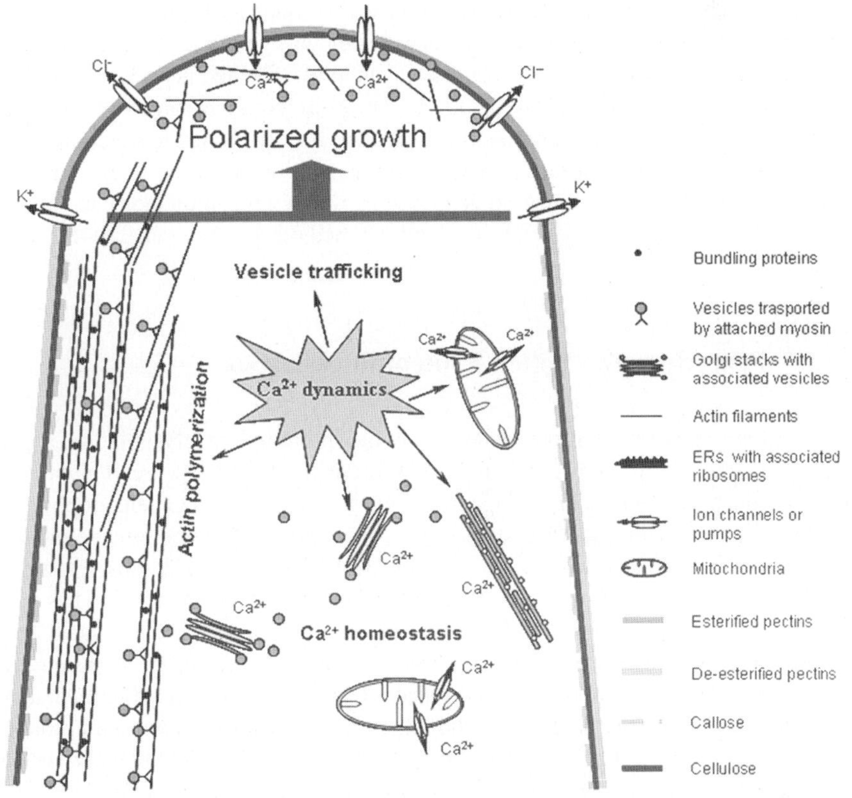

Fig. 17.1 Schematic illustration of pollen tube depicting main factors controlling tip growth. Calcium ions act as a primary regulator of cellular activities (e.g., actin polymerization, morphology and physiology of organelles, vesicle trafficking and fusion, and cell wall remodeling) in the growing pollen tube

Transcriptomic analysis of Arabidopsis pollen has identified some transcripts expressed preferentially in pollen (Honys and Twell 2003). Functional categorization of deduced proteins encoded by mRNAs expressed preferentially in pollen revealed that most are involved in signal transduction, cell wall biosynthesis and cytoskeleton remodeling, thus suggesting the crucial importance of these cellular processes in pollen development and function (Becker et al. 2003; Honys and Twell 2003; Pina et al. 2005). Previous studies, however, demonstrated a lack of a direct correlation between the transcriptional data and protein profile in a given cell or tissue (Greenbaum et al. 2003). The prevailing situation in eukaryotic cells is that one gene codes for several protein isoforms, which arise through alternative splicing and/or post-translational modifications (PTMs). These isoforms might have different physiological functions. Therefore, high-throughput proteomic studies are essential for a better understanding of pollen function.

Genomic sequence information and corresponding protein sequence databases provide the potential to identify proteins by correlating data from mass spectrometry (MS), i.e., masses of intact peptides (peptide mass fingerprinting) or their fragments (tandem mass spectrometry, MS/MS), with in silico sequences from database entries. Protein samples from several model plants with available genomic sequence information have been identified easily by direct matching of collected peptide signals with existing data in publicly available databases. However, conventional protein identification algorithms, such as Mascot and Sequest, are largely limited to exact matching of peptides to existing database sequences, thus hampering proteome characterization in plant species with non-sequenced genomes. Nowadays, this limitation can be overcome by sequence-similarity database searching methods (Liska et al. 2004).

17.2 Procedure for Pollen Proteomic Analysis

The common experimental procedure for pollen proteomic analysis includes gel-based separation of proteins, MS analysis, subsequent database searching and data mining (illustrated in Fig. 17.2). Up until now, two-dimensional gel electrophoresis (2-DE) combined with MS analysis has proved the most potent and reliable method for proteomic investigations (see Chap. 2 by Hurkman and Tanaka, this volume).

Most of the published proteomic work on pollen and pollen tubes so far has been based on total protein samples. Because mature pollen grains are often isolated from pooled pollen samples, protein samples from pollen grains should be considered as "average samples" and not biological replicates (Holmes-Davis et al. 2005). An ideal protocol for total protein preparation should reproducibly resolve almost all proteins in a proteome with a low contamination of other molecules. Additionally, proper purification and enrichment of proteins before sample loading are required for most samples derived from plant material. Due to the rigid nature of the pollen cell wall, efforts should be made to fully crush the solid wall and release the proteins before sample preparation. Total protein precipitation is achieved using TCA and acetone in commonly used protocols, which are based on protein denaturation under acidic and/or hydrophobic conditions to help concentrate proteins and remove contaminants.

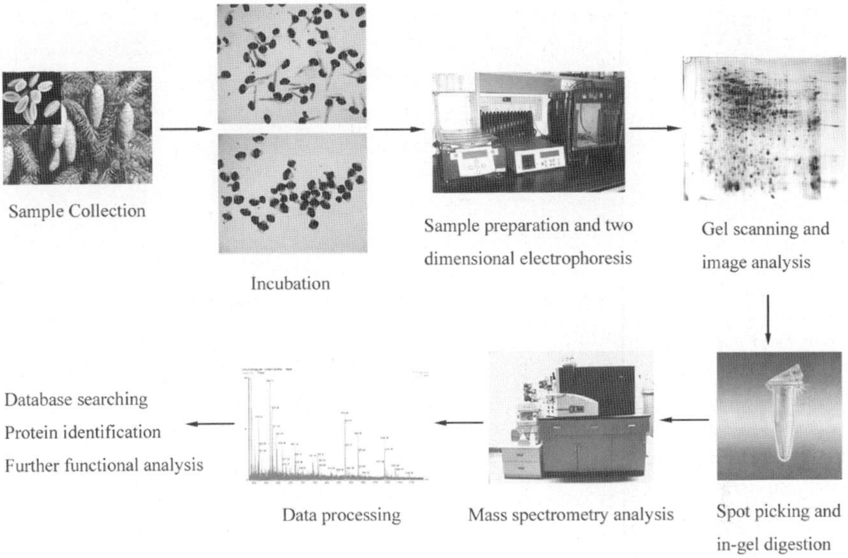

Fig. 17.2 Common procedure for proteomic analysis of pollen/pollen tube development

Most commonly, 2-DE techniques combined with MALDI-TOF (matrix-assisted laser desorption ionization / time of flight) MS or ESI Q-TOF (electrospray ionization quadrupole-TOF) MS/MS have been carried out to investigate proteomes during pollen development within the anther as well as proteomes of mature and germinated pollen in various plant species (Table 17.1). These proteomic studies have significantly improved our knowledge of the regulation of pollen and pollen tube development at the molecular level.

17.3 Proteomics Focused on Pollen Development

17.3.1 Angiosperm Pollen Proteomes

Pollen grains usually have a smaller and overall unique transcriptome (e.g., 6,587 genes are expressed in Arabidopsis pollen) with greater proportions of selectively expressed (11%) and enriched (26%) genes compared to vegetative tissues (Pina et al. 2005). Relative gene ontology (GO) categories in pollen and vegetative tissues revealed a shift of functional classification in the pollen transcriptome toward signaling, vesicle trafficking and cytoskeleton dynamics, suggesting a primary role of these cellular events in pollen germination and tube growth.

Protein synthesis during pollen germination and tube growth has been studied extensively in angiosperm species (Mascarenhas 1993). The complete release of genome information of model plants *Arabidopsis thaliana* and *Oryza sativa* has greatly accelerated the development of plant proteomics. However, proteomic

Table 17.1 Selected proteomic studies performed on male gametogenesis or pollen/pollen tube development. *2-DE* two-dimensional gel electrophoresis, *MALDI* matrix assisted laser desorption/ionisation, *TOF* time-of-flight, *ESI* electrospray ionisation, *Q* quadrupole, *MS* mass spectrometry, *MS/MS* tandem mass spectrometry

Organism	Material	Number of identified proteins	Proteomics methods	Reference
Oryza sativa L. cultivar Doogara	Anther at the young microspore stage	53	2-DE/MALDI-TOF MS	Imin et al. 2001
Arabidopsis thaliana	Pollen coat proteins	12	1-D SDS-PAGE/MS	Mayfield et al. 2001
Oryza sativa L. cultivar Doogara	Anther	40	2-DE/MALDI-TOF MS/ N-terminal sequencing	Kerim et al. 2003
Oryza sativa L. cultivar Limmaeus	Anther at the young microspore stage	70	2-DE/MALDI-TOF MS	Imin et al. 2004
Arabidopsis thaliana	Pollen grain	135	2-DE/ESI Q-TOF MS/MS 2005	Holmes-Davis et al.
Arabidopsis thaliana	Pollen grain	121	2-DE/MALDI-TOF MS/ ESI Q-TOF MS/MS	Noir et al. 2005
Pinus strobus	Pollen/pollen tube	57	2-DE/MALDI-TOF MS	Fernando, 2005
Picea meyeri	Pollen tube	53	2-DE/MALDI-TOF MS/ ESI Q-TOF MS/MS	Chen et al. 2006
Oryza sativa L. ssp japonica	Mature pollen grain	322	2-DE/MALDI-TOF MS/ ESI Q-TOF MS/MS/nano-LC ESI Q-TOF MS/MS	Dai et al. 2006
Arabidopsis thaliana	Pollen grain	110	2-DE/MALDI-TOF MS/ ESI Q-TOF MS/MS/nano-LC ESI Q-TOF MS/MS	Sheoran et al. 2006
Oryza sativa L. cultivars Doogara and HSC55	Anther	37	N-terminal Edman sequencing/ MALDI-TOF-MS and MS/ MS analysis	Imin et al. 2006

analyses focused on sexual reproduction, particularly male gametophyte development, remains scarce in higher plants. Only a limited number of proteomic studies to date have analyzed anthers and pollen during male gametogenesis and/or pollen germination, as summarized in Table 17.1.

17.3.1.1 Rice Pollen Proteomics

Rice (*Oryza sativa* L.) is a model organism for monocotyledonous plants, especially for members of the grass family. Numerous characteristics, such as small genome size, diploid nature, transformability, and establishment of genetic and molecular resources, make it a valuable system for plant biologists studying crop plants.

Imin et al. (2001) used 2-DE protein mapping and MS analysis to investigate the protein expression profile in rice (*O. sativa* L. cv. Doogara) anthers at the young microspore stage. The 53 anther protein spots identified were classified into several functional categories. This study provided the first comprehensive understanding of normal microspore development. Moreover, these techniques can be applied to determine responses to environmental stresses in future experiments. Later, Kerim et al. (2003) investigated changing patterns of protein synthesis during pollen development in rice anthers by means of proteomic analysis. A total of 40 differentially displayed proteins were characterized by MALDI-TOF MS and allocated into different functional categories, such as sugar metabolism, cell elongation and cell expansion. This differential proteomic analysis provided a new insight into the intricate process of anther development because most of the identified proteins were essential for maintenance of pollen activities and subsequent pollen germination. However, the above-mentioned studies did not discriminate among proteins expressed in gametophytic versus sporophytic tissues of the anther. It should be considered that anther sporophytic tissues, such as epidermis, endothecium, middle layers, and tapetum, undergo various developmental changes that are certainly accompanied by corresponding changes at the protein level during development of the male gametophyte (Fernando 2005). Therefore, more focused studies aimed at truly characterizing proteome of the male gametophyte (microspores and pollen) are required.

Recently, Dai et al. (2006) collected fractions of pollen-released and pollen-interior proteins and subsequently separated these samples by 2-DE; 322 proteins were identified by MALDI-TOF MS and/or ESI Q-TOF MS/MS. These proteins were associated with 14 distinct metabolic pathways involved in pollen functions, and most had not previously been reported to be present in the mature pollen of *O. sativa* L. ssp *japonica*. Moreover, 23% of these unique proteins were found to be represented by several isoforms, suggesting that PTM may be an important way to diversify the function of a particular protein in the haploid genome.

Male reproduction in rice is very sensitive to various forms of environmental stress. Thus, 2-DE combined with MALDI-TOF MS and/or MS/MS was also employed to characterize the proteins differentially expressed during cold treatment of anthers at the young microspore stage (Imin et al. 2004, 2006). The cold-responsive proteins identified are potentially involved in protein synthesis and folding, stress responses, lipid biogenesis, cell wall formation, protein degradation and energy metabolism. All of these functions are of crucial importance because their perturbation may lead to the destructive effects of cold treatment. Furthermore, the differentially displayed proteins could be used as molecular markers for breeding in order to improve cold tolerance in rice and possibly also in other crops. Importantly, these proteomic data provide new insights into microspore development and responses to environmental stress at the protein level – a result that could not be achieved by transcriptomic approaches alone.

17.3.1.2 Arabidopsis thaliana Pollen Proteomics

A. thaliana is one of the leading model organisms for genetic and molecular biology studies. Proteomic analysis on this model plant has been performed on different organs, tissues, organelles, and under various developmental stages and environmental conditions (see other chapters in this volume). However, investigations focused on development of the male gametophyte are still scarce in *A. thaliana* due to difficulties with sample collection and purification. First, Mayfield et al. (2001) used cyclohexane to extract proteins from the pollen coat of *A. thaliana* and identified 12 proteins, mainly oleosins and lipases. Later, Holmes-Davis et al. (2005) applied multiple protein extraction techniques followed by 2-DE and ESI-MS/MS, leading to the identification of 135 distinct proteins from a total number of 179 protein spots. Of these 135 proteins, most were involved in metabolism (20%), energy production (17%), or cell structure (12%). The authors quantified the protein spots in these three latter categories and compared normalized protein volumes to their corresponding mRNA signals in the transcriptome. The data revealed the surprising result that highly represented mRNAs corresponded to low-abundant proteins while mRNAs that were less abundant encoded some of the more abundant proteins, thus indicating an inverse relationship between mRNA and protein levels. Overall, the Arabidopsis pollen proteome preferentially contained cell wall and cytoskeletal proteins. These results are similar to those obtained in transcriptomic studies, and support the idea that, in addition to mRNAs, the mature pollen grain contains ready-made proteins that are necessary for germination and rapid pollen tube growth. Additionally, Noir et al. (2005) established a reference map of the *A. thaliana* mature pollen proteome by characterizing 145 proteins using MALDI-TOF MS and LC-MS/MS. Subsequent comparison with pollen transcriptomic data proved these results to be in accordance with the previous study by Holmes-Davis et al. (2005). Thus, it was concluded that a low transcript abundance may be sufficient to ensure accumulation of a significant amount of protein in some instances, for example proteins with a low turnover rate. Alternatively, some polypeptides might have been translated at an earlier stage of pollen development (e.g., in the bicellular stage) and persist longer in the mature pollen. Recently, Sheoran et al. (2006) completed studies on mature Arabidopsis pollen using ecotype Landsberg erecta. They identified 110 distinct proteins from 150 analyzed spots. Among them, 66 new proteins were identified that had not been reported in two previous studies on Arabidopsis. This was explained by the different ecotype used, and by differences in the methods of protein extraction and spot selection. The most abundant proteins were classified into functional categories including energy regulation, defense-related mechanisms, calcium-binding and signaling, cytoskeletal regulation, pollen allergens, glycine-rich proteins and late embryogenesis abundant proteins. Interestingly, some of them were similar to proteins found in the seed.

17.3.2 Proteomic Analysis of Gymnosperm Pollen and Pollen Tube Development

Studies on gymnosperm pollen tube development lag behind those in angiosperms. Only two proteomics reports on gymnosperm pollens and pollen tubes are available so far (Fernando 2005; Chen et al. 2006). Interestingly, the pollen tubes of gymnosperms differ from their angiosperm counterparts in several parameters, such as relatively slow rate and extended period of growth, extremely delayed gametogenesis, special characteristics of cell wall construction, and different control of cytoskeletal components (Fernando et al. 2005). These important differences point to the major evolutionary divergence in the development of male gametophytes in flowering plants, and to the primary importance of tip growth machinery for gymnosperm pollen tube growth and successful fertilization (Fernando et al. 2005; Wang et al. 2005; Chen et al. 2006).

Fortunately, gymnosperm species produce large amounts of pollen grains that can be collected free of contamination. Thus, they represent an ideal source of material for proteomic experiments. The protein expression profiles of ungerminated pollen and 2-day-old germinated pollen tubes of *Pinus strobus* were compared in order to better understand the changes in protein expression associated with pollen tube development (Fernando 2005). Proteomic analysis revealed that, overall, 645 and 647 protein spots were reproducibly resolved from pollen grains and pollen tubes, respectively. Among them, 57 differentially expressed proteins from mature pollen versus pollen tubes were identified by MALDI-TOF MS and categorized according to their cellular activities into the following groups: metabolism, stress/defense response, gene regulation, and cell wall formation. This study represents the first report describing a global analysis of differentially expressed proteins in the pollen tube versus pollen grain in seed plants, thus providing an important insight into the molecular basis that separates the development of pollen tubes from that of pollen grains.

Sophisticated signaling pathways are thought to interact with each other and to regulate pollen germination and subsequent tube elongation (Feijo et al. 2004; Malho et al. 2006). Tip-growing pollen tubes are particularly amenable to investigations focused on signals that are responsible for mediating changes to the actin cytoskeleton (Cheung and Wu 2004; Chen et al. 2006, 2007; Šamaj et al. 2006).

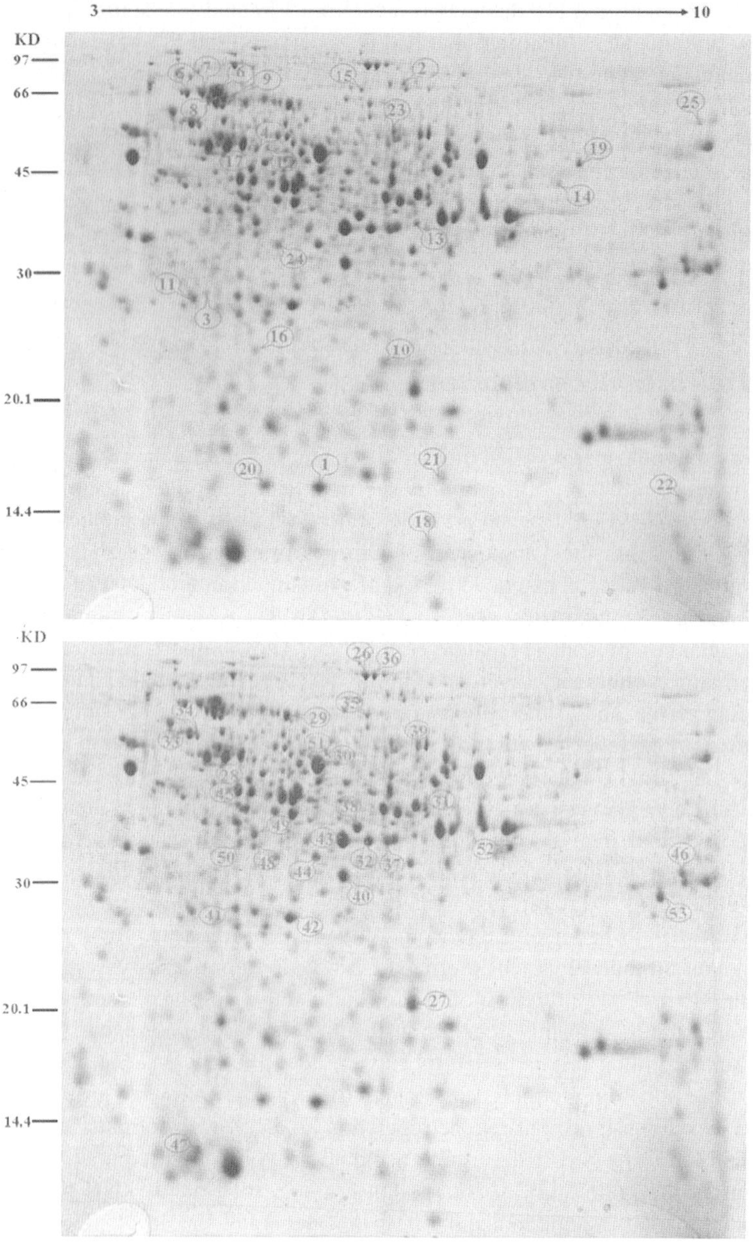

Fig. 17.3 Representative two-dimensional gel electrophoresis (2-DE) gels of pollen proteins. Pollen was treated with 20nM latrunculin B for 6h. Differentially expressed protein spots are indicated by *arrows*. *Upper image* Up-regulated protein spots ($n=25$); *lower image* down-regulated spots ($n=28$). Reproduced from Chen et al. (2006) with permission from Blackwell Publishing

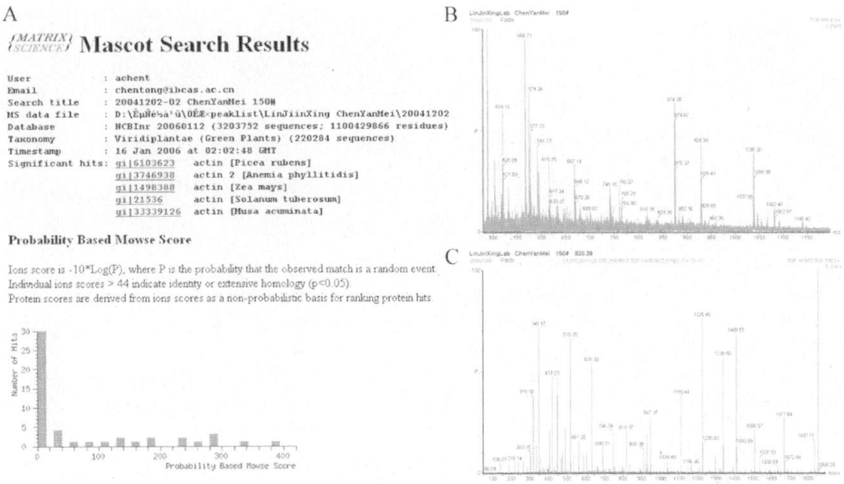

Fig. 17.4 Down-regulation of actin expression following latrunculin B treatment as identified by electrospray ionisation quadrupole time-of-flight tandem mass spectrometry (ESI Q-TOF MS/MS) and MASCOT search. **a** Result of Mascot search for one of the examined protein spots showing a positive match with actin in the database. **b** Representative MS spectrum of actin identified by nano-ESI Q-TOF MS/MS. **c** Picture showing one of the peaks with double charge (*m/z* 928.39)

It is widely accepted that the actin cytoskeleton helps to guide delivery of secretory vesicles to the pollen tube apex, but the exact mechanism remains enigmatic (Cheung and Wu 2004; Šamaj et al. 2006). Pharmacological treatments combined with proteomic and cytological analyses represent an efficient way of studying signaling pathways in plants. In order to investigate the role of the actin cytoskeleton in the pollen of *Picea meyeri*, a proteomic approach was used to analyze changes in protein expression profiles during pollen germination and subsequent development of pollen tubes upon actin disruption via latrunculin B (Chen et al. 2006). A total of 53 differentially expressed proteins were identified by means of 2-DE and ESI MS/MS at different time points and varying doses of latrunculin B (Fig. 17.3). These proteins were grouped into distinct functional categories including signaling, actin cytoskeleton, cell expansion and carbohydrate metabolism (Fig. 17.4). Moreover, pronounced changes in the morphology of Golgi stacks, mitochondria and amyloplasts were recorded (Fig. 17.5), along with differential expression of proteins involved in their functions. This report provides the first insight into the multiple effects of actin cytoskeleton disruption on global protein patterns and pollen tube cytoarchitecture in flowering plants.

17.4 Conclusions and Future Perspectives

The study of complex biological questions through comparative proteomics is becoming increasingly attractive to plant biologists as the rapidly expanding plant genomic and expressed sequence tag (EST) databases provide better opportunities

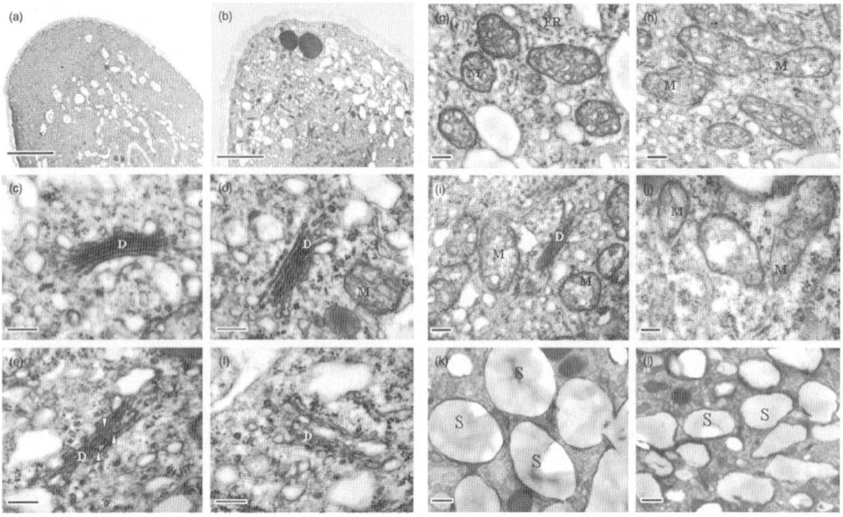

Fig. 17.5 Electron micrographs of control (**a**, **c**, **g**, **k**) and latrunculin B-treated (**b**, **d**–**f**, **h**–**j**, **l**) pollen tubes of *Picea meyeri*. Cytoplasmic organization and organelle morphology both undergo drastic changes following latrunculin B treatment (**a**, **b**). Obvious disruption of Golgi stacks (**c**–**f**; *arrowheads* in **e**) and mitochondria (**g**–**j**) can be observed. Synthesis of starch greatly decreases after inhibitor treatment (**k**, **l**). All of these data indicate that the actin cytoskeleton is closely related to both metabolic activity and energy production in pollen tubes. *ER* Endoplasmic reticulum, *M* mitochondria, *D* Golgi stacks, *S* starch granules. *Bars* **a**, **b** 5 µm; **c**–**j** 0.2 µm; **k**, **l** 1 µm. Reproduced from Chen et al. (2006) with permission from Blackwell Publishing

for precise protein identification. Thus, proteomics has proved a reliable and efficient method of complementing genomic tools in recent years.

Transcriptomic analysis and cytogenetic investigations have demonstrated that higher plants make a significant investment in gametophyte production. This is not surprising considering the fact that specialized structures such as pollen tubes, along with specialized molecular machinery regulating their rapid tip growth and cell–cell communication, have evolved during pollen–pistil interactions. Pollen tubes allow survival of the gametophyte free of the sporophyte, and the efficient delivery of sperm to the embryo sac. Angiosperm species are at the forefront of current proteomic research. On the other hand, conifer pollen tubes also represent an important, yet underdeveloped, experimental system in plant biology. Moreover, the first two proteomics studies on gymnosperm pollen have already contributed to a better understanding of pollen tube development and actin function in tip growth.

In the near future, proteomic investigations supported by genomic and genetic tools (mutants and transgenic technology), and by integrated cell biological approaches, will surely dissect new fascinating mechanisms of pollen development and pollen tube tip-growth in higher plants.

Acknowledgements This work was supported by the National Key Basic Research Program (2006CB910606) from MOST, by a general programme grant (30570100), and by a research grant from the Deutsche Forschungsgemeinschaft (DFG) to J.Š. (SA 1564/2-1).

References

Becker JD, Boavida LC, Carneiro J, Haury M, Feijo JA (2003) Transcriptional profiling of Arabidopsis tissues reveals the unique characteristics of the pollen transcriptome. Plant Physiol 133:713–725

Chen Y, Chen T, Shen S, Zheng M, Guo Y, Lin J, Baluška F, Šamaj J (2006) Differential display proteomic analysis of *Picea meyeri* pollen germination and pollen-tube growth after inhibition of actin polymerization by latrunculin B. Plant J 47:174–195

Chen T, Teng N, Wu X, Wang Y, Tang W, Šamaj J, Baluška F, Lin J (2007) Disruption of actin filaments by latrunculin B affects cell wall construction in *Picea meyeri* pollen tube by disturbing vesicle trafficking. Plant Cell Physiol 48:19–30

Cheung AY, Wu HM (2004) Overexpression of an Arabidopsis formin stimulates supernumerary actin cable formation from pollen tube cell membrane. Plant Cell 16:257–269

Dai S, Li L, Chen T, Chong K, Xue Y, Wang T (2006) Proteomic analyses of *Oryza sativa* mature pollen reveal novel proteins associated with pollen germination and tube growth. Proteomics 6:2504–2529

Feijo JA, Costa SS, Prado AM, Becker JD, Certal AC (2004) Signalling by tips. Curr Opin Plant Biol 7:589–598

Fernando DD (2005) Characterization of pollen tube development in *Pinus strobus* (Eastern white pine) through proteomic analysis of differentially expressed proteins. Proteomics 5:4917–4926

Fernando DD, Lazzaro MD, Owens JN (2005) Growth and development of conifer pollen tubes. Sex Plant Reprod 18:149–162

Greenbaum D, Colangelo C, Williams K, Gerstein M (2003) Comparing protein abundance and mRNA expression levels on a genomic scale. Gen Biol 4:117

Hepler PK, Vidali L , Cheung AY (2001) Polarized cell growth in higher plants. Annu Rev Cell Dev Biol 17:159–187

Holmes-Davis R, Tanaka CK, Vensel WH, Hurkman WJ, McCormick S (2005) Proteome mapping of mature pollen of *Arabidopsis thaliana*. Proteomics 5:4864–4884

Honys D, Twell D (2003) Comparative analysis of the Arabidopsis pollen transcriptome. Plant Physiol 132:640–652

Imin N, Kerim T, Weinman JJ, Rolfe BG (2001) Characterisation of rice anther proteins expressed at the young microspore stage. Proteomics 1:1149–1161

Imin N, Kerim T, Rolfe BG, Weinman JJ (2004) Effect of early cold stress on the maturation of rice anthers. Proteomics 4:1873–1882

Imin N, Kerim T, Weinman JJ, Rolfe BG (2006) Low temperature treatment at the young microspore stage induces protein changes in rice anther. Mol Cell Proteomics 5:274–292

Kerim T, Imin N, Weinman JJ, Rolfe BG (2003) Proteome analysis of male gametophytes development rice anthers. Proteomics 3:738–751

Liska A J, Shevchenko A, Pick U, Katz A (2004) Enhanced photosynthesis and redox energy production contribute to salinity tolerance in Dunaliella as revealed by homology-based proteomics. Plant Physiol 136:2806–2817

Malho R, Liu Q, Monteiro D, Rato C, Camacho L, Dinis A (2006) Signalling pathways in pollen germination and tube growth. Protoplasma 228:21–30

Mascarenhas JP (1993) Molecular mechanisms of pollen tube growth and differentiation. Plant Cell 5:1303–1314

Mayfield JA, Fiebig A, Johnstone SE, Preuss D (2001) Gene families from the *Arabidopsis thaliana* pollen coat proteome. Science 292:2482–2485

Noir S, Brautigam A, Colby T, Schmidt J, Panstruga R (2005) A reference map of the *Arabidopsis thaliana* mature pollen proteome. Biochem Biophys Res Commun 337:1257–1266

Pina C, Pinto F, Feijo JA , Becker JD (2005) Gene family analysis of the Arabidopsis pollen transcriptome reveals biological implications for cell growth, division control, and gene expression regulation. Plant Physiol 138:744–756

Šamaj J, Müller J, Beck M, Böhm N, Menzel D (2006) Vesicular trafficking, cytoskeleton and signalling in root hairs and pollen tubes. Trends Plant Sci 11:594–600

Sheoran IS, Sproule KA, Olson DJH, Ross ARS, Sawhney VK (2006) Proteome profile and functional classification of proteins in *Arabidopsis thaliana* (Landsberg erecta) mature pollen. Sex Plant Reprod 19:185–196

Wang Q, Kong L, Li Y, Lin J, Šamaj J, Baluška F (2005) Effects of Brefeldin A on pollen germination and tube growth: antagonistic effects on endocytosis and secretion. Plant Physiol 139:1692–1703

Chapter 18
Plant Proteomics Upon Fungal Attack

Frank Colditz, Franziska Krajinski, and Karsten Niehaus

Abstract Plant diseases caused by fungi and oomycetes are considered to represent a severe limiting factor to food production, with major economic impact. Therefore, it is of general concern to elucidate the details of fungal plant pathogenesis, including the molecular mechanisms of disease and the nature of the host cellular response. In recent years, proteomic analyses have emerged as a powerful approach to study effector proteins that derive from phytopathogens as well as defence-related proteins that are induced in infected plants to overcome disease. Proteomics now represents a valuable complement to genomic approaches. This chapter reviews recent advances in proteomic research, focussing on both the above-mentioned aspects, i.e. (1) the molecular mechanisms by which fungal plant pathogens establish infection, and (2) the defence-related strategies of plants to counteract or resist such challenges. Types or classes of proteins involved in these processes that have been investigated with proteomic or related methods are presented here. These include extracellular and cytoplasmic effector proteins, pathogen-associated molecular patterns, and inducible defence-related plant proteins such as pathogenesis-related proteins as well as phosphorylated proteins involved in disease signalling. Additional aspects of the responses to disease of plants, concerning specificity, systemic signal transduction and systemic acquired resistance, are also addressed.

18.1 Introduction

The ability of plants to defend themselves against potential pathogens depends on sensitive perception mechanisms and subsequent activation of effective defence responses. Generally, plants possess preformed as well as inducible mechanisms to

Frank Colditz, University of Bielefeld, Department 7 Proteome- and Metabolome Research, Universitätsstraße 25, 33615 Bielefeld, Germany, E-mail: fcolditz@cebitec.uni-bielefeld.de

Franziska Krajinski, University of Hannover, Institute for Plant Genetics, Department for Molecular Genetics, Herrenhäuser Straße 2, 30419 Hannover, Germany

Karsten Niehaus, University of Bielefeld, Department 7 Proteome- and Metabolome Research, Universitätsstraße 25, 33615 Bielefeld, Germany

J. Šamaj and J. Thelen (eds.), *Plant Proteomics*
© Springer 2007

resist pathogen invasion. Beyond the passive or basal protection at the plant surface provided by morphological barriers such as waxy cuticular ‘skin’ layers and preformed antimicrobial compounds (Dangl and Jones 2001), plant tolerance or resistance towards pathogen challenge involves a broad array of inducible defence responses.

Once contact with, or penetration into, the plant host cell has been established, released elicitors induce further defence mechanisms and patterns. Among these are (1) the accumulation of cell wall structural polymers such as callose, lignin or suberin for reinforcement of cell walls, (2) the synthesis of (secondary) metabolites (phytoanticipins) and antimicrobial compounds such as phytoalexins, and (3) the synthesis of various defence-related proteins (Corbin et al. 1987; Slusarenko et al. 2000). While initial chemosensoric recognition of a pathogen involves processes such as depolarisation of plasma membranes, modifications in ion fluxes, increasing cytosolic Ca^{2+} concentrations, production of reactive oxygen species (ROS), as well as co- or post-translational processing such as protein phoshorylation, glycosylation and sulfation, subsequent events such as the synthesis of inducible defence- or pathogenesis-related (PR-) proteins and enzymes (mainly chitinases and glucanases) and the accumulation of phytoalexins requires induction of gene expression (Dixon and Lamb 1990; Guy et al. 1994; Hammond-Kosack and Jones 1997; Yang et al. 1997). Among inducible defence-related proteins, extracellular proteins are thought to form the primary line of defence, before an area-wide penetration by the pathogen has taken place. As it takes time to accumulate these proteins, virulent pathogens may have already passed this barrier, and cellular defence-related proteins then function as a secondary line of defence (van Loon et al. 2006). Where tissue damage and disruption has occurred, lysozyme-like enzymes are released; thus, many fungal pathogens adopt a hemibiotrophic lifestyle in order to avoid damage to cells and tissues (van Loon et al. 2006). Regarding the establishment of systemic resistance to further invasion, besides the fundamental participation of typical signalling compounds such as salicylic acid (SA), jasmonic acid (JA) and ethylene (ET), inducible defence-related proteins are exclusively involved; indeed, the majority of the protein repertoire is thought to be located in the primary line of defence.

In recent years, considerable research interest has focused on plant genes expressed exclusively upon pathogenic interactions. Data in this field come mainly from microarray analyses. Compared to the relatively high number of published microarray studies, only a few proteomic studies – of varying quality – addressing the modified induction patterns of gene products as a plant responds to a pathogen challenge have so far been reported. This is quite remarkable considering that proteomic approaches more directly represent cellular status by monitoring the actual protein composition, which in turn reflects the presence of biochemical cellular pathways (Gygi et al. 1999). Moreover, the most significant advantage of proteomics is that it can monitor synthesis rates and expression levels as well as co- and post-translational protein modification (Xing et al. 2002). Such data provide additional information contributing to a fuller understanding of plant defence responses. Indeed, data gathered with proteomic tools can powerfully complement data sets obtained from transcriptomic approaches.

To conclude, this chapter presents an in-depth discussion of the application of proteomic and proteomic-related approaches to the seemingly obligatory and manifold modifications of protein induction patterns in plants due to pathogen infection. Evaluating changes in protein patterns in the host plant in response to a particular pathogenic interaction represents only one side of the story. Even better would be to also evaluate the corresponding changes in the proteome of the pathogen. However, only a few direct proteomic approaches have been applied to fungal plant pathogens to date, mainly due to a lack of available protein sequence data that can be used for reliable mass spectrometry (MS)-based protein identification. To address this issue, we have also included a section focussing on specific protein induction among fungal and oomycete pathogens during infection or penetration.

18.2 Types of Plant–Fungi Interactions

18.2.1 Mutualistic Plant–Fungi Interactions with Microsymbionts

All fungi are carbon-heterotrophic and therefore dependent on other organisms. Most fungi exhibit a saprophytic lifestyle, obtaining energy and molecules for their own metabolism from dead organisms. However, some fungi interact with other living organisms to gain access to energy and nutrients. These plant–microbe interactions vary from very loose, non-specific, associations to highly specialised interactions where the fungal partner is entirely dependent on its particular host. From the plant's point of view, the interaction can be beneficial, as in case of the mycorrhizal symbiosis, or fatal, as in the case of infection by a necrophytic pathogen.

Most plants in natural environments are infected by fungi that cause no disease. The outstanding example of a beneficial fungal–plant interaction is the mycorrhizal symbiosis. Arbuscular mycorrhiza (AM) symbiosis has a global impact on plant phosphorus nutrition and has become a model system for the study of the underlying molecular events (see Chap. 20 by Dumas-Gaudot and Recobert, this volume). Legumes, especially *Medicago truncatula* and *Lotus japonicus*, have been chosen as model plants since a direct comparison with the nitrogen-fixing *Rhizobium* symbiosis is possible (Harrison 2005; Parniske 2005; see also Chap 9 by Lei et al. and Chap. 19 by Sarma et al., this volume). Although the initial signals that initialise the mycorrhiza–plant interaction remain obscure, three plant signalling proteins required for signalling pathways used by AM fungi and rhizobia have been cloned. Both AM and rhizobial plant–microbe interactions have a strong effect on the plant proteome. In the *Sinorhizobium meliloti* symbiosis with *Medicago alba*, over 250 proteins were induced or up-regulated in the root nodule compared with uninfected roots, and over 350 proteins were down-regulated in the bacteroid form of the rhizobia compared with free living cells (Natera et al. 2000). These changes in the proteomes correlate with the morphological differentiation of the induced plant organ – the root nodule – and with physiological adaptations due to the nitrogen fixation and assimilation process. Proteomic approaches analysing AM symbiosis have paid special attention to membrane proteins. Upon infection of the root via apressoria formation, fungal hyphae infect the apoplastic space of the root cortex and penetrate single plants cells, where the fungus forms a highly branched filamentous structure called an arbusculum. The arbusculum is entirely covered by a plant membrane – the periarbuscular membrane – that forms the interface where nutrients, namely sugars and phosphate, are exchanged between both organisms. Recently, proteins of the periarbuscular membrane were analysed using two dimensional liquid chromatography and tandem mass spectrometry (2D-LC MS/MS), resulting in the identification of 87 proteins, of which 2 were differentially expressed (Valot et al. 2006). One of these proteins corresponded to an H(+)-ATPase, most probably involved in creating an electrochemical gradient required for the translocation of nutrients. H(+)-ATPases were also identified by a systematic proteome analysis of the peribacteroid membrane that surrounds the nitrogen fixing bacteroids in the *Rhizobium*–legume symbiosis (Wienkoop and Saalbach 2003).

18.2.2 Parasitic Plant–Fungi Interactions – Infection Strategies of Fungal Pathogens

The lifestyle of necrotrophs, e.g. the corn leaf blight fungus *Cochliobolus heterostrophus*, is optimised at the other end of the spectrum of plant–microbe interactions. These pathogens quickly kill plant cells and eventually the entire host organism, to feed as saprotrophs. Destruction of the host cells can involve the release of host-selective toxins, as in case of *Fusarium oxysporum*. The fungal protein Nep1 (necrosis- and ethylene-inducing peptide) disturbs membrane integrity and induces necrotic spots on *Arabidopsis thaliana* leaves treated with *F. oxysporum* Nep1 (Bae et al. 2006). Most interestingly, in many cases an active role of the invaded host is required for quick disruption. A striking example is the oxidative burst, a characteristic plant defence reaction directed against invading biotrophic microbes, which is involved in mediating a hypersensitive response (HR) including the death of the plant cell. In the case of a necrotroph such as *Botrytis cinerea*, plant cell death is beneficial to the pathogen and leads to susceptibility of the particular host plant (Govrin and Levine 2000). Although these pathogens follow a "brute force" strategy, extensive signalling is required during the early steps of the interaction. Whole-genome analysis of different pathogenic fungi revealed a huge set of two-component signal transduction genes, falling into 11 classes. A few of these classes are more prevalent in pathogenic fungi, suggesting that they contain paralogs required for virulence (Catlett et al. 2003). In addition, heterotrimeric G-proteins and mitogen-activated protein kinases (MAPKs) are essential components of the *C. heterostrophus* signal transduction chain leading to virulence (Ganem et al. 2004). Upon killing of the host cells, a complex cocktail of extracellular enzymes is secreted from the pathogen to decompose the plant materials, allowing the pathogen to develop and fulfill its life cycle.

Fungal biotrophy resembles the complex signal exchange of symbiotic interactions despite the fact that no obvious beneficial effect or rapid killing of host cells is observed. Characteristic of such biotrophic interactions is the tight interlocking of host and fungal life cycles. Strict biotrophs are highly specialised to particular host plants and entirely dependent on one or more hosts to complete their life cycle (Mendgen and Hahn 2002). Killing of the host cells is avoided and plant defence reactions might be suppressed, since these pathogens rely entirely on the metabolism of the host. For this reason, these pathogens cannot be cultivated on synthetic media. As a consequence, molecular biology tools gain only limited access to these interesting plant–microbe interactions. The infection strategy involves, following spore attachment and germination, the formation of a specific infection structure called the appressorium. The formation of an appressorium is essential for successful infection. Biotrophic pathogens, such as *Uromyces fabae*, infect via the stomatal pore and spread into the apoplastic space of the infected leaf. Later, starting from a haustorial mother cell, plant cells are invaded by fungal hyphae that differentiate into an organ specialised to extract nutrients from the still living host cell. This organ, the haustorium, is entirely covered by a membrane derived from the host cell's plasma membrane (Vögele 2006). Haustoria build the essential interface for sugar transport from the plant to the fungus. Infected tissues change dramatically in their metabolic capacity, with the fungus converting the absorbed sugars to mannitol (Vögele et al. 2005). Since the host plant is normally not able to synthesise or use sugar alcohols, the usage of mannitol seems an ideal strategy for the fungal pathogen. The complex adaptation of the plant to the fungal pathogen was shown by a recent microarray analysis of expressed sequence tags (ESTs) from haustoria (Jakupovic et al. 2006). Extensive proteome studies are hindered by limited genome information and the complex morphological differentiation of the pathogen during infection. Cooper et al. (2006) used 2-D nanoflow HPLC and MS/MS to identify more than 400 proteins from asexual uredospores from *Uromyces appendiculatus*. The challenging task will be to extend such approaches to later stages of the infection process.

In contrast to *U. fabae*, the hemibiotrophic infection of a plant by *Colletotrichum lindemuthianum* is not dependent on stomatal pores. The germinated spore generates a germ tube that forms an appressorium on the epidermal cell that is infected.

The primary infected plant cell must stay alive for the further development of the infection (Mendgen and Hahn 2002). If the plant recognises the infection, formation of papilla can block the penetration, or a hypersensitive cell death will kill the host cell as well as the invading pathogen (Schmelzer 2002). If the infection is effective, fungal hyphae spread to adjacent plant cells, which are infected and subsequently killed. The sequence of a transient biotrophic phase followed by cell killing is repeated and the pathogen feeds saprophytically on the damaged cells. Obviously, killing of plant cells and a saprophytic lifestyle do not favour colonisation of the plant. Freeman and Rodriguez (1993) reported a non-pathogenic mutant of *Colletitrichum magna* that grew throughout host tissues as an endophyte. Under these conditions, a positive effect of the infection could even be demonstrated.

Interestingly, plants infected by the non-pathogenic mutant are more resistant to infections by other pathogens (Redman et al. 1999).

18.2.3 Model Plant/Fungi Pathosystems

A number of pathogenic plant–fungus or plant–oomycete interactions have been selected as model systems for genome research. Sequencing projects have been initiated for the genomes of several phytopathogenic ascomycetes and basidiomycetes

Table 18.1 Model fungal and oomycete plant pathogens for which genome sequencing projects have been initiated, and the plant diseases they cause on their host plant(s)

Fungi	Subgroup	Plant diseases caused
Botrytis cinerea/ fuckeliana	Ascomycetes	Greymould of most vegetable and fruit crops, several tree varieties, shrubs, flowers, and weeds[a]
Fusarium verticillioides	Ascomycetes	Kernel and ear rot of maize[a]
Fusarium graminearum	Ascomycetes	Head blight (scab) of wheat and barley[a]
Magnaporthe grisea	Ascomycetes	Blast disease of rice and other cereals[a,b]
Fusarium solani	Ascomycetes	A variety of diseases in a number of crops including peas, soybean, potato, and ginger[a]
Phaeosphaeria nodorum	Ascomycetes	Leaf and glume blotch disease of wheat[a]
Phakopsora meibomiae	Basidiomycetes	Rust of soybean[a]
Phakopsora pachyrhizi	Basidiomycetes	Rust in over 30 legume species[a]
Phanerochaete chrysosporium	Ascomycetes	White-rot of wood[a]
Sclerotinia sclerotiorum	Ascomycetes	White mould of bean, head rot of sunflower, white mould of canola, and Sclerotinia stem rot of soybean[a]
Ustilago maydis	Basidiomycetes	Corn smut disease[a]
Oomycetes		
Phytophthora infestans	Peronosporales	(Late) blight disease of potato and tomato rot diseases of legumes[c]
Phytophthora ramorum	Peronosporales	Sudden oak death and diseases in economically and agri-culturally important dicot plants including crops, trees, and shrubs[a]
Phytophthora sojae	Peronosporales	Stem and root rot of legumes[a]

[a]http://www.ncbi.nlm.nih.gov/genomes
[b]Dean et al. 2005
[c]Paquin et al. 1997

as well as for some *Phytophthora* species. Table 18.1 gives an overview of model fungal and oomycete plant pathogens for which genome sequencing projects have been initiated, including fungi of economic relevance such as *B. cinerea*, *Fusarium* sp. and *Ustilago maydis*. Model oomycete plant pathogens of major economical impact include *Phytophthora infestans* and *Phytophthora sojae*.

18.3 The "Fungal Side" – Proteomics of Fungal Pathogens

Proteomic approaches using fungal and oomycete plant pathogens are now aimed mainly at detecting effector proteins, but also at monitoring stage-specific differences in protein patterns. In this context, two major challenges arise: (1) the diversity of fungal plant pathogens, and (2) the poor availability of genome sequence data, with only a few MS-based protein sequence references available (Padliya and Cooper 2006). This lack of MS-based protein references from fungi becomes strikingly apparent when carrying out proteomic studies; often, only a minority of proteins can be identified. Performing cross-species identification (CSI) using available protein sequence data from other related fungi is one possible way to bridge this gap. For example, protein identification from spores of the bean rust fungus *U. appendiculatus* initially led to only three identified proteins, but an enhanced number of proteins could be matched to data from related fungi, among them 25 proteins from the Basidiomycete *U. maydis* (Cooper et al. 2006). In cases where genome sequencing programs have been initiated, the situation appears more favourable. For oomycete pathogens from the species *Phytophthora*, initiated sequencing programmes (*Phytophthora* Genome Consortium; National Centre for Genome Resources, NCGR) have led to the identification of stage-specific as well as extracellular proteins from *P. infestans*, *P. palmivora* and *P. nicotianae* (Mitchel et al. 2002; Sheperd et al. 2003; Torto et al. 2003). Thus, several proteins were found to be stage-specific or specific to one or another spore type; their identification will help determine their specific functions during pathogenesis (van West et al. 2003).

The following section describes key effector proteins from phytopathogenic fungi and oomycetes, as consideration of these factors is crucial to thorough elucidation of plant–pathogen interactions.

18.3.1 Fungal and Oomycete Effector Proteins

It was first documented over 10 years ago that bacterial plant pathogens translocate effector proteins – including avirulence (AVR) proteins – into the host plant cytoplasm via the type III secretion system (Van den Ackerveken et al. 1996). The presence of corresponding factors involved in AVR protein perception in the plant cell was proposed, and indeed these were subsequently identified as resistance (R) proteins. This gene-for-gene interaction model is thought to mediate HR-associated resistance (Dangl and Jones 2001). In the intervening years, various R proteins, which function as "surveillance systems" to detect pathogen effectors, have been identified also for eukaryotic plant pathogens such as fungi and oomycetes (Martin et al. 2003; Birch et al. 2006).

The extensively studied interaction between the leaf mould pathogen *Cladosporium fulvum* and tomato (*Lycopersicon esculentum*) has been established as a model system for studying gene-for-gene resistance (Rivas and Thomas 2005). In contrast to bacterial plant pathogens, this interaction appears to be entirely extracellular: tomato Cf genes encode type I membrane-associated receptor-like proteins (RLPs) anchored in the plasma membrane that recognise fungal AVR peptides secreted into the leaf apoplast (Rivas and Thomas 2005; Ellis et al. 2006). However, as Cf proteins lack signal peptides, activation of HR cell death and the defence response is thought to be mediated through cellular interaction partners; several components of the Cf-9/Avr9-induced signalling pathway have been identified. Very recently, a Cf-9-interacting thioredoxin (CITRX) was identified via a yeast two-hybrid screen as a putative adaptor protein connecting Cf-9 and a Ser/Thr protein kinase (ACIK1) mediating the Cf-9/Avr9-induced defence response in tomato (Nekrasov et al. 2006). Gonzales-Lamothe et al. (2006) discovered that two Avr9/Cf-9 rapidly elicited (ACRE) genes essential for Cf-9- and Cf-4-dependent HR production in *Nicotiana benthamiana* encode putative U-box E3 ubiquitin ligase components and that the ubiquitination pathway is involved in the Cf-9/Avr9 interaction.

18.3.1.1 Cytoplasmic Effector Proteins

In contrast to the above-described extracellular AVR/R-interaction, the situation for the majority of known fungal and oomycete plant–pathogen interactions is different (and comparable to those involving bacterial pathogens), indicating that their AVR proteins enter the plant cell, where they bind mainly to cytoplasmic nucleotide binding site leucine-rich repeat (NBS-LRR) proteins (Ellis et al. 2006). The rice Pita-R protein corresponding to the AVR protein AVR-Pita from the rice blast fungus *Magnaporthe grisea* represents such an NBS-LRR protein, where direct interaction of the R/AVR-couple has been demonstrated via the yeast two-hybrid system (Jia et al. 2000). Hence, these AVR proteins represent cytoplasmic effectors and although they initially function within the host cells, it is mainly macroscopic disease symptoms such as lesion formation, tissue rotting or chlorosis, that become apparent after infection. Various other biotrophic pathogens [e.g. rust and mildew fungi; oomycete pathogens that cause devastating plant diseases like the agents of downy mildew and white rust (*Plasmopara* and *Perenospora* species); the genus *Phytophthora*, including the most destructive pathogens of dicots causing late blight, root rot and black pod diseases; and *Aphanomyces* species], deliver their AVR proteins into the host cell through specialised infection structures termed haustoria that invaginate host cells but remain isolated by a separating double membrane (Vogele and Mendgen 2003; Ellis et al. 2006; Birch et al. 2006). Thus,

AvrL567 from the flax rust fungus *Melampsora lini* is recognised in the host cytoplasm; cloning of flax rust *R* genes encoding proteins with typical NBS-LRR motifs suggests an almost entirely cytoplasmic perception system (Dodds et al. 2004; Catanzariti et al. 2006). The same is true for the Uf-RTP1 protein secreted from the rust fungi *U. fabae* (Kemen et al. 2005). Recently, the first oomycete AVR proteins were identified: Avr1b-1 from the soybean pathogen *P. sojae* (Shan et al. 2004), Avr3a from *P. infestans* (Armstrong et al. 2005) and ATR13 and ATR1^{NdWsB} from *Hyaloperenospora parasitica*, which infects Arabidopsis (Allen et al. 2004; Rehmany et al. 2005). Interestingly, sequence alignment of oomycete AVR proteins revealed a conserved amino acid signature, termed the 'RxLR' motif, occurring within 32 residues of the N-terminal signal peptide, which is related to a translocation motif in the virulence proteins of the malaria disease agent *Plasmodium fulciparum* responsible for their secretion into the host erythrocytes (Hiller et al. 2004; Marti et al. 2004; Rehmany et al. 2005). Bioinformatic analyses revealed that more than 40 genes from the *P. infestans* genome encode secretion proteins with this typical 'RxLR' motif (Rehmany et al. 2005). Considering the ever increasing amount of available genome sequences for oomycetes, one can expect the discovery of many more such proteins in the future (Ellis et al. 2006). Other cytoplasmic effector molecules derived from oomycetes have been investigated via a combination of functional in planta expression screens and subsequent computational predictions for putative candidate secreted proteins of *P. infestans* (Pex) as well as proteomic evaluation. Examples of effector molecules identified by such methods include the crinkling- and necrosis-inducing CRN protein family (Torto et al. 2003). Again, analogies can be made with bacterial (cytoplasmic) effectors, where obvious macroscopic effects (chlorosis, discolorations) appear after expression in the host cell.

18.3.1.2 Apoplastic Effector Proteins

In addition to cytoplasmic effectors, several classes of apoplastic effector molecules deriving from fungi or oomycetes have also been identified. These proteins function mainly as counter-defence mechanisms against hydrolytic enzymes from plants – usually PR proteins such as glucanases, chitinases and proteases. Hence, they are termed according to their function, e.g. glucanase inhibitor proteins (GIPs) or protease inhibitors (PIs). Other apoplastic effector molecules include the large class of small cysteine-rich proteins containing typical elicitins, AVR proteins, and the necrosis- and ethylene-inducing (NEP1)-like protein family as well as fungal cell wall glycoproteins such as transglutaminases (TGases) or the cellulose binding elicitor lectin (CBEL) from oomycete pathogens (Kamoun 2005). Some of these proteins represent general elicitors, or so-called "pathogen-associated molecular patterns" (PAMPs) (see below).

The 24-kDa necrosis-and ethylene-inducing protein Nep1 was originally identified in *F. oxysporum* f. sp. *erythroxyli* (Bailey et al. 1997) and Nep1-like proteins (NLPs) have been discovered subsequently in diverse species, particularly plant-associated bacteria, fungi and oomycetes (Pemberton and Salmond 2004). NLPs share remarkable kingdom-wide sequence similarities, and are capable of inducing necrosis and cell death in several dicotyledonous plants (Pemberton and Salmond 2004). For plant pathogens that switch between biotrophy and necrotrophy, NLP gene activation was found to occur particularly at the initiation of, or during transition to, the necrotrophic phase, which is likely to facilitate colonisation of the plant by the pathogen (Qutob et al. 2002). Recently, one member of the NLP family (PiNPP1.1) from the oomycete *P. infestans*, which induces cell death in *Nicotiana* spp., was found to interact with the ubiquitin ligase-associated protein SGT1 and the heat shock protein HSP90 in the host cell (Kanneganti et al. 2006). Interestingly, cell death triggered by PiNPP1.1 was dependent on specific signalling proteins distinct from those induced by the elicitin INF1. Combined expression of PiNPP1.1 and INF1 resulted in enhanced cell death, also suggesting distinct signalling requirements and a synergistic interplay between the two cell death responses (Kanneganti et al. 2006).

Many of the AVR proteins encoded by *Avr* genes of eukaryotic fungi and oomycetes represent small (<150 amino acids) cysteine-rich proteins that are secreted by plant pathogens and capable of inducing defence responses (Kamoun 2005). This group comprises the *C. fulvum* AVR2, AVR4 and AVR9 proteins as well as typical oomycete elicitins such as the large family of INF effector proteins isolated from *P. infestans* (Kamoun et al. 1993, 1998), the PcF and PcF-like proteins from *P. cactorum* (Orsomando et al. 2001; Liu et al. 2005), and the PPAT proteins from the *A. thaliana* pathogen *H. parasitica* (Bittner-Eddy et al. 2003). The most notable similarities between these small cysteine-rich proteins are their secretion into the plant apoplast and the formation of disulfide bridges between two cysteine residues, which are presumably essential for avirulence function and defence response stimulation and which are thought to protect against proteolytic degradation (Luderer et al. 2002). In addition to a core motif of six conserved cysteine residues, members of the *Phytophthora* INF family (for example) possess high frequencies of serine, threonine, alanine and proline residues, suggesting O-glycosylation and linkage to the plant cell wall. In contrast, *H. parasitica* PPAT proteins share no sequence similarities to other known cysteine-rich proteins (Kamoun 2005).

18.3.2 Pathogen-Associated Molecular Patterns

In recent years, remarkable similarities and distinctions have been discovered among general plant defence response mechanisms triggered by microbial compounds called elicitors and those induced by the so-called innate immune response in animals (Gomez-Gomez and Boller 2000; Nürnberger et al. 2004; Zipfel and Felix 2005). These inducible defence responses are activated by recognition of effector molecules termed pathogen-associated molecular patterns (PAMPs), which are conceptually equivalent to general elicitors (Nürnberger and Lipka 2005). Referring to plants, this term takes into account the broader character of basal perception systems for patterns that are characteristic of entire groups or classes of microorganisms.

In contrast to an *R*-gene-dependent gene-for-gene interaction, responses to PAMPs do not always result in HR-associated cell death (Zipfel et al. 2004).

One of the best-analysed PAMPs is the hepta-β-glucoside elicitor of *P. sojae* (*megasperma*). Oligosaccharides that induce the formation of phytoalexins were first detected in culture filtrates of this pathogen (Ayers et al. 1976). It turned out that these glucosides are breakdown products of oomycete cell walls. Interestingly, only a very defined oligosaccharide structure exhibited biological activity. The smallest elicitor-active oligosaccharide purified from the mixture of oligosaccharides released from mycelial cell walls is a branched hepta-β-glucoside with a backbone of five 1→6-linked β-glucosyl residues, and with two terminal β-glucosyl residues attached at C-3 of the second and fourth backbone glucose rings (Sharp et al. 1984). Even minor modification of this structure results in drastic or total loss of biological activity. Pioneering work using radioactively labelled hepta-glucan elicitor revealed the presence of high affinity binding sites in host plant membranes (Schmidt and Ebel 1987). Although elicitor-induced signal transduction could be analysed to some extent (Mithöfer et al. 2005), the receptor remains unknown.

Recent results define cellulose-binding domains (CBDs) in the cell wall glycoprotein CBEL from the oomycete pathogen *P. parasitica* var. *nicotianae* (*Ppn*) as a novel class of PAMP (Gaulin et al. 2006). CBEL is widespread in the genus *Phytophthora* and maintains a dual function: (1) it is a potent elicitor in the *Ppn* host plant tobacco (but also in non-host plants such as *Arabidopsis*), inducing HR-like necrosis and expression of defence genes; (2) it contains two Cys-rich domains with similarities to the fungal type I CBD consensus pattern of several fungal cellulases, and is required for binding to plant cellulosic substrates (Gilkes et al. 1991; Khatib et al. 2004; Kamoun 2005). Silencing of the CBEL gene in *Ppn* resulted in the loss of the ability of the strain to bind to cellophane membranes, although it retained its ability to infect tobacco. Leaf infiltration assays with a recombinant protein containing the PAMP-CBD domains are both necessary and sufficient to stimulate defence responses (Gaulin et al. 2002, 2006).

Another cell wall glycoprotein identified in various *Phytophthora* species is the 42-kDa Ca^{2+}-dependent TGase GP42, which triggers a multifaceted defence response in parsley (Sacks et al. 1995). GP42 contains a 13-amino-acid peptide fragment (Pep-13) found to be invariant in all studied *Phytophthora* TGases, which again is both necessary and sufficient to activate defence responses in parsley, but also cell death in potato (Halim et al. 2004). Hence, this GP42 epitope conforms to PAMP character (Brunner et al. 2002).

18.4 The "Plant Side" – Plant Proteomics Upon Fungal Attack

Several direct proteomic approaches to the study of plant responses to fungal pathogens have been published recently. These studies comprise typical plant/fungi pathosystems such as *Oryza sativa*/*M. grisea* (Kim et al. 2004), *L. esculentum*/*F. oxysporum* (Rep et al. 2002), *Triticum aestivum*/*Fusarium graminearum*

(Wang et al 2005; Zhou et al. 2006), *A. thaliana/Fusarium monoliforme* (Ndimba et al. 2003) and *Zea mays/F. verticillioides* (Campo et al. 2004), but also oomycete pathosystems such as *M. truncatula/Aphanomyces euteiches* (Colditz et al. 2004, 2005). These approaches share the common theme that MS-based methods were used to identify plant proteins specifically induced upon pathogen infection.

This section presents an overview of the major groups of inducible defence-related plant proteins in pathogenic interactions with fungi and oomycetes that have been investigated via proteomic or related methods. Systemic aspects of the disease response as well as protein phosphorylation patterns due to pathogenic infection are also described in more detail.

18.4.1 Inducible Defence-Related Proteins and their Occurrence in Infected Plants

Inducible defence-related proteins have been extensively described in many plant species that are infected by fungal pathogens, oomycetes, bacteria or viruses. Pathogenesis related (PR-) proteins were first described in tobacco exhibiting HR in response to infection with tobacco mosaic virus (TMV). The majority of such proteins were termed PR proteins or "PR-like" proteins (PRLs) due to their specific induction in pathological or related situations. Since constitutive expression of PRs in non-pathogenesis affected plant tissue is also observed, and since, over time, the term was extended to include defence-related enzymes such as phenylalanine ammonia-lyase (PAL) and peroxidases, "PR proteins" now represents a collective term for all microbe-induced plant proteins and their homologues (van Loon et al. 2006). PR proteins, most of which can be considered as direct antimicrobial defence proteins, are currently classified into 17 families (Stintzi et al. 1993).

The most important PR families induced in response to interactions with fungi are PR-2 proteins (hydrolytic enzymes such as β-1,3-endoglucanases) and classes PR-3, -4, -8, and -11 comprising different types of endochitinases; these PR proteins exhibit inhibitory effects on the growth of fungi while depolymerising polysaccharides of mycelial walls. In addition, other classes of PR include proteins exhibiting broad antibacterial or antifungal activities. For example, antioomycete activity has been demonstrated for members of the PR-1, -5, and -10 classes (Ham et al. 1997; Fagoaga et al. 2001; Park et al. 2004). Generally, hydrolytic enzymes may exert a dual function in defence responses, encompassing both direct antimicrobial activity and/or bacterial induction. Thus, these enzymes accelerate the disease resistance response by eliciting HR.

Unfortunately, the molecular methods applied to the study of these enzymes are often not sensitive enough to monitor the qualitative specificities of the participating isozymes (Tuzun and Somanchi 2006). However, proteomic approaches have now been used to characterise induction of many PRs in plant–fungal pathosystems. Zhou et al. (2005) showed up-regulation of a PR-2 protein in wheat spikes upon infection with *F. graminearum*, which causes *Fusarium* head blight (FHB) disease. Very recently, this pathogenic interaction was comprehensively investigated using proteomic methods: 30 (Wang et al. 2005) and 41 (Zhou et al. 2006) differentially regulated proteins were identified, including PR proteins and those involved in antioxidant and JA signalling pathways. Consistently, eight identified fungal proteins have putative functions as antioxidants, requiring carbon from glycolysis, while proteins related to photosynthetic pathways were less abundant (Zhou et al. 2006). Two β-1,3-glucanases were found to be induced in rice leaves upon infection with the rice blast fungus *M. grisea*, in addition to other inducible defence-related proteins such as peroxidases, a thaumatin-like protein (PR-5), a PR-10, and a probenazole-inducible protein (Kim et al. 2004). Generally, several classes and types of β-1,3-glucanases are constitutively expressed in plants and specific induction upon antimicrobial activity has been proven for only a few PR-2-type proteins (Grover and Gowthaman 2003). PR-2 glucanase isozymes exhibit variations of up to 250-fold in specific activity with various substrates, indicating quite diverse properties in planta (Tuzun and Somanchi 2006). Ham et al. (1997) observed that phytopathogenic fungi are capable of overcoming the challenge of these enzymes for a certain time period and establishing resistance to β-1,3-glucanases via secretion of inhibitory proteins (glucanase inhibitor proteins, GIPs). A combination of glucanases and chitinases is apparently more effective in degrading fungal cell walls than each enzyme group alone (Broekaert et al. 2000; Grover and Gowthaman 2003). Moreover, the chitin layers of several fungal cell walls are buried in β-glucans, which results in inaccessibility to chitinases without prior hydrolysis with β-1,3-glucanases (Benhamou et al. 1990). Accordingly, induction of (endo-) glucanases and chitinases was found to occur simultaneously upon diverse plant–microbe interactions (Ham et al. 1997; Ndimba et al. 2003; Castillejo et al. 2004; Casado-Vela et al. 2006).

Concerning chitinases, including PR-3, -4, -8, and -11 (-type) proteins, various experiments have focussed mainly on transgenic plants that constitutively express these enzymes in order to enhance resistance levels upon colonisation by fungal pathogens (Broekaert et al. 2000; Grover and Gowthaman 2003). Although suppressive effects upon fungal colonisation could be observed for the targeted pathogen or the specific parasitic interaction investigated, enhanced tolerance or resistance often disappear upon expression in other plant species or inoculation with different pathogen strains. Thus, the effectiveness of such efforts depends heavily on the transgenic source, plant species and pathogen sensitivity (Van Loon et al. 2006). Moreover, these results indicate that there are no unitary paths in the specification of plant defence reponses. Several studies have highlighted the accumulation of PR proteins including chitinases in plant fruits and berries (Menu-Bouaouiche et al. 2003; Casado-Vela et al. 2006). It is postulated that they might contribute to basal resistance to pathogen challenge. Ferreira et al. (2004) found mainly chitinases and thaumatin-like (PR-5-type) proteins as prevalent proteins in grape berries, and a proteomic screening of six different grapevine cultivars revealed accumulation of dehydrin and invertase in addition to PR proteins (Sarry et al. 2004).

Thaumatins with sequence similarities to a protein from *Thaumatococcus danielli* (Edens et al. 1982) and thaumatin-like PR-5 proteins belong to a larger multi-gene family of proteins with antifungal and anti-oomycete properties that cause cytoplasmic leakage in hyphae (Vigers et al. 1991; Fagoaga et al. 2001; Van Loon et al. 2006). The antifungal action was thought to be directly correlated with β-1,3-glucan binding and endo-β-1,3-glucanase activity of some PR-5 (-type) proteins, but since there are also documented examples where such a correlation has not been observed, the evidence for a stringent correlation of these two properties is inconsistent (Grenier et al. 1999; Menu-Bouaouiche et al. 2003). However, cell wall binding activity is thought to enhance the effectiveness of osmotin (PR-5c, originally from tobacco) and other PR-5-type proteins contributing to basal resistance to fungal pathogens (Yun et al. 1997; Narasimhan et al. 2003). Two proteins induced in chickpea (*Cicer arietinum*) upon infection with *Ascochyta rabei* were identified as PR-5a and PR-5b proteins (Hanselle et al. 2001). In a proteomic approach, a new member of the PR-5 family was identified in tomato xylem sap after infection with the vascular wilt fungus *F. oxysporum* (Rep et al. 2002).

A complex antioxidant system comprising enzymes such as superoxide dismutase (SOD), peroxidases and catalases protect plant cells against ROS, which are generated during the HR of plant defence. The existence and induction of multiple isoenzymes with antioxidant activity favours the view that each enzyme performs a separate function in addition to that of the key enzyme SOD (Tuzun and Somanchi 2006). Antioxidant PR proteins – classified as PR-9 proteins – represent peroxidase isoenzymes and as the recently added PR-15 and -16 proteins of monocots, which include germin-like oxalate oxidases and oxalate oxidase-like proteins with SOD activity (Van Loon et al. 2006). A putative novel family (PR-18) was founded by isolating from higher plants a class of carbohydrate oxidases that are inducible by SA and fungal pathogens and that produce hydrogen peroxide as a by-product of the catalysed reaction (Custers et al. 2004).

The ribonuclease-like PR-10/PR-10-type proteins have also been described extensively to be induced in plants upon infection with fungal pathogens. Representatives of this family have been identified in more than 70 species of both mono- and dicotyledonous flowering plants (Markovic-Housley et al. 2003). Since their amino acid sequences lack signal peptides, PR-10 proteins represent the sole family of PRs where all members are thought to be located strictly in the cytoplasm. Thus, they are easily detected in proteomic approaches. Interestingly, PR10-1 and homologues were found to represent the most abundant proteins in the *M. truncatula* root proteome (and in those of suspension and embryonic cultures), appearing with four isoforms (Mathesius et al. 2001; Imin et al. 2004). PR10-1 expression in *M. truncatula* was found to be constitutive, functioning in phytohormone homeostasis by exhibiting binding affinities to cytokinin and brassinosteroids (Imin et al. 2004). However, a group of PR10-like proteins including proteins annotated as ABA-responsive proteins (ABR17) were found to be significantly induced in 2-D maps derived from *M. truncatula* roots infected with the oomycete root pathogen *A. euteiches* (Colditz et al. 2004). As their induction level correlates with the in planta infection level evaluated in different ecotypes, they can act as a marker for the cellular defence state upon this particular interaction (Colditz et al. 2005). Recently, a PR-10 member of maize was found to be involved in host resistance against fungal infection and aflatoxin production by *Aspergillus flavus* and *A. parasiticus* (Chen et al. 2006). In another interesting study, immunohistochemical analysis led to the localisation of two PR-10 proteins, one induced in mesophyll cells of rice leaves under the attachment site of appressoria of the rice blast fungus *M. grisea*, and the other localised mainly in vascular tissue (Kim et al. 2004). Two proteome analyses revealed induction of PR-10 occurring also under abiotic stress situations: in rice seedlings upon ozone stress (Agrawal et al. 2002) and in salt-tolerant peanut calli (Jain et al. 2006).

The thionin PR-13 family comprises small (~5 kDa) sulfur-rich proteins, located in either the vacuole or the cell wall, that exhibit antibacterial and antifungal activity (Bohlmann 1994) but their low molecular weight makes them difficult to detect with standard 2-DE methods. Using an immunogold labelling technique, the subcellular location of thionins and hydroxyproline-rich glycoproteins (HRGPs) was determined in *F. culmorum*-infected wheat spikes (Kang and Buchenauer 2003). These proteins were found to accumulate in the cell walls of infected lemma, ovary and rachis, but were increased in an infected resistant cultivar compared to a susceptible one. In a very interesting approach, a molecular model for phospholipid membrane lysis by plant thionins was postulated; two highly positive charged thionins were found to bind a negatively charged phosphate ion and a glycerol molecule (Stec et al. 2004). The in vitro interaction of phospholipids results in the formation of proteolipid complexes, causing solubilisation and lysis of the membrane. This model proposes basic structural principles of protein–membrane interactions. Two γ-thionins, representing the first plant defence protein class described in the literature (Peligrini and Franco 2005), were purified from maize using RP-HPLC chromatography and Edman degradation. ES/MS analysis revealed remarkable homologies to sorghum α-amylase isoinhibitors (Castro et al. 1996). A recombinant *Pisum sativum* defensin (rPsd1) was found to be processed post-translationally to the same mature protein as the native Psd1 exhibiting similar antifungal activity against *A. niger* (Cabral et al. 2003).

One aspect of major interest might be to monitor either synergistic or antagonistic effects of inducible defence-related proteins among the defence responses of plants. Proteomic approaches are particularly suited to this purpose. Very recently, antagonistic induction of *M. truncatula* PR-5b proteins in an RNAi silencing approach targetted at PR-10-type proteins was documented via proteomic methods; enhanced PR-5b induction led to increased tolerance in vitro upon infection with the oomycete root pathogen *A. euteiches* (Colditz et al 2007).

18.4.2 Induction of Systemic Acquired Resistance by Fungal Pathogens

As discussed for hemibiotrophic pathogens, infection of the plant may result in enhanced resistance to other pathogens (Redman et al. 1999). The phenomenon of an entire plant locally infected by one pathogen developing a broad spectrum of resistance against subsequent infection is called systemic acquired resistance (SAR). Initially, SAR was characterised in tobacco plants infected with TMV (Ross 1961), but the effect of "plant immunisation" had been observed long before that. SAR is an active process that requires local defence reactions to a primary infection, and the generation of a mobile signal that induces elevated resistance in other parts of the plant. A general observation in plants that exhibit SAR is the expression of PR genes in tissues that are not affected by the initial infection (van Loon et al. 2006). The elevated levels of PR proteins could be correlated with increased resistance in certain specific plant–microbe interactions (Alexander et al. 1993). In other cases over-expression of PR genes in transgenic plants had no significant effect on the resistance towards pathogens, indicating that other mechanisms must also be involved. Infected tissues produce salicylic acid (SA), and SA is able to induce PR genes and SAR. SAR requires both local and systemic SA accumulation as well as induction of a subset of PR genes. Nevertheless, SA is not the translocated signal responsible for induction of SAR, as shown by grafting of transgenic plants that are no longer able to accumulate SA (Vernooij et al. 1994). Arabidopsis mutants helped to identify the central signalling components that are responsible for the establishment of SAR. Non-expressor of pathogenesis related genes 1 (NPR1) is a central positive regulator of SAR that is controlled by SA-mediated redox signalling (Fobert and Despres 2005). Under oxidative conditions, NPR1 and the transcription factor TGA1 are oxidised and inactive, while under reducing conditions conserved cysteines in both proteins become reduced, enabling direct protein–protein interactions to promote a complex that is competent to activate SAR-regulated genes (Fobert and Despres 2005).

18.4.3 Mapping of Protein Phosphorylation Patterns Triggered by Fungal Pathogens or Elicitors

Along with the control of enzyme activity, transcriptional events, cell division and subcellular protein localisation, post-translational modification by phosphorylation/dephosphorylation is a general mechanism in the perception and transduction of signals originated from pathogens and is thought to represent the key regulating processes in elicitor-stimulated early plant defence responses (Viard et al. 1994; Stone and Walker 1995; Grant and Mansfield 1999). Treatment of plant cell cultures with elicitors derived from (fungal) pathogens has been shown to induce rapid changes in phosphoprotein profiles (Dietrich et al. 1990; Felix et al. 1991, 1994; Viard et al. 1994; Lecourieux-Ouaked et al. 2000; Peck et al. 2001). In contrast, inhibition of protein phoshorylation results in cessation of medium alkalinisation, generation of ROS, induction of defence gene expression, induction of phytoalexin synthesis, and HR-like cell death (Zhang et al. 2000). Hence, proteins differentially phosphorylated in response to microbial elicitation are very likely components of signal transduction pathways following microbe perception. Therefore, identification of protein kinases and protein phosphatases as well as their targeted substrates is of crucial importance in the elucidation of the early steps of plant-microbe interactions.

Different techniques, such as ^{32}P radio-pulse-labelling or immunoblot analyses detecting phosphoamino acids, have been developed for mapping phosphoproteins on 2D gels (Guy et al. 1994). The first results obtained by Dietrich et al. (1990) and Felix et al. (1991) upon infection with fungal pathogens or stimulation with fungal elicitors demonstrated that protein kinase-mediated phosphorylation occurs within minutes after elicitor treatment and that it is essential to maintain this state while initiating further transduction of elicitor signals.

Beyond the involvement in early plant defence responses, phosphorylation and dephosphorylation are aligned events embedded in a complex regulatory network. For example, treatment of tomato cells with elicitors from *C. fulvum* results in a cascade of phosphorylation, de- and re-phosphorylation of the host plasma membrane H^+ ATPase, where at least two different kinases, a protein kinase C (PKC) and a Ca^{2+}-calmodulin-dependent kinase (CaMK), are involved (Xing et al. 1997). A 2-DE approach was applied to elucidate in vitro upstream phosphorylation pathways in tobacco after exposure to the oomycete *Phytophthora* spp. elicitin cryptogein (Lecourieux-Ouaked et al. 2000). The interplay of protein kinases and phosphatases in tobacco defence reactions was well characterised, but the phosphorylated polypeptides after elicitation revealed no similarities to known proteins. Peck et al. (2001) established a comprehensive phosphoproteome map for *Arabidopsis* cell cultures in response to bacterial and fungal elicitors via ^{32}P radio-pulse-labeling and MS-based protein identification. Interestingly, one of the de novo induced proteins, AtPhos43, was found to be phosphorylated exclusively after exposure to bacterial flagellin but not after treatment with fungal chitin. Thus, specificity in phosphorylation-mediated signal transduction according to the quality or source of the stimulus was identified.

Besides distinct phoshorylation pathways, the identification of phosphorylated cell wall proteins and of a putative extracellular receptor-like kinase revealing no transmembrane domain in Arabidopsis cell cultures upon induction with fungal elicitors implicate the existence of an upstream extracellular phosphorylation network (Ndimba et al. 2003). Thus, next to the detection and identification of single proteins phosphorylated during plant defence, targeting the interacting members (proteins/DNA) of the pathways represents a major challenge. Protein–protein and protein–DNA interactions in plant defence pathways have been identified using the yeast two-hybrid system (Xing et al. 2001). Recently, de la Fuente van Bentem et al. (2005) showed, via yeast two-hybrid screening, that the tomato R protein I-2, which confers resistance to the fungal pathogen *F. oxysporum*, interacts with the protein phosphatase 5 (PP5) to form a multi-protein complex with cytosolic heat shock protein 90 (HSP90). Concerning the role of dephoshorylation in plant defence regulation, silencing of subfamily I of protein phosphatase PP2Ac in *N. benthaminana* resulted in the activation of a plant defence response associated with increased resistance to bacterial and fungal pathogens (*C. fulvum*) and accelerated HR (He et al. 2004). Moon et al. (2005) isolated a homologue of the UBP6 deubiquitinating enzyme family exhibiting a Ca^{2+}-dependent calmodulin

(CaM)-binding domain in *A. thaliana* by screening an expression library with a calmodulin2 (AtCaM2) antibody, demonstrating that CaM is also involved in ubiquitin-mediated protein turnover.

Although many kinases involved in plant signalling are regulated at the post-translational level, it should also be mentioned here that transcriptional regulation of kinases is thought to be of similar importance. Hence, both processes are essential and are thought to be well coordinated for stringent kinase-mediated defence signalling. For the tobacco *WIPK* gene, which encodes a wound-induced MAPK, activation due to fungal elicitation could be documented at multiple levels during induction of HR-like cell death (Zhang et al. 2000).

18.5 Induction of Defence-Related Proteins in Non-Pathogenic Interactions

Plants are able to protect themselves not only against most pathogenic microorganisms and fungi, but also against various stresses such as wounding, application of chemicals including phytohormones and heavy metals, air pollutants such as ozone, ultraviolet rays, and harsh growing conditions. Induction of defence-related genes has been observed under all these latter conditions (Bowles 1990).

In addition, the induction of plant-defence-related genes also occurs during the mutualistic plant–fungus AM symbiosis (reviewed in Harrison 1999), particularly during the early stages of AM development (Gianinazzi-Pearson et al. 1996). In some cases, induction of paralogous genes of classical defence-related proteins has also been observed in *M. truncatula*. Thus, a new member of the chitinase gene family is exclusively expressed in mycorrhizal roots, whereas other members of this gene family are predominantly expressed during parasitic infections (Salzer et al. 2000). Other than this activation of specific members of families of defence-related proteins, the induction of defence-related genes seems to be rather weak and transient (Lambais and Mehdy 1993). Furthermore, over-expression of PR genes in transgenic tobacco plants did not reduce infection or colonisation by the AM fungus *Glomus mosseae* (Vierheilig et al. 1995). Hence, the role of defence-related genes during the AM symbiosis remains unclear.

18.6 Conclusions and Future Perspectives

Proteomic and related approaches have become crucial factors in the characterisation of pathogenic interactions between fungal and oomycete pathogens and their host plants. The detailed elucidation and a better understanding of the molecular aspects of plant diseases caused by phytopathogenic fungi is becoming ever more important due to the notable increase of pathogen spread caused by considerable changes in the global environment. The lack of (or lack of access to) genome and protein sequence information for these phytopathogens might be one explanation why researchers have focussed mainly on the identification of induced defence responses in the corresponding host plants. Nevertheless, progressing genome sequencing programmes for important plant pathogens will redress this situation, as evidenced by the steadily increasing number of recent investigations performed on virulent fungi and oomycetes. For example, several thousand ESTs are currently available for the important oomycete plant pathogens *P. infestans* and *P. sojae*, provided by the *Phytophthora* Genome Consortium, the Consortium for the Functional Genomics of Microbial Eukaryotes (COGEME), Syngenta or also via individual processed EST libraries (Kamoun et al. 1999; Waugh et al. 2000; van West et al 2003). The increasing amount of available sequence data makes it worthwhile to perform functional genomic approaches on fungal pathogens comprising the characterisation of effector molecules, AVR genes and proteins, and PAMPs, as well as stage-specific expression patterns. In addition, a comparative analysis of protein diversity among pathogenic fungi and oomycetes will be possible, hopefully contributing to our understanding of more fundamental aspects such as host specificity, pathogenicity, susceptibility and resistance.

Evaluating the field of plant proteomics with respect to challenges by fungal or oomycete pathogens, although access to genome and protein sequence data for the host plants appears much better in many cases, it is not entirely satisfactory and some difficulties are evident. Here, one major point concerns the fact that many genes encoding inducible defence-related proteins, particularly PR proteins, comprise broad evolutionarily conserved families whose members differ widely in appearance and activity (van Loon et al. 2006). Thus, these multi-gene families encode proteins with diverse functions, which comprise not only their specific induction due to a pathogenic interaction but also their occurrence during other stress situations or even at particular developmental stages. In addition, the induction of plant defence-related genes and gene products has been reported also in symbiotic interactions (see below). Even the well-investigated PR-1 proteins, which are often used as molecular markers for an enhanced defensive state or pathogen-induced SAR (van Loon et al. 2006), still remain elusive in their stringent induction in connection with pathogenesis or at least SAR.

Regarding SAR, very interesting and promising results were described by Xia et al. (2004) in terms of the identification of an SA-dependent *constitutive disease resistance* (*CDR1*) gene in *Arabidopsis* encoding an apoplastic aspartic protease. Induction of *CDR1* results in resistance to virulent *Pseudomonas syringae* and in the generation of a small mobile peptide signal involved in the activation of inducible resistance mechanisms (Xia et al. 2004). Previously, the involvement of a lipid transfer protein was suggested in the case of *Arabidopsis* SAR, implying that the mobile signal consists of a lipid moiety (Maldonado et al. 2002).

Regarding early phosphorylation events of proteins during pathogenic challenges, a direct relation appears very obvious and is thought to be proven in many cases, but many technical challenges still remain to be overcome concerning the detection and evaluation of protein phosphorylation patterns.

Acknowledgments The authors thank Dr. Matthew McIntosh, University of Bielefeld, Institute of Genome Researc1h, for proof-reading the manuscript and for helpful discussions.

References

Agrawal GK, Rakwal R, Yonekura M, Kubo A, Saji H (2002) Proteome analysis of differentially displayed proteins as a tool for investigating ozone stress in rice (*Oryza sativa* L.) seedlings. Proteomics 2:947–959

Alexander D, Goodman RM, Gut-Rella M, Glascock C, Weymann K, Friedrich L, Maddox D, Ahl-Goy P, Luntz T, Ward E (1993) Increased tolerance to two oomycete pathogens in transgenic tobacco expressing pathogenesis-related protein 1a. Proc Natl Acad Sci USA 90:7327–7331

Allen R, Bittner-Eddy PD, Grenville-Briggs LJ, Meitz JC, Rehmany AP, Rose LE, Beynon JL (2004) Host-parasite coevolutionary conflict between *Arabidopsis* and downy mildew. Science 306:1957–1960

Armstrong MR, Whisson SC, Pritchard L, Bos JI, Venter E, Avrova AO, Rehmany AP, Bohme U, Brooks K, Cherevach I, Hamlin N, White B, Fraser A, Lord A, Quail MA, Churcher C, Hall N, Berriman M, Huang S, Kamoun S, Beynon JL, Birch PR (2005) An ancestral oomycete locus contains late blight avirulence gene *Avr3a*, encoding a protein that is recognized in the host cytoplasm. Proc Natl Acad Sci USA 102:7766–7771

Ayers AR, Ebel J, Finelli F, Berger N, Albersheim P (1976) Host-pathogen interactions. IX. Quantitative assays of elicitor activity and characterization of the elicitor present in the extracellular medium of cultures of *Phytophthora megasperma* var. *sojae*. Plant Physiol 57:751–759

Bae H, Kim MS, Sicher RC, Bae HJ, Bailey BA (2006) Necrosis- and ethylene-inducing peptide from *Fusarium oxysporum* induces a complex cascade of transcripts associated with signal transduction and cell death in *Arabidopsis*. Plant Physiol 141:1056–1067

Bailey BA, Jennings JC, Anderson JD (1997) The 24-kDa protein from *Fusarium oxysporum* f. sp. *erythroxyli*: occurrence in related fungi and the effect of growth medium on its production. Can J Microbiol 43:45–55

Benhamou N, Joosten MHAJ, de Wit PJGM (1990) Subcellular localization of chitinase and of its potential substrate in tomato root tissues infected by *Fusarium oxysporum* f. sp. *radicis-lycopersici*. Plant Physiol 92:1108–1120

Birch PRJ, Rehmany AP, Pritchard L, Kamoun S, Beynon JL (2006) Trafficking arms: oomycete effectors enter host plant cells. Trends Microbiol 14:8–11

Bittner-Eddy PD, Allen RL, Rehmany AP, Birch P, Beynon JL (2003) Use of suppression subtractive hybridization to identify downy mildew genes expressed during infection of *Arabidopsis thaliana*. Mol Plant Pathol 4:501–507

Broekaert WF, Terras FRG, Cammue BPA (2000) Induced and performed antimicrobial proteins. In: Slusarenko AJ, Fraser RSS, Van Loon LC (eds) Mechanisms of resistance to plant diseases, Kluwer, Dordrecht

Bohlmann H (1994) The role of thionins in plant protection. Crit Rev Plant Sci 13:1–16

Bowles DJ (1990) Defence-related proteins in higher plants. Annu Rev Biochem 59:873–907

Brunner F, Rosahl S, Lee J, Rudd JJ, Geiler C, Kauppinen S, Rasmussen G, Scheel D, Nürnberger T (2002) Pep-13, a plant defence-inducing pathogen-associated pattern from *Phytophthora* transglutaminases. EMBO J 21:6681–6688

Cabral KMS, Almeida MS, Valente AP, Almeida FCL, Kurtenbach E (2003) Production of the active antifungal *Pisum sativum* defensin 1 (Psd1) in *Pichia pastoris*: overcoming the inefficiency of the STE13 protease. Prot Expr Purif 31:115–122

Campo S, Carrascal M, Coca M, Abian J, San Segundo B (2004) The defence response of germinating maize embryos against fungal infection: a proteomics approach. Proteomics 4:383–396

Casado-Vela J, Sellas S, Martinez RB (2006) Proteomic analysis of tobacco mosaic virus-infected tomato (*Lycopersicon esculentum* M.) fruits and detection of viral coat protein. Proteomics 6:S196–206

Castillejo M, Amiour N, Dumas-Gaudot E, Rubiales D, Jorrin JV (2004) A proteomic approach to studying plant response to crenate broomrape (*Orobanche crenata*) in pea (*Pisum sativum*). Phytochemistry 65:1817–1828

Castro MD, Fontes W, Morphy L, Bloch C (1996) Complete amino acid sequences of two gamma-thionins from maize (*Zea mays* L.) seeds. Protein Peptide Lett 3:267–274

Catanzariti AM, Dodds PN, Lawrence GJ, Ayliffe MA, Ellis JG (2006) Haustorially expressed secreted proteins from flax rust are highly enriched for avirulence elicitors. Plant Cell 18:243–256

Catlett NL, Yoder OC, Turgeon BG (2003) Whole-genome analysis of two-component signal transduction genes in fungal pathogens. Eukaryotic Cell 2:1151–1161

Chen ZY, Brown RL, Rajasekaran K, Damann KE, Cleveland TE (2006) Identification of a maize kernel pathogenesis-related protein and evidence for its involvement in resistance to *Aspergillus flavus* infection and aflatoxin production. Phytopathology 96:87–95

Colditz F, Nyamsuren O, Niehaus K, Eubel H, Braun H-P, Krajinski F (2004) Proteomic approach: identification of *Medicago truncatula* proteins induced in roots after infection with the pathogenic oomycete *Aphanomyces euteiches*. Plant Mol Biol 55:109–120

Colditz F, Braun H-P, Jacquet C, Niehaus K, Krajinksi F (2005) Proteomic profiling unravels insights into the molecular background underlying increased *Aphanomyces euteiches*-tolerance of *Medicago truncatula*. Plant Mol Biol 59:387–406

Colditz F, Niehaus K, Krajinski F (2007) Silencing of PR-10-like proteins in *Medicago truncatula* results in an antagonistic induction of other PR proteins and in an increased tolerance upon infection with the oomycete *Aphanomyces euteiches*. Planta (in press) DOI: 10.1007/s00425-006-0466-y

Cooper B, Garett WM, Campbell KB (2006) Shotgun identification of proteins from uredospores of the bean rust *Uromyces appendiculatus*. Proteomics 6:2477–2484

Corbin DR, Sauer N, Lamb CJ (1987) Differential regulation of a hydroxyproline-rich glycoprotein gene family in wounded and infected plants. Mol Cell Biol 7:4337–4344

Custers JHHV, Harrison SJ, Sela-Buurlage MB, Van Deventer E, Lageweg W, Howe PW, van der Meijs PJ, Ponstein AS, Simons BH, Melchers LS, Stuiver MH (2004) Isolation and characterisation of a class of carbohydrate oxidases from higher plants, with a role in active defence. Plant J 39:147–160

Dangl JL, Jones JDG (2001) Plant pathogens and integrated defence responses to infection. Nature 411:826–833

Dean RA, Talbot NJ, Ebbole DJ, Farman ML, Mitchell TK, Orbach MJ, Thon M, Kulkarni R, Xu JR, Pan H, Read ND, Lee YH, Carbone I, Brown D, Oh YY, Donofrio N, Jeong JS, Soanes DM, Djonovic S, Kolomiets E, Rehmeyer C, Li W, Harding M, Kim S, Lebrun MH, Bohnert H, Coughlan S, Butler J, Calvo S, Ma LJ, Nicol R, Purcell S, Nusbaum C, Galagan JE, Birren BW (2005) The genome sequence of the rice blast fungus *Magnaporthe grisea*. Nature 434:980–986

De la Fuente van Bentem S, Vossen JH, de Vries KJ, van Wees S, Tameling WI, Dekker HL, de Koster CG, Haring MA, Takken FL, Cornelissen BJ (2005) Heat shock protein 90 and its co-chaperone protein phosphatase 5 interact with distinct regions of the tomato I-2 disease resistance protein. Plant J 43:284–298

Dietrich A, Mayer JE, Hahlbrock K (1990) Fungal elicitor triggers rapid, transient, and specific protein phosphorylation in parsley cell suspension cultures. J Biol Chem 265:6360–6365

Dixon RA, Lamb CJ (1990) Molecular communication in plant–microbial pathogen interactions. Annu Rev Plant Physiol Plant Mol Biol 41:339–367

Dodds PN, Lawrence GJ, Catanzariti AM, Ayliffe MA, Ellis JG (2004) The *Melampsora lini AvrL567* avirulence genes are expressed in haustoria and their products are recognized inside plant cells. Plant Cell 16:755–768

Edens L, Hselinga L, Klok R, Ledeboer AM, Maat J, Toonen MY, Visser C, Verrips C (1982) Cloning of cDNA encoding the sweet tasting plant protein thaumatin and its expression in *E. coli*. Gene 18:1–12

Ellis J, Catanzariti A-M, Dodds P (2006) The problem of how fungal and oomycete avirulence proteins enter plant cells. Trends Plant Sci 11:61–63

Fagoaga C, Rodrigo I, Conejero V, Hinarejos C, Tuset JJ, Arnau J, Pina JA, Navarro L, Peña L (2001) Increased tolerance to *Phytophthora citrophthora* in transgenic orange plants constitutively expressing a tomato pathogenesis related protein PR-5. Mol Breed 7:175–185

Felix G, Grosskopf DG, Regenass M, Boller T (1991) Rapid changes of protein phosphorylation are involved in transduction of the elicitor signal in plant cells. Proc Natl Acad Sci USA 88:8831–8834

Felix G, Regenass M, Spanu P, Boller T (1994) The protein phosphatase inhibitor calyculin A mimics elicitor action in plant cells and induced rapid hyperphosphorylation of specific protein as revealed by pulse labelling with [^{32}P] phosphate. Proc Natl Acad Sci USA 91:952–956

Ferreira RB, Montiero SS, Piçarra-Pereira MA, Teixeira AR (2004) Engineering grapevine for increased resistance to fungal pathogens without compromising wine stability. Trends Biotechnol 22:168–173

Fobert PR, Despres C (2005) Redox control of systemic acquired resistance. Curr Opin Plant Biol 8:378–382

Freeman S, Rodriguez RJ (1993) Genetic conversion of a fungal plant pathogen to a nonpathogenic, endophytic mutualist. Science 260:75–78

Ganem S, Lu SW, Lee BN, Chou DY, Hadar R, Turgeon BG, Horowitz BA (2004) G-protein beta subunit of *Cochliobolus heterostrophus* involved in virulence, asexual and sexual reproductive ability, and morphogenesis. Eukaryotic Cell 3:1653–1663

Gaulin E, Jauneau A, Villalba F, Rickauer M, Esquerré-Tugayé M-T, Bottin A (2002) The CBEL glycoprotein of *Phytophthora parasitica* var *nicotianae* is involved in cell wall deposition and adhesion to cellulosic substrates. J Cell Sci 115:4565–4575

Gaulin E, Dramé N, Lafitte C, Torto-Alalibo T, Martinez Y, Ameline-Torregrosa C, Khatib M, Mazarguil H, Villalba-Mateos F, Kamoun S, Mazars C, Dumas B, Bottin A, Esquerré-Tugayé M-T, Rickauer M (2006) Cellulose binding domains of a *Phytophthora* cell wall protein are novel pathogen-associated molecular patterns. Plant Cell 18:1766–1777

Gianinazzi-Pearson V, Dumas-Gaudot E, Gollotte A, Tahhiri-Alaoui A, Gianinazzi S (1996) Cellular and molecular defence-related root response to invasion by arbuscular mycorrhizal fungi. New Phytol 133:45–57

Gilkes NR, Henrissat B, Kilburn DG, Miller RCJ, Warren RA (1991) Domains in microbial beta-1,4-glycanases: sequence conservation, function, and enzyme families. Microbial Rev 55:303–315

Gomez-Gomez L, Boller T (2000) FLS2: an LRR receptor-like kinase involved in the perception of the bacterial elicitor flagellin in *Arabidopsis*. Mol Cell 5:1003–1011

Gonzalez-Lamothe, R, Tsitsigiannis DI, Ludwig, AA, Panicot M, Shirasu K, Jones JDG (2006) The U-box protein CMPG1 is required for efficient activation of defence mechanisms triggered by multiple resistance genes in tobacco and tomato. Plant Cell 19:1067–1083

Govrin EM, Levine A (2000) The hypersensitive response facilitates plant infection by the necrotrophic pathogen *Botrytis cinerea*. Curr Biol 10:751–757

Grant M, Mansfield J (1999) Early events in host-pathogen interactions. Curr Opin Plant Biol 2:312–319

Grenier J, Potvin C, Trudel J, Asselin A (1999) Some thaumatin-like proteins hydrolyse polymeric β-1,3-glucans. Plant J 19:473–480

Grover A, Gowthaman R (2003) Strategies for development of fungus-resistant transgenic plants. Curr Sci 84:330–340

Guy GR, Philip R, Tan YH (1994) Analysis of cellular phosphoproteins by two-dimensional gel electrophoresis: applications for cell signalling in normal and cancer cells. Electrophoresis 15:417–40

Gygi SP, Rochon Y, Franza BR, Aebersold R (1999) Correlation between protein and mRNA abundance in yeast. Mol Cell Biol 19:1720–1730

Halim VA, Hunger A, Macioszek V, Landgraf P, Nürnberger T, Scheel D, Rosahl, S (2004) The oligopeptide elicitor Pep-13 induces salicylic acid-dependent and -independent defence reactions in potato. Physiol Mol Plant Pathol 64:311–318

Ham KS, Wu SC, Darvill AG, Albersheim P (1997) Fungal proteins secrete an inhibitor protein that distinguishes isoforms of plant pathogenesis-related endo-beta-1,3-glucanases. Plant J 11:169–179

Hammond-Kosack KE, Jones JDG (1997) Plant disease resistance genes. Annu Rev Plant Physiol Plant Mol Biol 48:575–607

Hanselle T, Ichinoseb Y, Barz W (2001) Biochemical and molecular biological studies on infection (*Ascochyta rabiei*)-induced thaumatin-like proteins from chickpea plants (*Cicer arietinum L.*). Z Naturforsch 56:1095–1107

Harrison MJ (1999) Molecular and cellular aspects of the arbuscular mycorrhizal symbiosis. Annu Rev Plant Physiol Plant Mol Biol 50:361–389

Harrison MJ (2005) Signalling in the arbuscular mycorrhizal symbiosis. Annu Rev Microbiol 59:19–42

He X, Anderson JC, del Pozo O, Gu YQ, Tang X, Martin GB (2004) Silencing of subfamily I of protein phosphatase 2A catalytic subunits results in activation of plant defence responses and localized cell death. Plant J 38:563–577

Hiller NL, Bhattacharjee S, van Ooij C, Liolios K, Harrison T, Lopez-Estrano C, Haldar K (2004) A host-targeting signal in virulence proteins reveals a secretome in malarial infection. Science 306:1934–1937

Imin N, De Jong F, Mathesius U, van Noorden G, Saeed NA, Wang X-D, Rose RJ, Rolfe BG (2004) Proteome reference maps of *Medicago truncatula* embryonic cell cultures generated from single protoplasts. Proteomics 4:1883–1896

Jain S, Srivastava S, Sarin NB, Kav NN (2006) Proteomics reveals elevated levels of PR10 proteins in saline-tolerant peanut (*Arachis hypogaea*) calli. Plant Physiol Biochem 44:253–259

Jakupovic M, Heintz M, Reichmann P, Mendgen K, Hahn M (2006) Microarray analysis of expressed sequence tags from haustoria of the rust fungus *Uromyces fabae*. Fungal Genet Biol 43:8–19

Jia Y, McAdams SA, Bryan GT, Hershey HP, Valent B (2000) Direct interaction of resistance gene and avirulence gene products confers rice blast resistance. EMBO J 19:4004–4014

Kamoun S (2005) A catalogue of the effector secretome of plant pathogenic oomycetes. Annu Rev Phytopathol 44:1–20

Kamoun S, Young M, Glascock C, Tyler BM (1993) Extracellular protein elicitors from *Phytophthora*: host-specificity and induction of resistance to fungal and bacterial phytopathogens. Mol Plant Microbe Interact 6:15–25

Kamoun S, van West P, Vleeshouwers VG, de Groot KE, Govers F (1998) Resistance of *Nicotiana benthamiana* to *Phytophthora infestans* is mediated by the recognition of the elicitor protein INF1. Plant Cell 10:1413–1426

Kamoun S, Hraber P, Sobral B, Nuss D, Govers F (1999) Initial assessment of gene diversity for the oomycete pathogen *Phytophthora infestans* based on expressed sequences. Fungal Genet Biol 28:94–106

Kang Z, Buchenauer H (2003) Immunocytochemical localization of cell wall-bound thionins and hydroxyproline-rich glycoproteins in *Fusarium culmorum*-infected wheat spikes. J Phytopathol 151:120–129

Kanneganti TD, Huitema E, Cakir C, Kamoun S (2006) Synergistic interaction of the plant cell death pathways induced by *Phytophthora infestans* Nep1-like protein PiNPP1.1 and INF1 elicitin. Mol Plant Microbe Interact 19:854–863

Kemen E, Kemen AC, Rafiqi M, Hempel U, Mendgen K, Hahn M, Voegele RT (2005) Identification of a protein from rust fungi transferred from haustoria into infected plant cells. Mol Plant Microbe Interact 18:1130–1139

Khatib M, Lafitte C, Esquerré-Tugayé M-T, Bottin A, Rickauer M (2004) The CBEL elicitor of *Phytophthora parasitica* var. *nicotianae* activates defence in *Arabidosis thaliana* via three different signalling pathways. New Phytol 162:501–510

Kim ST, Kim SG, Hwang DH, Kang SY, Kim HJ, Lee BH, Lee JJ, Kang KY (2004) Proteomic analysis of pathogen-responsive proteins from rice leaves induced by rice blast fungus, *Magnaporthe grisea*. Proteomics 4:3569–3578

Lambais MR, Mehdy MC (1993) Suppression of endochitinase, β-1,3-endoglucanase and chalcone isomerase expression in bean vesicular arbuscular mycorrhizal roots under different soil phosphate conditions. Mol Plant Microbe Interact 6:75–83

Lecourieux-Ouaked F, Pugin A, Lebrun-Garcia A (2000) Phosphoproteins involved in the signal transduction of cryptogein, an elicitor of defence reactions in tobacco. Mol Plant Microbe Interact 13:821–829

Liu Z, Bos JIB, Armstrong M, Whisson SC, da Chunha L, Torto-Alalibo T, Win J, Avrova AO, Wright F, Birch PRJ, Kamoun S (2005) Patterns of diversifying selection in the phytotoxin-like *scr74* gene family of *Phytophthora infestans*. Mol Biol Evol 22:659–672

Luderer R, de Kock MJD, Dees RHL, de Wit PJM, Joosten MH (2002) Functional analysis of cysteine residues of ECP elicitor proteins of the fungal tomato pathogen *Cladosporium fulvum*. Mol Plant Pathol 2:91–95

Maldonado AM, Doerner P, Dixon RA, Lamb CJ, Cameron RK (2002) A putative lipid transfer protein involved in systemic resistance signalling in *Arabidopsis*. Nature 419:399–403

Markovic-Housley Z, Degano M, Lamba D, von Roepenack-Lahaye E, Clemens S, Susani M, Ferreira F, Scheiner O, Breitender H (2003) Crystal structure of a hypoallergenic isoform of the major birch pollen allergen Bet v 1 and its likely biological function as a plant steroid carrier. J Mol Biol 325:123–133

Marti M, Good RT, Rug M, Knuepfer E, Cowman AF (2004) Targeting malaria virulence and remodeling proteins to the host erythrocyte. Science 306:1930–1933

Martin GB, Bogdanove AJ, Sessa G (2003) Understanding the functions of plant disease resistance proteins. Annu Rev Plant Biol 54:23–61

Mathesius U, Keijzers G, Natera SH, Weinman JJ, Djordjevic MA, Rolfe BG (2001) Establishment of a root proteome reference map for the model legume *Medicago truncatula* using the expressed sequence tag database for peptide mass fingerprinting. Proteomics 1:1424–1440

Mendgen K, Hahn M (2002) Plant infection and the establishment of fungal biotrophy. Trends Plant Sci 7:352–356

Menu-Bouaouiche L, Vriet C, Peumans WJ, Barre A, Van Damme EJM, Rougé P (2003) A molecular basis for the endo-β1,3-glucanase activity of the taumatin-like proteins from edible fruits. Biochemie 85:123–131

Mitchel HJ, Kovac KA, Hardham AR (2002) Characterisation of *Phytophthora nicotianae* zoospore and cyst membrane proteins. Mycol Res 106:1211–1223

Mithöfer A, Ebel J, Felle HH (2005) Cation fluxes cause plasma membrane depolarization involved in beta-glucan elicitor-signalling in soybean roots. Mol Plant Microbe Interact 18: 983–990

Moon BC, Choi MS, Kang YH, Kim MC, Cheong MS, Park CY, Yoo JH, Koo SC, Lee SM, Lim CO, Cho MJ, Chung WS (2005) Arabidopsis ubiquitin-specific protease 6 (AtUBP6) interacts with calmodulin. FEBS Lett 579:3885–3890

Narasimhan ML, Lee H, Damsz B, Singh NK, Ibeas JI, Matsumoto TK, Woloshuk CP, Bressan RA (2003) Over-expression of a cell wall glycoprotein in *Fusarium oxysporum* increases virulence and resistance to a plant PR-5 protein. Plant J 36:390–400

Natera SH, Guerreiro N, Djordjevic MA (2000) Proteome analysis of differentially displayed proteins as a tool for the investigation of symbiosis. Mol Plant Microbe Interact 13:995–1009

Ndimba BK, Chivasa S, Hamilton JM, Simon WJ, Slabas AR (2003) Proteomic analysis of changes in the extracellular matrix of Arabidopsis cell suspension cultures induced by fungal elicitors. Proteomics 3:1047–1059

Nekrasov V, Ludwig AA, Jones JDG (2006) CITRX thioredoxin is a putative adaptor protein connecting Cf-9 and the ACIK1 protein kinase during the Cf-9/Avr9-induced defence response. FEBS Lett 580:4236–4241

Nürnberger T, Lipka V (2005) Non-host resistance in plants: new insights into an old phenomenon. Mol Plant Pathol 6:335–345

Nürnberger T, Brunner F, Kemmerling B, Piater L (2004) Innate immunity in plants and animals: striking similarities and obvious differences. Immunol Rev 198:249–266

Orsomando G, Lorenzi M, Raffaelli N, Dalla Rizza M, Mezzetti B, Ruggieri S (2001) Phytotoxic protein PcF, purification, characterization, and cDNA sequencing of a novel hydroxyproline-containing factor secreted by the strawberry pathogen *Phytophthora cactorum*. J Biol Chem 276:21578–21584

Padliya ND, Cooper B (2006) Mass spectrometry-based proteomics for the detection of plant pathogens. Proteomics 6:4069–4075

Paquin B, Laforest MJ, Forget L, Roewer I, Wang Z, Longcore J, Lang BF (1997) The fungal mitochondrial genome project: evolution of fungal mitochondrial genomes and their gene expression. Curr Genet 31:380–95

Park C-J, Kim K-J, Shin R, Park JM, Shin Y-C, Paek K-H (2004) Pathogenesis-related protein 10 from hot pepper functions as a ribonuclease in an antiviral pathway. Plant J 37:186–198

Parniske M (2005) Plant fungal associations: cue for the branching connection. Nature 435:750–751

Peck SC, Nühse TS, Hess D, Iglesias A, Meins F, Boller T (2001) Directed proteomics identifies a plant-specific protein rapidly phosphorylated in response to bacterial and fungal elicitors. Plant Cell 13:1467–1475

Peligrini PB, Franco OL (2005) Plant gamma-thionins: novel insights on the mechanisms of action of a multi-functional class of defence proteins. Int J Biochem Cell Biol 37:2239–2253

Pemberton CL, Salmond GPC (2004) The Nep1-like proteins: a growing family of microbial elicitors of plant necrosis. Mol Plant Pathol 5:353–359

Qutob D, Kamoun S, Gijzen M (2002) Expression of a *Phytophthora sojae* necrosis-inducing protein occurs during transition from biotrophy to necrotrophy. Plant J 32:361–373

Redman RS, Freeman S, Clifton DR, Morrel J, Brown G, Rodriguez RJ (1999) Biochemical analysis of plant protection afforded by a nonpathogenic endophytic mutant of *Colletotrichum magna*. Plant Physiol 119:795–804

Rehmany AP, Gordon A, Rose LE, Allen RL, Armstrong MR, Whisson SC, Kamoun S, Tyler BM, Birch PR, Beynon JL (2005) Differential recognition of highly divergent downy mildew avirulence gene alleles by *RPP1* resistance genes from two *Arabidopsis* lines. Plant Cell 17:1839–1850

Rep M, Dekker HL, Vossen JH, de Boer AD, Houterman PM, Speijer D, Back JW, de Koster CG, Cornelissen BJC (2002) Mass spectrometric identification of isoforms of PR proteins in xylem sap of fungus-infected tomato. Plant Physiol 130:904–917

Rivas S, Thomas CM (2005) Molecular interactions between tomato and the leaf mold pathogen *Cladosporium fulvum*. Annu Rev Phytopathol 43:395–436

Ross AF (1961) Systemic acquired resistance induced by localized virus infections in plants. Virology 14:340–358

Sacks W, Nurnberger T, Hahlbrock K, Scheel D (1995) Molecular characterization of nucleotide sequences encoding the extracellular glycoprotein elicitor from *Phytophthora megasperma*. Mol Gen Genet 166:1517–1527

Salzer P, Bonanomi A, Beyer K, Vogeli-Lange R, Aeschbacher RA, Lange J, Wiemken A, Kim D, Cook DR, Boller T (2000) Differential expression of eight chitinase genes in *Medicago truncatula* roots during mycorrhiza formation, nodulation, and pathogen infection. Mol Plant Microbe Interact 13:763–777

Sarry JE, Sommerer N, Sauvage FX, Bergoin A, Rossignol M, Albagnac G, Romieu C (2004) Grape berry biochemistry revisited upon proteomic analysis of the mesocarp. Proteomics 4:201–215

Schmelzer E (2002) Cell polarization, a crucial process in fungal defence. Trends Plant Sci 7:411–415

Schmidt WE, Ebel J (1987) Specific binding of a fungal glucan phytoalexin elicitor to membrane fractions from soybean *Glycine max*. Proc Natl Acad Sci USA 84:4117–4121

Shan W, Cao M, Leung D, Tyler BM (2004) The *Avr1b* locus of *Phytophthora sojae* encodes an elicitor and a regulator required for avirulence on soybean plants carrying resistance gene *Rps1b*. Mol Plant Microbe Interact 17:394–403

Sharp JK, McNeil M, Albersheim P (1984) The primary structures of one elicitor-active and seven elicitor-inactive hexa(β-D-glucopyranosyl)-D-glucitols isolated from the mycelial walls of *Phytophthora megasperma* var. *sojae*. J Biol Chem 259:11321–11336

Sheperd SJ, van West P, Gow NAR (2003) Proteomic analysis of asexual development of *Phytophthora palmivora*. Mycol Res 107:395–400

Slusarenko AJ, Fraser RSS, Van Loon LC (2000) Mechanisms of resistance to plant diseases. Kluwer, Dordrecht

Stec B, Markman O, Rao U, Heffron G, Henderson S, Vernon LP, Brumfeld V, Teeter MM (2004) Proposal for molecular mechanisms of thionins deduced from physico-chemical studies of plant toxins. J Pept Res 64:210–224

Stintzi A, Heitz T, Prasad V, Wiedemann-Merdinoglu S, Kauffmann S, Geoffroy P, Legrand M, Fritig B (1993) Plant pathogenesis-related proteins and their role in defence against pathogens. Biochemie 75:687–706

Stone JM, Walker JC (1995) Plant protein kinase families and signal transduction. Plant Physiol 108:451–457

Torto T, Li S, Styer A, Huitema E, Testa A, Gow NAR, van West P, Kamoun S (2003) EST mining and functional expression assays identify extracellular effector proteins from the plant pathogen *Phytophthora*. Genome Res 13:1675–1685

Tuzun S, Somanchi A (2006) The possible role of PR proteins in multigenic and induced systemic resistance. In: Tuzun S, Bent E (eds) Multigenic and induced systemic resistance in plants. Springer, New York, pp 112–142

Valot B, Negroni L, Zivy M, Gianinazzi S, Dumas-Gaudot E (2006) A mass spectrometric approach to identify arbuscular mycorrhiza-related proteins in root plasma membrane fractions. Proteomics 6:145–155

Van den Ackerveken G, Marois E, Bonas U (1996) Recognition of the bacterial avirulence protein AvrBs3 occurs inside the host plant cell. Cell 87:1307–1316

Van Kan JA (2006) Licensed to kill: the lifestyle of a necrotrophic plant pathogen. Trends Plant Sci 11:247–253

Van Loon LC, Rep M, Pieterse CMJ (2006) Significance of inducible defence-related proteins in infected plants. Annu Rev Phytopathol 44:1–28

Van West P, Appiah AA, Gow NAR (2003) Advances in research on oomycete root pathogens. Physiol Mol Plant Pathol 62:99–113

Vernooij B, Friedrich L, Morse A, Reist R, Kolditz-Jawhar R, Ward E, Uknes S, Kessmann H, Ryals J (1994) Salicylic acid is not the translocated signal responsible for inducing systemic acquired resistance but is required in signal transduction. Plant Cell 6:959–965

Viard M-P, Martin F, Pugin A, Ricci P, Blein J-P (1994) Protein phosphorylation is induced in tobacco cells by the elicitor cryptogein. Plant Physiol 104:1245–1249

Vierheilig H, Alt M, Lange J, Gut-Rella M, Wiemken A, Boller T (1995) Colonization of transgenic tobacco constitutively expressing pathogenesis-related proteins by the vesicular-arbuscular mycorrhizal fungus *Glomus mosseae*. Appl Environ Microbiol 61:3031–3034

Vigers AJ, Roberts WK, Selitrennikoff CP (1991) A new family of plant antifungal proteins. Mol Plant Microbe Interact 4:315–323

Vögele RT (2006) *Uromyces fabae*: development, metabolism, and interactions with its host *Vicia faba*. FEMS Microbiol Lett 259:165–173

Vögele RT, Mendgen K (2003) Rust haustoria: nutrient uptake and beyond. New Phytol 159:93–100

Vögele RT, Hahn M, Lohaus G, Link T, Heiser I, Mendgen K (2005) Possible roles for mannitol and mannitol dehydrogenase in the biotrophic plant pathogen *Uromyces fabae*. Plant Physiol 137:190–198

Wang Y, Yang L, Xu H, Li Q, Ma Z, Chu C (2005) Differential proteomic analysis of proteins in wheat spikes induced by *Fusarium graminearum*. Proteomics 5:4496–4503

Waugh M, Hraber P, Weller J, Wu Y, Chen G, Inman J, Kiphart D, Sobral B (2000) The *Phytophthora* genome initiative database: informatics and analysis for distributed pathogenomic research. Nucleic Acids Res 28:87–90

Wienkoop S, Saalbach G (2003) Proteome analysis. Novel proteins identified at the peribacteroid membrane from *Lotus japonicus* root nodules. Plant Physiol 131:1080–1090

Xia Y, Suzuki H, Borevitz J, Blount J, Guo Z, Patel K, Dixon RA, Lamb C (2004) An extracellular aspartic protease functions in *Arabidopsis* disease resistance signalling. EMBO J 23:980–988

Xing T, Higgins VJ, Blumwald E (1997) Race-specific elicitors of *Cladosporium fulvum* promote translocation of cytosolic components of NADPH oxidase to the plasma membrane of tomato cells. Plant Cell 9:249–259

Xing T, Rampitsch C, Mauthe W, Oullet T, Miki B (2001) Proteomic identification of protein–protein and protein–DNA interactions in MAP kinase pathways that are involved in plant defence reactions. Plant Physiol 125[Suppl]:40

Xing T, Ouellet T, Miki BL (2002) Towards genomic and proteomic studies of protein phosphorylation in plant–pathogen interactions. Trends Plant Sci 7:224–230

Yang Y, Shah J, Klessig DF (1997) Signal perception and transduction in plant defence responses. Genes Dev 11:1621–1639

Yun D-J, Zhao Y, Pardo JM, Narasimhan ML, Damsz B, Lee H, Abad LR, D'Urzo MP, Hasegawa PM, Bressan RA (1997) Stress proteins on the yeast cell surface determine resistance to osmotin, a plant antifungal protein. Proc Natl Acad Sci USA 94:7082–7087

Zhang S, Liu Y, Klessig DF (2000) Multiple levels of tobacco WIPK activation during the induction of cell death by fungal elicitins. Plant J 23:339–347

Zipfel C, Felix G (2005) Plants and animals: a different taste for microbes? Curr Opin Plant Biol 8:353–360

Zipfel C, Robatzek S, Navarro L, Oakeley EJ, Jones JDG, Felix G, Boller T (2004) Bacterial disease resistance in *Arabidopsis* through flagellin perception. Nature 428:764–767

Zhou W, Kolb FL, Riechers DE (2005) Identification of proteins induced or upregulated by *Fusarium* head blight infection in the spikes of hexaploid wheat (*Triticum aestivum*). Genome 48:770–780

Zhou W, Eudes F, Laroche A (2006) Identification of differentially regulated proteins in response to a compatible interaction between the pathogen *Fusarium graminearum* and its host, *Triticum aestivum*. Proteomics 6:4599–4609

Chapter 19
Metabolic Intricacies of the Symbiotic Association between Soybean and *Bradyrhizobium japonicum*: A Proteomic Outlook

Annamraju D. Sarma, Nathan W. Oehrle, and David W. Emerich

Abstract Symbiotic nitrogen fixation, the primary pathway by which inorganic nitrogen is made available for living organisms, requires complex communication between the bacterial microsymbiont and the host plant, beginning in the soil and ending at nodule senescence. The symbiosis takes the form of a highly complex structure referred to as a nodule, which appears as a tumor-like growth on the roots of certain leguminous plants. Both partners exchange signals and change metabolically and morphologically in response to their fellow symbiont. These changes by necessity must be coordinated and complementary. Proteomic analysis has revealed extensive changes in the proteomes of each organism during symbiosis.

19.1 Introduction

Plant–microbe interactions involve complex signaling systems and metabolic responses with varying levels of complexity. Perception of a signal from the interactive partner results in activation of the appropriate signal transduction pathway. The response proceeds through signal transduction proteins and often results in de novo protein synthesis. Some responses are transient whereas others lead to longer term metabolic and morphological changes. Molecular biology has been the primary tool for identification of the known signal perception and transmission events. Proteomics offers the technology for identification of the downstream events that lead to longer term metabolic and morphological changes.

Annamraju D. Sarma, Department of Biochemistry, 117 Schweitzer Hall, University of Missouri, Columbia, MO 65211, USA

Nathan W. Oehrle, Department of Biochemistry, 117 Schweitzer Hall, University of Missouri, Columbia, MO 65211, USA

David W. Emerich, Department of Biochemistry, 117 Schweitzer Hall, University of Missouri, Columbia, MO 65211, USA, E-mail: EmerichD@Missouri.edu

J. Šamaj and J. Thelen (eds.), *Plant Proteomics*
© Springer 2007

19.2 Symbiotic Nitrogen Fixation

Symbiotic nitrogen fixation – the primary pathway by which inorganic nitrogen is made available for living organisms – requires complex communication between the bacterial microsymbiont and the host plant, beginning in the soil and ending at nodule senescence. The symbiosis takes the form of a highly complex structure referred to as a nodule, which appears as a tumor-like growth on the roots of certain leguminous plants. Both partners exchange signals and change metabolically and morphologically in response to their fellow symbiont. These changes by necessity must be coordinated and complementary. Proteomic analysis performed on only one of the symbionts would not reveal the coordinated and complementary nature of the symbiosis.

The complexity of the symbiosis extends from pre-infection and continues throughout the lifetime of the symbiosis. Specific rhizobium species infect specific leguminous host plants, for example, *Bradyrhizobium japonicum* infects soybean but not other leguminous plants. The nodules on soybean are located predominately at the root crown and at maturity are up to 4 mm in diameter. The signals to initiate symbiosis are believed to be specific flavonoids excreted by the plant roots and recognized by a rhizobial receptor protein: NodD or NodD-like protein. NodD, in the presence of the specific flavonoid(s), transcriptionally activates other rhizobial *nod* genes (Dénarié and Debellé 1996). The *nod* gene products participate in the synthesis and export of compounds called Nod factors, which are lipochitin-oligosaccharide molecules, substituted with sulfate, acetate or fucosyl residues in a species-specific manner (Dénarié and Debellé 1996). The tips of the newly emerging root hairs of the plant root are the preferred sites of rhizobial attachment (Sprent 1989). After the rhizobia bind, the root hairs curl and the bacteria invade the root hair encased within a plant-derived membrane called the infection thread. Transcriptomic analysis has been employed to identify additional genes involved in early steps of infection (Bontemps et al. 2005; Capela et al. 2005). Recently, Wan et al. (2005) used proteomics to identify proteins in root hairs after infection by *B. japonicum*. They found proteins previously identified as being produced in response to rhizobium inoculation, such as lipoxygenases, agglutinin, actin, peroxidase and phenylalanine ammonia lyase, but also novel proteins such as phospholipase D and phosphoglucomutase. Phospholipase D functions in lipid signaling pathways and may be an early step in a signal transduction pathway (den Hartog et al. 2001). Phosphoglucomutase plays a role in the partitioning of stored carbon in plants and thus may be involved in provision of energy for infection thread elongation. Chitinase, class I, and stress-induced gene H4, were found exclusively in the root hairs whereas CDPK (calmodulin-like domain protein kinase isoenzyme β), phosphoenolpyruvate carboxylase and ascorbate peroxidase 2 were more abundant in root hairs than in roots.

The infection thread elongates into the central cortical region of the root, branches, then penetrates meristematic cells depositing bacteria within membrane sacs called symbiosomes (also called the peri-bacteroid membrane). The plant cells differentiate into uninfected and infected cells, which multiply and expand to form a nodule. The mitotic phase of soybean nodule development, a period of rapid cell division that continues for 12–18 days after inoculation, is characterized by declining soluble protein content and slowly rising total protein content per nodule (Anthon and Emerich 1990; Newcomb 1981). The period of cell expansion that follows is characterized by relatively constant soluble protein content (Anthon and Emerich 1990). The rapid rise in total protein coincides with the appearance of nitrogenase activity. The bacteria within the symbiosomes of the infected cells differentiate morphologically and biochemically to a form referred to as a bacteroid, which can derepress nitrogenase, the enzyme complex that reduces atmospheric N_2 to ammonium.

Determinant, or spherical, nodules, such as those found on soybean formed by infection with *B. japonicum*, do not have persistent meristems; the vascular system becomes closed, conferring a continuous system of vascular branches, and there is little or no involvement of infection threads in the distribution of bacteria into nodule cells. Determinate nodules are developmentally synchronized, i.e., they contain only one specific developmental form, which is dictated by the age of the nodule. Furthermore, the infection process is controlled such that all the nodules within the root crown are of the same developmental age. In contrast, indeterminant nodules, such as those formed by *Sinorhrizobium meliloti* and alfalfa produce longitudinal nodules with persistent meristems containing tissues and bacteria/bacteroids of graded age from the growing point, which contains the youngest material, to the base of the nodule, which contains the oldest.

The nodule is a complex structure with different cell types and different physiological and functional zones. These zones can be categorized physiologically based on the metabolic activities of the component cells. Oxygen concentration plays a role in establishing these zones and consequently their metabolic activities (Hunt and Layzell 1993; Wycoff et al. 1998). The oxygen concentration in the outer layer of the nodule is at normal concentrations for plant material. Surrounding the central infected zone in the cortex is an endodermis layer that has suberized walls. This layer acts as a gas diffusion barrier thereby reducing the partial pressure of oxygen within the infected region of the nodule to about 10 nM O_2. Leghemoglobin, a myoglobin-like protein found in the infected region of the nodule, acts as an oxygen buffer maintaining a low partial pressure of oxygen to permit the functioning of the oxygen-labile nitrogenase component proteins within the bacteroids.

The central portion of a determinant nodule is notable for the abundant presence of leghemoglobin, resulting in a red-colored interior. This portion of a nodule is interspersed with smaller infected and larger uninfected cells, resulting in a ration of approximately 1.6 uninfected cells for each infected cell (Newcomb 1981). Plasmodesmata connect infected and uninfected cells within the central core, and cortical cells surrounding the central region. The structural complexity underlies the metabolic complexity of the nodule. Infected and uninfected cells have different metabolic roles in symbiotic nitrogen fixation. Infected and uninfected nodule cell protoplasts have been separated from each other and analyzed for enzymatic activities (Copeland et al. 1989; Kouchi et al. 1988; Shelp et al. 1983; Suganuma et al. 1987).

Glycolytic enzymes were found predominately in the outer cortical and uninfected cells. Enzymes of nitrogen assimilation, such as glutamine synthetase, glutamate synthase, aspartase and alanine aminotransferase, were several-fold greater in abundance in uninfected cells than in infected cells. Ureide metabolism is found primarily in uninfected cells. Determinate nodules export primarily ureides as the final nitrogen assimilation product to the plant, whereas indeterminate nodules export the amide amino acids, asparagine and glutamine. Proteomic and metabolite analysis have not been fully exploited to further our understanding of functional activities within specific cell types of the nodule.

19.3 Mitochondrial Proteomics

Mitochondria within infected cells are cristae-rich due to intensive folding of the inner mitochondrial membrane, and are found at the cell periphery adjacent to intercellular air spaces presumably to take advantage of the available oxygen. Hoa et al. (2004) analyzed proteins from the mitochondria of soybean nodules via two dimensional electrophoresis (2-DE) and matrix-assisted laser desorption ionisation (MALDI) analysis and detected phosphoserine aminotransferase, flavanone 3-hydroxylase, coproporphyrinogen III oxidase, and a ribonucleoprotein as being found exclusively in nodule mitochondria and absent in root mitochondria. Proteins that were up-regulated in the nodule mitochondria as compared to root mitochondria were: heat shock protein 70, glycine dehydrogenase, ATP synthase D chain, ATP synthase δ chain and the 27 kDa subunit of NADH dehydrogenase. Proteins down-regulated relative to root mitochondria were ATP synthase α chain, ATP synthase β chain, and dihydrolipoamide dehydrogenase. The presence of phosphoserine aminotransferase and glycine dehydrogenase is consistent with ureide production and the transport of ureides as the major carrier of fixed nitrogen to the shoots. Coproporphyrinogen III oxidase catalyzes the conversion of coproporphyrinogen III to protoporphyrinogen IX in the heme biosynthetic pathway, which could be used as the prosthetic group of leghemoglobin, one of the more prominent proteins within a nodule. Lee et al. (2004) also identified transcripts for coproporphyrinogen III oxidase, demonstrating the critical need for heme during symbiosis. Flavanone 3-hydroxylase was identified as specific to nodule mitochondria and is thought to participate in modifying the flavonoid and isoflavonoid pool (Charrier et al. 1995).

19.4 Nodule Cytosol Proteomics

A whole proteome analysis of the soybean nodule cytosol was undertaken in our laboratory using 2-DE protein separation followed by peptide mass fingerprinting (PMF; Oehrle 2005) and identification using the soybean UniGene database (Mooney et al. 2004). Selection of proteins for analysis was random but biased toward the most highly abundant proteins of the soybean nodule cytosol proteome. The largest category of proteins was that defined as being involved in carbon metabolism (~28%) followed by nitrogen metabolism (~12%) and oxygen protection (~12%). These three categories define the primary metabolic activities of the nodule and, as such, together constitute more than half of all the proteins identified. Proteins known to be present more in either infected cells or uninfected cells were found. For example, sucrose synthase has been reported to be present in greater abundance in infected cells whereas glutamine synthetase and aspartate aminotransferase were more abundant in uninfected cells (Kouchi et al. 1988). Urate oxidase has been localized to the peroxisomes of infected cells (Nguyen et al. 1985). Ascorbate peroxidase (Dalton et al. 1993) was found in the cytoplasm of both infected and uninfected nodule cells.

A number of proteins were represented both in the proteome, as found by Oehrle (2005), and as mRNAs in the transcriptome, as reported by Lee et al. (2004). Sucrose synthase, coproporphyrinogenase, leghemoglobin-A and -C and some members of the 14-3-3 protein family, peroxidases and resistance proteins were identified both as proteins and as transcripts. The fact that these were identified both in the proteome (Oehrle 2005) and in the transcriptome (Lee et al. 2004) demonstrates the critical role these components perform in nodule function. For example, sucrose synthase catalyzes the cleavage of sucrose to fructose and UDP-glucose. Anthon and Emerich (1990) previously described the operation of the sucrose synthase pathway in soybean nodules. The advantage of the sucrose synthase pathway in nodule metabolism is that it requires three molecules of ATP for metabolism of sucrose to triose-phosphates as opposed to the four ATPs required by the invertase pathway (Huber and Akazawa 1986).

Protein related to oxygen protection constituted about 12% of the total protein population identified. Leghemoglobins-A and -C, which maintain the low oxygen partial pressure within the nodule, and ferric leghemoglobin reductase-2, which maintains the iron in leghemoglobin in its active ferrous state (Lin et al. 1991), were also identified in both the proteome (Oehrle 2005) and the transcriptome (Lee et al. 2004), which attests to the critical roles of oxygen protection and supply within the nodule. Glutamate-semialdehyde aminotransferase, which catalyzes the conversion of [S]-4-amino-5-oxopentanoate to 5-aminolevulinate, an early step in heme biosynthesis, was identified, as were the heme-containing proteins catalase and ascorbate peroxidase. Ascorbate peroxidase, catalase, Fe-superoxide dismutase and Mn-superoxide dismutase, which perform roles in scavenging active oxygen species were all found in the nodule cytosol (Oehrle 2005). In contrast, a single Fe-superoxide dismutase was identified in the *B. japonicum* bacteroid proteome (Sarma and Emerich 2005).

Glutamine synthetase, aspartate aminotransferase, and inosine monophosphate dehydrogenase – all involved in ureide biosynthesis – were prominent proteins identified in the soybean nodule cytosol. Ureides are the major nitrogen transport product of soybean nodules. Several additional enzymes of amino acid metabolism were identified in the proteome, such as serine hydroxymethyltransferase and methionine synthase. In contrast, soybean nodule bacteroids were shown to have a considerable ability to synthesize a number of amino acids (Sarma and Emerich 2005). The ability of both symbionts to metabolize amino acids provides support for nutrient exchange cycles between the symbionts (Kahn et al. 1985; Lodwig et al. 2003).

Lipoxygenase and fatty acid epoxide hydroxylase were both highly expressed proteins in nodules (Oehrle 2005). Lipoxygenase had been documented in soybean root nodules previously (Junghans et al. 2004; Mohammadi and Karr 2003) but fatty acid epoxide hydroxylase had not. The expression of lipoxygenase and fatty acid epoxide hydrolase provides an alternative metabolic route for hydroperoxides formed by lipoxygenase instead of traumatic acid or jasmonic acid, which are involved in cell division and senescence, respectively. Traumatic acid is a member of a large class of alkenals that are toxic natural products of fatty acid oxidation (Farmer 1994). Thus, fatty acid epoxide hydrolase may perform a role in developmental synchrony and senescence of soybean nodules. Fatty acid epoxide hydrolase acts after the branch point at which 13-hydroperoxylinolenic acid is either cleaved by hydroperoxide lyase to cis-3-hexenal and 12-oxo-cis-dodencenoic acid – the latter being isomerized to traumatic acid (13-oxo-trans-10-dodecenoic acid) – or converted by allene oxide synthase to 12,13-epoxylinolenic acid. This latter compound can be metabolized through a series of rearrangement, cyclization, reduction and β-oxidation steps to form jasmonic acid.

Plant-genome-encoded proteins that are found only in the root nodules and not in any other parts of the host plant are called nodulins. Nodulins were originally identified by comparing in vivo translation products from nodule and root mRNA and by identifying cDNA clones by differential screening. Investigations have shown that several nodulins have been found in other portions of the plant, thus requiring the original definition to be modified to include genes/proteins highly up-regulated during symbiosis. "Early" nodulins are generally defined as being expressed during the infection and invasion process. The "late" nodulins are involved in nodule function and maintenance and are generally defined as coinciding with the beginning of measurable nitrogen fixation activity. The two general classes of late nodulins are metabolic nodulins and symbiosome membrane nodulins. Metabolic nodulins include leghemoglobin, uricase (a key enzyme of uriede biosynthesis), glutamine synthetase (which catalyzes the first step in ammonium assimilation), and sucrose synthase (which catalyzes the cleavage of sucrose to begin the pathway that produces carbon metabolites for bacteroid energy production). Proteomic analysis may identify additional nodulins.

A number of known nodulins were identified in recent proteomic investigations. Sucrose synthase, urate oxidase, ferric leghemoglobin reductase, leghemoglobin A and leghemoglobin C were identified by Oehrle (2005). Panter et al. (2000) identified Nodulin 53 as one of 17 proteins associated with the symbiosome membrane of soybean root nodules. Nodulin 53 is not predicted to have a transmembrane domain, but it may be anchored to the symbiosome membrane via myristoylation (Winzer et al. 1999). Panter et al. (2000) noted that Nodulin 53 was identified as a 31 kDa protein presumably after proteolysis and speculated that it was a processed product with a discrete function. The 31 kDa protein similar to Nodulin 53 was also identified by Oehrle (2005), confirming the observations of Panter et al. (2000).

A number of proteins requiring nucleosides and nucleotides were found in the whole nodule proteome: nucleoside diphosphate kinase, ATP synthase, ATPase β chain, and apyrase (Oehrle 2005). The nucleoside diphosphate kinase identified here was identical to that first reported by Krishnan et al. (1999) in the exudate from imbibing seed, and matches the nucleoside diphosphate kinase (NDPK) type I clones in the soybean databases. The energy demands of nitrogen fixation would require a capacity to maintain ATP concentrations via ATP synthase and NDPK. The presence of NDPK type I in both germinating seeds and nodules substantiates the extensive roles this enzyme performs in energy metabolism.

Vegetative storage proteins of 27, 29 and 94 kDa were prominent protein in the soybean nodule cytosol (Oehrle 2005). These proteins are lysine-rich glycoproteins first identified in the leaves of soybean (Wittenbach 1982) and have subsequently been found in other plant organs and ascribed potential metabolic roles (Meuriot et al. 2004; Penheiter et al. 1997; Tranbarger et al. 1991) in addition to nitrogen storage. In soybean, the lower molecular weight vegetative storage proteins have homology to acid phosphatases (Staswick 1989; Penheiter et al. 1997) and the high molecular weight vegetative storage proteins have similarity to lipoxygenase (Tranbarger et al. 1991). The low molecular weight vegetative storage proteins differ most strikingly from the acid phosphatases in the substitution of a serine in place of the conserved nucleophilic aspartate residue in the N terminus, which renders it a phosphatase. Leelapon et al. (2004) reported that a single amino acid substitution in soybean vegetative storage protein α increased its acid phosphatase activity by about 20-fold. Bona fide acid phosphatases that are closely related to vegetative storage proteins are also found in soybean nodules (Penheiter et al. 1997). In *Medicago sativa* taproots, methyl jasmonate alters expression of the low molecular weight vegetative storage proteins (Meuriot et al. 2004). As mentioned above, fatty acid epoxide hydrolase may partition 13-hydroperoxylinolenic acid to form jasmonic acid and thus regulate vegetative storage proteins in soybean nodules. In addition, β-amylase was identified in the soybean nodule proteome (Oehrle 2005) and in alfalfa taproots (Gana et al. 1998), suggesting it may function as a storage reserve.

Three proteins from *B. japonicum* were identified among the selected nodule cytosol proteins: the β-chain of nitrogenase MoFe protein, and the chaperonin GroEL, which are both abundant proteins within the microsymbiont and blr1670, an unknown protein with similarity to adenylate cyclases. Catalano et al. (2004) found 23 bacterial proteins among the symbiosome membrane proteins of *Medicago truncatula* root nodules. Wienkoop and Saalbach (2003) analyzed the proteome of symbiosome membranes from *Lotus japonicus* root nodules and found the Fe protein of nitrogenase and 13 other bacteroid proteins. Panter et al. (2000) found a number of unknown proteins associated with the symbiosome membrane of soybean root nodules. Their work was published prior to the sequencing of the *B. japonicum* genome. Analysis of the seven proteins they found with no homology to known proteins revealed that one was homologous to a *B. japonicum* gene that is annotated as an outer membrane protein. These results emphasize the technical difficulties in separating the plant and bacterial symbionts and their component macromolecules for analysis.

An alternative explanation to the presence of *B. japonicum* proteins in the plant nodule cytosol is that they are actively exchanged between the symbiotic partners. Clathrin and ADP-ribosylation factor were identified in the whole nodule proteome (Oehrle 2005), which permits speculation of active exchange of macromolecules and membrane proteins between the infected and uninfected nodule cells, between infected cells and the symbiosome, and perhaps between the symbiosome and the bacteroid. Clathrin is the major protein of the polyhedral coat of coated pits and vesicles formed during endocytosis of materials at the surface of cells (Brodsky et al. 2001; Jürgens 2004). Two different adaptor protein complexes link the clathrin lattice either to the plasma membrane or to the trans-Golgi network. ADP-ribosylation factors are 20 kDa GTP-binding proteins involved in protein trafficking that are members of the Ras superfamily of regulatory GTP-binding proteins (Memon 2004) and regulate metabolism via 14-3-3 gene activation (Zuk et al. 2003). In addition, enolase – a prominent protein found in the whole nodule proteome (Oehrle 2005) – has been reported to activate homotypic vacuole fusion and protein transport to the vacuole in yeast (Decker and Wickner 2006).

Another possible example of glycolytic enzymes that have cellular functions in addition to their roles in glycolysis is glyceraldehyde 3-phosphate dehydrogenase – a prominent protein in the soybean nodule proteome (Oehrle 2005). Glyceraldehyde 3-phosphate dehydrogenase has been shown to bind selenium in *Escherichia coli* (Lacourciere et al. 2002) and humans (Ogasawara et al. 2005). In addition, Oehrle (2005) identified a selenium-binding protein believed to bind to selenium without containing selenocysteine. Although its physiological role remains to be identified, it has recently attracted interest due to its participation in the late stages of intra-Golgi protein transport (Ishida et al. 2002). An abundant selenium-binding protein was also reported in young nodules of *L. japonicus* (Flemetakis et al. 2002).

The largest group of identified proteins in the whole soybean nodule proteome represented proteins involved in carbon metabolism (Oehrle 2005). Besides enolase and glyceraldehyde 3-phosphate dehydrogenase, phosphoglycerate kinase, NADP-isocitrate dehydrogenase, malate dehydrogenase, sucrose synthase, isoflavone reductase, 4-coumarate-CoA ligase, α-carboxyltransferase, and β-amylase were identified. NADP-dependent isocitrate dehydrogenase would be anticipated to be localized within the mitochondria, but it was not detected in the mitochondrial proteome (Hoa et al. 2004). Miller et al. (1998) identified a cytosolic and a nodule-enhanced malate dehydrogenase in alfalfa, and Appels and Haaker (1988) found multiple malate dehydrogenase activities in the cytosol of pea nodules. Hoa et al. (2004) reported two different malate dehydrogenases found both in root and nodule mitochondria of soybean.

Three out of the four enzymes in the 3-carbon segment of glycolysis, glyceraldehyde 3-phosphate dehydrogenase, phosphoglycerate kinase and enolase, were among the most abundant proteins in the plant fraction of soybean nodules (Oehrle 2005).

Phosphoglycerate mutase, the remaining enzyme of the 3-carbon segment of glycolysis was not identified, but it is possible that it is one of a number of proteins not well represented in annotated genomes and libraries. The activities of all four enzymes correlated with nitrogen fixation, with glyceraldehyde 3-phosphate dehydrogenase showing the greatest increase in activity during peak nitrogen fixation (Oehrle 2005). The production of phosphoenolpyruvate provides the substrate for phosphoenolpyruvate carboxylase, resulting in the production of oxaloacetate, which is converted by malate dehydrogenase to malate for transport to the bacteroid (Wadham et al. 1996).

Phosphoenolpyruvate is also a substrate for the biosynthesis of shikimic acid, which is a central precursor for phenylpropanoids and benzoates, of which the enzymes isoflavone reductase and 4-coumarate-CoA ligase were identified in the soybean nodule cytosol (Oehrle 2005). Isoflavone reductase is an enzyme involved in the biosynthesis of the phytoalexin medicarpin, which is found constitutively in roots and nodules and is inducible in leaves in *Medicago* (Lopez-Meyer et al. 2002). 4-Coumarate-CoA ligase is the last enzyme of general phenylpropanoid metabolism, supplying substrates for lignin formation and other individual phenylpropanoid pathways.

19.5 Symbiosome Proteomics

Each infected soybean nodule cell may contain up to 20,000 bacteroids that occur in groups of up to 20 within a single symbiosome membrane. The bacteroids never come into direct contact with the host cell cytoplasm. The symbiosome membrane acts as a selective permeability barrier and thus regulates the transfer of metabolites between the infected plant cell and the bacteroids. The flow of metabolites and signal molecules from the plant must pass through the infected plant cell, across the symbiosome membrane, the symbiosome space (the volume outside the bacteroids inside the symbiosome), the outer bacteroid membrane, the periplasmic space and the inner bacteroid membrane. The flow of metabolites from the bacteroid to the plant would follow the reverse pathway. In soybean root nodules, almost 30 times more membrane is generated in the form of symbiosome membrane than plasma membrane (Winzer et al. 1999). The symbiosome membrane is derived from the host plasma membrane that surrounds the infection thread at the time of release of bacteria into the host cell. Following endocytosis, this membrane undergoes significant changes in composition (Verma and Hong 1996). Proteins are delivered to the symbiosome membrane and symbiosome space via vesicles from the Golgi (Cheon et al. 1993; Kinnback et al. 1987). During the mitotic phase of soybean nodule development, when most of the symbiosome membranes are formed, the soluble protein content of the nodule declines whereas the total protein content of the nodule increases (Anthon and Emerich 1990; Newcomb 1981). This change in protein content reflects the tremendous increase in proteins associated with the symbiosome membrane.

The symbiosome can be considered as a special form of lytic compartment of the infected plant cell (Mellor 1989). The symbiosome membrane is produced from the host plant by membrane components from the Golgi and from the endoplasmic reticulum. The fatty acid composition of the membrane in soybean nodules is more similar to that of the endoplasmic reticulum than that of the Golgi (Bassarab et al. 1989). The proteins identified in soybean symbiosome membranes by biochemical (Werner 1992) and proteomic (Panter et al. 2000) analysis include: Nodulin 24, Nodulin 26, Nodulin 53, vacuolar ATPase, thiol protease, subtilisin protease, protein disulfide isomerase, heat shock protein 60 and BiP protein. The BiP protein is a widely distributed and highly conserved endoplasmic reticulum luminal protein that has been implicated in co-translational folding of nascent polypeptides and in the recognition and disposal of misfolded polypeptides. Panter et al. (2000) suggested that, due to the presence of several proteins associated with the symbiosome membrane that function in protein folding, symbiosomes may import proteins directly from the infected cell cytoplasm. Simonsen and Rosendahl (1999) demonstrated that isolated pea symbiosomes can import plant-translated nodule proteins.

The symbiosome space has a much lower density than the infected plant cell cytoplasm. Proteins identified in the soybean symbiosome space via biochemical methods include: acid phosphatase, α-mannosidase II, proteases, protease inhibitors, α-glucosidase, aspartate aminotransferase and trehalase (Werner 1992). Saalbach et al. (2002) used proteomic methods to identify 46 proteins in the symbiosome space of pea nodules. Among the proteins found were chaperonins, heat shock proteins, protein disulfide isomerase, and ATP synthase components, as would be expected from reports on the symbiosome membrane. However, the symbiosome space contained a number of enzymes of the citric acid cycle, such as aconitase, succinyl-CoA synthetase and malate dehydrogenase. Succinate semialdehyde dehydrogenase was also found, which, together with the identified citric acid cycle enzymes, suggests that the symbiosome space may utilize organic acids. However, both nitrogenase proteins were found in the symbiosome space and the majority of the identified proteins (21/27) were from *Rhizobium leguminosarum* bacteroids, the endophyte of pea. This again reflects the inherent difficulty of separating symbionts and their macromolecules. Nevertheless, combined biochemical and proteomic analyses demonstrate that the symbiosome membrane and symbiosome space perform highly specialized functions essential for symbiotic functioning.

19.6 Bacteroid Proteomics

The global proteomic analysis of *B. japonicum* bacteroids has revealed a dominant and elaborate protein network for nitrogen and carbon metabolism that is closely dependent on plant-supplied metabolites complemented by a selective group of bacteroid transporter proteins (Sarma and Emerich 2005). Earlier, Djordjevic et al. (2003) presented a global analysis of *S. meliloti* bacteroids, which result from an indeterminant symbiosis. Both analyses matched very well with previous biochemical and genetic reports and clearly showed the inter-connection between several metabolic pathways that meet the needs of the bacteroid. There was significant overlap of proteins in each functional category but the majority of identified proteins were unique to each organism. For example, chaperonins were particularly abundant in *B. japonicum* bacteroids compared to *S. meliloti* bacteroids, whereas many more regulatory proteins were identified in *S. meliloti* bacteroids.

In *B. japonicum* bacteroids, the largest categories based on general functions were transcription, translation and protein folding; carbon metabolism; nitrogen metabolism; and solute transporters and membrane proteins (Sarma and Emerich 2005). Among the transcription, translation and protein folding proteins were six amino acid tRNA synthetases plus a tRNA modification GTPase, elongation factor Tu, translation factor Ts, 30S ribosomal proteins S1 and S6, 50S ribosomal protein L9, chaperonins GroEL and GroES and heat shock protein 70. Thus, *B. japonicum* bacteroids apparently expend considerable energy for protein synthesis and maintenance. A typical growing cell expends ~80% of its energy toward protein synthesis (Lehninger 1965). Since *B. japonicum* bacteroids are non-growing cells, this expenditure in protein synthesis and maintenance is all the more remarkable.

Among the more abundant proteins in bacteroids of both *B. japonicum* and *S. meliloti* are the nitrogenase proteins, which provide the source of reduced nitrogen for ultimate assimilation into plant matter. Nitrogenase requires considerable amounts of energy to reduce atmospheric dinitrogen – at least 16 molecules of ATP per N_2. The energy for the reduction of atmospheric dinitrogen is derived ultimately from photosynthetic carbon products translocated to the roots. Nitrogenase has a relatively slow turnover rate and thus the bacteroid compensates by producing higher levels of these proteins to supply the needs of the plant.

A number of amino acid aminotransferases and the enzymes for the biosynthesis of the amino acids alanine, arginine, asparagine, aspartate, glutamate, glycine, isoleucine, leucine, lysine, methionine, proline, threonine, tryptophan and valine were found in *B. japonicum* bacteroids (Sarma and Emerich 2005). Of these, alanine and aspartate were notable as they may serve as carriers of reduced atmospheric dinitrogen to the plant (Lodwig et al. 2003; Rosendahl et al. 2001; Waters et al. 1998). In particular there were a relatively large number of enzymes expressed in the lysine biosynthetic pathway. Streeter (1987) reported earlier that *B. japonicum* bacteroids possessed a higher concentration of lysine than the surrounding plant nodule cytosol. At least 19 proteins involved in the transport of amino acids and carbon compounds were identified, demonstrating that bacteroids have the ability to transport a multitude of compounds.

Malate is believed to be the primary compound transported to the bacteroid from the infected plant cell (Stumpf and Burris 1979). Exogenous malate, succinate and fumarate actively support nitrogen fixation in bacteroid suspensions. However, recent evidence has questioned the functioning of a classic citric acid cycle in bacteroids during the phase of active nitrogen fixation (Green and Emerich 1997; Green et al. 2000; Shah and Emerich 2006). Proteomic analysis identified malate dehydrogenase, fumarase and components of succinate dehydrogenase and of α-ketoglutarate dehydrogenase. Interestingly, a number of proteins participating in benzoate and phenylpropanoid metabolism were identified in *B. japonicum* bacteroids, including anthranilate phosphoribosyl transferase, 2-hydroxyhepta-2,4-diene-1,7-dioate isomerase, phenol hydroxylase, phospho-2-dehydro-3-deoxyheptonate aldolase and several cytochrome P450 oxygenases. When considered with regard to the proteins in benzoate and phenylpropanoid metabolism identified by Oehrle (2005) in the soybean nodule proteome, this suggests that benzoates and phenylpropanoids may represent a major carbon source for bacteroid metabolism, perhaps providing energy for symbiotic nitrogen fixation. Curiously, bacteroids possess a number of enzymes of the omega oxidation pathway of fatty acids: NADH-enoyl acyl carrier protein, acyl aldehyde dehydrogenase, acyl alcohol dehydrogenase and acyl-CoA dehydrogenases. A primary product of the omega oxidation pathway of fatty acids is succinate. Considering the stimulatory effect of succinate on nitrogen fixation activity, the omega oxidation pathway may be another in vivo source of this carbon metabolite.

19.7 An Integrated Proteomic View of the Soybean Nodule

The central theme of symbiotic nitrogen fixation is the provision of carbon to the non-growing bacteroid by the photosynthetic plant partner in exchange for the nitrogen fixed by the prokaryotic microsymbiont. The large energy requirement of the bacteroid nitrogen fixation process requires a vast amount of energy from the plant in terms of carbon. Recent results have left in doubt the prevailing notion that the bacteroid citric acid cycle is the main provider of energy for symbiotic nitrogen fixation. Proteomics has revealed several possible alternative metabolic sources for the energy needed by the bacteroid. Unexpectedly, proteins for benzoate and phenylpropanoid metabolism were identified in both the soybean nodule cytosol and in the bacteroid. The logistics and energetics of benzoate and phenylpropanoid biosynthesis by the plant nodule cells, their subsequent transport to the bacteroid, and the extraction of energy from these compounds by the bacteroid had not previously been considered. However, many of the required proteins are present, and the membranes of both the symbiosome and the bacteroid have a wealth of transporters that could transport not only benzoates and phenylpropanoids but also many other compounds not previously identified or considered as central to the symbiotic process. The omega oxidation pathway of fatty acids may serve as another source of energy for bacteroid metabolism. Other notable proteomic discoveries include the abundance of enzymes of amino acid biosynthesis in the bacteroid as well as potential amino acid transporters, all of which could be key components of more elaborate nutrient exchange cycles (Kahn et al. 1985; Lodwig et al. 2003). Proteomics has expanded the possibilities of nodule function and provided clues as to the large number and types of interactions that occur between symbiotic partners.

References

Appels MA, Haaker H (1988) Identification of cytoplasmic nodule-associated forms of malate dehydrogenase involved in symbiosis between *Rhizobium leguminosarum* and *Pisum sativum*. Eur J Biochem 171:515–522

Anthon GE, Emerich DW (1990) Developmental regulation of enzymes of sucrose and hexose metabolism in effective and ineffective soybean nodules. Plant Physiol 92:346–351

Bassarab S, Schenk SU, Werner D (1989) Fatty acid composition of the peribacteroid membrane and the ER in nodules of *Glycine max* varies after infection by different strains of the microsymbiont . Bot Acta 102:196–201

Bontemps C, Golfier G, Gris-Liebe C, Carrere S, Talini L, Boivin-Masson C (2005) Microarray-based detection and typing of the Rhizobium nodulation gene nodC: potential of DNA arrays to diagnose biological functions of interest. Appl Environ Microbiol 71:8042–8048

Brodsky FM, Chen CY, Knuehl C, Towler MC, Wakeham DF (2001) Biological basket weaving: formation and function of clathrin-coated vesicles. Annu Rev Cell Dev Biol 17:517–568

Capela D, Carrere S, Batut J (2005) Transcriptome-based identification of the *Sinorhizobium meliloti* NodD1 regulon. Appl Environ Microbiol 71:4910–4913

Catalano, CM, Lane WS, Sherrier DJ (2004) Biochemical characterization of symbiosome membrane proteins from *Medicago truncatula* root nodules. Electrophoresis 25:519–531

Charrier B, Coronado C, Kondorosi A, Ratet P (1995) Molecular characterization and expression of alfalfa (*Medicago sativa* L.) flavanone-3-hydroxylase and dihydroflavonol-4-reductase encoding genes. Plant Mol Biol 29:773–786

Cheon C-I, Lee N-G, Siddique ABM, Bal AK, Verma DPS (1993) Roles of plant homologs of Rab7p in biogenesis of peribacteroid membrane, a subcellular compartment formed de novo during root nodule symbiosis. EMBO J 12:4125–4135

Copeland L, Vella J, Hong Z (1989) Enzymes of carbohydrate metabolism in soybean nodules. Phytochemistry 28:57–61

Dalton DA, Baird LM, Langeberg L, Taugher CY, Anyan WR, Vance CP, Sarath G (1993) Subcellular localization of oxygen defense enzymes in soybean (*Glycine max* [L.] Merr.) root nodules. Plant Physiol 102:481–489

Decker BL, Wickner WT (2006) Enolase activates homotypic vacuole fusion and protein transport to the vacuole in yeast. J Biol Chem 281:14523–14528

Dénarié J, Debellé F (1996) *Rhizobium* lipooligosaccharide nodulation factors: signaling molecules mediating recognition and morphogenesis. Annu Rev Biochem 65:503–535

Den Hartog M, Musgrave A, Munnik T (2001) Nod factor-induced phosphatidic acid and diacylglycerol pyrophosphate formation: a role for phospholipase C and D in root hair deformation. Plant J 25:55–65

Djordjevic MA, Chen HC, Natera S, Van Noorden G, Menzel C, Taylor S, Renard C, Geiger O, the Sinorhizobium DNA sequencing consortium, Weiller GF (2003) A global analysis of protein expression profiles in *Sinorhizobium meliloti*: discovery of new genes for nodule occupancy and stress adaptation. Mol Plant Microbe Interact 16:508–524

Farmer EE (1994) Fatty acid signaling in plants and their associated microorganisms. Plant Mol Biol 26:1423–1437

Flemetakis E, Agalou A, Kavroulakis N, Dimou M, Martsikovaskaya A, Slater A, Spaink HP, Roussis A, Katinakis P (2002) *Lotus japonicus* gene Ljsbp is highly conserved among plants and animals and encodes a homologue to the mammalian selenium-binding proteins. Mol Plant Microbe Interact 15:313–322

Gana JA, Kalengamaliro NE, Cunningham SM, Volenec JJ (1998) Expression of β-amylase from alfalfa taproots. Plant Physiol 118:1495–1506

Green LS, Emerich DW (1997) *Bradyrhizobium japonicum* does not require α-ketoglutarate dehydrogenase for growth on succinate or malate. J Bacteriol 179:194–201

Green LS, Emerich DW, Bergersen FJ, Day DA (2000) Catabolism of α-ketoglutarate by a *suc*A mutant of *Bradyrhizobium japonicum*: evidence for an alternative tricarboxylic acid cycle. J Bacteriol 182:2838–2844

Hoa LT–P, Nomura M, Kajiwara H, Day DA, Tajima S (2004) Proteomic analysis on symbiotic differentiation of mitochondria in soybean nodules. Plant Cell Physiol 45:300–308

Huber SC, Akazawa T (1986) A novel sucrose synthase pathway for sucrose degradation in cultured sycamore cells. Plant Physiol 81:1008–1013

Hunt S, Layzell DB (1993) Gas exchange of legume nodules and the regulation of nitrogenase activity. Annu Rev Plant Physiol Mol Biol 44:483–511

Ishida T, Ishii Y, Yamada H, Oguri K (2002) The induction of hepatic selenium-binding protein by aryl hydrocarbon (Ah)-receptor ligands in rats. J Health Sci 48:62–68

Junghans TG, de Almeida-Oliveira MG, Moreira MA (2004) Lipoxygenase activities during development of root and nodule of soybean. Pesq Agropec Bras 39:625–630

Jürgens, G (2004) Membrane trafficking in plants. Annu Rev Cell Dev Biol 20:481–504

Kahn ML, Kraus J, Somerville JE (1985) A model of nutrient exchange in the rhizobium-legume symbiosis. In: Evans HJ, Bottomley PJ, Newton WE (eds) Nitrogen fixation research progress. Nijhoff, Dordrecht, pp 193–199

Kinnback A, Mellor R, Werner D 1987 α-Mannosidase II isoenzyme in the peribacteroid space of *Glycine max* root nodules. J Exp Bot 38:1373–1377

Kouchi H, Fukai K, Katagin H, Minamisawa K, Tajima S (1988) Isolation and enzymological characterization of infected and uninfected cell protoplasts from root nodules of *Glycine max*. Physiol Planta 73:327–334

Krishnan HB, Karr DB, Emerich DW (1999) Purification of an autophosphorylating protein from imbibing soybean (*Glycine max* L.) seed exudate and its identification as a nucleoside diphosphate kinase. J Plant Physiol 154:584–590

Lacourciere GM, Levine RL, Stadtman TC (2002) Direct detection of potential selenium delivery proteins by using an *Escherichia coli* strain unable to incorporate selenium from selenite into proteins. Proc Natl Acad Sci USA 99:9150–9153

Lee H, Hur C-G, Oh CJ, Ho BK, Park S-Y, An CS (2004). Analysis of the root nodule-enhanced transcriptome in soybean. Mol Cells 18:53–62

Leelapon O, Sarath G, Staswick PE (2004) A single amino acid substitution in soybean VSPα increases its acid phosphatase activity nearly 20-fold. Planta 219:1071–1079

Lehninger, AL (1965) Bioenergetics: the molecular basis of biological energy transformations, Benjamin, New York

Lin J, Wood S, Becana M, Klucas RV (1991) Purification and characterization of soybean root nodule ferric leghemoglobin reductase. Plant Physiol 96:32–37

Lodwig EM, Hosie AHF, Bourdes A, Findlay K, Allaway D, Karunakaran R, Downie JA, Poole PS (2003) Amino-acid cycling drives nitrogen fixation in the legume–rhizobium symbiosis. Nature 422:722–726

Lopez-Meyer M, Paiva NL (2002) Immunolocalization of vestitone reductase and isoflavone reductase, two enzymes involved in the biosynthesis of the phytoalexin medicarpin. Physiol Mol Plant Pathol 61:15–30

Mellor RB (1989) Bacteroids in the *Rhizobium*-legume symbiosis inhabit a plant internal lytic compartment: implications for other microbial endosymbioses. J Exp Bot 40:831–839

Memon AR (2004) The role of ADP-ribosylation factor and SAR1 in vesicular trafficking in plants. Biochim Biophys Acta 1664:9–30

Meuriot F, Noquet C, Avice J-C, Volenec JJ, Cunningham SM, Sors TG, Caillot S, Ourry A (2004) Methyl jasmonate alters N partitioning, N reserves accumulation and induces gene expression of a 32-kDa vegetative storage protein that possesses chitinase activity in *Medicago sativa* taproots. Physiol Planta 120:113–123

Miller SS, Driscoll BT, Gregerson RG, Gantt JS, Vance CP (1998) Alfalfa malate dehydrogenase (MDH): molecular cloning and characterization of five different forms reveals a unique nodule-enhanced malate dehydrogenase. Plant J 15:173–184

Mohammadi M, Karr AL (2003) Induced lipoxygenases in soybean root nodules. Plant Sci 164:471–479

Mooney BP, Krishnan HB, Thelen JJ (2004) High-throughput peptide mass fingerprinting of soybean seed proteins: automated workflow and utility of UniGene expressed sequence tag databases for protein identification. Phytochemistry 65:1733–1744

Newcomb EH (1981) Nodule morphogenesis and differentiation. Int Rev Cytol Suppl 13:247–297

Nguyen T, Zelechowska M, Foster V, Bergmann H, Verma DPS (1985) Primary structure of the soybean nodulin-35 gene encoding uricase II localized in the peroxisomes of uninfected cells of nodules. Proc Natl Acad Sci USA 82:5040–5044

Oehrle NW (2005) Physiologic and metabolic interactions in the soybean/*Bradyrhizobium japonicum* symbiosis. PhD Thesis, University of Missouri

Ogasawara Y, Lacourciere GM, Ishii K, Stadtman TC (2005) Characterization of potential selenium-binding proteins in the selenophosphate synthetase system. Proc Natl Acad Sci USA 102:10112–1016

Panter S, Thomson R, de Bruxelles G, Laver D, Trevaskis B, Udvardi M (2000) Identification of proteomics of novel proteins associated with the peribacteroid membrane of soybean root nodules. Mol Plant Microbe Interact 13:325–333

Penheiter AR, Duff SMG, Sarath G (1997) Soybean root nodule acid phosphatase. Plant Physiol 114:597–604

Rosendahl L, Mouritzen P, Rudbeck A (2001) Nitrogen transfer in the interface between the symbionts in pea roots. Plant Soil 230:31–37

Saalbach G, Erik P, Wienkoop S (2002) Characterisation by proteomics of peribacteroid space and peribacteroid membrane preparations from pea (*Pisum sativum*) symbiosomes. Proteomics 2:325–337

Sarma AD Emerich DW (2005) Global protein expression pattern of *Bradyrhizobium japonicum* bacteroids: a prelude to functional proteomics. Proteomics 5:4170–4184

Shah RJ, Emerich DW (2006) Characterization of an isocitrate dehydrogenase mutant of *Bradyrhizobium japonicum*. J Bacteriol 188:7600–7608

Shelp BJ, Atkins CA, Storer PJ, Canvin DT (1983) Cellular and subcellular organization of pathways of ammonia assimilation and ureide synthesis in nodules of cowpea (*Vigna unguiculata* L. Walp). Arch Biochem Biophys 224:429–441

Simonsen ACW, Rosendahl L (1999) Origin of de novo synthesized proteins in the different compartments of pea–*Rhizobium* sp. symbiosomes. Mol Plant Microbe Interact 12:319–327

Sprent JI (1989) Which steps are essential for the formation of functional legume nodules? New Phytol 111:129–153

Staswick PE (1989) Preferential loss of an abundant storage protein from soybean pods during seed development. Plant Physiol 90:1252–1255

Streeter JG (1987) Carbohydrate, organic acid, and amino acid composition of bacteroids and cytosol from soybean nodules. Plant Physiol 85:768–773

Stumpf DK, Burris RH (1979) A micromethod for the purification and quantification of organic acids of the tricarboxylic acid cycle in plant tissues. Anal Biochem 95:311–315

Suganuma N, Kitou M, Yamamoto Y (1987) Carbon metabolism in relation to cellular organization of soybean root nodules and respiration of mitochondria aided by leghemoglobin. Plant Cell Physiol 28:113–122

Tranbarger TJ, Franceschi VR, Hildebrand DF, Grimes HD (1991) The soybean 94-kilodalton vegetative storage protein is a lipoxygenase that is localized in the paravenial mesophyll cell vacuoles. Plant Cell 3:973–987

Verma DPS, Hong Z (1996) Biogenesis of the peribacteroid membrane in root nodules. Trends Microbiol 4:364–368

Wadham C, Winter H, Schuller KA (1996) Regulation of soybean nodule phosphoenolpyruvate carboxylase in vivo. Physiol Planta 97:531–535

Wan J, Torres M, Ganapathy A, Thelen J, DaGue BB, Mooney B, Xu D, Stacey G (2005) Proteomic analysis of soybean root hairs after infection by *Bradyrhizobium japonicum*. Mol Plant Microbe Interact 18:458–467

Waters JK, Hughes BL, Purcell LC, Gerhardt KO, Mawhinney TP, Emerich DW (1998) Alanine, not ammonia is excreted from N_2-fixing soybean nodule bacteroids. Proc Natl Acad Sci USA 95:12038–12042

Werner D (1992) Physiology of nitrogen-fixing legume nodules: compartments and functions. In: Stacey G, Burris RH, Evans HJ (eds) Biological nitrogen fixation. Chapman and Hall, New York, pp 399–431

Wienkoop S, Saalbach G (2003) Proteome analysis. Novel proteins identified at the peribacteroid membrane from *Lotus japonicus* root nodules. Plant Physiol 131:1080–1090

Winzer T, Bairl A, Linder M, Linder D, Werner D, Muller P (1999) A novel 53-kDa nodulin of the symbiosome membrane of soybean nodules, controlled by *Bradyrhizobium japonicum*. Mol Plant Microbe Interact 12:218–226

Wittenbach VA (1982) Effect of pod removal on leaf senescence in soybeans. Plant Physiol 70:1544–1548

Wycoff KL, Hunt S, Gonzales MB, VandenBosch KA, Layzell DB, Hirsch AM (1998) Effects of oxygen on nodule physiology and expression of nodulins in alfalfa. Plant Physiol 117:385–395

Zuk M, Prescha A, Keprzynski J, Szopa J (2003) ADP ribosylation factor regulates metabolism and antioxidant capacity of transgenic potato tubers. J Agric Food Chem 51:288–294

Chapter 20
Proteomes in Arbuscular Mycorrhizal Symbiosis

Eliane Dumas-Gaudot and Ghislaine Recorbet

Abstract Since proteins are key effectors of plant responses to environmental cues, including recognition, signalling, transport and defence reactions, much interest has focussed on characterising proteins involved in the establishment and functioning of arbuscular mycorrhizal (AM) symbiosis. The recent development of high-throughput techniques in the model species *Medicago truncatula* is providing, step-by-step, a large-scale analysis of symbiosis-related proteins. Depending on the symbiotic stage and the abundance of mycorrhizal material, different proteomic strategies, which can be combined with other large-scale approaches (transcriptomic and metabolomic), can be used and are presented in this chapter. In addition to providing an update on AM related-proteomics research, our aim is to highlight proteomic strategies available to study AM symbiosis. Finally, further developments that can be expected from the most recent technical improvements in plant proteomics are discussed for this particular research area.

20.1 Introduction

Plant roots undergo various interactions with soil microbes. These interactions fall into three categories: (1) mutualistic, where both the host and the microbe benefit from the association; (2) commensal, where the microbe benefits and the host is unharmed; and (3) pathogenic (or parasitic), where the microbe benefits and the host is harmed by the association. The arbuscular mycorrhizal (AM) symbiosis formed between more than 80% of plant species and fungi belonging to the phylum of *Glomeromycota* (Schussler et al. 2001) are included within the first category. Mycorrhiza develops in plant roots where the AM fungi colonise the cortical cells to obtain carbon from the plant, while assisting the plant with the acquisition of

Eliane Dumas-Gaudot, UMR 1088 INRA/CNRS 5184/UB Plante-Microbe-Environnement, INRA/CMSE, BP 86510, 21065 Dijon Cedex, France, E-mail: dumas@epoisses.inra.fr

Ghislaine Recorbet, UMR 1088 INRA/CNRS 5184/UB Plante-Microbe-Environnement, INRA/CMSE, BP 86510, 21065 Dijon Cedex, France

J. Šamaj and J. Thelen (eds.), *Plant Proteomics*
© Springer 2007

phosphate from the soil (Smith and Read 1997). Thus, AM symbiosis is not only important from an ecological viewpoint, but also has largely proved its efficiency for improving plant health in a wider agronomical context (Auge 2001; Azcón-Aguilar and Barea 1996; Jeffries and Barea 1994; Gianinazzi et al. 2002; Bethlenfalvay 1992). Nowadays, *in vitro* mass production of AM fungi (Adholeya et al. 2005) together with adapted farm management practices (Gosling et al. 2006) are opening the way for massive agricultural use of these beneficial microorganisms. Mycorrhiza develops according to a well-orchestrated and finely tuned process in which both partners undergo morpho-physiological modifications (Bonfante and Perotto 1995; Gianinazzi-Pearson 1996; Harrison 1997). Briefly, starting from spore germination and root recognition – a process recently shown to be dependent on exchange of molecules (Matusova et al. 2005; Akiyama et al. 2005; Kosuta et al. 2003; Olah et al. 2005) – fungal hyphae penetrate the root epidermis through an appressorium, grow through the cortical cell layers and finally develop into haustorial structures called "arbuscules" within inner cortex cells. At no timepoint, however, does fungus enter the plant cytoplasm. In arbuscule-containing cells, a periarbuscular membrane that surrounds the arbuscule is formed. The fungal cell wall becomes progressively thinner as the arbuscule develops. This leads to the establishment of an extensive intracellular interface between the two symbionts, where phosphate, and possibly also carbon, is transferred between the plant and the fungus. Despite its central role in mycorrhiza function, the life span of an arbuscule is short; the arbuscule collapses and becomes degraded after only a few days (Alexander et al. 1988). Contrary to the 'Paris' type of mycorrhiza, in which arbuscule development is very restricted, the 'Arum' type is characterised by abundant highly branched intracellular fungal structures (Smith and Read 1997). Finally, the AM fungus achieves its life cycle by producing intraradicular spores and vesicles (in some cases – some AM fungal strains do not form vesicles), as well as, to a much larger extent, extraradicular mycelium harbouring disseminating spores. Although comprehensive cytological and molecular studies (Bonfante and Perotto 2000; Genre et al. 2005) have provided a thorough description of AM morphological aspects, our understanding of the physiological events underlying the development of AM symbiosis remains limited. Recent integrated biochemical, molecular and genetic approaches have provided new insights into various aspects of AM development (Strack et al. 2003; Parniske 2004; Harrison 2005; Hause and Fester 2005). In the last 4 years, advances have been achieved in our understanding of the signals required for appressorium formation (Requena et al. 2002, Weidmann et al. 2004), the transcriptional changes occurring inside roots during AM colonisation (Liu et al. 2003, 2004; Brechenmacher et al. 2004; Hohnjec et al. 2005), and the signal transduction pathways common to both AM and nitrogen fixing symbioses (Kistner et al. 2005; Akiyama and Hayashi 2006). These recent findings profit from the use of large-scale approaches such as transcriptomics along with screening of vast numbers of mutants in two model plants, namely *Medicago truncatula* (Journet et al. 2001; Colebatch et al. 2002; Cook 1999) and *Lotus japonicum* (Sandal et al. 2006; Udvardi et al. 2005). *Arabidopsis* is not suitable for the study of beneficial plant–microbe interactions because this plant does not form any symbiosis with either AM fungus or nitrogen-fixing bacteria. Macro- and micro-array transcriptional profiling experiments have enabled the identification of gene sets that show significant alterations in their expression during AM symbiosis. Many of the induced genes were novel and predicted to be involved in signalling, gene regulation, cell-wall biosynthesis and modification as well as in plant secondary metabolism (Liu et al. 2003, 2004; Brechenmacher et al. 2004; Hohnjec et al. 2005). Although these analyses provided a wealth of information about AM-related genes, it must be remembered that one gene does not equal one transcript and that one transcript does not equal one protein (Gygi et al. 1999; Peck 2005). Thus, post-genomic studies should consider the fact that mRNA and protein abundance are not clearly correlated, that low-copy number mRNAs are not measured as readily as abundant mRNAs, and that gene expression studies do not provide information about either subcellular protein localisation or post-translational protein modifications that may be essential for its function, transport and activation (Gygi et al. 1999; Peck 2005). The relatively recent scientific strategy of proteomics, i.e. the large-scale analysis of proteins in biological systems at precise timepoints, has emerged to bridge this gap. Proteomic methodology is undergoing rapid development due to major breakthroughs in genomic, bioinformatic and transcriptomic disciplines. Because it addresses genuine cell effectors, proteome analysis is the strategy of choice for the investigation and dissection of mechanisms involved in any physiological process accompanying plant development and/or adaptation to environmental factors.

Several reviews have pointed out the efficiency of proteomics for studying plant–microbe interactions, including AM symbiosis (Rolfe et al. 2000; Xing et al. 2002; Canovas et al. 2004; Bestel-Corre et al. 2004a; Thurston et al. 2005; Jorrin et al. 2006; Padliya and Cooper 2006; see also Chap. 9 by Lei et al. and Chap. 19 by Sarma et al., this volume). In this chapter, besides providing an update of AM-related proteomics research, we will discuss the further developments that can be expected from the more recent technical developments in plant proteomics.

20.2 Trends in AM Symbiosis Towards Simultaneous "...omics" Strategies

20.2.1 Combining Protein, Transcript and Metabolite Extractions for Investigation of AM Root/Fungus Interaction

Since the first reports on the induction by AM fungi of additional root proteins, referred to as 'endomycorrhizins" (Dumas et al. 1989; Wyss et al. 1990), and the purification of one of the first mycorrhiza-related proteins (Slezack et al. 2001), the detection, visualisation and, most importantly, identification of mycorrhiza-related proteins have strongly benefitted from recent developments in proteomic devices and tools, including immobilised pH gradient (IPG) strips, larger gels, multi-cuve systems, staining detection, image analysis software, mass spectrometry (MS), bioinformatics, etc. The general strategy is based on protein extraction and separation – two-dimensional gel electrophoresis separation (2-DE) being the most frequent – followed by image analysis and spot excision. Protein spots are then submitted to enzymatic digestion (very often trypsin) and are further analysed by MS. The two most frequent MS devices used in protein identification utilise different methods for sample ionisation and generate different information data: matrix assisted laser desorption ionisation (MALDI) time-of-flight (TOF) is used to generate peptide mass fingerprints (PMFs), while electrospray ionisation (ESI) tandem mass spectrometry (MS/MS), usually combined with high performance liquid chromatography (HPLC), is used to generate sequence data (Lahm and Langen 2000; Nyman 2001; Aebersold and Mann 2003; Newton et al. 2004). One of the first prerequisites for developing proteomics in AM symbiosis is the availability of sequence databases for subsequent MS analysis. This has been partly achieved thanks to the promotion of barrel medic (*Medicago truncatula*) among model plants to study symbiosis-related molecular events – a major goal that cannot be achieved using *Arabidopsis thaliana* (Cook 1999; Journet et al. 2001). Nowadays, five cDNA libraries isolated from AM roots and more than 218,980 annotated expressed sequence tag (ESTs) corresponding to tissues/organs and plant responses to various biological situations are available (latest release 23 August 2006 of the TIGR *M. truncatula* gene index; http://www.tigr.org/tdb/tgi). Additionally, the production of 25,000 ESTs identified within the frame of a large-scale project targetted at specifying genes expressed during both mycorrhization and nodulation (MtBC library; Journet et al. 2002), was complemented by proteomic investigations (Dumas-Gaudot et al. 2001; Bestel-Corre et al. 2002). Figure 20.1 illustrates this general strategy as well as the methodological approaches used in the case of *M. truncatula*–AM symbiosis. The first improvement was provided by a protocol designed to obtain both RNA and proteins from the same root sample (Dumas-Gaudot et al. 2004a), which has proved very valuable for studying the early steps of the AM symbiosis for which only a limited amount of root material is available (Dumas-Gaudot et al. 2007). Assuming a protein can be identified by MS, this protocol is very helpful in investigating the abundance of the corresponding transcript. Recently, another technical improvement has allowed the same root tissues to be studied at both the proteomic and metabolomic levels (Fig. 20.1). Two categories of secondary metabolite are of interest in AM symbiosis: (1) phytoalexins and/or isoflavonoïds, for their role in abiotic and biotic stress responses as well as antimicrobial molecules (Morandi 1996); and (2) carotenoids and their precursors, in regard to their specific accumulation during AM colonisation (Lohse et al. 2005, 2006; Fester et al. 2005; Walter et al. 2000; Strack et al. 2003). A protocol combining HPLC quantification of phytoalexins and/or isoflavonoïds and apocarotenoids in methanol-extracted roots together with quantification of mycorrhizal infection has been designed. The usefulness of this protocol in monitoring phytoalexin and/or isoflavonoïd accumulation in *M. truncatula* roots either colonised or not with an AM fungus is currently being tested. Additionally, the effect of cadmium stress is being studied (A. Aloui et al., manuscript in preparation). This protocol is also being tested for

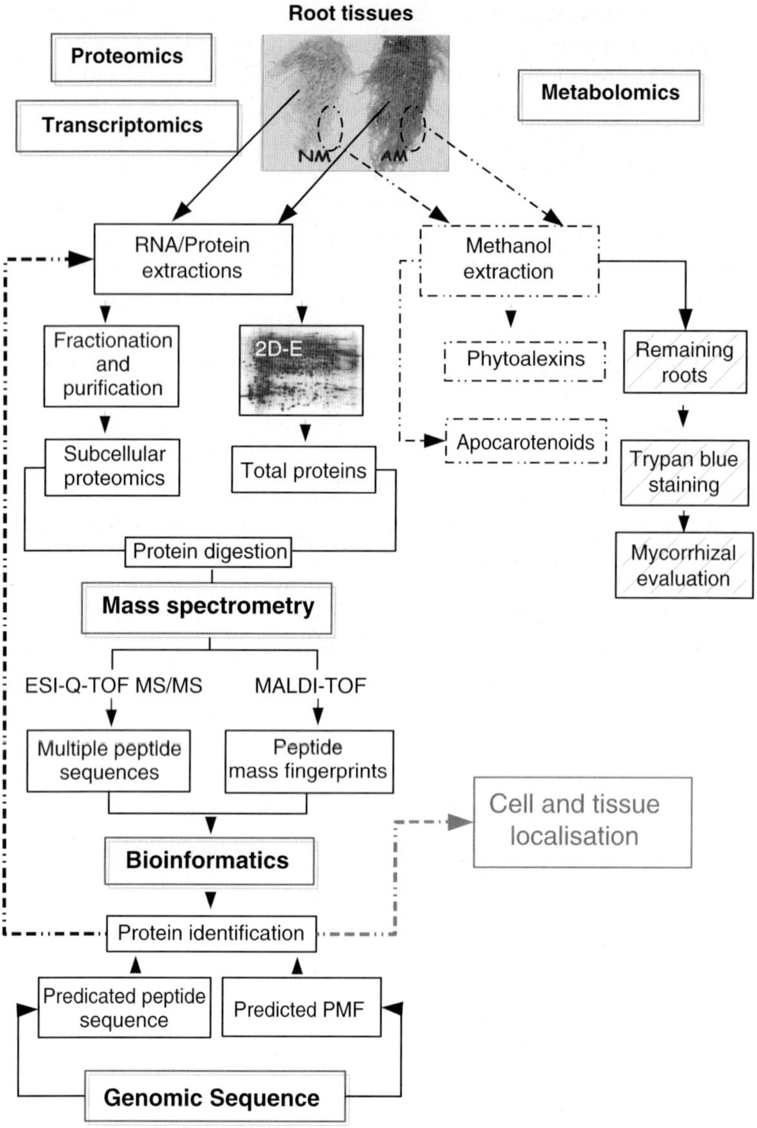

Fig. 20.1 Potential strategies for studying arbuscular mycorrhizal (AM) symbiosis. The upper photograph shows non-inoculated (*NM*) and *Glomus intraradices*-colonised (*AM*) *Medicago truncatula* roots (7 weeks; Z. Daher, unpublished data). The various steps in the proteomic strategy as well as links with transcriptomics are presented on the left. The recent experimental set up we designed to analyse metabolome on the same root samples along with mycorrhiza quantification is shown on the right (Z. Daher and A. Aloui, unpublished data). The 2-DE profile of total proteins corresponds to 5-week-old *M. truncatula* roots inoculated with *G. intraradices* (B. Valot, unpublished data). Further potential functional analyses are indicated in *grey*

apocarotenoid quantification carried out in parallel to a sub-cellular proteomic approach focusing on plastids (Z. Daher et al. in preparation).

20.2.2 Experimental Plant Design

Investigations into the molecular mechanisms regulating development of AM symbiosis have been hampered by the obligatory biotrophic nature of the fungus, and the asynchronous development of the colonisation process. Although several experimental systems allowing more synchronous AM infections have been designed (Rosewarne et al. 1997; Hause et al. 2002), none were fully satisfactory. Moreover, proteomics has its own constraints, such as the necessity to have a reasonable amount of root material for protein extraction due to the lack of an amplification system for proteins equivalent to PCR-based DNA amplification. So far, proteomic approaches have focussed on three well-determined stages in AM symbiosis: (1) the extraradicular fungal material represented by a mixture of spores and mycelium, (2) the early stage at which the fungus develops appressoria at the root surface, and (3) the late stage corresponding to a fully functioning symbiosis (Fig. 20.2). Although the two first stages are amenable to proteomic analysis by the use of particular experimental systems, the third stage represents a compromise since arbuscule-enriched roots also harbour appressoria and intraradical hyphae. Harvesting of extraradicular fungal material has been facilitated greatly by the use of the dual in vitro compartmentalised system first designed by St-Arnaud et al. (1996), while early stages have become more accessible by performing early root sampling according to Weidmann et al. (2004). This latter system, although providing very limited amount of elicited root material, has proved very useful for subsequent proteomic analysis (Amiour et al. 2006; Dumas-Gaudot et al. 2007). Recently, such a system has been improved by pre-culturing plantlets for a couple of weeks before their transfer into an AM–soil mix substrate (D. Morandi, personal communication). This will allow more root material to be obtained to complete studies on the first contact stages between the two symbiotic partners. Concerning the late AM stage, in the absence of better control of its development, only laser dissection microscopy coupled with subsequent proteomic analysis will help us focus more specifically on the arbuscular compartment and, maybe, to discriminate between young, fully mature and senescing arbuscules.

20.3 Proteome Profiling During AM Symbiosis

Figure 20.2 and Table 20.1 summarise the proteomic profiling studies realised to date and aimed at improving our knowledge of AM symbiosis.

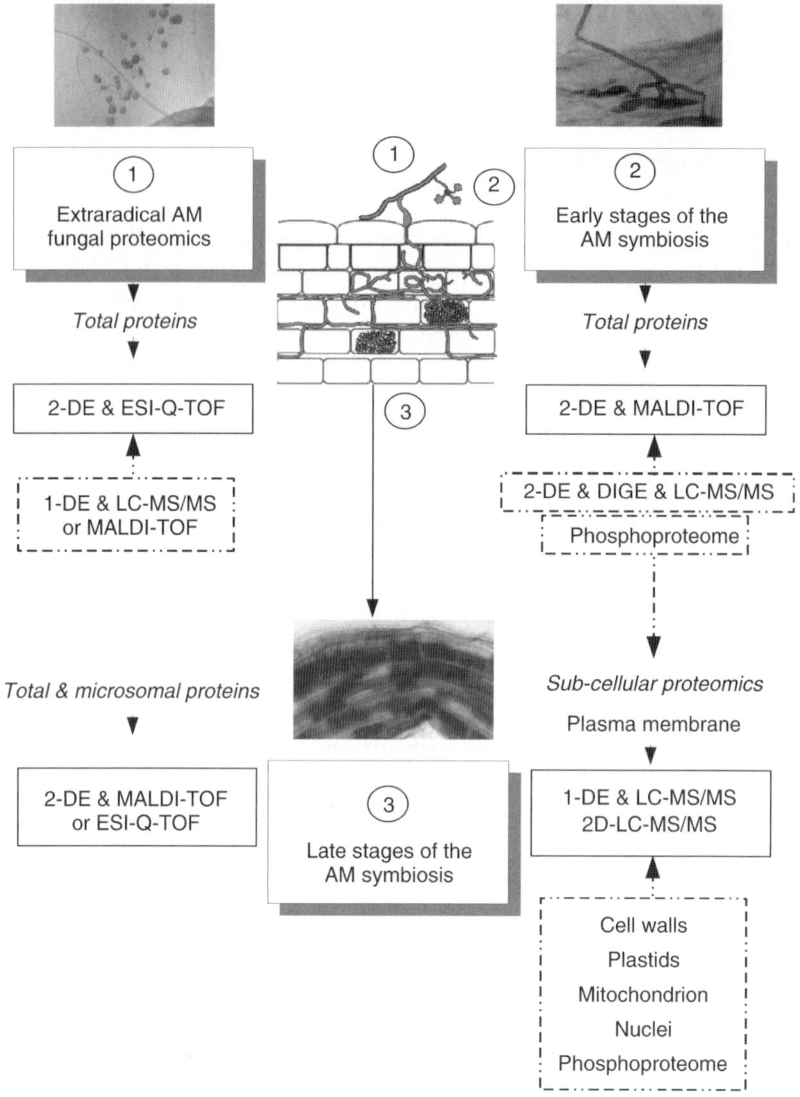

Fig. 20.2 Schematic representation of the AM fungal life cycle indicating the proteomes that have been studied so far (*solid boxes and arrows*) and those requiring further investigation (*dotted boxes and arrows*). Photographs: *upper left* extraradicular fungal structures (spores and mycelium) from the dual compartment of *G. intraradices*-colonised Ri-T-DNA-transformed roots of carrot (E. Dumas-Gaudot, unpublished); *upper right Glomus mosseae* appressorium stained with Trypan Blue at the root surface of *M. truncatula* (Bestel-Corre et al. 2002); *centre* Trypan Blue-stained arbuscules in *G. intraradices*-colonised *M. truncatula* roots (Z. Daher, unpublished data)

Table 20.1 *Medicago truncatula* root proteomes during arbuscular mycorrhizal (AM) symbiosis. *MS* mass spectrometry, *2-DE* two-dimensional gel electrophoresis, *ESI-Q-TOF* electrospray ionisation quadrupole time-of-flight, *MALDI-TOF* matrix assisted laser desorption ionisation time-of-flight, *MS/MS* tandem MS, *PM* plasma membrane

Proteome	Type of proteomic separation and MS analysis	Observed modifications	Proteins identifications (*n*)	Reference
Total root proteins during *Glomus mosseae* colonisation of *M. truncatula* J5	2-DE ESI-Q-TOF	55	8	Bestel-Corre et al. 2002
Total root proteins of *M. truncatula* J5	2-DE ESI-Q-TOF	34	2	Bestel-Corre et al. 2004
Microsomal root proteins of *M. truncatula* J5	Adapted 2-DE MALDI-TOF	Good resolution of 440 spots on silver-stained 2-DE	115	Valot et al. 2004
Microsomal root proteins in response to *Glomus intraradices* colonisation of *M. truncatula* J5	Adapted 2D-E ESI-Q-TOF/ MALDI-TOF	36	36	Valot et al. 2005
PM-enriched fraction of *M. truncatula* J5	1D-MS/MS 2D-MS/MS	78	78	Valot et al. 2006
PM-enriched fraction in response to *G. intraradices* colonisation of *M. truncatula* J5	1D-MS/MS 2D-MS/MS	2	2	Valot et al. 2006
Total root proteomics of the early stage of the symbiosis of:				
M. truncatula J5/ *G. intraradices*		0		
M. truncatula sunn TR122/ *G. intraradices*	2-DE/MALDI-TOF	9	13	Amiour et al. 2006
M. truncatula dmi3 TRV25/ *G. intraradices*		8		
M. truncatula J5/ *G. mosseae*		19	10	Dumas-Gaudot et al. 2007

20.3.1 Proteomics of Total Proteins

20.3.1.1 Plant Proteomics

The first proteomic AM symbiosis investigation aimed to compare the protein profiles of *M. truncatula* J5 roots in response to infection with either the mycorrhizal fungus *Glomus mosseae* (BEG12) or the nitrogen-fixing bacteria *Sinorhizobium meliloti* (Bestel-Corre et al. 2002). This study was conducted as a comparative time-course analysis carried out on total proteins extracted with a phenol-based protocol adapted from Hurkman and Tanaka (1987), and further separated on 13 cm 3–10 NL IPG strips. Although at that time no differentially displayed proteins were found to be common to the two symbioses, as had been expected from molecular studies (Gianinazzi-Pearson and Dénarié 1997; Kistner et al. 2005; Marsh and Schultze 2001; Stracke et al. 2002; Mathesius 2003; Sanchez et al. 2004; Borisov et al. 2004), this first proteomic approach revealed 55 proteins (apparently novel proteins – both up- and down-regulated constitutive proteins) whose accumulation level varied during the AM colonisation process (Bestel-Corre et al. 2002). Since very limited genomic information with which to identify *M. truncatula* proteins induced/modulated in response to AM colonisation was accessible, and virtually no MS data was available for fungal proteins, we used mainly ESI-Q-TOF MS to identify novel proteins. A similar 2-DE strategy was used to follow changes in protein accumulation during AM symbiosis of *M. truncatula* plants cultivated on sewages sludges spiked with pollutants (Bestel-Corre et al. 2004b). In this case, due to the use of larger IPG strips and multi-tank 2-DE device, we were able to reveal more protein modifications in AM roots collected at an equivalent time of colonisation than in our previous study. However, only two more proteins could be identified following de novo sequencing.

It has to be emphasised that no fungal protein was detected in these two reports. Such a result is not so surprising if one considers that generally no more than 60% of a root system is colonised with a proportion of fungal RNA not exceeding 8.1% (Liu et al. 2003). Therefore, it is likely that no AM fungal protein would be identified by non-targeted proteomics unless special care was taken in the initial selection of the starting biological material.

Protein modifications accompanying the early stages of AM symbiosis were also evaluated on total proteins separated by 2-DE, a study in which a genotype response to AM inoculation was included, leading to the identification of sets of appressoria-responsive proteins modified in a genotype-dependent manner (Amiour et al. 2006). Although the number of *M. truncatula* ESTs was considerably larger at the time of this study than had previously been the case, MALDI-TOF MS analyses led to the identification of only 13 proteins, some of which were also found when the *M. truncatula* Jemalong J5 genotype was inoculated with a different fungal strain, i.e. *G. mosseae* BEG 12 (Dumas-Gaudot et al. 2007).

20.3.1.2 Fungal Proteomics

Until now, two strategies to identify AM symbiosis-related fungal proteins have been proposed; the first is based on a preliminary harvest of extraradical fungal material, the second on the development of subcellular proteomics on AM-colonised roots. Although some authors have reported the possible isolation of extraradical AM mycelium from pot cultures (Dodd 1994), the use of *in vitro* mycorrhizal root organ cultures possesses several advantages over soil-based studies. These advantages have made the root organ culture system very attractive for investigating the genetics, cell biology and physiology of AM fungi (reviewed in Declerck et al. 2005). A few years ago, using such a system, we launched the beginning of AM fungal proteomics. Although a limited number of proteins could be identified after tandem MS, the first fungal 2-DE reference map, displaying 438 spots from *G. intraradices*, was set up together with a comparative analysis of 2-DE protein profiles of transformed carrot roots either colonised or not with this fungus (Dumas-Gaudot et al. 2004b). Nowadays, ongoing sequencing of the *G. intraradices* genome (Lammers et al. 2004; Martin et al. 2004) will completely change the situation, making AM fungal proteomics both easier and cheaper. The availability of AM fungal genomic data in public databases will allow not only MALDI-TOF-based protein identification of 2-DE separated proteins but also the identification of 2-DE recalcitrant membrane proteins after either 1D-liquid chromatography (LC) MS/MS or 2D-LC MS/MS.

20.3.2 Sub-Cellular Proteomics

Taking into account that a typical plant cell likely contains more than 20,000 different polypeptide species, with 2-DE typically able to resolve and detect a maximum of only 3,000 spots (Lilley et al. 2002), it was logical to embark upon sub-cellular proteomics in order to reveal more proteins of which accumulation is modulated during AM symbiosis (Fig. 20.2, Table 20.1). Considering that the establishment of AM symbiosis is accompanied by drastic and profound reorganisation of several cell compartments (Harrison 1997; Genre et al. 2005), and assuming that the complexity of proteome analysis will be somewhat reduced by focussing on the proteomes of sub-cellular compartments and/or organelles, research has been orientated towards proteomics of the membrane compartments (Valot et al. 2005, 2006) and, more recently, on proteomics of root plastids (Z. Daher, personal communication). One of the most important morphological changes occurring in AM roots is the invagination of the plant plasma membrane (PM) around the fungal arbuscular structures, resulting in formation of the periarbuscular membrane (Bonfante and Perotto 1995). Firstly, with the aim of gaining access to total membrane-associated proteins, a sub-cellular proteomics study targetted root microsomes of *M. truncatula* (Valot et al. 2004). After setting up the purification process, microsomal proteins were separated by 2-DE. Of the 440 well-resolved proteins (some of which showed several isoforms), 96 were identified by MALDI-TOF PMF. This protocol was then employed to reveal membrane protein modifications in relation to AM symbiosis (Valot et al. 2005). Such a comparative proteomic approach between non-inoculated and *G. intraradices*-inoculated roots of *M. truncatula* revealed that 36 spots were differentially displayed in response to fungal colonisation including 15 induced, 3 up-regulated and 18 down-regulated proteins; among the latter, 7 proteins were found to be commonly down-regulated in AM-colonised and phosphate-fertilised roots. Of these 36 spots, 25 could be identified by MALDI-TOF and/or MS/MS analyses. With the exception of an acid phosphatase and a lectin, none of them were previously reported as being regulated during AM symbiosis. Interestingly, for the first time, this sub-cellular proteomic approach allowed the authors to identify several AM fungal proteins *in planta*. Indeed, several isoforms of ATP synthase α and β subunits, as well as a protein of unknown function but matching to a *Glomus intraradices* extra-radical cDNA, were identified (Valot et al. 2005).

The second approach aimed to characterise root PM proteins of *M. truncatula*. First, membrane extracts were enriched in PM using a discontinuous sucrose gradient method. The resulting PM fractions were further analysed with (1) 2D LC MS/MS and (2) 1D-SDS-PAGE combined with systematic LC-MS/MS analysis. A total of 78 proteins, including hydrophobic proteins, could be reproducibly identified in the PM fraction from non-inoculated roots, representing the first survey of the *M. truncatula* root PM proteome (Table 20.1). Comparative proteomics between non-inoculated and *G. intraradices*-inoculated roots revealed two proteins that were present only in the mycorrhizal root PM fraction (Table 20.1). These corresponded to an H^+-ATPase (Mtha1) and a predicted glycosylphosphatidylinositol (GPI)-anchored blue copper-binding protein (MtBcp1) (Valot et al. 2006). Both proteins were previously reported to be localised mainly in arbuscule-containing cortical cells (Krajinski et al. 2002; Hohnjec et al. 2005). Data gained from this sub-cellular proteomic approach indicated the possible localisation of these two proteins in the periarbuscular membrane; however, this needs to be clearly demonstrated by *in situ* immunolocalisation. Such analysis is in progress for the MtBcp1 protein, for which an exact role in the AM symbiosis remains to be determined.

Plastids constitute another root compartment that was recently chosen for investigation by sub-cellular proteomics. Although chloroplasts have been the subject of a large number of proteomic studies (reviewed in Taylor et al. 2003; Van Wijk 2004; Newton et al. 2004; see Chap. 14 by Jarvis, this volume), much less is known about the root plastid compartment. During AM symbiosis, it is obvious that plastids undergo drastic re-organisation characterised by plastid proliferation and shape modification, as well as network formation through stromules throughout AM development, which is accompanied by concomitant biosynthesis of apocarotenoids (Lohse et al. 2006). Although this phenomenon appears to be common to several plant species in response to AM colonisation, and biochemical mechanisms have been well described (Fester et al. 2002, 2005; Walter et al. 2000), the ongoing proteomic approach will certainly give rise to important information on both soluble and membrane-associated proteins of root plastids.

20.4 Further Insights

It is obvious that we can expect to identify more and more proteins involved in AM symbiosis, and that we will continue to benefit from protocol modifications and the development of new procedures in proteomics. Still, there are a number of issues that need to be addressed in order to improve AM symbiotic proteomics.

Firstly, the release of the genome sequences of both partners of AM symbiosis, i.e. *M. truncatula* and *G. intraradices*, will undoubtedly make the MS part of any proteomics strategy more efficient and less expensive. For example, it is noteworthy that even for the model plant species *M. truncatula*, for which genome sequence information is now available in public databases, the percentages of successful protein identification following MALDI-TOF analysis and PMF search are rather limited when proteomics was directed to total proteins from either various plant organs (55% in Watson et al. 2003) or different biological situations including AM symbiosis (Mathesius et al. 2002; Imin et al. 2004; Amiour et al. 2006). More success (83%) was, however, reached in the case of a sub-cellular proteomics approach (Valot et al. 2004).

The description of the AM fungal proteome has been strongly hampered by the low amount of fungal biomass within roots, difficulties in isolating the fungal partner free of host tissues, and the very limited genomic data available (http://darwin.nmsu.edu/~plammers/; Sawaki and Saito 2001). We can really expect a breakthrough when the *G. intraradices* fungal genome sequence is released. Moreover, it is obvious that AM fungal proteomics will also support the AM genome annotation process. In addition, this will open the way for more environmental studies aimed at identifying AM proteins responsive to various abiotic stresses (heavy metals, polycyclic hydrocarbons, etc.). Furthermore, the limitations we faced in identifying fungal proteins expressed *in planta* could be avoided by using laser-capture microdissection techniques that can be coupled to proteome profiling (Inada and Wildermuth 2005), allowing enrichment of samples in mycorrhizal structures such as arbuscules and appressoria. Such an approach may also be carried out with fluorescent labelling of proteins extracted from microdissected tissues prior to 2-DE separation, although this technique has not yet been applied in the plant kingdom (Kondo et al. 2003).

To further investigate the AM initiation process, more studies should focus on the formation of the plant pre-penetration apparatus before visible AM fungus infection as well as on changes in the cell wall that likely occur at this stage (Genre et al. 2005). Recent reports revealed improvements in cell wall extraction methods that can be applied to the study of the early stages of AM (Feiz et al. 2006; Jamet et al. 2006). Additionally, progress has been made in the analysis of phosphoproteins and phosphopeptides (Stasyk et al. 2005; Wolschin et al. 2005; Zhang et al. 2005; Kalume et al. 2003). Thus, it should be possible to detect membrane proteins that are regulated through phosphorylation during establishment of AM symbiosis. Moreover, we may expect that the use of methods such as fluorescence DIGE (2-D difference gel electrophoresis) (Lilley and Dupree 2006; Amme et al. 2006;

Komatsu et al. 2006) and /or ICAT (isotope coded affinity tag) labelling (Dunkley et al. 2004; Newton et al. 2004; Komatsu et al. 2006) will further improve comparative analysis between mycorrhizal and non-mycorrhizal roots including a true quantitative ICAT comparison. Moreover, subcellular proteomics based on gel-free separations (MUDPIT; multidimensional protein identification technology, also called peptidomics) will benefit from both the technical progress in 2D LC and data set queries using special algorithms (America et al. 2006). This will be very helpful for analysing qualitatively and, more interesting, quantitatively, the data acquired from other sub-cellular compartments such as root plastids, mitochondria and/or nuclei, the response of all of which in response to AM symbiosis needs to be investigated. Indeed, sub-cellular proteomics will be essential to a better understanding of organelle function and regulation as well as for detecting dynamic changes that may occur during the different phases of AM infection. In relation to implementation of genomic data, we expect that queries on orphan PM proteins previously analysed by 1D and 2D-LC MS/MS (Valot et al. 2006) should be repeated in order to retrieve more proteins differentially expressed in response to AM colonisation. Finally, proteomics will be more relevant when combined with other functional genomic tools and approaches.

20.5 Conclusions and Future Perspectives

Although less developed than transcriptomics, proteome profiling has provided new insights into the molecular events that govern AM processes. Proteins not detected by transcriptomics, as well as proteins unravelled by previous molecular data, have been found using proteomic approaches. Nevertheless, besides the obvious necessity to identify and characterise proteins differentially displayed in response to AM colonisation, we should focus on those proteins whose post-transcriptional modifications (phosphorylation, glycosylation, etc.) could play essential roles during AM interaction. Additionally, rare proteins should also be detected and identified, since they may have essential roles. In parallel, functional analysis of mycorrhiza-related proteins is required to confirm crucial involvement of these proteins in AM establishment and/or functioning, and also to determine exactly their role(s). Such functional analysis may be achieved through genetic studies, e.g. using RNA interference (RNAi) lines with expression of specific genes down-regulated or knocked-out (Delalande et al. 2005; Ott et al. 2005), and through the screening of TILLING (targeting induced local lesions in genome) collections (Slade and Knauf 2005), either alone or complemented with tissue and cellular localisation and transcript accumulation experiments. Although most recent proteomic studies have focussed on the model plant species *M. truncatula* (see Chap. 9 by Lei et al., this volume), other plant species and their AM fungi could also provide nice complementary molecular information, as was recently shown in a study of ericoid mycorrhizal symbiosis in response to cadmium (S. Perotto, personal communication). Finally, proteomics will certainly be a valuable tool for investigating non-legume plant species undergoing AM symbiosis. In this respect, tomato (*Solanum esculentum*) is attractive because several mutants affected at different stages in mycorrhiza development are already available (Gao et al. 2001, 2004; Barker et al. 1998; David-Schwartz et al. 2001, 2003; Gadkar et al. 2003), along with extensive genomic data (http://www.sgn.cornell.edu/solanaceae-project/index.pl). Rice (*Oryza sativa*) might also be a nice model because of its central position as an important crop plant, the recent completion of its genome sequence (http://rgp.dna.affrc.go.jp/IRGSP/) and, last but not least, because rice proteomics is well developed already (Komatsu and Yano 2006; see Chap. 7 by Hirano, this volume).

Acknowledgments The authors thank the Council of Burgundy for financial support (04 516 CP09 5327). Z. Daher, A. Aloui, and B. Valot are acknowledged for kindly sharing unpublished pictures or data.

References

Adholeya A, Tiwari P, Singh R (2005) Large-scale inoculum production of arbuscular mycorrhizal fungi on root organs and inoculation strategies. In: Declerck S, Strullu D-G, Fortin JA (eds) *In vitro* culture of mycorrhizas. Springer, Berlin, pp 315–338

Aebersold R, Mann M (2003) Mass spectrometry-based proteomics. Nature 422:198–207

Akiyama K, Hayashi H (2006) Strigolactones: chemical signals for fungal symbionts and parasitic weeds in plant roots. Ann Bot 97:925–931

Akiyama K, Matsuzaki K, Hayashi H (2005) Plant sesquiterpenes induce hyphal branching in arbuscular mycorrhizal fungi. Nature 435:824–827

Alexander T, Meier R, Toth R, Weber HC (1988) Dynamics of arbuscule development and degeneration in mycorrhizas of *Triticum aestivum* L. and *Avena sativa* L. with reference to *Zea mays* L. New Phytol 110:363–370

America AHP, Cordewener JHG, van Geffen MHA, Lommen A, Vissers JPC, Bino RJ, Hall RD (2006) Alignment and statistical difference analysis of complex peptide data sets generated by multidimensional LC-MS. Proteomics 6:641–653

Amiour N, Recorbet G, Robert F, Gianinazzi S, Dumas-Gaudot E (2006) Mutations in *DMI3* and *SUNN* modify the appressorium-responsive root proteome in arbuscular mycorrhiza. Mol Plant Microbe Interact 19:988–997

Amme S, Matros A, Schlesier B, Mock HP (2006) Proteome analysis of cold stress response in *Arabidopsis thaliana* using DIGE-technology. J Exp Bot 57:1537–1546

Auge RM (2001) Water relations, drought and vesicular-arbuscular mycorrhizal symbiosis. Mycorrhiza 11:3–42

Azcón-Aguilar C, Barea JM (1996) Arbuscular mycorrhizas and biological control of soil-borne plant pathogens – an overview of the mechanisms involved. Mycorrhiza 6:457–464

Barker SJ, Stummer B, Gao L, Dispain I, O'Connor PJ, Smith SE (1998) A mutant in *Lycopersicon esculentum* Mill. with highly reduced VA mycorrhizal colonization: isolation and preliminary characterisation. Plant J 15:791–797

Bestel-Corre G, Dumas-Gaudot E, Poinsot V, Dieu M, Dierick JF, van Tuinen D, Remacle J, Gianinazzi-Pearson V, Gianinazzi S (2002) Proteome analysis and identification of symbiosis-related proteins from *Medicago truncatula* Gaertn. by two-dimensional electrophoresis and mass spectrometry. Electrophoresis 23:122–137

Bestel-Corre G, Dumas-Gaudot E, Gianinazzi S (2004a) Proteomics as a tool to monitor plant–microbe endosymbioses in the rhizosphere. Mycorrhiza 14:1–10

Bestel-Corre G, Gianinazzi S, Dumas-Gaudot E (2004b) Impact of sewage sludges on *Medicago truncatula* symbiotic proteomes. Phytochemistry 65:1651–1659

Bethlenfalvay GW (1992) Mycorrhizae and crop productivity. In: Bethenthalvay GJ, Linderman RG (eds) Mycorrhizae in sustainable agriculture. American Society of Agronomy, Madison, pp 1–27

Bonfante P, Perotto S (1995) Strategies of arbuscular mycorrhizal fungi when infecting host plants. New Phytol 130:3–21

Bonfante P, Perotto S (2000) Outside and inside the roots: cell-to-cell interactions among arbuscular mycorrhizal fungi, bacteria and host plants. In: Podila GK, Douds DD (eds) Current advances in mycorrhiza. American Phytopathological Society, St Paul, pp 141–155

Borisov AY, Danilova TN, Koroleva TA, Naumkina TS, Pavlova ZB, Pinaev AG, Shtark OY, Tsyganov VE, Voroshilova VA, Zhernakov AI, Zhukov VA, Tikhonovich IA (2004) Pea (*Pisum sativum* L.) regulatory genes controlling development of nitrogen-fixing nodule and arbuscular mycorrhiza: fundamentals and application. Biologia 59:137–144

Brechenmacher L, Weidmann S, van Tuinen D, Chatagnier O, Gianinazzi S, Franken P, Gianinazzi-Pearson V (2004) Expression profiling of up-regulated plant and fungal genes in early and late stages of *Medicago truncatula–Glomus mosseae* interactions. Mycorrhiza 14:253–262

Canovas FM, Dumas-Gaudot E, Recorbet G, Jorrin J, Mock HP, Rossignol M (2004) Plant proteome analysis. Proteomics 4:285–298

Colebatch G, Trevaskis B, Udvardi M (2002) Symbiotic nitrogen fixation research in the postgenomics era. New Phytol 153:37–42

Cook DR (1999) *Medicago truncatula* – a model in the making! Curr Opin Plant Biol 2:301–304

David-Schwartz R, Badani H, Smadar W, Levy AA, Galili G, Kapulnik Y (2001) Identification of a novel genetically controlled step in mycorrhizal colonization: plant resistance to infection by fungal spores but not extra-radical hyphae. Plant J 27:561–569

David-Schwartz R, Gadkar V, Wininger S, Bendov R, Galili G, Levy AA, Kapulnik Y (2003) Isolation of a premycorrhizal infection (*pmi2*) mutant of tomato, resistant to arbuscular mycorrhizal fungal colonization. Mol Plant Microbe Interact 16:382–388

Declerck S, Strullu DG, Fortin JA (2005) In vitro culture of mycorrhizas. Springer, Berlin

Delalande F, Carapito C, Brizard JP, Brugidou C, Van Dorsselaer A (2005) Multigenic families and proteomics: extended protein characterisation as a tool for paralog gene identification. Proteomics 5:450–460

Dodd JC (1994) Approaches to the study of the extraradical mycelium of arbuscular mycorrhizal fungi. In: Gianinazzi S, Schüepp H (eds) Impact of arbuscular mycorrhizas on sustainable agriculture and natural ecosystems. Birkhäusser, Basel, pp 147–166

Dumas E, Gianinazzi-Pearson V, Gianinazzi S (1989) Production of new soluble proteins during VA endomycorrhiza formation. Agric Ecosyst Environ 29:111–114

Dumas-Gaudot E, Bestel-Corre G, Gianinazzi S (2001) Proteomics, a powerful approach towards understanding functional plant root interactions with arbuscular mycorrhizal fungi. In: Pandalai SG (ed) Recent research developments in plant biology. Research Signpost, Trivandrum, India, pp 95–104

Dumas-Gaudot E, Amiour N, Weidmann S, Bestel-Correl G, Valot B, Lenogue S, Gianinazzi-Pearson V, Gianinazzi S (2004a) A technical trick for studying proteomics in parallel to transcriptomics in symbiotic root–fungus interactions. Proteomics 4:451–453

Dumas-Gaudot E, Valot B, Bestel-Corre G, Recorbet G, St-Arnaud M, Fontaine B, Dieu M, Raes M, Saravanan RS, Gianinazzi S (2004b) Proteomics as a way to identify extra-radicular fungal proteins from *Glomus intraradices*-RiT-DNA carrot root mycorrhizas. FEMS Microb Ecol 48:401–411

Dumas-Gaudot E, Recorbet G, Robert F, Valot B, Amiour N (2007) Functional genomic of arbuscular mycorrhizal symbiosis: why and how using proteomics. In: Varma A (ed) Mycorrhizal manual. Uttar Pradesh, in press.

Dunkley TPJ, Dupree P, Watson RB, Lilley KS (2004) The use of isotope-coded affinity tags (ICAT) to study organelle proteomes in *Arabidopsis thaliana*. Biochem Soc Trans 32:520–523

Feiz L, Irshad M, Pont-Lezica RF, Canut H, Jamet E (2006) Evaluation of cell wall preparations for proteomics: a new procedure for purifying cell walls from *Arabidopsis* hypocotyls. Plant Methods 2:2–13

Fester T, Schmidt D, Lohse S, Walter MH, Giuliano G, Bramley PM, Fraser PD, Hause B, Strack D (2002) Stimulation of carotenoid metabolism in arbuscular mycorrhizal roots. Planta 216:148–154

Fester T, Wray V, Nimtz M, Strack D (2005) Is stimulation of carotenoid biosynthesis in arbuscular mycorrhizal roots a general phenomenon? Phytochemistry 66:1781–1786

Gadkar V, David-Schwartz R, Nagahashi G, Douds DD Jr, Wininger S, Kapulnik Y (2003) Root exudate of *pmi* tomato mutant M161 reduces AM fungal proliferation in vitro. FEMS Microbiol Lett 223:193–198

Gao LL, Delp G, Smith SE (2001) Colonization patterns in a mycorrhiza-defective mutant tomato vary with different arbuscular-mycorrhizal fungi. New Phytol 151:477–491

Gao LL, Knogge W, Delp G, Smith FA, Smith SE (2004) Expression patterns of defense-related genes in different types of arbuscular mycorrhizal development in wild-type and mycorrhiza-defective mutant tomato. Mol Plant Microbe Interact 17:1103–1113

Genre A, Chabaud M, Timmers T, Bonfante P, Barker DG (2005) Arbuscular mycorrhizal fungi elicit a novel intracellular apparatus in *Medicago truncatula* root epidermal cells before infection. Plant Cell 17:3489–3499

Gianinazzi S, Schüepp H, Haselwandter K, Barea J (2002) Mycorrhizal technology in agriculture: from genes to bioproducts. Birkhäuser, Basel

Gianinazzi-Pearson V (1996) Plant cell responses to arbuscular mycorrhizal fungi: getting to the roots of the symbiosis. Plant Cell 8:1871–1883

Gianinazzi-Pearson V, Dénarié J (1997) Red carpet genetic programmes for root endosymbioses. Trends Plant Sci 2:371–372

Gosling P, Hodge A, Goodlass G, Bending GD (2006) Arbuscular mycorrhizal fungi and organic farming. Agric Ecosyst Environ 113:17–35

Gygi SP, Rochon Y, Franza BR, Aebersold R (1999) Correlation between protein and mRNA abundance in yeast. Mol Cell Biol 19:1720–1730

Harrison MJ (1997) The arbuscular mycorrhizal symbiosis: an underground association. Trends Plant Sci 2:54–60

Harrison MJ (2005) Signaling in the arbuscular mycorrhizal symbiosis. Annu Rev Microbiol 59:19–42

Hause B, Fester T (2005) Molecular and cell biology of arbuscular mycorrhizal symbiosis. Planta 221:184–196

Hause B, Maier W, Miersch O, Kramell R, Strack D (2002) Induction of jasmonate biosynthesis in arbuscular mycorrhizal barley roots. Plant Physiol 130:1213–1220

Hohnjec, Vieweg MF, Pühler A, Becker A, Küster H (2005) Overlaps in the transcriptional profiles of *Medicago truncatula* roots inoculated with two different *Glomus* fungi provide insights into the genetic program activated during arbuscular mycorrhiza. Plant Physiol 137:1283–1301

Hurkman WJ, Tanaka CK (1987) The effect of salt on the pattern of protein synthesis in barley roots. Plant Physiol 83:517–524

Imin N, De Jong F, Mathesius U, van Noorden G, Saeed NA, Wang XD, Rose RJ, Rolfe BG (2004) Proteome reference maps of *Medicago truncatula* embryogenic cell cultures generated from single protoplasts. Proteomics 4:1883–1896

Inada N, Wildermuth MC (2005) Novel tissue preparation method and cell-specific marker for laser microdissection of *Arabidopsis* mature leaf. Planta 221:9–16

Jamet E, Canut H, Boudart G, Pont-Lezica RF (2006) Cell wall proteins: a new insight through proteomics. Trends Plant Sci 11:33–39

Jeffries P, Barea JM (1994) Biogeochemical cycling and arbuscular mycorrhizas in the sustainability of plant-soil systems. In: Gianinazzi S, Schüepp H (eds) Impact of arbuscular mycorrhizas on sustainable agriculture and natural ecosystems. Birkhaüser, Basel, pp 101–115

Jorrin JV, Rubiales D, Dumas-Gaudot E, Recorbet G, Maldonado A, Castillejo MA, Curto M (2006) Proteomics: a promising approach to study biotic interaction in legumes. A review. Euphytica 147:37–47

Journet EP, Carreau V, Gouzy J, Thoquet P, Rosenberg C, Barker D, Huguet T, Denarie J, Gamas P (2001) The model legume *Medicago truncatula*: recent advances and perspectives in genomics. Oleagineux Corps Gras Lipides 8:478–484

Journet EP, van Tuinen D, Gouzy J, Crespeau H, Carreau V, Farmer MJ, Niebel A, Schiex T, Jaillon O, Chatagnier O, Godiard L, Micheli F, Kahn D, Gianinazzi-Pearson V, Gamas P (2002) Exploring root symbiotic programs in the model legume *Medicago truncatula* using EST analysis. Nucleic Acids Res 30:5579–5592

Kalume DE, Molina H, Pandey A (2003) Tackling the phosphoproteome: tools and strategies. Curr Opin Chem Biol 7:64–69

Kistner C, Winzer T, Pitzschke A, Mulder L, Sato S, Kaneko T, Tabata S, Sandal N, Stougaard J, Webb KJ, Szczyglowski K, Parniske M (2005) Seven *Lotus japonicus* genes required for transcriptional reprogramming of the root during fungal and bacterial symbiosis. Plant Cell 17:2217–2229

Komatsu S, Yano H (2006) Update and challenges on proteomics in rice. Proteomics 6:4057–4068

Komatsu S, Zang X, Tanaka N (2006) Comparison of two proteomics techniques used to identify proteins regulated by gibberellin in rice. J Proteome Res 5:270–276

Kondo T, Seike M, Mori Y, Fujii K, Yamada T, Hirohashi S (2003) Application of sensitive fluorescent dyes in linkage of laser microdissection and two-dimensional gel electrophoresis as a cancer proteomic study tool. Proteomics 3:1758–1766

Kosuta S, Chabaud M, Lougnon G, Gough C, Denarie J, Barker DG, Becard G (2003) A diffusible factor from arbuscular mycorrhizal fungi induces symbiosis-specific *MtENOD11* expression in roots of *Medicago truncatula*. Plant Physiol 131:952–962

Krajinski F, Hause B, Gianinazzi-Pearson V, Franken P (2002) *Mtha1*, a plasma membrane H+-ATPase gene from *Medicago truncatula*, shows arbuscule-specific induced expression in mycorrhizal tissue. Plant Biol 4:754–761

Lahm HW, Langen H (2000) Mass spectrometry: a tool for the identification of proteins separated by gels. Electrophoresis 21:2105–2114

Lammers P, Tuskan GA, DiFazio SP, Podila GK, Martin F (2004) Mycorrhizal symbionts of *Populus* to be sequenced by the United States Department of Energy's Joint Genome Institute. Mycorrhiza 14:63–64

Lilley KS, Dupree P (2006) Methods of quantitative proteomics and their application to plant organelle characterisation. J Exp Bot 57:1493–1499

Lilley KS, Razzaq A, Dupree P (2002) Two-dimensional polyacrylamide gel electrophoresis: recent advances in sample preparation, detection and quantification. Curr Opin Chem Biol 6:46–50

Liu JY, Blaylock LA, Endre G, Cho J, Town CD, VandenBosch KA, Harrison MJ (2003) Transcript profiling coupled with spatial expression analyses reveals genes involved in distinct developmental stages of an arbuscular mycorrhizal symbiosis. Plant Cell 15:2106–2123

Liu JY, Blaylock LA, Harrison MJ (2004) cDNA arrays as a tool to identify mycorrhiza-regulated genes: identification of mycorrhiza-induced genes that encode or generate signaling molecules implicated in the control of root growth. Can J Bot 82:1177–1185

Lohse S, Schliemann W, Ammer C, Kopka J, Strack D, Fester T (2005) Organization and metabolism of plastids and mitochondria in arbuscular mycorrhizal roots of *Medicago truncatula*. Plant Physiol 139:329–340

Lohse S, Hause B, Hause G, Fester T (2006) FtsZ characterisation and immunolocalization in the two phases of plastid reorganization in arbuscular mycorrhizal roots of *Medicago truncatula*. Plant Cell Physiol 47:1124–1134

Marsh JF, Schultze M (2001) Analysis of arbuscular mycorrhizas using symbiosis-defective plant mutants. New Phytol 150:525–532

Martin F, Tuskan GA, DiFazio SP, Lammers P, Newcombe G, Podila GK (2004) Symbiotic sequencing for the *Populus* mesocosm. New Phytol 161:330–335

Mathesius U (2003) Conservation and divergence of signalling pathways between roots and soil microbes – the *Rhizobium*–legume symbiosis compared to the development of lateral roots, mycorrhizal interactions and nematode-induced galls. Plant Soil 255:105–119

Mathesius U, Imin N, Chen HC, Djordjevic MA, Weinman JJ, Natera SHA, Morris AC, Kerim T, Paul S, Menzel C, Weiller GR, Rolfe BG (2002) Evaluation of proteome reference maps for cross-species identification of proteins by peptide mass fingerprinting. Proteomics 2:1288–1303

Matusova R, Rani K, Verstappen FWA, Franssen MCR, Beale MH, Bouwmeester HJ (2005) The strigolactone germination stimulants of the plant-parasitic *Striga* and *Orobanche* spp. are derived from the carotenoid pathway. Plant Physiol 139:920–934

Morandi D (1996) Occurrence of phytoalexins and phenolic compounds in endomycorrhizal interactions, and their potential role in biological control. Plant Soil 185:241–251

Newton RP, Brenton AG, Smith CJ, Dudley E (2004) Plant proteome analysis by mass spectrometry: principles, problems, pitfalls and recent developments. Phytochemistry 65:1449–1485

Nyman TA (2001) The role of mass spectrometry in proteome studies. Biomol Eng 18:221–227

Olah B, Briere C, Becard G, Denarie J, Gough C (2005) Nod factors and a diffusible factor from arbuscular mycorrhizal fungi stimulate lateral root formation in *Medicago truncatula* via the DMI1/DMI2 signalling pathway. Plant J 44:195–207

Ott T, van Dongen JT, Gunther C, Krusell L, Desbrosses G, Vigeolas H, Bock V, Czechowski T, Geigenberger P, Udvardi MK (2005) Symbolic leghemoglobins are crucial for nitrogen fixation in legume root nodules but not for general plant growth and development. Curr Biol 15:531–535

Padliya ND, Cooper B (2006) Mass spectrometry-based proteomics for the detection of plant pathogens. Proteomics 6:4069–4075

Parniske M (2004) Molecular genetics of the arbuscular mycorrhizal symbiosis. Curr Opin Plant Biol 7:414–421

Peck SC (2005) Update on proteomics in *Arabidopsis*. Where do we go from here? Plant Physiol 138:591–599

Requena N, Mann P, Hampp R, Franken P (2002) Early developmentally regulated genes in the arbuscular mycorrhizal fungus *Glomus mosseae*: identification of *GmGIN1*, a novel gene with homology to the C-terminus of metazoan hedgehog proteins. Plant Soil 244:129–139

Rolfe BG, Mathesius U, Djordjevic M, Weinman JJ, Guerreiro N, Natera S, Morris AC (2000) The use of the genus *Trifolium* for the study of plant–microbe interactions, root development and plant defence responses. In: Triplett EW (ed) Prokaryotic nitrogen fixation. Horizon, Wymondham, pp 643–656

Rosewarne GM, Barker SJ, Smith SE (1997) Production of near-synchronous fungal colonization in tomato for developmental and molecular analyses of mycorrhiza. Mycol Res 101:966–970

Sanchez L, Weidmann S, Brechenmacher L, Batoux M, van Tuinen D, Lemanceau P, Gianinazzi S, Gianinazzi-Pearson V (2004) Common gene expression in *Medicago truncatula* roots in response to *Pseudomonas fluorescens* colonization, mycorrhiza development and nodulation. New Phytol 161:855–863

Sandal N, Petersen TR, Murray J, Umehara Y, Karas B, Yano K, Kumagai H, Yoshikawa M, Saito K, Hayashi M, Murakami Y, Wang XW, Hakoyama T, Imaizumi-Anraku H, Sato S, Kato T, Chen WL, Hossain MS, Shibata S, Wang TL, Yokota K, Larsen K, Kanamori N, Madsen E, Radutoiu S, Madsen LH, Radu TG, Krusell L, Ooki Y, Banba M, Betti M, Rispail N, Skot L, Tuck E, Perry J, Yoshida S, Vickers K, Pike J, Mulder L, Charpentier M, Muller J, Ohtomo R, Kojima T, Ando S, Marquez AJ, Gresshoff PM, Harada K, Webb J, Hata S, Suganuma N, Kouchi H, Kawasaki S, Tabata S, Hayashi M, Parniske M, Szczyglowski K, Kawaguchi M,

Stougaard J (2006) Genetics of symbiosis in *Lotus japonicus*: recombinant inbred lines, comparative genetic maps, and map position of 35 symbiotic loci. Mol Plant Microbe Interact 19:80–91

Sawaki H, Saito M (2001) Expressed genes in the extraradical hyphae of an arbuscular mycorrhizal fungus, *Glomus intraradices*, in the symbiotic phase. FEMS Microbiol Lett 195:109–113

Schussler A, Schwarzott D, Walker C (2001) A new fungal phylum, the Glomeromycota: phylogeny and evolution. Mycol Res 105:1413–1421

Slade A, Knauf V (2005) TILLING moves beyond functional genomics into crop improvement. Transgenic Res 14:109–115

Slezack S, Negrel J, Bestel-Corre G, Dumas-Gaudot E, Gianinazzi S (2001) Purification and partial amino acid sequencing of a mycorrhiza-related chitinase isoform from *Glomus mosseae*-inoculated roots of *Pisum sativum* L. Planta 213:781–787

Smith SE, Read DJ (1997) The symbionts forming VA mycorrhizas. In: Smith SE, Read DJ (eds) Mycorrhizal symbiosis. Academic, San Diego, pp 9–33

St-Arnaud M, Hamel C, Vimard B, Caron M, Fortin A (1996) Enhanced hyphal growth and spore production of the arbuscular mycorrhizal fungus *Glomus intraradices* in an in vitro system in the absence of host roots. Mycol Res 100:328–332

Stasyk T, Morandell S, Bakry R, Feuerstein I, Huck CW, Stecher G, Bonn GK, Huber LA (2005) Quantitative detection of phosphoproteins by combination of two-dimensional difference gel electrophoresis and phosphospecific fluorescent staining. Electrophoresis 26:2850–2854

Strack D, Fester T, Hause B, Schliemann W, Walter MH (2003) Arbuscular mycorrhiza: biological, chemical, and molecular aspects. J Chem Ecol 29:1955–1979

Stracke S, Kistner C, Yoshida S, Mulder L, Sato S, Kaneko T, Tabata S, Sandal N, Stougaard J, Szczyglowski K, Parniske M (2002) A plant receptor-like kinase required for both bacterial and fungal symbiosis. Nature 417:959–962

Taylor SW, Fahy E, Ghosh SS (2003) Global organellar proteomics. Trends Biotechnol 21:82–88

Thurston G, Regan S, Rampitsch C, Xing T (2005) Proteomic and phosphoproteomic approaches to understand plant–pathogen interactions. Physiol Mol Plant Pathol 66:3–11

Udvardi MK, Tabata S, Parniske M, Stougaard J (2005) *Lotus japonicus*: legume research in the fast lane. Trends Plant Sci 10:222–228

Valot B, Gianinazzi S, Dumas-Gaudot E (2004) Sub-cellular proteomic analysis of a *Medicago truncatula* root microsomal fraction. Phytochemistry 65:1721–1732

Valot B, Dieu M, Recorbet G, Raes M, Gianinazzi S, Dumas-Gaudot E (2005) Identification of membrane-associated proteins regulated by the arbuscular mycorrhizal symbiosis. Plant Mol Biol 59:565–580

Valot B, Negroni L, Zivy M, Gianinazzi S, Dumas-Gaudot E (2006) A mass spectrometric approach to identify arbuscular mycorrhiza-related proteins in root plasma membrane fractions. Proteomics 6:145–155

Van Wijk KJ (2004) Plastid proteomics. Plant Physiol Biochem 42:963–977

Walter MH, Fester T, Strack D (2000) Arbuscular mycorrhizal fungi induce the non-mevalonate methylerythritol phosphate pathway of isoprenoid biosynthesis correlated with accumulation of the 'yellow pigment' and other apocarotenoids. Plant J 21:571–578

Watson BS, Asirvatham VS, Wang LJ, Sumner LW (2003) Mapping the proteome of barrel medic (*Medicago truncatula*). Plant Physiol 131:1104–1123

Weidmann S, Sanchez L, Descombin J, Chatagnier O, Gianinazzi S, Gianinazzi-Pearson V (2004) Fungal elicitation of signal transduction-related plant genes precedes mycorrhiza establishment and requires the *DMI3* gene in *Medicago truncatula*. Mol Plant Microbe Interact 17:1385–1393

Wolschin F, Wienkoop S, Weckwerth W (2005) Enrichment of phosphorylated proteins and peptides from complex mixtures using metal oxide/hydroxide affinity chromatography (MOAC). Proteomics 5:4389–4397

Wyss P, Mellor RB, Wiemken A (1990) Vesicular-arbuscular mycorrhizas of wild-type soybean and non-nodulating mutants with *Glomus mosseae* contain symbiosis-specific polypeptides (mycorrhizins), immunologically cross-reactive with nodulins. Planta 182:22–26

Xing T, Ouellet T, Miki BL (2002) Towards genomic and proteomic studies of protein phosphorylation in plant–pathogen interactions. Trends Plant Sci 7:224–230

Zhang HX, Zhang CJ, Lajoie GA, Yeung KKC (2005) Selective sampling of phosphopeptides for detection by MALDI mass spectrometry. Anal Chem 77:6078–6084

Chapter 21
Plant Proteome Responses to Abiotic Stress

Delphine Vincent and Michel Zivy

Abstract Proteomic studies of plant response to abiotic stress include analyses of the effects of water deficit, salt excess, low and high temperatures, high light, and presence of toxic chemicals such as herbicides or heavy metals in the environment. In several instances, proteomics has allowed the identification or novel genes and the characterisation of their regulation and function. In other instances, already known function of proteins found to be regulated by stress have allowed the identification of cellular processes involved in the response. Although many studies have analysed total soluble proteins, only a few have analysed particular subcellular compartments such as chloroplasts or mitochondria. Not only variations in protein abundance but also regulation by post-translational modifications and changes in the subcellular localisation of certain proteins have been observed in response to abiotic stress. Regardless of the species or the environmental conditions studied, some common responses are reported, such as up-regulation of heat shock proteins, stimulation of the antioxidant system or RuBisCO degradation, but it is still difficult to decipher the multiple proteome responses, which vary according to the stress characteristics and to the tissue (differentiation, plant developmental stage). Several studies have shown that the combination of proteomics with metabolomics and genetics may allow a better understanding of integrated plant responses to abiotic stresses. Linking proteome variations to physiological and phenotypic changes will make possible the identification of genes and alleles of interest for the selection of plants able to maintain crop yield as high as possible in unfavourable environments.

Delphine Vincent, UMR de Génétique Végétale, Institut National de la Recherche Agronomique, Centre National de la Recherche Scientifique, Université Paris-Sud, Institut National Agronomique Paris-Grignon, la Ferme du Moulon, 91190 Gif-sur-Yvette, France

Michel Zivy, UMR de Génétique Végétale, Institut National de la Recherche Agronomique, Centre National de la Recherche Scientifique, Université Paris-Sud, Institut National Agronomique Paris-Grignon, la Ferme du Moulon, 91190 Gif-sur-Yvette, France, E-mail: zivy@moulon.inra.fr

J. Šamaj and J. Thelen (eds.), *Plant Proteomics*
© Springer 2007

21.1 Introduction

The first studies on the effect of abiotic stresses on the plant proteome were published in the late 1980s. Although at that time most proteins remained anonymous, these studies could address questions such as the impact of the stress on de novo protein synthesis (Hurkman and Tanaka 1987), the existence of changes in different membrane fractions (Hurkman et al. 1988), and the genetic variability of the response (Zivy 1987). Edman microsequencing then allowed the first lists of proteins induced by abiotic stresses to be established (e.g. Costa et al. 1998; Riccardi et al. 1998). The emergence of mass spectrometry (MS) techniques for protein identification opened new perspectives. Chang et al. (2000) were probably the first to exploit these novel techniques to analyse responses to abiotic stress; they studied the protein changes induced by anoxia in maize root tips. Matrix assisted laser desorption ionisation time-of-flight (MALDI-TOF) MS analyses allowed them to explore mechanisms other than the well-known induction of glycolytic enzymes, which alone cannot explain acclimatisation. In addition to the use of high throughput protein identification tools, more and more articles now present sophisticated designs in which the response to stress is better characterised by taking into account the kinetics of the response (several time points), stress levels (e.g. dose effects), differentiation (organs and tissues, stages of development), or by analysing the individual response of different organelles. In general, quantitative analyses are now undertaken, sometimes allowing the classification of proteins according to their different patterns of expression. The effect of abiotic stresses on post-translational events has also been analysed in several studies. This includes not only post-translational modifications and turnover but also protein migration from one cell compartment to another.

Several abiotic stresses share similar consequences at the cellular level. Water deficit and saline stress decrease osmotic potential. Chilling stress also induces changes in cell water status. Abscisic acid (ABA), as well as being implicated in several mechanisms of transcriptional control, is commonly involved in the responses to these stresses. Proteomic analyses have contributed by exposing common features. However, responses to the same stress can differ according to the factors mentioned above (stress intensity and duration, differentiation, etc.) such that it is hazardous to draw general conclusions from analyses of different stresses on different species in different laboratories. Therefore, in this review each abiotic stress will be presented separately.

21.2 Water Deficit

Water is a major limiting factor for plant growth and development. Responses to drought are many and varied and depend on the organ and on the stage of development, e.g. stomatal closure and decreased photosynthesis for aerial parts, growth reduction for young aerial organs, growth continuation for roots, and accelerated senescence for older leaves. Carbon metabolism and relations between sink and source organs are perturbed, as well as the metabolism of elements that are normally absorbed with water. Cellular responses include osmotic adjustment, regulation of water circulation (aquaporins), protection or degradation of proteins, and protection against oxidative stress. These responses are partly triggered by ABA.

Rey et al. (1998) found a protein (CDSP32) induced by severe drought stress in potato chloroplasts. The three micro-sequences obtained showed no homology with any known protein. The cDNA was identified by screening an expression library with an antiserum against the N-terminal sequence. The gene was shown to encode a new form of thioredoxin. Its targets, which include peroxiredoxins, were later identified by affinity chromatography and immunoprecipitation (Rey et al. 2005). This study is a good example of gene discovery and analysis of function using different proteomic techniques.

Rather than focussing the effort on characterisation of a single protein, the following studies aimed at identifying proteins and metabolic pathways involved in the response to drought. In several of these studies the objective was also to explore the genetic variability of the proteome responses, and to identify candidate proteins.

Costa et al. (1998) followed, over the course of several months, the effects of severe and moderate water deficit, including rewatering, on needle proteins from 2-year-old maritime pine seedlings. By using a model of analysis of variance including both genetic background and seasonal variations, and measuring water potential to assess the drought level, they isolated 38 drought-responsive protein spots. Two down-regulated proteins were identified as actin and ribulose 1,5-bisphosphate carboxylase/oxygenase (RuBisCO) activase. Five up-regulated proteins corresponded to two glutathione peroxidase isoforms, a caffeoyl CoA-*O*-methyltransferase, a low molecular weight heat shock protein (HSP) and a Cu/Zn superoxide dismutase. The three other spots showing increased intensity corresponded to degradation products of the RuBisCO large subunit, and thus were symptomatic of degradation of RuBisCO upon drought.

Salekdeh et al. (2002a) compared the responses to water deficit of two rice cultivars contrasting in osmotic adjustment and root features. Plants were subjected to water deficit for 23 days, and then rewatered for 10 days. Leaf samples were collected twice during water deficit and twice after the beginning of rewatering. Control plant leaves were collected at the same dates. Forty-two proteins showed a quantitative change in response to gradual drought stress. All changes were fully or substantially reversed upon rewatering. Although both cultivars presented comparable leaf water status, they differed at the proteomic level. Fifteen changes were common to both genotypes. The majority of the differences involved proteins that were up-regulated only in the tolerant line or down-regulated only in the sensitive line. This could be the result of a more active response in the tolerant line and of a more affected metabolism in the sensitive line. The up-regulated proteins included a translation elongation factor EF-Tu, two RuBisCO active isoforms, a fructose 1,6-bisphosphate aldolase, a cytosolic triosephosphate isomerase, a GSH-dependent dehydroascorbate reductase, an actin depolymerising factor, two Cu/Zn superoxide dismutase isoforms and a nucleoside disphosphate kinase. An isoflavone reductase-like protein and two isoforms of a chloroplast Rieske Fe-S protein were identified among the down-regulated proteins. Two of the most up-regulated proteins corresponded to isoforms of S-like RNase, whose cDNA was isolated and sequenced based on peptide sequence information.

Hajheidari et al. (2005) subjected two sugar beet genotypes to a progressive mild drought by reducing the water supply during the growing season. The originality of this study is that the samples were taken from field-grown plants subjected to progressive water deficit. Most of the drought-responsive leaf proteins separated by two-dimensional electrophoresis (2-DE) were shared by both genotypes. Forty-five proteins were down-regulated and 35 were up-regulated under drought stress. A putative nascent polypeptide-associated complex alpha-chain was down-regulated, and four breakdown products of RuBisCO large subunit accumulated. A Cu/Zn superoxide dismutase, a nucleoside diphosphate kinase, a 2-Cys peroxiredoxin, two low molecular weight HSPs, a putative oxidoreductase, an osmotin-like protein and two cyclophilin isoforms were up-regulated.

An original experiment was conducted on grapevine plants in which the reduction of water potential caused by withholding water was reproduced by exposure to increasing concentrations of NaCl (Vincent et al. 2007). Two grapevine cultivars were subjected in parallel to water deficit and salinity over a 16 day time period. Interestingly, among stress-related proteins such as mitochondrial peroxiredoxin, part of the antioxidant system, or pathogenesis-related (PR) proteins, ten were up-regulated by stress only in the most susceptible cultivar. Most of the proteins responding to water deficit were also affected by salt stress. The greatest increase in abundance was observed for a nuclear matrix constituent protein 1 and ribosomal protein L39.

Riccardi et al. (1998) used two maize inbred lines differing in their ability to sustain yield under water-deprived conditions, and their hybrid, to isolate drought-responsive proteins. After withholding water for 10 days, the elongating zone of leaves were collected and analysed by 2-DE. Out of 78 drought-responsive protein spots, 23 were down-regulated, 40 were up-regulated, 10 were visible only in stressed samples and 5 showed only a significant genotype x treatment interaction. Sixteen proteins showing increased abundance were subsequently identified: three β-glucosidase isoforms, a ferritin, a putative cytoplasmic malate dehydrogenase, a cytosolic triose phosphate isomerase, a glutamate-1-semialdehyde 2,1-aminotransferase, an ABA-induced OSR40 protein, a chloroplastic fructose bisphosphate aldolase, a cysteine synthase, an ABA/stress/ripening responsive protein (ASR), a soluble inorganic pyrophosphatase, an enolase, a phosphoribulokinase, a dehydrin, and a caffeate-*O*-methyltransferase (COMT). The expression kinetics of these proteins was then analysed up to the 14th day of water deficit (Riccardi et al. 2004). Relative water content, water potential and ABA concentrations were measured on the same plants, thus allowing analysis of the relationship between protein regulation and physiological and biochemical plant characteristics. Different types of responses were observed. For example, the relative quantity of some proteins (e.g. a protein similar to OSR40) was highly correlated with that of ABA, while a plateau was observed for other proteins. Differences between the two inbred lines were of special interest to isolate "candidate proteins". For instance, this analysis showed that differences in expression of ZmASR1 between the inbred lines previously observed on a specific sampling date were not related to a delay in the response of the susceptible line, but rather to a specific expression pattern of the protein in the tolerant line. For another protein (similar to the barley aleurone ABA45 protein), this analysis showed that the higher level in the tolerant line was most likely related to a higher level of ABA in leaves rather than to a genetic variation in the control of this specific protein.

Further proteomic experiments were conducted to analyse the response of maize leaves to water deficit as a function of cell differentiation and development, indicating the co-regulation of enzymes involved in lignin synthesis (Vincent et al. 2005). Because the meristem (elongation zone) is located at the base of the leaf, a gradient of tissue age exists along the maize leaf. Sampling maize leaf fragments along the leaf allowed profiles of protein expression as a function of tissue age to be studied. Comparisons between profiles from leaves subjected to different levels of water deficit and sampled at different stages of development suggested that there is a lignification zone above the elongation zone, and that water deficit not only induces a shift of this zone toward the basal region of the maize leaf, but also causes a down-regulation of the whole lignin pathway (Fig. 21.1). These hypotheses were based on the profiles of various isoforms of enzymes involved in the synthesis of monolignols (e.g. COMT, caffeoyl-CoA 3-O-methyltransferase), enzymes in the phenylpropanoid pathway (e.g. phenylalanine ammonia lyase) and in the synthesis of S-adenosyl-L-methionine (SAM), i.e. the donor of the methyl group in methylation reactions that is widely used during monolignol synthesis (SAM synthase, methio-

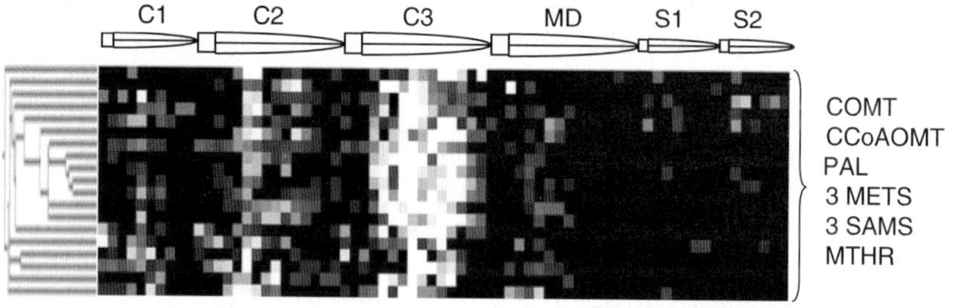

Fig. 21.1 Down-regulation of enzymes related to lignin synthesis in maize leaves upon water deficit. Maize plants were submitted to mild (*MD*) or severe (*S1* and *S2*) water deficit. Controls were collected during the linear growth of leaf 6 (*C1* and *C2*) and when growth decreased (*C3*). Samples were collected along leaf 6. Sample position along the leaf is symbolised by the schematic above the graph. Hierarchical classification allowed the identification of this cluster, which includes proteins involved in monolignol biosynthesis (see text). High and low protein amounts are represented by *bright* and *dark* squares, respectively. In control leaves, the amount of these proteins reaches a maximum by the end of growth (C3) in a region situated on top of the elongation zone (10–20 cm from the insertion point). When plants were submitted to water deficit, these proteins were down-regulated (data from Vincent et al. 2005)

nine synthase and methylenetetrahydrofolate reductase). A metabolomic analysis of lignin compounds along the leaf confirmed that the region where lignin is synthesised corresponds to the region where enzymes involved in lignin biosynthesis display a maximum abundance; under water deficit conditions, this region shifts toward the base of the leaf and lignin accumulation is reduced. Thus the increase of COMT abundance observed at the base of the maize leaves in Riccardi et al. (1998) was not due to the up-regulation of COMT, but was in fact related to the shift toward the base of the lignification zone. This study is an example of discovery of a physiological response by using a proteomic approach. It also emphasises the importance of comparing tissues at similar developmental stages.

One of the proteins shown to be induced by water deficit in maize leaves was ZmASR1. This protein was induced in one of the two analysed inbred lines, but was not produced by the other line. With the aim of looking for candidate proteins, i.e. proteins whose genetic variation of expression could be the cause of the variation of phenotypic responses to water deficit, a proteomic analysis of a population of recombinant inbred lines obtained from the cross between these two lines was carried out. The locus determining the expression/non expression of ZmASR1 was mapped to chromosome 10, at the same location as the encoding gene, which suggests that the polymorphism responsible for the variation of ZmASR1 expression was within the gene encoding it. In the same chromosomal region were located quantitative trait loci (QTLs) for anther-silking interval and leaf senescence under conditions of water deficit. The co-localisation of phenotypic QTLs with a QTL of ZmASR1 protein expression (a PQL, protein quantity locus) made it possible to hypothesise that the cause of variation of these phenotypic traits was the variation in expression of this protein (de Vienne et al. 1999). This hypothesis was further supported by plant transformation; results with sense and antisense constructs confirmed the effect of ZmASR1 expression on leaf senescence (Jeanneau et al. 2002).

21.3 Salt Stress

Salt excess causes not only osmotic shock, but also ionic stress. Cells remove sodium from the cytoplasm, where it is toxic, and sequester it in the vacuole. In addition to studies similar to those on responses to drought, studies on responses to salt stress include the analysis of response to exogenous ABA and the study of specific cellular compartments.

Moons et al. (1995, 1997) analysed the proteome response of rice roots to different concentrations of ABA. Three cultivars were compared, and the most tolerant to salinity showed the highest up-regulation of a dehydrin, a late embryogenesis abundant (LEA) protein of group 3, and an unknown protein, whose cDNA was then isolated and sequenced thanks to the availability of peptide sequences. A new family of genes encoding highly hydrophilic proteins with duplicated domains (*osr40* genes) was thus discovered.

Salekdeh et al. (2002b) compared the effects of progressive saline stress (7 days at 50 mM followed by 7 days at 100 mM NaCl) in two rice cultivars. They identified three proteins that were up-regulated by this stress in roots. Interestingly, an ASR-like protein was induced in the tolerant line, while no change was observed in the sensitive line. It is interesting to notice that a relation between ASR expression and plant response to drought, previously demonstrated in maize, also seems possible in rice. A protein similar to an ascorbate peroxidase was found induced in both genotypes, but was more abundant in the tolerant genotype in the absence of stress. Thus the tolerant variety might be constitutively better protected against oxidative stress than the sensitive variety. This highlights the fact that genetic variations for constitutive levels of expression can be responsible for phenotypic differences between lines.

Parker et al. (2006) applied a low concentration (50 mM) of NaCl to rice seedlings, and studied the short-term (1 day) and long-term (7 days) effects of this treatment on the leaf proteome. Proteins from the blade of the fourth leaf were separated by 2-DE. Interestingly, the authors analysed the technical and biological reproducibility of their system. The 2D technique introduced a coefficient of variation (CV) of 0.26 on average. The biological variation was relatively low, since it added only 0.05 to the average spot CV. Nine protein spots showed a short-term response, and 31 showed a long-term response. One protein, a chloroplastic phosphoglycerate kinase, was transiently induced. A RuBisCO activase, a ferritin isoform, and an ATP synthase responded to both short- and long-term NaCl application. A superoxide dismutase and another isoform of ferritin were up-regulated, and a porphobilinogen deaminase (synthesis of tetrapyrroles), the translation initiation factor 5A and two isoforms of SAM synthase were down-regulated only in samples taken after 7 days of salt stress.

Yan et al. (2005) analysed the kinetics of the rice root proteome response to saline stress. A higher (150 mM) concentration of NaCl was supplied for 24 h, 48 h and 72 h. Of the 58 protein spots affected by salinity, 12 were identified as being involved in regulation of carbohydrate, nitrogen and energy metabolism, reactive oxygen species (ROS) scavenging, mRNA and protein processing, as well as cytoskeleton stability.

Askari et al. (2006) opted for a halophyte species, *Sueda aegiptiaca*, to unravel salt adaptation in plants. They applied increasing concentrations of NaCl (0, 150, 300, 450 and 600 mM) for 30 days and observed the 2-D patterns obtained from leaf samples. Most of the protein changes occurred at the highest NaCl concentration. Different methods of classification allowed the clustering of the 102 significantly varying spots into 12 different groups according to their response to the different doses. Twenty seven up-regulated proteins were identified, including many proteins involved in ROS scavenging. Other proteins were involved in glycine betaine synthesis, cytoskeleton remodelling, photosynthesis, ATP production, protein degradation and protection, and cyanide detoxification.

Difference in-gel electrophoresis (DIGE) was used by (Ndimba et al. 2005) to analyse the effect of salinity and osmotic stress on *Arabidopsis* cell cultures.

The identification of 76 proteins showed the involvement of many cellular processes, although the percentage change for most identified proteins was lower than 30%.

To identify novel signalling components of response to salinity stress, Lee et al. (2004) carried out a proteomic analysis of the *Arabidopsis thaliana* root microsomal fraction. Seedlings were grown for 2 weeks in MS–sucrose liquid medium and were treated with 250 mM NaCl for 2 h. Two of the up-regulated spots were identified as two forms of the same annexin, AnnAt1. Annexins participate in essential cellular processes in animal cells and are known to interact with Ca^{2+}, which is a second messenger in the response of plants to stress. AnnAt1 spots were further characterised by immunodetection methods, which allowed the identification of two additional forms. Two-dimensional blots revealed that the accumulation in the microsomal fraction was induced not only by salt stress but also by ABA, PEG and mannitol. The cytosolic fraction was also studied. Interestingly, AnnAt1 protein accumulation was highly increased in the microsome after 2 h of salt treatment and returned to the control level after 24 h, while, on the contrary, the protein was abundant in the cytoplasm at time 0, down-regulated drastically at 2 h, and recovered to control levels at 24 h of treatment. As the protein quantity in the cytosol was estimated to represent more than 99% of the total, this cannot be explained by a simple translocation of the protein from the cytosol to the membranes: protein turnover in the cytosol is likely involved. In vitro experiments showed that the association of AnnAt1 with the microsomal fraction is strengthened by Ca^{2+} and weakened by EGTA. In vivo experiments showed that Ca^{2+} depletion almost cancelled out the effect of NaCl on both fractions, suggesting that the salt stress-induced response of AnnAt1 is regulated by Ca^{2+}. This study is a nice example of the ability of proteomic studies to analyse phenomena such as protein translocation between cellular compartments and protein turnover, which are typically out of the reach of transcriptomic analyses. Interestingly, the RNA level of *AnnAt1* was unaffected by the salt treatment.

The plant cell apoplast – with its primary role in signal perception and cell nutrition – is a compartment of interest for the analysis of response to stresses. The apoplast is also involved in cell wall elongation, which is determinant for plant growth. Dani et al. (2005) focussed their observation on the apoplastic compartment in leaves from tobacco plants treated with 100 mM NaCl for 20 days, which reduced the growth rate. Apoplastic proteins were recovered following a vacuum infiltration procedure and separated by 2-DE. Approximately 150 spots were resolved, 7 of which were up-regulated by salinity, whereas 13 spots were down-regulated. Two of the salt-induced proteins corresponded to different isoforms of a lipid transfer protein and the other up-regulated proteins were identified as a β-galactosidase, an acidic endochitinase Q, a chitinase PR-P and a germin-like protein. All of the down-regulated proteins were identified as peroxidases. The observation that the levels of cell wall-associated peroxidases were decreased is interesting because down-regulation of proteins acting in the synthesis of cell wall might be a consequence of the lower rate of leaf elongation.

21.4 Cold Stress

Low temperature is an important limiting factor for plant growth and photosynthesis. Cold acclimatisation is associated with metabolic changes involving, in particular, carbon metabolism and membrane lipid composition. This has led several teams to focus their analyses specifically on chloroplast and membrane proteins.

Yan et al. (2006) deciphered the cold shock response in rice seedlings by exposing them to 6°C for 6 h or 24 h, and to 6°C for 24 h followed by 24 h recovery. Leaf proteins were separated by 2-DE. Of the 1,000 reproducible spots, 65 were significantly up-regulated and 31 down-regulated by cold treatment. Most of them were common to both 6 h and 24 h time-points. About 50% of the down-regulated proteins were no longer visible after 24 h of recovery, while approximately two-thirds of up-regulated proteins were still observed. Eighty five of these proteins could be identified by MS. In addition to proteins involved in the response to ROS (peroxidases) and the control of redox homeostasis (thioredoxins), the functional category most affected was photosynthesis (35%), followed by carbon/nitrogen/sulfur metabolism (12%), protein synthesis (10.5%), energy metabolism (8%), photorespiration (7%) and signal transduction (5%). Interestingly, this study also showed that chilling stress enhances protein degradation: 19 fragments of the large subunit of RuBisCO were identified in this study. Evidence for the degradation of five other photosynthetic proteins was found; amounts of their intact form decreased while that of fragments increased. This might be symptomatic of the susceptibility of the photosynthetic apparatus to chilling stress.

Amme et al. (2006) used a DIGE strategy to study cold response in *Arabidopsis*. Seedlings were exposed to 6°C or 10°C for a week before being allowed to recover for another week. The majority of the cold-responsive proteins returned to control levels upon cold stress relief. Most of the 14 identified up-regulated proteins, such as a RNA-binding protein, a glycine-rich protein, a dehydrin ERD 10 and low-temperature-induced protein 78, were already known to be induced by cold stress.

Imin et al. (2004) studied cold responses in rice anthers because of the high susceptibility of microspores to low temperature, which can lead to male sterility. Cold treatment was mild and consisted of a 4-day exposure to 12°C at the young microspore stage. Plants were then placed at control temperatures for 10 days, which allowed them to reach the trinucleate microspore stage. At this stage the treatment had a visible impact on anther size and pollen structure. Of the 3,000 revealed spots, 70 showed a greater than twofold variation in response to cold treatment. Peptide mass fingerprint (PMF) analyses allowed the identification of 18 cold-responsive proteins. An interesting point of this study was the observation of different responses of isoforms produced by the same gene. For example, six isoforms of ascorbate peroxidase were identified, corresponding to the products of two different genes. Three of the four isoforms detected for the first gene were up-regulated while the remaining isoform was down-regulated. Likewise, one isoform encoded by the other gene was up-regulated while the other was down-regulated. These results show that one must be cautious when drawing conclusions regarding protein induction, since the existence of other isoforms with different regulation cannot be excluded. Another interesting point of this study was that degradation products of seven different proteins accumulated under cold treatment, indicating that cold treatment triggered the cleavage or partial degradation of these proteins. Interestingly, the peptides identified in these breakdown products allowed the authors to identify them as N- or C-terminal products.

Cui et al (2005) subjected rice seedlings to a 3 day-long progressive decrease of temperature from 28°C (control) to 5°C. The objective of the progressive decrease of temperature was to distinguish cold response from possible responses to cellular injuries caused by a cold shock. Leaf tissues were analysed by 2-DE at different steps of the temperature treatment. Different types of kinetics were detected. Some proteins showed a progressive increase, while others displayed a specific increase at a particular step. Of the 60 proteins up-regulated at least once during the experiment, 38 were increased by a factor greater than 2. Many cold-responsive proteins were identified as being involved in protein synthesis, protection and degradation. Other up-regulated proteins were directly or indirectly related to lignin synthesis, response to oxidative stress and detoxification, and energy metabolism. More than 40% were predicted to be located in the chloroplast, underlining the fact that chloroplasts are specifically affected by cold stress.

Goulas et al. (2006) focused on chloroplastic proteins of *Arabidopsis* leaves. Mature plants were placed at 5°C for either a short- (24 h) or long-term (10 and 40 days) exposure. After chloroplast isolation, proteins from the stroma and lumen fractions were recovered. The DIGE technique was used for protein visualisation. No change in abundance was observed after 24 h of cold stress, while approximately 10% (50 spots) of proteins were affected after 10 or 40 days of cold in each fraction. Most stromal proteins involved in the Calvin cycle showed a significant decrease. Several subunits of the extrinsic catalytic (CF1) portion of the ATP synthase complex were shown to increase in the stroma after 10 days at 5°C. These proteins are normally coupled to the thylakoid membrane. Their release into the stroma during the early phase of cold acclimatisation may indicate that the coupling factor is affected by cold stress. Interestingly, leaves that grew during the 40 days of cold stress did not exhibit this phenomenon, suggesting that acclimatisation allowed for protection of ATP synthase. The same phenomenon was observed for subunits of the core photosynthetic complexes, indicative of rearrangements in the thylakoid membrane. Several induced stromal proteins were involved in detoxification and response to oxidative stress. Interestingly, the observed decrease in the amount of 2-Cys Prx peroxiredoxins, which act as peroxidases under control conditions, is consistent with their association to thylakoid membrane to form a chaperone complex upon cold stress. Changes were also observed for proteins involved in signalling, e.g. a protein involved in the synthesis of jasmonic acid and a PII uridylyl transferase involved in the control of nitrogen and carbon metabolism.

Kawamura and Uemura (2003) analysed plasma membranes isolated from leaves of *Arabidopsis* plants placed at 2°C for up to 7 days. This temperature induced increased freezing tolerance after 1 day, with maximum tolerance being attained after 7 days. Proteins associated with plasma membranes were divided into a soluble fraction that was analysed by 2-DE and an insoluble fraction that was analysed by SDS-PAGE. Twenty-seven soluble membrane-associated proteins responded to freezing treatments, most of which were up-regulated after 1 day of freezing stress. Insoluble protein fractions analysed by 1-DE allowed 15 freezing-responsive proteins to be added to the previous list. The function of several of the up-regulated proteins could be related to acclimatisation. Proteins ERD10 and ERD14 are SK-type acidic dehydrins: the association of this type of dehydrin with plasma membranes has been observed before. A lipoprotein-like protein appeared to be a member of the lipokalin family. It has been suggested that membrane-anchored lipokalin proteins might play a role in membrane biogenesis and repair. Other proteins could play a role in signal transduction: a tobacco DREPP-like protein may be associated with Ca^{2+}; a synaptotagmin-like protein and a phospholipase D may bind phospholipids in a Ca^{2+}-dependent manner, and PLD-induced phosphatidic acid acts as a second messenger.

Bae et al. (2003) studied how nucleoproteins responded to cold in *Arabidopsis*. Seedlings were exposed to 4°C for 6h and leaf nuclear proteins were recovered using Percoll density gradients. Both 4–7 and 6–9 pI ranges were used for 2-DE analyses. Of the 184 identified proteins, 40 were up-regulated and 14 were down-regulated by cold treatment. These proteins were involved in multiple cellular functions, among which were, as expected, transcriptional factors, RNA-binding proteins and other proteins involved in signalling.

21.5 Heat Stress

Plants exposed to non-lethal temperatures acquire thermotolerance, i.e. the ability to withstand subsequent higher temperatures. Heat shock proteins (HSPs) are chaperones that are induced while other proteins are down-regulated upon elevated temperatures eliciting thermotolerance. As HSPs are defined by these expression patterns, they could be analysed by 2-DE even before precise identification of the genes encoding them. The aim of several of studies (Zivy 1987; Krishnan et al. 1989; Süle et al. 2004), was to study the genetic variability of HSP expression and to correlate it to thermotolerance. A high level of genetic variability was detected for low molecular weight HSPs in wheat (Zivy 1987). Wang and Luthe (2003) showed that a chloroplastic HSP is genetically correlated to thermotolerance in bentgrass. Banzet et al. (1998) reported that some small HSPs are also induced by oxidative stress. HSPs are developmentally regulated; small HSPs accumulate in developing seeds. Several studies analysed by 2-DE the expression of HSPs in seeds of rice and wheat as well as in cork oak somatic embryos, and observed their up-regulation by heat (Lin et al. 2005; Majoul et al. 2003; Puigderrajols et al. 2002). When plants are submitted to day/night temperature variations, the proteome response is not as simple as the induction of HSPs and down-regulation of other proteins. In fact, Ferreira et al. (2006) observed up-regulation of 20% of leaf proteins spots when hydroponically grown poplar plants were transferred from

30°C/20°C day/night temperature to 42°C/37°C for 3 days. Many functions were found to be up-regulated in addition to HSPs.

Phosphorylation of maize HSP22 was proved by ^{32}P-ATP labelling of isolated mitochondria. LC-MS experiments allowed the authors not only to identify the phosphorylated serine, but also to detect two forms of HSP22 that are produced by alternative intron splicing (Lund et al. 2001).

Heat shock provokes a rapid reprogramming of gene expression to favour translation of HSPs in which translation factors could play a role. Gallie et al. (1997) analysed translation initiation factors in wheat heat-shocked seedlings and during seed development. Immunoblots revealed that modifications of the pI of different eukaryotic initiation factors (EIFs) occur in response to heat shock as well as during development. Modifications during seed germination could be correlated to the resumption of translation. Phosphatase assays were used to show that the modification originated from phosphorylation. Regulation of phosphorylation of EIF-4A and EIF-4B in particular was associated with heat stress.

21.6 High Light

When light exceeds photosynthetic capacity, ROS are generated in the chloroplasts and cause oxidative damage. Plant responses to high light include synthesis of carotenoids to dissipate light energy in the form of heat, and the synthesis of proteins involved in ROS scavenging. Chloroplasts are the primary targets of damage caused by high light.

A proteomic approach was used to study high light stress in ginseng leaves by exposing them to 1,700 µmol photons m^{-2} s^{-2} for 1–4 h (Nam et al. 2003). Proteins were analysed by 2-DE after 1 h, 2 h and 4 h of stress. A small number of proteins were shown to be significantly affected by this stress. Four isoforms of cytosolic ascorbate peroxidase, two isoforms of a small HSP, and two isoforms of a major latex-like protein were up-regulated under high light conditions. A Rieske Fe/S protein and a 3-dehydroxysteroid dehydrogenase/isomerase, possibly involved in the biosynthesis of ginsenosides, were down-regulated. It is interesting to note that the response to oxidative stress was represented only by cytosolic ascorbate peroxidases, while for example different forms of superoxide dismutase (SOD) were also identified in these 2-D gels.

Phee et al. (2004) analysed the effect of high light on *Arabidopsis* chloroplasts. Most of the 35 proteins showing reduced expression, such as photosystem-related proteins and ATP synthase-related proteins, were involved in photosynthetic pathways. Ribosomal proteins were also found to decrease. Of the 17 up-regulated proteins, most were chaperones and proteins involved in ROS scavenging. Among other induced proteins were a glycerol-3-phosphate acyltransferase involved in the metabolism of membrane lipids, and a NADP-dependent oxidoreductase involved in photosynthetic electron transport.

Giacomelli et al. (2006) further narrowed down their focus to thylakoid membranes within chloroplastic structures by deciphering the effect of high light stress on the ascorbate-deficient *Arabidopsis* mutant *vtc2-2*. Following a 1,000 μmol photons m^{-2} s^{-2} exposure for 0, 1, 3, or 5 days, rosettes were collected and intact thylakoid membranes were recovered using Percoll gradients. 2-DE separated proteins were stained with the fluorescent dye Sypro Ruby. This is a nice example of a quantitative study where an equilibrated design was built with biological as well as technical replicates at each of the time points, allowing the detection of significant variations as a function of treatment, genotype and interaction between genotype and treatment by analysis of variance. A total of 155 reproducible spots was tested for statistical significance. Most of them were identified, including proteins either affected or not by high light or ascorbate deficiency. Six of the proteins most highly induced by light were major components of plastoglobules (PGs), namely four members of the fibrillin family, a fructose-biphosphate aldolase and a flavin reductase. PGs are thylakoid-associated lipid-rich particles that contain quinones and α-tocopherol, as well as proteins likely involved in the cleavage of lipids and carotenoids. Such proteins are believed to play a role in the response to oxidative stress. YCF37 is another protein found increased by high light in thylakoids, which can be related to accelerated turnover of PSI. Several proteins known to participate in the response to oxidative stress or to be involved in redox homeostasis or detoxification were not induced (e.g. peroxiredoxins A and E). Interestingly, ascorbate deficiency induced responses in the mutant that were not observed in the wild type and vice versa. For example, a putative glyoxylase I was down-regulated by high light in the wild type only and HSPs were induced only in the mutant. Thus, this analysis effectively allowed the study of compensation mechanisms for the weakening of one of the detoxification systems.

21.7 Metal Stress and Herbicides

Contamination of soil by toxic heavy metals is a major environmental hazard. Elevated metal concentrations induce several defence mechanisms in plants, including the chelation of metals by phytochelatins (PCs) or metallothioneins. Metal stress causes oxidative stress, and ROS scavenging mechanisms are induced. The analysis of the response to metal stress is often combined with the response to other oxidative stress, such as the action of a herbicide. PR proteins are also induced in response to metals.

Sarry et al. (2006) analysed the effect of the application to *Arabidopsis* cell cultures of several cadmium (Cd) concentrations (0.5, 2, 5, 20, 50 and 200 mM) for 24 h. The experiment was performed on two different culture mediums because previous results had shown an effect of the culture medium on protein expression. Soluble proteins were separated by 2-DE. The number of spots significantly affected varied from 10% for 0.5 mM $CdCl_2$ to 25% for 200 mM $CdCl_2$. Many enzymes involved in energy (especially glycolysis), nitrogen, and sulfur metabolic pathways were increased in response to Cd, as well as proteins related to the response to oxidative stress, detoxification and protein destination. Different forms of response to oxidative stress were observed, including a catalase, a monodehydroascorbate reductase, peroxiredoxins, a NADPH oxido-reductase, a phospholipid hydroperoxide and a glutathione peroxidase. Products of degradation were detected for several proteins, indicating an increased proteolytic activity, which is consistent with the increased amount of several proteins involved in protein degradation. In parallel to the proteomic analysis, the authors carried out a metabolomic analysis, which revealed that the major response to Cd at the metabolite level was the increased accumulation of different families of PCs. PCs have the general formula (γGlu-Cys)n-Gly and are synthesized from glutathione (GSH). With this knowledge, and possibly enhanced by the proteomic analysis, the authors were able to construct interesting hypotheses regarding the roles of the different pathways. Thus the increased synthesis of enzymes involved in glycolysis could be related not only to a demand for ATP and reducing power, but also to a demand for carbon skeletons required for the synthesis of glutathione. The increased accumulation of enzymes involved in the biosynthesis of glutamate, cysteine, serine and glycine could be directly related to the synthesis of GSH and PCs. The increased amounts of proteins of C1 metabolism could be indirectly involved. GSH is probably not only used for the synthesis of PCs, since several GSTs were also found to be induced: these might conjugate metabolites damaged by the oxidative stress provoked by Cd. This study is a nice example of the complementarity of metabolomics and proteomics studies.

Le Lay et al. (2006) analysed the effect of caesium (Cs) in *Arabidopsis* cells exposed to 1 mM CsCl for 24 h. As Cs accumulates in plants as a potassium analogue, they analysed the proteome response in the presence and absence of potassium and found that 92 proteins were significantly modified by at least one of the treatments. K^+ shortage actually caused more changes than the addition of Cs to the culture medium: 32 proteins were affected by K^+-depletion treatment while 18 were affected by Cs in the presence of K^+. The presence of Cs in the absence of K^+ induced 40 protein changes. Interestingly, both Cs and K^+ treatments induced the overexpression of antioxidant enzymes.

A proteomic analysis of arsenic (As) toxicity on maize roots showed that most As-responsive proteins were enzymes involved in ROS scavenging (Requejo and Tena 2005).

Cuypers et al. (2005) reported the accumulation of PR proteins in *Phaseolus vulgaris* roots in response to Cu excess. Rakwal et al. (1999) also found the up-regulation of PR proteins in response to Cu, as well as to jasmonate acid and UV irradiation of rice leaves. A strong effect of different metals on RuBisCO in rice leaves was observed through the increased amount of degradation products (Hajduch et al. 2001).

Fecht-Christoffers et al. (2003) looked for changes in soluble proteins extracted from the apoplastic washing fluid of cowpea leaves after 5 days of exposure to manganese (50 mM $MnSO_4$). Blue-native (BN)-PAGE was used to reveal peroxidase activity. The treatment induced a large increase in the concentration of proteins in the apoplastic washing fluid that was apparently related not specifically to peroxidases but also to a number of PR proteins.

Most proteins induced by fluxioxazin (a herbicide) in grapevine roots and shoots were PR class 10 isoforms (Castro et al. 2005), thus providing another example of induction of PR proteins by oxidative stress.

Taylor et al. (2005) compared the effects of different stresses, i.e. drought, chilling, and herbicide, on the mitochondrial proteome of pea. The herbicide induced severe oxidative stress. Thirty-three spots were shown to change significantly in abundance in response to at least one stress treatment. In several instances, breakdown products were identified and their accumulation was negatively correlated to the accumulation of the intact product, i.e. their quantities increased as that of the intact protein decreased. Mechanisms common to the different analysed stresses were observed. In particular, the degradation of glycine decarboxylase and serine hydroxymethyltransferase was provoked by the three treatments, although with quantitative differences. Interestingly, the treatments induced very different responses for various classes of HSPs: a HSP90 was found to be induced by drought and chilling but not by the herbicide; a HSP70 decreased in response to the herbicide only; HSP22 was induced by all three treatments. In an earlier paper, Sweetlove et al. (2002) analysed the specific effect of oxidative stress on mitochondrial proteins in more depth.

Many studies, such as the latter, focus on a particular compartment to analyse the effect of a stress. In contrast, the following study focussed rather on a particular function. Glutathione *S*-transferases (GSTs) are involved in detoxification and protection against oxidative stress. The regulation of GSTs in response to Cu and to a herbicide safener was investigated in *Arabidopsis* seedlings using gel-free tools (Smith et al. 2004). Proteins were applied to a GSH-agarose affinity column to purify GSTs. Eight GSTs out of the 53 reported in the *Arabidopsis* genome were identified. Quantification was performed by differential isotopic labelling; after digestion by trypsin, peptides were labelled with either acetate (control) or trideuteroacetate (treated) before being combined and analysed by tandem MS (MS/MS). The effect of the herbicide safener and Cu was analysed for seven GSTs, for which at least three peptides could be quantified. Only one GST was induced by the herbicide safener, while four were induced by Cu. One GST was decreased by both treatments, and two GSTs were unaffected.

21.8 Conclusions

A number of proteome responses were found common to many stresses. For example, the induction of chaperones and of enzymes involved in ROS scavenging was observed in nearly all studies. Degradation of the large subunit of RuBisCO was also often observed through the increased abundance of breakdown products. However, numerous differences also exist, even for the induction of these relatively common responses. Superoxide dismutases were found induced in some analyses and constant in others. Although the induction of HSPs is relatively general, obviously they do not all respond in the same manner to different treatments. Other functions are often found to be involved in the response to stress, but not necessarily in the same way. For example, SAM synthetases were found increased in some studies and decreased in others. In addition, isoforms differing due to post-translational modifications can present different patterns of expression, as noted for ascorbic peroxidase isoforms.

In general, a large number of protein functions are induced but it must be acknowledged that, in most cases, an integrated interpretation is still difficult. A few studies have shown however that the parallel analysis of proteomic and metabolomic changes allow a better interpretation of physiological responses to environmental changes. A careful characterisation of the environmental conditions and of plant stress is also necessary. Almost all analyses of plant proteome responses to abiotic stresses have been carried out using 2-DE. This has allowed the identification of many responding proteins, and it should be noted that only the analysis of full length proteins, as opposed to the analysis of peptides, allows the identification of processes such as protein degradation and post-translational modification. However, there is no doubt that additional information on the regulation of intrinsic membrane proteins would be of great interest for the analysis of most abiotic stresses. Few studies have combined proteomics with the quantitative analysis of transcripts. In general, analysis of transcripts only partially confirms the results observed with proteins. In addition to problems relating to the existence of multigene families, discordance can be related to differences in turnover between RNAs and proteins. This can be especially important when dynamic processes such as the study of response to stress are analysed. Another possible cause of divergence is the occurrence of post-translational modifications. The ability to analyse the dynamics of post-translational modifications is a huge advantage of proteomics, but it also makes it difficult to analyse correlations with mRNAs, since it is difficult to be sure that the analysed protein spots represent the totality of protein isoforms.

In conclusion, proteomics studies have contributed to the identification of a number of cellular responses to abiotic stresses. However, we are still far from being able to relate all observed changes to physiological adaptations and to the variation of plant phenotypes. Combination of proteomics with metabolomics and transcriptomics should allow a better understanding of integrated plant responses to abiotic stresses and the identification of genes and alleles of interest for the selection of plants able to maintain yields as high as possible in unfavourable environments.

References

Amme S, Matros A, Schlesier B, Mock HP (2006) Proteome analysis of cold stress response in *Arabidopsis thaliana* using DIGE-technology. J Exp Bot 57:1537–1546

Askari H, Edqvist J, Hajheidari M, Kafi M, Salekdeh GH (2006) Effects of salinity levels on proteome of *Suaeda aegyptiaca* leaves. Proteomics 6:2542–2554

Bae MS, Cho EJ, Choi EY, Park OK (2003) Analysis of the Arabidopsis nuclear proteome and its response to cold stress. Plant J 36:652–663

Banzet N, Richaud C, Deveaux Y, Kazmaier M, Gagnon J, Triantaphylides C (1998) Accumulation of small heat shock proteins, including mitochondrial HSP22, induced by oxidative stress and adaptive response in tomato cells. Plant J 13:519–527

Castro AJ, Carapito C, Zorn N, Magne C, Leize E, Van Dorsselaer A, Clement C (2005) Proteomic analysis of grapevine (*Vitis vinifera* L.) tissues subjected to herbicide stress. J Exp Bot 56:2783–2795

Chang WW, Huang L, Shen M, Webster C, Burlingame AL, Roberts JK (2000) Patterns of protein synthesis and tolerance of anoxia in root tips of maize seedlings acclimated to a low-oxygen environment, and identification of proteins by mass spectrometry. Plant Physiol 122:295–318

Costa P, Bahrman N, Frigerio JM, Kremer A, Plomion C (1998) Water-deficit-responsive proteins in maritime pine. Plant Mol Biol 38:587–596

Cui S, Huang F, Wang J, Ma X, Cheng Y, Liu J (2005) A proteomic analysis of cold stress responses in rice seedlings. Proteomics 5:3162–3172

Cuypers A, Koistinen KM, Kokko H, Karenlampi S, Auriola S, Vangronsveld J (2005) Analysis of bean (*Phaseolus vulgaris* L.) proteins affected by copper stress. J Plant Physiol 162:383–392

Dani V, Simon WJ, Duranti M, Croy RR (2005) Changes in the tobacco leaf apoplast proteome in response to salt stress. Proteomics 5:737–745

Fecht-Christoffers MM, Braun HP, Lemaitre-Guillier C, VanDorsselaer A, Horst WJ (2003) Effect of manganese toxicity on the proteome of the leaf apoplast in cowpea. Plant Physiol 133:1935–1946

Ferreira S, Hjerno K, Larsen M, Wingsle G, Larsen P, Fey S, Roepstorff P, Salome Pais M (2006) Proteome profiling of *Populus euphratica* Oliv. upon heat stress. Ann Bot 98:361–377

Gallie DR, Le H, Caldwell C, Tanguay RL, Hoang NX, Browning KS (1997) The phosphorylation state of translation initiation factors is regulated developmentally and following heat shock in wheat. J Biol Chem 272:1046–1053

Giacomelli L, Rudella A, van Wijk KJ (2006) High light response of the thylakoid proteome in arabidopsis wild type and the ascorbate-deficient mutant vtc2-2. A comparative proteomics study. Plant Physiol 141:685–701

Goulas E, Schubert M, Kieselbach T, Kleczkowski LA, Gardestrom P, Schroder W, Hurry V (2006) The chloroplast lumen and stromal proteomes of *Arabidopsis thaliana* show differential sensitivity to short- and long-term exposure to low temperature. Plant J 47:720–734

Hajduch M, Rakwal R, Agrawal GK, Yonekura M, Pretova A (2001) High-resolution two-dimensional electrophoresis separation of proteins from metal-stressed rice (*Oryza sativa* L.) leaves: drastic reductions/fragmentation of ribulose-1,5-bisphosphate carboxylase/oxygenase and induction of stress-related proteins. Electrophoresis 22:2824–2831

Hajheidari M, Abdollahian-Noghabi M, Askari H, Heidari M, Sadeghian SY, Ober ES, Salekdeh GH (2005) Proteome analysis of sugar beet leaves under drought stress. Proteomics 5:950–960

Hurkman WJ, Tanaka CK (1987) The effect of salt on the pattern of protein synthesis in barley roots. Plant Physiol 83:517–524

Hurkman WJ, Tanaka CK, Dupont FM (1988) The effects of salt stress on polypeptides in membrane fractions from barley roots. Plant Physiol 88:1263–1273

Imin N, Kerim T, Rolfe BG, Weinman JJ (2004) Effect of early cold stress on the maturation of rice anthers. Proteomics 4:1873–1882

Jeanneau M, Gerentes D, Foueillassar X, Zivy M, Vidal J, Toppan A, Perez P (2002) Improvement of drought tolerance in maize: towards the functional validation of the Zm-Asr1 gene and increase of water use efficiency by over-expressing C4-PEPC. Biochimie 84:1127–1135

Kawamura Y, Uemura M (2003) Mass spectrometric approach for identifying putative plasma membrane proteins of Arabidopsis leaves associated with cold acclimation. Plant J 36:141–154

Krishnan M, Nguyen HT, Burke JJ (1989) Heat shock protein synthesis and thermal tolerance in wheat. Plant Physiol 90:140–145

Le Lay P, Isaure MP, Sarry JE, Kuhn L, Fayard B, Le Bail JL, Bastien O, Garin J, Roby C, Bourguignon J (2006) Metabolomic, proteomic and biophysical analyses of *Arabidopsis thaliana* cells exposed to a caesium stress. Influence of potassium supply. Biochimie 88:1533–1547

Lee S, Lee EJ, Yang EJ, Lee JE, Park AR, Song WH, Park OK (2004) Proteomic identification of annexins, calcium-dependent membrane binding proteins that mediate osmotic stress and abscisic acid signal transduction in Arabidopsis. Plant Cell 16:1378–1391

Lin SK, Chang MC, Tsai YG, Lur HS (2005) Proteomic analysis of the expression of proteins related to rice quality during caryopsis development and the effect of high temperature on expression. Proteomics 5:2140–2156

Lund AA, Rhoads DM, Lund AL, Cerny RL, Elthon TE (2001) In vivo modifications of the maize mitochondrial small heat stress protein, HSP22. J Biol Chem 276:29924–29929

Majoul T, Bancel E, Triboi E, Ben Hamida J, Branlard G (2003) Proteomic analysis of the effect of heat stress on hexaploid wheat grain: characterisation of heat-responsive proteins from total endosperm. Proteomics 3:175–183

Moons A, Bauw G, Prinsen E, Van Montagu M, Van der Straeten D (1995) Molecular and physiological responses to abscisic acid and salts in roots of salt-sensitive and salt-tolerant Indica rice varieties. Plant Physiol 107:177–186

Moons A, Gielen J, Vandekerckhove J, Van der Straeten D, Gheysen G, Van Montagu M (1997) An abscisic-acid- and salt-stress-responsive rice cDNA from a novel plant gene family. Planta 202:443–454

Nam MH, Heo EJ, Kim JY, Kim SI, Kwon KH, Seo JB, Kwon O, Yoo JS, Park YM (2003) Proteome analysis of the responses of *Panax ginseng* C. A. Meyer leaves to high light: use of electrospray ionization quadrupole-time of flight mass spectrometry and expressed sequence tag data. Proteomics 3:2351–2367

Ndimba BK, Chivasa S, Simon WJ, Slabas AR (2005) Identification of Arabidopsis salt and osmotic stress responsive proteins using two-dimensional difference gel electrophoresis and mass spectrometry. Proteomics 5:4185–4196

Parker R, Flowers TJ, Moore AL, Harpham NV (2006) An accurate and reproducible method for proteome profiling of the effects of salt stress in the rice leaf lamina. J Exp Bot 57:1109–1118

Phee BK, Cho JH, Park S, Jung JH, Lee YH, Jeon JS, Bhoo SH, Hahn TR (2004) Proteomic analysis of the response of Arabidopsis chloroplast proteins to high light stress. Proteomics 4:3560–3568

Puigderrajols P, Jofre A, Mir G, Pla M, Verdaguer D, Huguet G, Molinas M (2002) Developmentally and stress-induced small heat shock proteins in cork oak somatic embryos. J Exp Bot 53:1445–1452

Rakwal R, Agrawal GK, Yonekura M (1999) Separation of proteins from stressed rice (*Oryza sativa* L.) leaf tissues by two-dimensional polyacrylamide gel electrophoresis: induction of pathogenesis-related and cellular protectant proteins by jasmonic acid, UV irradiation and copper chloride. Electrophoresis 20:3472–3478

Requejo R, Tena M (2005) Proteome analysis of maize roots reveals that oxidative stress is a main contributing factor to plant arsenic toxicity. Phytochemistry 66:1519–1528

Rey P, Pruvot G, Becuwe N, Eymery F, Rumeau D, Peltier G (1998) A novel thioredoxin-like protein located in the chloroplast is induced by water deficit in *Solanum tuberosum* L. plants. Plant J 13:97–107

Rey P, Cuine S, Eymery F, Garin J, Court M, Jacquot JP, Rouhier N, Broin M (2005) Analysis of the proteins targeted by CDSP32, a plastidic thioredoxin participating in oxidative stress responses. Plant J 41:31–42

Riccardi F, Gazeau P, de Vienne D, Zivy M (1998) Protein changes in response to progressive water deficit in maize. Quantitative variation and polypeptide identification. Plant Physiol 117:1253–1263

Riccardi F, Gazeau P, Jacquemot MP, Vincent D, Zivy M (2004) Deciphering genetic variations of proteome responses to water deficit in maize leaves. Plant Physiol Biochem 42:1003–1011

Salekdeh GH, Siopongco J, Wade LJ, Ghareyazie B, Bennett J (2002a) Proteomic analysis of rice leaves during drought stress and recovery. Proteomics 2:1131–1145

Salekdeh GH, Siopongco J, Wade LJ, Ghareyazie B, Bennet J (2002b) A proteomic approach to analysing drought- and salt-responsiveness in rice. Field Crop Res 76:199–219

Sarry JE, Kuhn L, Ducruix C, Lafaye A, Junot C, Hugouvieux V, Jourdain A, Bastien O, Fievet JB, Vailhen D, Amekraz B, Moulin C, Ezan E, Garin J, Bourguignon J (2006) The early responses of *Arabidopsis thaliana* cells to cadmium exposure explored by protein and metabolite profiling analyses. Proteomics 6:2180–2198

Smith AP, DeRidder BP, Guo WJ, Seeley EH, Regnier FE, Goldsbrough PB (2004) Proteomic analysis of Arabidopsis glutathione S-transferases from benoxacor- and copper-treated seedlings. J Biol Chem 279:26098–26104

Süle A, Vanrobaeys F, Hajos G, Van Beeumen J, Devreese B (2004) Proteomic analysis of small heat shock protein isoforms in barley shoots. Phytochemistry 65:1853–1863

Sweetlove LJ, Heazlewood JL, Herald V, Holtzapffel R, Day DA, Leaver CJ, Millar AH (2002) The impact of oxidative stress on Arabidopsis mitochondria. Plant J 32:891–904

Taylor NL, Heazlewood JL, Day DA, Millar AH (2005) Differential impact of environmental stresses on the pea mitochondrial proteome. Mol Cell Proteomics 4:1122–1133

Vienne D de, Leonardi A, Damerval C, Zivy M (1999) Genetics of proteome variation for QTL characterisation: application to drought-stress responses in maize. J Exp Bot 50:303–309

Vincent D, Lapierre C, Pollet B, Cornic G, Negroni L, Zivy M (2005) Water deficits affect caffeate O-methyltransferase, lignification, and related enzymes in maize leaves. A proteomic investigation. Plant Physiol 137:949–960

Vincent D, Ergül A, Bohlman M, Tattersall A, Tillett R, Wheatley M, Woolsey R, Quilici D, Joets J, Schlauch K, Schooley D, Cushman J, Cramer G (2007) Proteomic analysis reveals differences between *Vitis vinifera* L. cv. Chardonnay and cv. Cabernet Sauvignon and their responses to water deficit and salinity. J Exp Bot 58:1873–1892

Wang D, Luthe DS (2003) Heat sensitivity in a bentgrass variant. Failure to accumulate a chloroplast heat shock protein isoform implicated in heat tolerance. Plant Physiol 133:319–327

Yan S, Tang Z, Su W, Sun W (2005) Proteomic analysis of salt stress-responsive proteins in rice root. Proteomics 5:235–244

Yan SP, Zhang QY, Tang ZC, Su WA, Sun WN (2006) Comparative proteomic analysis provides new insights into chilling stress responses in rice. Mol Cell Proteomics 5:484–496

Zivy M (1987) Genetic variability for heat shock proteins in common wheat. Theor Appl Genet 74:209–213

Index